Antarctica

Secrets of the Southern Continent

Antarctica
Secrets of the Southern Continent

Chief Consultant David McGonigal

FIREFLY BOOKS

A FIREFLY BOOK

Published by Firefly Books Ltd. 2008

Publisher Cataloging-in-Publication Data (U.S.)

McGonigal, David.
    Antarctica : secrets of the southern continent / David
McGonigal.
[400] p. : col. ill., photos. (chiefly col.); cm.
Includes index.
Summary: Includes geology and geography, flora and fauna,
current scientific research, conservation issues, and the impact
of global warming. Stories of polar exploration and scientific
enterprise, through Amundsen's conquest of the South Pole in
1911, the ratification of the Antarctic Treaty of 1959, and the
subsequent cooperative effort to preserve Antarctica as a
continent for peace and science.
ISBN-13: 978-1-55407-398-6
ISBN-10: 1-55407-398-7
1. Antarctica. I. Title.
919.8/9 dc22   G860.M34 2008

Library and Archives Canada Cataloguing in Publication

McGonigal, David, 1950–
    Antarctica : secrets of the southern continent / David
McGonigal.
Includes index.
ISBN-13: 978-1-55407-398-6
ISBN-10: 1-55407-398-7
    1. Antarctica. I. Title.
G860.M333 2008   919.8'9   C2008-902624-1

Published in the United States by
Firefly Books (U.S.) Inc.
P.O. Box 1338, Ellicott Station
Buffalo, New York 14205

Published in Canada by
Firefly Books Ltd.
66 Leek Crescent
Richmond Hill, Ontario L4B 1H1

Film separation Pica Digital Pte Ltd, Singapore
Printed in China by SNP Leefung Printers Limited

Developed by Global Book Publishing
Global Book Publishing
Level 8, 15 Orion Road, Lane Cove,
NSW 2066, Australia
Ph: (612) 9425 5800   Fax: (612) 9425 5804
Email: rightsmanager@globalpub.com.au

Managing Director Chryl Campbell
Publishing Director Sarah Anderson
Art Director Kylie Mulquin
Project Manager Mary Trewby
Chief Consultant David McGonigal
Consultant Lynn Woodworth
Contributors Don Adamson, Ian Allison,
Thomas Bauer, Dana Bergstrom,
Gary Burns, Louise Crossley,
Arthur Ford, Robert Hill, Julia
Jabour, Paul Lehmann, Peter
Lemon, David McGonigal,
Harvey Marchant, Rob Massom,
Gary Miller, Stephen Nicol, Hugh
Pennington, Kim Pitt, Patrick
Quilty, Stephen Rintoul, Patricia
Selkirk, Michael Stoddart, Patrick
Toomey, Lynn Woodworth
Editors Maggie Aldhamland, Derek Barton,
Anna Cheifetz, Dannielle Doggett,
Fiona Doig, Kate Etherington,
Vanessa Finney, Janet Healey,
Denise Imwold, Margaret McPhee,
Margaret Olds, Judith Simpson,
Mary Trewby, Tracy Tucker
Editorial Assistant Christine Leonards
Map Editors Vanessa Finney, Oliver Laing,
Valerie Marlborough, Janet
Parker, Judith Simpson,
Jan Watson,
Cover Design Erin R. Holmes
Designers Jo Buckley, Cathy Campbell, Avril
Makula, Jacqueline Richards
Junior Designer Althea Aseoche
Design Concept Kylie Mulquin
Cartographer John Frith/Flat Earth Mapping
Additional Cartography Robert Taylor
Map Contour Artwork Oliver Rennert
Wildlife Distribution Maps Lynn Woodworth
Picture Research Gordon Cheers, Joanna Collard,
Heather McNamara,
Christine Leonards
Principal Photographer David McGonigal
Photo Library Alan Edwards
Illustrations Glen Vause
Diagrams Mike Gorman
Index Jon Jermey
Proofreader Elizabeth Connolly
Production Ian Coles
Contracts Alan Edwards
Foreign Rights Kate Hill
Publishing Assistant Christine Leonards

Contributors

CHIEF CONSULTANT
David McGonigal

David McGonigal is an award-winning travel writer and author/editor of more than a dozen books. In 1997 he rode his motorcycle in Antarctica (then to Alaska and across Siberia) on the world's first seven-continent motoring journey. A graduate in arts/law and a fellow of the Royal Geographical Society, David has convened university courses on Antarctica and publishes and broadcasts widely. He has made over 100 journeys to the polar regions as expedition leader, lecturer, adventurer, and photographer.

CONSULTANT AND CONTRIBUTOR
Lynn Woodworth

Dr Lynn Woodworth has a background in conservation biology. She completed her Ph.D., investigating the unique genetic problems of small populations of endangered species, in 1996. She has managed a genetics research laboratory and coordinated conferences for the Society for Conservation Biology. Lynn's first Antarctic voyage was in 1995, and she has been there every year since, spending up to eight months a year on polar vessels as expedition leader, Zodiac driver, and wildlife lecturer.

CONTRIBUTORS
Don Adamson

The late Dr Don Adamson, formerly a senior research fellow in biology at Macquarie University in Sydney, spent 23 years as an Antarctic researcher. He undertook summer fieldwork in the Vestfold Hills, the Prince Charles Mountains, and near Mawson Station in continental Antarctica, as well as on sub-Antarctic Macquarie Island. His research interests spanned geomorphology, vegetation–landscape interactions, and environmental history.

Ian Allison

Ian Allison is a research scientist with the Australian Antarctic Division, where he is leader of the Ice, Ocean, Atmosphere and Climate Program. He also works with the Antarctic Climate and Ecosystems Cooperative Research Centre (CRC). He has studied the Antarctic for 40 years, participated in or led more than 25 research expeditions, and published over 100 papers on Antarctic science. His current research interests include: ice shelf–ocean interaction; Antarctic weather and climate; sea ice; and the mass budget of the Antarctic Ice Sheet. Ian was a lead author of the chapter in the IPCC *Fourth Assessment Report* dealing with changes to snow and ice, and was co-chair of the Joint Committee for the International Polar Year 2007–08.

Thomas Bauer

Thomas Bauer is an assistant professor of tourism in the School of Hotel and Tourism Management at the Hong Kong Polytechnic University. During a 16-year affiliation with Antarctica, he has participated in tourism research, lectured aboard cruise ships, and driven Zodiacs. He is the photographer and co-producer of the *Voyage to Antarctica* CD-ROM and video. His writings include *Tourism in the Antarctic: Opportunities, Constraints, and Future Prospects.* Thomas is a rainforest guardian of the Australian Rainforest Foundation, and the co-owner of Dundee Park, an environmental education facility in Mission Beach, in far north Queensland, Australia.

Dana Bergstrom

Dr Dana Bergstrom is a senior research scientist at the Australian Antarctic Division. An ecologist, she has worked in the Antarctic region, and particularly in the sub-Antarctic, for 25 years. Her specialty areas incorporate studying the impact of change: both climate change and change bought about by human activity on Antarctic ecosystems.

Gary Burns

Dr Gary Burns is a principal research scientist at the Australian Antarctic Division. He has wintered at Casey Station and on Macquarie Island, and spent a summer at Davis Station overseeing the equipment used to study aurora and the upper atmosphere. Gary is presently monitoring the Antarctic atmosphere for indicators of climate change. He has published articles on pulsating aurora, the auroral green line, and auroral linkages between the hemispheres.

Louise Crossley

Louise Crossley is scientist, historian, university lecturer, museum director, broadcaster, futurist, Antarctic expeditioner, and writer. She has spent three winters and four summers in Antarctica with ANARE, and has been a guide/lecturer on icebreaker cruises to the Antarctic Peninsula, the Weddell Sea, the Ross Sea, and on three semi-circumnavigations. Louise is the author of *Explore Antarctica* (1995), a general introduction to the continent, and the editor of *Trial by Ice: The Antarctic Journals of Captain John King Davis* (1997).

Arthur Ford

For nearly three decades, Dr Arthur Ford led or participated in numerous United States Geological Survey expeditions into many areas of Antarctica, and has also undertaken geological research in Alaska. Since retiring in 1995, he has lectured on Antarctic cruise ships. The chapter on Antarctica in *Encyclopædia Britannica* is among his more than 200 publications. A recipient of the United States Antarctica Service Medal and president of The Antarctican Society, Arthur is a fellow of the Royal Geographical Society, the Explorers Club, and the Geological Society of America.

Robert Hill

Professor Robert Hill spent 20 years at the University of Tasmania, Hobart, researching the plant macrofossil record in Tasmania and Antarctica over the past 40 million years. Now executive dean of the Faculty of Sciences at the University of Adelaide and the head of science at the South Australian Museum, he is investigating the impact of Australia's drying climate on vegetation over the past 30 million years. Robert is professor emeritus at the University of Tasmania and professor at the University of Adelaide.

Julia Jabour

Dr Julia Jabour has been studying, researching, writing, and lecturing on Antarctic law and policy for more than 16 years at the University of Tasmania, Hobart. She has visited Antarctica five times and has been on the Australian delegation to Antarctic Treaty Consultative Meetings. Julia has a B.A. in politics, philosophy, and sociology, and a Graduate Diploma (honors) in Antarctic and Southern Ocean studies. Her Ph.D. investigated the changing nature of sovereignty in the Arctic and Antarctic in response to global environmental interdependence.

Paul Lehmann

Dr Paul Lehmann received a Ph.D. in upper atmospheric physics at the University of Melbourne, Australia, followed by research fellowships in atmospheric physics at the University of Illinois and the Max Planck Institute, Germany. In 1984–86 he held a faculty position in the Physics Department at the University of Melbourne, and in 1986–89 received a National Research Scholarship for an appointment at CSIRO. In 1989 he was appointed senior physicist in the Bureau of Meteorology, Melbourne, in the area of stratospheric ozone depletion. Currently, he is a research scientist in the Centre for Australian Weather and Climate Research at the Bureau of Meteorology. He has represented Australia at UN meetings on ozone depletion.

Peter Lemon

Peter Lemon has been connected with Australia's Peregrine Adventures since 1983, undertaking four trips to the Antarctic and many more to Nepal and southern Africa. He is especially interested in wildlife and landscape photography. Peter is also fascinated by vintage and veteran aircraft, and Australian commercial aviation history, particularly that between the two world wars. He enjoys flying in aircraft—the older the better, within reason.

Harvey Marchant

Since his retirement from the Australian Antarctic Division in 2006 as head of Australia's Antarctic biology program, Dr Harvey Marchant has been a visiting fellow at the Australian National University, Canberra. He has an honors degree in science from the University of Adelaide and a Ph.D. from the Australian National University. As a senior Fulbright scholar, he worked at the University of California, Santa Cruz, in 1987. In 2005 he was awarded the Royal Society of Tasmania Medal. Harvey is an international authority on Antarctic biology, with a specific interest in aquatic microorganisms. He is author of over 100 scientific papers and book chapters, senior editor of *Australian Antarctic Science: the first 50 years of ANARE*, co-editor of *Antarctic Marine Protists*, and co-author of *Antarctic Fishes*. He is a member of the editorial boards of two international journals and has been a lead author of the IPCC Assessments in 1995, 2001, and 2007. Harvey has been to Antarctica 12 times with Australian, Japanese, and United States programs.

Rob Massom

Rob Massom has been involved in both Antarctic and Arctic research since 1980, and participated in ten Antarctic and three Arctic sea-ice research campaigns. After receiving a Ph.D. from the Scott Polar Research Institute, University of Cambridge, in 1989, he spent three years in the Oceans and Ice Branch of NASA Goddard Space Flight Center, before joining the Antarctic Cooperative Research Centre (CRC) in Hobart in 1992 as a sea-ice research scientist. He is currently a senior research scientist with the Australian Antarctic Division and Antarctic Climate and Ecosystems CRC in Hobart. Rob's current research interests include: remote sensing of sea ice and ice sheets (he is author of three books on satellite remote sensing of polar regions); physical properties of sea ice and their ecological significance; interactions between sea ice and the Antarctic Ice Sheet; and the impact of modes of large-scale anomalous atmospheric circulation on sea-ice distribution and ecology and ice shelf break-up.

Gary Miller

Dr Gary Miller is a visiting research fellow at the University of Western Australia and a research assistant professor at the University of New Mexico. More than 30 years ago he began working in the polar regions, studying the behaviour of polar bears and the distribution of bowhead whales in the Arctic. In Antarctica, he has studied the behaviour, ecology, genetics, and diseases of penguins and skua. Gary has participated in United States, New Zealand, and Australian Antarctic programs, as well as lecturing and leading tourist ships in Antarctica.

Stephen Nicol

Dr Stephen Nicol has worked on krill since 1979. He joined the Australian Antarctic Division in 1987 to lead the krill research team, and has made eight voyages to the Antarctic. Since 1999 Stephen has been program leader for the division's Antarctic Marine Living Resources and Southern Ocean Ecosystems programs. He has been a member of Australia's delegation to the Commission for the Conservation of Antarctic Marine Living Resources (CCAMLR) since 1987.

Hugh Pennington

Professor Hugh Pennington graduated in medicine and gained his Ph.D. from St Thomas's Hospital Medical School, London. After posts there, at the University of Wisconsin, and at the University of Glasgow, in 1979 he was appointed to the chair of bacteriology at Aberdeen University, where he became emeritus professor in 2003. His research—sometimes involving British Antarctic Survey medical officers—focused on developing molecular fingerprinting methods for bacteria. Hugh is a fellow of the Royal Society of Edinburgh and the Academy of Medical Sciences.

Kim Pitt

Kim F. Pitt, A.M., joined the Australian Antarctic Division in October 1997 and is its general manager of operations. He is responsible for providing the infrastructure (station and field operations, shipping and air operations, engineering and construction programs) that support Australia's interests in the Antarctic and sub-Antarctic. Before this, he spent 32 years in the Royal Australian Navy serving in submarines, the intelligence community, and the Navy's Support Command.

Patrick Quilty

Professor Patrick Quilty was chief scientist with the Australian Antarctic Division and is now honorary research professor at the University of Tasmania. He has degrees from the universities of Western Australia and Tasmania. He has worked in academia, in industry, and federal government. He has participated in many marine science programs and has been, and is, involved in many learned committees. He has published over 190 scientific papers. In 1997 he was awarded membership of the Order of Australia (A.M.) and was made inaugural Distinguished Alumnus from the University of Tasmania. He received the US Antarctic Services Medal, Royal Society of Tasmania Medal, was distinguished lecturer for the Petroleum Exploration Society of Australia, and was invited speaker in the North American Speaker Series for 1998–99. He has five species, a range of nunataks, and a bay named in his honor. He is a patron of the University of Western Australia Geoscience Foundation. He first visited Antarctica in 1965–66 with the University of Wisconsin and since then he has been on many tourist ventures to the region.

Stephen Rintoul

Dr Stephen R. Rintoul studies the influence of Southern Ocean currents on the Earth's climate. For the past 15 years, he has led the Southern Ocean program at Australia's CSIRO Marine and Atmospheric Research and the Antarctic Cooperative Research Centre. He received his Ph.D. from the Woods Hole Oceanographic Institution and Massachusetts Institute of Technology. An active sea-going oceanographer, Stephen has led ten expeditions to the Southern Ocean. His numerous awards include the Priestley Medal from the Australian Meteorological and Oceanography Society, the M. R. Banks Medal from the Royal Society of Tasmania, and the Georg Wüst Prize from the German Society of Marine Research. He was elected a fellow of the Australian Academy of Science in 2006 and appointed a CSIRO fellow, CSIRO's highest honour, in 2007.

Patricia Selkirk

Dr Patricia Selkirk is a senior research fellow in biology at Macquarie University in Sydney, Australia. Over 29 years she has spent many summers undertaking Antarctic fieldwork on Ross Island, in Victoria Land, and near Casey Station, and sub-Antarctic fieldwork on Macquarie and Heard islands and Iles Kerguelen. Her research interests include plant–environment interactions, environmental history, biogeography, and evolutionary biology.

Michael Stoddart

Michael Stoddart is chief scientist to Australia's Antarctic program. He has occupied the position since 1998 after serving four years as deputy vice-chancellor at the University of New England in Armidale, NSW. Previously he was professor of zoology at the University of Tasmania, Hobart, and prior to that reader and lecturer in zoology at the University of London. His main research is on the olfactory biology of mammals. He is the author of *The Scented Ape: the biology and culture of human odour* (1990). He holds the degrees of B.Sc., Ph.D. and D.Sc. from the University of Aberdeen, Scotland.

Captain Patrick Toomey

Retired from the Canadian Coast Guard, Captain Patrick Toomey is presently a consultant ice navigation specialist. He has piloted Russian icebreakers in Arctic waters, made transits of both the Northwest Passage and Siberian Northern Sea Route, and reached the North Pole. He serves as ice pilot and lecturer on Arctic and Antarctic voyages and has circumnavigated the Antarctic continent. Patrick specialises in nautical journalism, training of ice navigators, expert testimony during litigation proceedings, and development of international regulations concerning ice navigation.

Contents

FOREWORD   10

The End of the Earth   14

Antarctic Regions   72

Antarctic Wildlife   136

Antarctic Exploration   260

Antarctica Today   354

Resources   382

GAZETTEER   389

INDEX   393

ACKNOWLEDGMENTS   400

Foreword

There is a book in my life that has led me to where I am today. As an eight-year-old, I visited the home of the Norwegian polar explorer, scientist, and Nobel Peace Prize winner Fridtjof Nansen. When we got back home, my father gave me Nansen's book about his traverse of the Greenland icecap. It was written in language that was too hard for me at that age. But in the school library the next day I found a book for kids about Roald Amundsen's expedition to the South Pole—and that book inspired my childhood dream of skiing to the South Pole. The visit to the school library also opened my eyes to the fact that books could take me everywhere—and teach me everything I wanted to know.

Christmas Eve 1994—more than 30 years after I had read that first book about Antarctic exploration—I reached the South Pole. I had skied, solo and unsupported, over 750 miles (1,200 km). It took 50 days. Books about the old explorers had mentally prepared me for a harsher journey than I had. As the days passed on my way to the South Pole, I felt that I got more and more joy and energy. The undulating Antarctic landscape of white and blue and grey, the patterns in the snowdrifts and in the skies all mixed with the poetry I read at night—making it as much a mental expedition as a physical one. Even though I was pulling a sled weighing 220 lb (100 kg), the hard work is not what I remember. It was the feeling of being at one with nature—of knowing why I was there, what life is, who I am.

Antarctica is the world's last true wilderness. Each visit is an adventure for all who have the privilege to go there. Each visit gives us new experiences and knowledge. The southern continent is home to a rich wildlife, and it is a uniquely beautiful and harsh environment. It is a continent where diverse—and sometimes warring—nations coexist and cooperate, and where the environment is protected. Perhaps reading the text of the Antarctic Treaty can inspire international cooperation to solve the problems we are struggling with in the rest of the world—after all, so far its peaceful conflict resolution process has managed to preserve Antarctica as "a continent for peace and science."

This book reveals the secrets of the southern continent in a way that will create new explorers of all ages and in all arenas. It tells the complete story of the unique Antarctic zone, and is full of wonderful photographs that inspire us to find out more about this fascinating region. We can read how Antarctica has taught us the world's history through its ancient layers of ice. The book features—in a reader-friendly way—the latest information about the region's geology and geography, flora and fauna, scientific research, conservation issues, and the impact of global warming. Knowledge about Antarctica is important for humankind and this book teaches us why. My hope is that this will be the most opened book all around the world.

Liv Arnesen
Norwegian lecturer, educator, and polar explorer.
The first woman to ski solo to the South Pole (1994); since then
she has crossed the Antarctic landmass with Ann Bancroft.

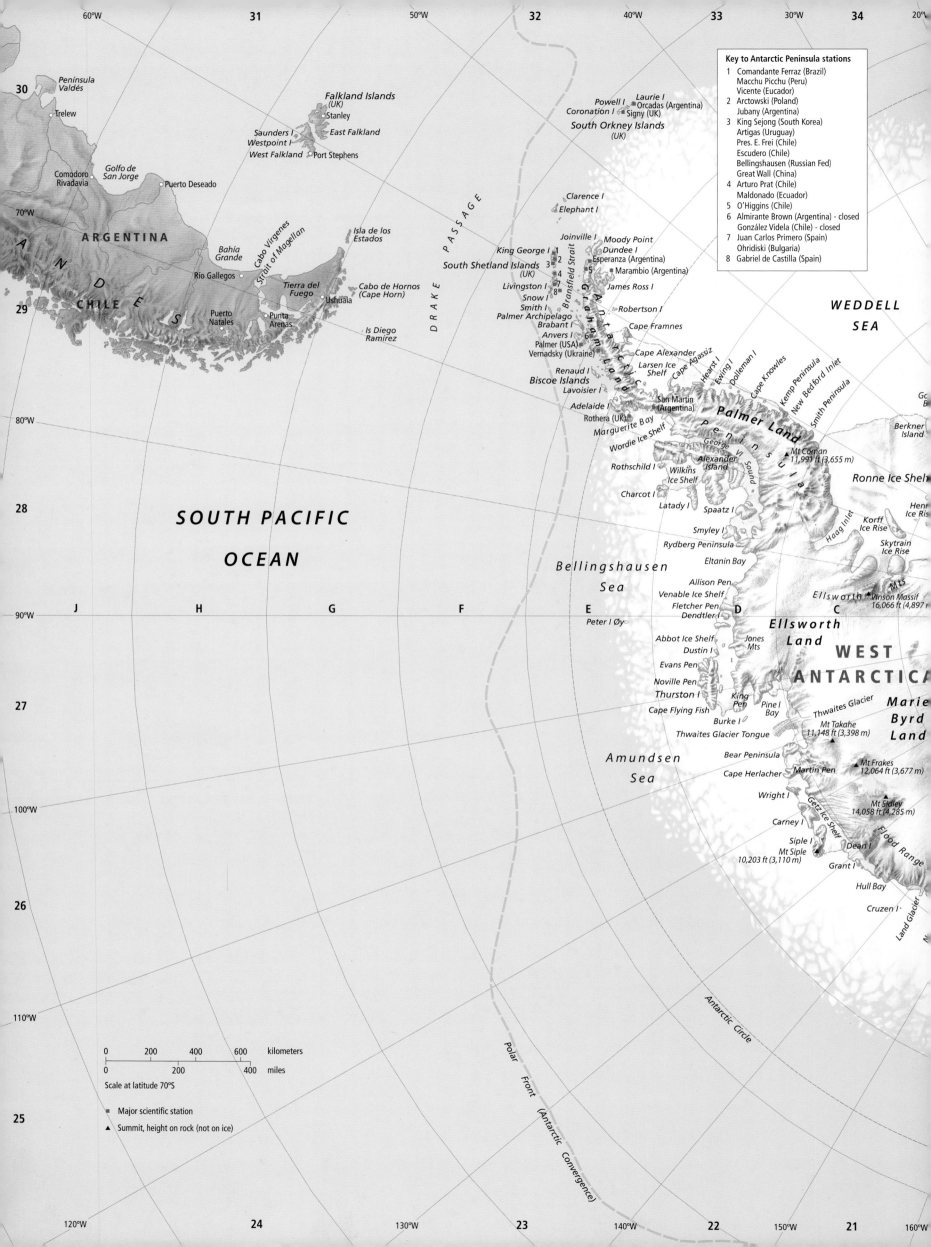

30

Península
Valdés

Trelew

ARGENTINA

Comodoro
Rivadavia *Golfo de*
San Jorge

70°W Puerto Deseado

A
N
D
E
S

CHILE

Bahía
Grande

Río Gallegos

Cabo Vírgenes
Strait of Magellan

Puerto
Natales

Punta
Arenas

Ushuaia

Tierra del
Fuego

Isla de los
Estados

Cabo de Hornos
(Cape Horn)

Is Diego
Ramírez

29

Falkland Islands
(UK)

Stanley

Saunders I
Westpoint I
West Falkland Port Stephens

East Falkland

D
R
A
K
E

P
A
S
S
A
G
E

Powell I Laurie I
Coronation I Orcadas (Argentina)
Signy (UK)
South Orkney Islands
(UK)

Clarence I
Elephant I

Joinville I Moody Point
Dundee I
King George I Esperanza (Argentina)
South Shetland Islands Marambio (Argentina)
(UK)
Livingston I James Ross I
Snow I
Smith I Robertson I
Palmer Archipelago
Brabant I Cape Framnes
Anvers I
Palmer (USA)
Vernadsky (Ukraine)

Bransfield Strait

G
r
a
h
a
m

L
a
n
d

Cape Alexander
Larsen Ice Cape Agassiz
Shelf Hearst I
Ewing I
Dolleman I Cape Knowles
Kemp Peninsula
New Bedford Inlet
Smith Peninsula

Renaud I
Biscoe Islands
Lavoisier I

San Martin
(Argentina)

WEDDELL

SEA

Berkner
Island

28

SOUTH PACIFIC

OCEAN

J **H** **G** **F** **E** **D** **C**

Adelaide I
Rothera (UK)

Marguerite Bay

Wordie Ice Shelf

Rothschild I

Charcot I

Latady I

Smyley I

Rydberg Peninsula

Eltanin Bay

A
n
t
a
r
c
t
i
c

P
e
n
i
n
s
u
l
a

George VI Sound

Wilkins
Ice Shelf

Alexander
Island

Spaatz I

Palmer Land

Mt Coman
11,991 ft (3,655 m)

Haag Inlet

Henr
Ice Ris

Korff
Ice Rise

Skytrain
Ice Rise

Ronne Ice Shel

M
ts

Ellsworth Vinson Massif
16,066 ft (4,897 m)

Bellingshausen

Sea

Peter I Øy

90°W

Allison Pen

Venable Ice Shelf
Fletcher Pen
Dendtler I

Abbot Ice Shelf
Dustin I

Evans Pen

Noville Pen

Thurston I

Cape Flying Fish

Burke I

Thwaites Glacier Tongue

Jones
Mts

King
Pen Pine I
Bay

Ellsworth

Land

WEST

ANTARCTICA

Thwaites Glacier

Mt Takahe
11,148 ft (3,398 m)

Marie

Byrd

Land

27

26

25

Amundsen

Sea

Bear Peninsula

Cape Herlacher Martin Pen

Wright I

Carney I

Mt Siple
10,203 ft (3,110 m)

Dean I

Grant I

Mt Frakes
12,064 ft (3,677 m)

Mt Sidley
14,058 ft (4,285 m)

Getz Ice Shelf

Hull Bay

Cruzen I

Flood Range

Land Glacier

N

100°W

110°W

120°W **24** 130°W **23** 140°W **22** 150°W **21** 160°W

80°W

70°W

Antarctic Circle

Polar
Front
(Antarctic Convergence)

Scale at latitude 70°S

0	200	400	600	kilometers
0	200	400	miles	

■ Major scientific station

▲ Summit, height on rock (not on ice)

Key to Antarctic Peninsula stations

1 Comandante Ferraz (Brazil)
   Macchu Picchu (Peru)
   Vicente (Euacdor)
2 Arctowski (Poland)
   Jubany (Argentina)
3 King Sejong (South Korea)
   Artigas (Uruguay)
   Pres. E. Frei (Chile)
   Escudero (Chile)
   Bellingshausen (Russian Fed)
   Great Wall (China)
4 Arturo Prat (Chile)
   Maldonado (Ecuador)
5 O'Higgins (Chile)
6 Almirante Brown (Argentina) - closed
   González Videla (Chile) - closed
7 Juan Carlos Primero (Spain)
   Ohridiski (Bulgaria)
8 Gabriel de Castilla (Spain)

The End of the Earth

About the Polar Regions

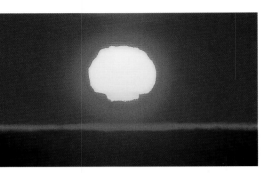

Magnificent Sunrise
Its image distorted in accordance with the laws of atmospheric physics, the sun rises over the Ross Ice Shelf. South of the Antarctic Circle, the sun never rises above the horizon on some winter days; on an equal number of summer days the sun never sets below it.

From space, the Earth is a blue ball capped with white at the poles, but few pause to wonder why the poles are cold enough to maintain permanent ice fields while the tropics are always hot. The energy generated by the sun is capable of heating the whole planet to 57°F (14°C), and the sun is 93 million miles (150 million km) away—so why should it heat one part of the Earth more than another?

The main reason is that the Earth is spherical, with the Equator pointing directly at the sun but the poles tilting away: therefore, sunlight strikes the surface at an oblique angle. At 30° north or south of the Equator (roughly at the latitude of Florida or Sydney), this cuts the amount of sunlight received to about 86 percent of that falling on the Equator. At 60° (about the latitude of Oslo or the sub-Antarctic South Sandwich Islands), the intensity of the sunlight is reduced to 50 percent, and by 80° (about the latitude of the northern coast of Greenland or the edge of the Ross Ice Shelf in West Antarctica) it has fallen to 17.4 percent. From the two

poles, the sun is lower in the sky and its rays are less warming. If the Earth were not tilted, the poles would receive no sunlight at all, and a polar observer would see the sun traveling around the horizon, and never rising above that.

Two other factors help deprive the poles of the sun's warmth. The greater angle means that the sun's rays must penetrate more of the Earth's atmosphere before reaching the surface. More importantly, at least 85 percent of the solar radiation that does reach the Earth's surface is reflected back into the atmosphere by the ice and snow covering the poles.

Every 24 hours, the planet Earth rotates through 360 degrees, so that alternate heating and cooling moderate the impact of the energy emanating from the sun; if it were not so, the side facing the sun would be too hot to sustain life, and the area in shadow too cold. The poles form the axis of this rotation, so move very little, whereas points on the Equator move at 1,038 mph (1,670 kph) to complete a rotation in 24 hours.

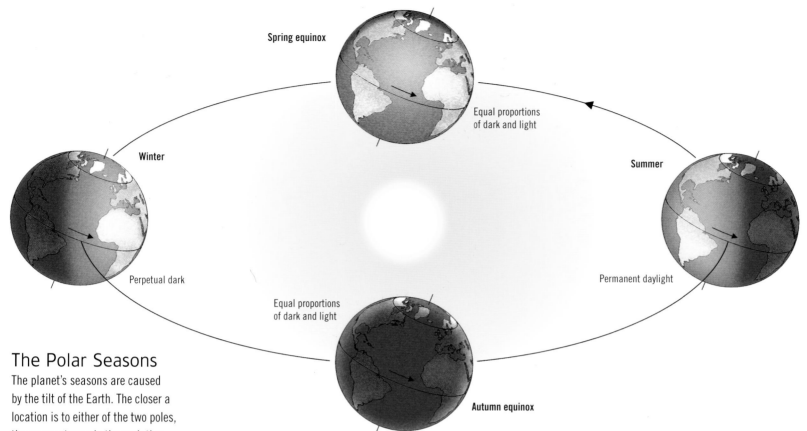

Spring equinox

Equal proportions
of dark and light

Winter

Perpetual dark

Summer

Permanent daylight

Equal proportions
of dark and light

Autumn equinox

The Polar Seasons

The planet's seasons are caused by the tilt of the Earth. The closer a location is to either of the two poles, the more extreme is the variation between summer and winter there. As the illustration shows, during winter in the Antarctic, the South Pole is tilted away from the sun and lies in perpetual darkness. The South Pole is tilted more toward the sun at midsummer, and thus is bathed in perpetual daylight. The North Pole, which is tilted in the opposite direction, experiences the opposite season: summertime in Antarctica is winter at the North Pole.

Resting on Ice

Midnight—but the sun is still high in the summer sky above the snow-free surface of a frozen tarn on Ross Island. Ice of such Antarctic ponds and lakes is frequently beautifully patterned. At this latitude the sun never sets between late October and late February. Continuous sunlight tempts many visitors to push themselves to their physical limits.

The Earth also travels around the sun. If the poles were perpendicular to the sun's rays, every day the whole of the Earth would have 12 hours of daylight followed by 12 hours of darkness; there would be no seasons—and no easily definable year.

The patterns of night and day, and of the seasons, occur because the Earth's axis of rotation inclines from the perpendicular somewhat, so that for half the year the South Pole leans toward the sun and experiences summer, while for the other half it is the North Pole that is inclined toward the sun. At the two solstices—in June and December—the Earth is at the points in its orbit where the North Pole or the South Pole, respectively, are most inclined toward the sun. At these solstice times, one of the poles is bathed in 24-hour daylight and receives more sunlight than anywhere else on the Earth; the other is bathed in darkness.

Midway between the solstices are the September and March equinoxes, when radiation from the sun falls vertically at the Equator. The name "equinox" does suggest that everywhere on the Earth has equal hours of light and darkness on those days. However, sunrise and sunset are judged not from the angle of the sun but from the appearance of its first and last rays, so that different locations experience equal hours of light and darkness several days on either side of the true equinox. (The time also varies, defined by established time zones, so midday may not be midway between sunrise and sunset.)

The equinoxes also mark the point when the poles graduate from six months of sunshine to six months of darkness, and vice versa. The Earth does not orbit the sun in a perfect circle: when the South Pole is leaning toward the sun, the Earth is about 3 percent closer to

the sun. Nevertheless, Antarctica is much colder than the Arctic, mainly because of the dominant effect of its high polar ice sheets, but also because it is a landmass that blocks the moderating influence of the ocean.

The Earth's Poles

In reference to the Earth, a pole is defined as either end of an axis around which the planet revolves. The Earth has no fewer than seven poles. Some move over time; others are fixed, although the markers left by humans move as the West Antarctic Ice Sheet slides toward the sea and the Arctic ice shifts with ocean currents.

Geographic poles are fixed poles at latitudes 90°N and 90°S. Halfway between them—about 6,220 miles (10,000 km) from each—lies the Equator. In the late eighteenth century, there was an attempt to define the meter as one 10-millionth of this distance.

Poles of rotation are moving poles forming the axis of the Earth's rotation. They are located within about 65 ft (20 m) of the geographic poles, around which they rotate over a period of 435 days. The exact locations of the geographic poles are calculated by taking the average measurement of the rotational poles.

Celestial poles are moving poles at the positions in the sky that are occupied by an imaginary line extended infinitely through the geographic poles. The reason these poles move is because the Earth wobbles on its axis over a variety of cycles—lasting 100,000, 41,000, or 23,000 years—as it orbits the sun.

Magnetic poles are moving poles located where the magnetic field is at right angles to the Earth's surface. These poles fluctuate daily, under the influence of solar winds, in an oval and have moved around 1,240 miles (2,000 km) in the last four centuries. Since 1841, the

South Magnetic Pole has shifted northwestward at an average of 6 miles (9 km) each year; in 2007, it was at 64°29′49″S, 137°41′2″E, and the North Magnetic Pole was at 78°35′42″N, 104°11′54″W.

Geomagnetic poles are calculations of where the magnetic poles would be if the Earth's magnetic field worked like a simple bar magnet; they are situated at 78°39′N, 69°W, and 78°30′S, 111°E. These poles are also the outer limits of the Earth's geomagnetic field and the centers of auroral activity.

The poles of inaccessibility are differently defined in the two polar regions: in the Arctic, it is the position that is equidistant from the surrounding landmasses—84°03′N, 174°51′W; in the Antarctic, it is the position, on average, farthest from the sea—85°50′S, 65°47′E.

Defining the Polar Regions

The Arctic and Antarctic Circles, at about 66°33′, are the farthest latitudes from the poles where there is at least one day when the sun does not sink below the horizon at midsummer or rise above it at midwinter. Because the Earth wobbles on its axis—making its tilt from the perpendicular vary—these circles slowly shift back and forth latitudinally. This occurs over one of the periods known as the Milankovitch cycles. These can last 23,000, 41,000, and 100,000 years. Currently, the

Time Zones

Each time zone occupies 15 degrees of longitude. The first figure for each zone indicates the number of hours it is ahead or behind Universal Time (UT). Those figures in brackets show the time in each zone when it is 12 noon UT.

Frozen Mass

In winter, the sea ice in Antarctica (below) extends beyond the Antarctic Circle (66°33′S). If it were permanent, the ice would make Antarctica the third largest continent.

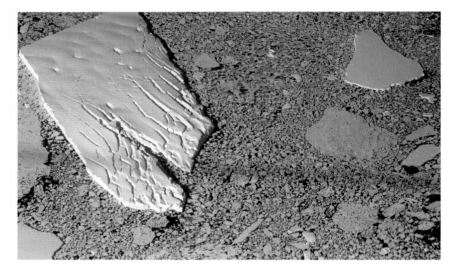

Polar Contrasts

Although the two polar regions share extremes of daylight and darkness, in other respects they differ widely.

ANTARCTIC

- Continent surrounded by ocean.
- South Pole: 9,300 ft (2,836 m) above sea level; bedrock 100 ft (30 m) above sea level.
- Deep, narrow continental shelf; restricted ice-free frozen ground; no tree line; no tundra; no native population.
- Polar ice sheets covering 98 percent of land.
- Icebergs from glaciers and shelf ice measured in cubic miles.
- Sea ice mainly annual, salty, and less than 7 ft (2 m) thick.
- Mean annual temperature at South Pole: −58°F (−50°C).
- Marine mammals (whales, seals); no terrestrial mammals.
- Less than 20 bird species between latitudes 70°S and 80°S.
- Lichens at latitude 82°S.

ARCTIC

- Ocean surrounded by continents.
- North Pole: 3 ft (1 m) of sea ice; bedrock 1,400 ft (427 m) below sea level.
- Shallow, extensive continental shelf; extensive frozen ground; clear tree line; well-defined tundra; circumpolar native populations.
- Limited land ice.
- Icebergs from glaciers measured in cubic yards.
- Sea ice mainly multi-year, low in salinity, and more than 7 ft (2 m) thick.
- Mean annual temperature at North Pole: 0°F (−18°C).
- Terrestrial mammals (reindeer, wolf, musk ox, hare, lemming, fox); marine mammals (whales, seals, polar bears).
- More than 100 bird species between latitudes 75°N and 80°N.
- About 90 flowering plant species at latitude 82°N.

Scenic Serenity on a Grand Scale

Cradled in a natural amphitheater, a vessel lies securely at anchor in Horseshoe Harbor in Mac.Robertson Land. Lightly crevassed slopes lead gently onto the plateau where the peaks of Mount Henderson (left) and the Northern Masson Range (right) pierce the blue continental ice.

For the Birds

The humpbacked shape of Cuverville Island marks the northern end of the Errera Channel in the Antarctic Peninsula. Gentoo penguins, shags, and sheathbills nest on the available low land. Petrels, gulls, terns, and skuas claim spots higher up on the island's steep, lichen-covered slopes.

The Geographic Center
The South Pole is the southern axis of the Earth's rotation. Its position is marked by the Ceremonial Pole, a chromium globe, surrounded by flags of the 12 original signatories to the Antarctic Treaty. Behind is the "Dome," until recently the main Amundsen–Scott Base building.

circles are moving closer to the poles and, being about halfway to the turning point of a 41,000-year cycle (where the angle of the Earth's axis is inbetween 22° and 24.5°), they are moving at close to the maximum rate of about 47½ ft (14.5 m) per year. In 1996, the polar circles were at 66°33′37″; at the present time, they lie at about 66°33′39″; they will reach 68° about 10,000 years from now.

It would be convenient to define the polar regions as those lying within the polar circles, but conditions at the two circles differ so much that such a definition is meaningless. In fact, each region has several different definitions, and the criteria also differ between the two polar regions. The Antarctic Peninsula extends beyond the Antarctic Circle, for example, and almost all winter sea ice develops north of the circle.

"Antarctica" is defined as the Antarctic continent. "The Antarctic" has a variety of meanings but loosely is described as the area south of 60°S. An alternative definition of the Antarctic is of the region lying south of the natural boundary formed by the Polar Front— the convergence of oceanic waters that encircles the continent roughly between latitudes 40°S and 60°S. There is also a legal definition, which is embodied in the Antarctic Treaty and its associated protocols: this describes Antarctica as everywhere south of 60°S.

The Arctic, on the other hand, is not a landmass but rather is a frozen sea surrounded by land; even in the winter, sea ice stops forming well north of the Arctic Circle. One defining criterion for the Arctic region is the northern limit of the tree line: in Scandinavia, there are farms and towns that lie on the Arctic Circle, and trees will survive even farther north. Another definition, which effectively covers the same area, is based on temperature: the Arctic is described as a Northern Hemisphere region where the average temperature during the warmest month remains under 50°F (10°C).

Cold Seas
The Southern Ocean's freezing waters encircle and define the Antarctic region. The ocean was officially named by the International Hydrographic Organization in 2000 and given the boundary of 60ºS, coincident to the Antarctic Treaty's legal limit of latitude.

Time at the Ends of the World
At the Amundsen–Scott Base, sited at the South Pole, a popular exercise is to take a quick walk around the world. This is possible because the station is located at the geographic South Pole, where all the meridians of longitude converge, so you can cross all the world's time zones as fast as you can run in a circle.

Until the end of the nineteenth century, towns and cities operated on the time for their specific longitude: when it was noon in New York City, it was 12:12 PM in Boston. But the rapid development of transport and communications made some standardization of time essential; in October 1884, the meridian of longitude running through Greenwich, England, became known as the Prime Meridian, or 0 degrees, and Greenwich Mean Time (GMT), now known as Universal Time (UT), became the world's fixed time reference.

The Prime Meridian marks the International Date Line, running through the Bering Sea between eastern

Russia and Alaska, and then across the largely landless Pacific Ocean, with some deviations so that the sparse population does not have to manage the problem of neighboring towns operating a day apart. At the poles, the 0° and 180° meridians meet, dividing Antarctica into East and West Antarctica, and generating some confusion in one of the two parts of the world where "east" and "west" are meaningless concepts: at the geographic South Pole every direction is north.

In the polar regions, the sun provides few temporal clues, and therefore makes clocks important for human routine. At the Equator, each single degree of longitude occupies about 69 miles (111 km) and each time zone covers 15 degrees of longitude (1,037 miles/1,670 km), but at the polar circles, each of the time zones occupies only 413 miles (665 km). On the Antarctic continent, where longitude means little during 24 hours of daylight or darkness, stations often use the same time as their home country for convenience. DM

Measuring Antarctica

The rock and permanent ice of the Antarctic landmass cover around 5½ million sq miles (14 million sq km), making this the fifth largest continent, and considerably larger than Europe. If the ice melted, it would consist of the East Antarctica continent and the West Antarctica archipelago and be the smallest continent, at about half its present size. The winter sea ice roughly doubles the area of Antarctica; if it were permanent, Antarctica would be third in size after Asia and Africa. At its deepest point, the dome of the polar ice sheet is 15,800 ft (4,800 m) thick, and the South Pole stands on 1¾ miles (2.8 km) of ice. The average elevation of Antarctica is 7,100 ft (2,160 m)—Asia, the next highest continent, is about 3,280 ft (1,000 m).

Going North
Although it is over 10 degrees north of the Antarctic Circle, in the mountainous South Georgia—in the middle of the Polar Front—glaciers and ice fields cover around 60 percent of the island, and the bays around the coast freeze in the wintertime.

The Polar Landscape

Antarctica, sometimes called the Crystal Desert, has two faces: a visible one largely of ice, and a masked one of bedrock. Ice averaging 1½ miles (2.3 km) thick covers over 98 percent of the continent. In other continents, "land" are those parts lying above water; in the Antarctic, "land" is predominantly ancient bedrock clothed in crystalline water, except rarely in midsummer. Snow, equivalent to about ¾ in (18 mm) of water, accumulates over much of the interior. Major mountain ranges, 10,000 ft (3,000 m) or more in height, such as the subglacial Gamburtsev Mountains of East Antarctica, lie entirely hidden in places. The Inuit name *nunatak* is given to an Antarctic "island," the bedrock of a mountain peak or hill encircled by ice.

The Masked Face of Antarctica

Scattered nunataks and ice-bathed mountain ranges provide clues to the geological framework of Antarctica's approximately 5,500,000 sq miles (14,200,000 sq km). Extensive geophysical surveys using seismic soundings, rock magnetism, radio echo-sounding, and gravity have established the features of the continent's buried face. Comparisons with continents to which Antarctica was joined until 130 million years ago as the hub of a giant southern continent, Gondwanaland, also provide useful geological information.

Antarctica is roughly pear-shaped: the large and bulbous East Antarctic region lies mainly in the eastern longitudes; the smaller West Antarctica, with the spine

Icy Landscape

A frozen meltpool tops the Ross Ice Shelf at Cape Crozier on the eastern tip of Ross Island. The island is a volcanic complex that began forming about 5 million years ago. It contains the few Antarctic volcanoes that are considered to be active.

Obstacle Course

Commonwealth Bay, an open bay in East Antarctica about 30 miles (48 km) wide, is shaped by wind-driven waves. The first geological investigations of the area were carried out by the expedition led by Douglas Mawson in 1911–14. Decades later, when some samples collected by the expedition were further investigated, geochronological and petrological results were obtained.

of the Antarctic Peninsula stretching in the direction of South America, is in western longitudes. The Ross Sea at the south end of the Pacific Ocean and the Weddell Sea south of the Atlantic create deep indentations in the continent's outline. In the nineteenth and early twentieth centuries, geologists had speculated that these two seas might be connected beneath the ice of the interior, but surveys during the International Geophysical Year (IGY) of 1957–58 clearly showed that a ridge, ice-buried but above sea level, links the Ellsworth Mountains in West Antarctica to the great Transantarctic Mountains dividing West and East Antarctica.

A Mountainous Continent

Antarctica contains several of the major mountain belts on the Earth. The largest is the belt of the Transantarctic Mountains, a chain of nunataks, mountains, and ranges around 2,000 miles (3,200 km) long, extending—with some interruptions—from Cape Adare on the southern Pacific side to Coats Land on the Atlantic side. These mountains reach to within about 310 miles (500 km) of the South Pole, and form a topographic marker between East and West Antarctica.

A second major mountain system is formed by the Antarctic Peninsula. It stretches about 1,000 miles (1,600 km) south from Drake Passage to disappear as a chain of nunataks beneath the ice sheet in Ellsworth Land. An early ship's captain, noting the great similarity between the rocks in the peninsula and those of South America's Andes, called the belt the Antarctandes. The Andes mountains in Chile and numerous rocks from the Antarctic Peninsula show evidence of copper deposits, for example. Where this belt disappears—in Ellsworth Land—its structures begin bending westward; they seem to extend largely beneath ice, before reappearing along the coasts of the Bellingshausen and Amundsen seas.

During the Cretaceous period (144–65 million years ago), when ice-free Antarctica was connected to Australia and the dinosaurs lived, the mountains extended the Andes around the south Pacific to the eastern rim of Australia. Much later, the rim broke off and moved eastward during the opening and widening of the Tasman Sea to form New Zealand.

The spectacular Ellsworth Mountains, at the head of the Weddell Sea, where the highest point in Antarctica—Vinson Massif—rises to 16,066 ft (4,897 m), remain a geological enigma. Sandstone, shale, limestone, and glacial deposits of Paleozoic age (543–248 mya) reveal a geological history that is more like that of East than West Antarctica. These mountains in West Antarctica contain the long-extinct *Glossopteris*, a late Paleozoic tree and a fossil characteristic not only of East Antarctica but of all Gondwana. Geologists think these mountains probably formed in the Transantarctic Mountains and later became displaced.

Geophysically, the average thickness of the Earth's crust in Antarctica matches that of other continents: it is roughly 19 to 20 miles (30 to 32 km) thick in West Antarctica and about 25 miles (40 km) thick in East Antarctica. A sharp change in the thickness along the Transantarctic Mountains front may indicate a deep crustal fault system. The overall crustal stability of the Antarctic of today is confirmed by the absence of any significant earthquake activity, which is recorded only sporadically in a few volcanoes. AF

Rocks of Ancient Lineage
Nunataks of the Framnes Mountains in the vicinity of Amery Inlet on the Mawson Coast occupy an area that is geologically a part of the Precambrian shield of East Antarctica.

The Formation of Antarctica

Geologists use the term "plate tectonics" for movements of the Earth's large, rigid, crustal plates, and for their deformation by folding and breaking (faulting) of rocks under compression or by tension at places where, moving differently, the plates touch. Lavas—melted rocks—erupt along mid-ocean ridges as the ocean floors widen and continents separate. Along the collision zones, surface rocks can be carried to great depths and ranges of mountains squeezed up. At great depths, the rocks change under high pressures and temperatures, and in places melt. Lavas rise along conduits to form volcanic arcs where the plates collide: the Andes, for instance, resulted from the Pacific and South American plates colliding.

The geological map of Antarctica reveals a very long history of plate collisions. Some geologists hypothesize that the northern extremity of Victoria Land, near Cape Adare, was an island mass that collided with Antarctica by plate-tectonic movements. Because this originated away from the Antarctic continent, it is known as an "exotic terrane" of rocks.

Making Mountains

An episode of mountain building resulting from folding under compression, faulting, or upthrusting is known as an orogeny. Fossils in the rocks and radioactive dating of materials, such as volcanic rock and granites, formed during such events are used to determine which period ancient, long-eroded mountains were formed.

Like the other continents, Antarctica has a tectonic framework consisting of a long-stable core of rocks of Precambrian age (older than 543 million years) called a shield, adjoined by orogenic belts of younger rocks. East and West Antarctica are geological provinces of greatly different character, with West Antarctica having a record of much younger crustal movement. The

Land Shaped by Ice

The rugged west coast of the Antarctic Peninsula vividly encapsulates what a powerful force ice can be in sculpting mountains into a landscape of valleys and hills. Valley glaciers spill over the vast cliffs into the sea and, in doing so, they spawn huge icebergs.

The Edge of the Sea
Sunlight reflects on broken sea ice, just off Cape Royds (right), deep in the Ross Sea. Tidal movements break up the ice where it meets the land, providing breathing holes for seals and access into and out of the water for penguins.

Life at the Summit
Mount Melbourne (bottom right), a little-dissected stratovolcanic cone, rises 8,963 ft (2,732 m) above the western shore of the Ross Sea. It was named by James Clark Ross in 1841 after the British prime minister of the time. At the summit is a rare Antarctic habitat: a few hundred square yards of steam-warmed, ice-free ground where thin coverings of algae, mosses, and liverworts can grow.

Antarctic Peninsula formed as a volcanic arc during the Mesozoic era (248–65 mya), and was an active arc up to the beginning of the Cenozoic era (65–1.8 mya). By 30 million years ago, in the mid-Cenozoic, South America stretched northward from Antarctica, became disrupted, and formed the small, new plate of the Scotia Sea. The very active volcanoes of the South Sandwich Islands at the easternmost end of the Scotia Plate mark its collision with an oceanic plate to the east.

Continents Adrift
The geological history of Antarctica is largely the record of ancient plate movements. Pangaea, a supercontinent that encompassed most of the Earth's continental crust from more than 300 million to about 200 million years ago, rifted apart to form Laurasia, a northern landmass that encompasses today's Europe, North America, and Asia, as well as Gondwana, which later broke up into the southern continents. These ancient lands fractured along faults and split apart, perhaps many times. The Atlantic seems to have opened once, then closed, and then reopened again into today's ocean.

During the 1950s, there was controversy in some Northern Hemisphere universities over "continental drift," which resolved into the theory of plate tectonics by the late 1960s; once a plausible mechanism was identified, the lateral motion of continents became

scientifically acceptable. Some geologists had long known how continents could be refitted—for example, by matching the Brazilian bulge with Africa's great Ivory Coast indentation by the closing of the Atlantic Ocean. Other geophysicists had calculated that this was a mathematical impossibility, and many geologists believed them, but no one argues now against the oceans opening and closing, nor against Antarctica's key geological role in the configuration of Gondwana.

It is accepted that Gondwana broke apart along a number of rifts that developed into mid-ocean ridge

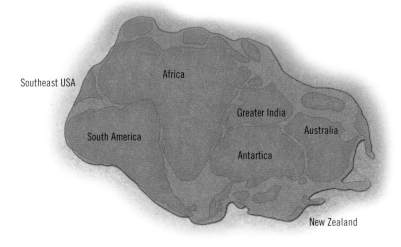

Gondwana
The supercontinent Gondwana, as it was about 180 million years ago. It comprised all today's southern continents and began to show signs of rifting at this time. A landmass consisting of Africa, South America, and India was first to separate; Australasia split off later.

Salt Antifreeze

Don Juan Pond, a pool of water lying in isolation in the south fork of the Wright Valley, is considered the saltiest body of water on the Earth. It is so salty, in fact, that it never freezes, even in the depths of the Antarctic winter. The pond has no outlet; it is fed by groundwater and, for a few weeks of each year, by snow melt.

Layers of History

Large boulders at the Upper Wright Glacier, in the Dry Valleys. The lighter rocks are sandstone; the darker rocks are dolerite. Bands of these totally different rock types are visible in the mountains behind. Sandstone was deposited by rivers and lakes during a geologically quiet period; dolerite is formed by intrusion of molten rock.

systems, and that new oceans carrying fragments of continents spread ever wider.

Antarctica forms one of the small number of the Earth's rigid crustal plates. Today's plates formed as they broke and rifted apart along faultlines of some earlier plate, under crustal stresses that were produced by thermal plumes rising from the Earth's interior. Early stages in the isolation of Antarctica can be witnessed in the rifts that are beginning to tear East Africa from West Africa, a process proceeding south from the Red Sea into the Rift Valleys. Numerous active volcanoes mark the zones of heat from the interior along these rifts in Kenya and Tanzania. In the Red Sea region, which is at a more advanced stage, rifting has developed into an incipient oceanic ridge system, with sea-floor spreading and separation of Arabia from Africa.

Antarctica seems to be at an early stage of interior splitting. The Ross Sea region and Marie Byrd Land contain active volcanoes—including Mount Erebus, Mount Terror, and Mount Melbourne—with the same unusual alkaline chemistry as those of the East African rifts. Geologists have found evidence in the area for a major zone of rifting, which they have called the West Antarctic Rift System.

East Antarctica's Precambrian shield was part of a much larger rock shield that included much of India, Madagascar, west Australia, and southern Africa. Some of the oldest rocks yet known—around 3,800 million years old—are reported from such places. Original sedimentary and igneous rocks became deeply buried in the crust, and tectonic processes again and again transformed them by metamorphism into many types of gneiss and schist. Eventually, late in the Precambrian era, a plate of an ancestral Pacific Ocean developed by volcanism that was associated with crustal rifting, and by sea-floor spreading.

As it opened, this ancestral plate pushed against, and collided with, the adjoining continental plate of Pangaea, and was dragged beneath the continent in the process that geologists call subduction. Where the subducted crust was dragged sufficiently deep, parts melted under the high temperatures, forming molten silicate rock material, called magma, which was then injected upward to crystallize as granite or extruded in volcanoes. The Pacific crust collided more than once along the edge of "paleo-Antarctica," which is marked by today's Transantarctic Mountains. Ancient mountains formed and eroded. The processes successively added to the continent, which grew toward the Pacific Ocean. Sediments were deposited in marine or terrestrial basins formed ahead of, and behind, volcanic arcs. Granites crystallized from materials melted at depth, and lavas were forced out where magmas coursed up fractures to the surface. The youngest belt thus formed is the mountain range system of the Antarctic Peninsula and its continuation along coastal West Antarctica. Ancient New Zealand and the Andes in Chile were extensions of this system. New Zealand, which was once part of West Antarctica, rifted away about 85 million years ago.

The Beacon Supergroup

Antarctica's geological evolution followed a course that was generally similar to that of the other Gondwanan continents. Most of the geological record through the early Mesozoic era, up to about 242 million years ago, is well displayed in the Transantarctic Mountains, in the rock sequence called the Beacon Sandstone (now the Beacon Supergroup) by the early explorers. To the excitement of palaeontologists at the British Museum, pioneer geologists of the expeditions led by Scott and Shackleton from 1904 to 1912 collected plant fossils and coal samples of the late Paleozoic era from the Beardmore Glacier at the edge of the Polar Plateau, as well as fossils of Devonian-age fish (about 380 million years old) from the quartzitic sandstones near McMurdo Sound. The coals were found to be extensive and were considered a potential mineral resource. Fossils of late Paleozoic amphibians and dinosaurs have now been found, and also of the late Paleozoic leaf, *Glossopteris*, a hallmark of Gondwana in the Permian period. This is compelling evidence that the southern lands were once connected, and separate from the northern continents.

The Beacon Supergroup has since been mapped throughout the Transantarctic Mountains and at places far within East Antarctica. It generally forms flat beds of Devonian to Jurassic periods, overlying an eroded surface on strongly folded, trilobite-bearing Cambrian limestone and other bedrock. Such structural relations, combined with an absence of sedimentary rocks aged between the Cambrian and Devonian periods, are evidence that strong deformation and mountain building occurred after the Cambrian and before Devonian times. Granites formed also during this event, known as the Ross Orogeny.

Effects of that orogeny and of an older one of the late Precambrian, the Beardmore Orogeny, are known throughout the Transantarctic Mountains. They are believed to result from collisions between an ancestral Pacific plate and the East Antarctic shield. During the late Mesozoic and early Cenozoic eras, the continent continued to enlarge by subduction (underthrusting) of the Pacific Ocean crust, accretion (the addition of new material), and the formation of volcanic arcs along the Antarctic Peninsula and its extensions, in an event called the Andean Orogeny.

Cold Deserts

The Dry Valleys in the Transantarctic Mountains, seen from the top of the Olympus Range. The largest snow- and ice-free area in Antarctica, these "cold deserts" contain many clues to the formation of Antarctica.

Gondwanan Folding

In *Our Wandering Continents*, published in 1937, the South African Alex du Toit predicted that folded rocks similar to those of South Africa's Cape Fold Belt near Cape Town must have once extended into Antarctica at the head of the Weddell Sea.

Lincoln Ellsworth had seen high mountains in the area on his 1935 transcontinental flight. The later-named Ellsworth Mountains were geologically wholly unknown until a United States exploration party visited them in 1958. They found that du Toit had predicted correctly, based on how he thought the continents had to refit. The tectonic event, in which rocks as young as Permian were strongly folded during the earliest part of the Mesozoic era, affected the Antarctic regions only around the margin of the Weddell Sea, including the Antarctic Peninsula, the Pensacola Mountains, and the Ellsworth Mountains. The Cape Fold Belt extended west into ancestral Argentina. When the Atlantic Ocean rifted open, one small fragment—today's Falkland Islands—was dragged out from Africa and left on the continental shelf of Argentina. Unlike the ocean-continent collisions, it seems that the Gondwanan Orogeny occurred within a continent by some little-understood mechanism.

In the middle and late Mesozoic era, heat currents ("plumes") rising through the mantle of the Earth led to successive crustal rifting, accompanied by volcanism on an immense scale. Gondwana was dismembered, and the new, smaller plates of today's continents were formed. In the mountains of Victoria Land, the events are marked by the conspicuous black dolerite, also known as diabase, that intrudes into the horizontally layered, white, quartz-rich Beacon Supergroup rocks. Some of these intrusions, known as sills, parallel the sandstone beds; others, called dykes, cut across them. Rifting began around 190 million years ago, early in

the Jurassic period. Vast piles of flood-lavas across the Transantarctic Mountains also mark this as one of the most dramatic volcanic events in the Earth's history. An exceptionally large body of igneous rock—the Dufek, a layered igneous complex near the southeastern Weddell Sea—may mark a mantle plume that is associated with the rifting of Africa from Antarctica in an early phase of the breakup of Gondwana.

Isolation Sets In

Plant and animal migration routes that had apparently freely interconnected all the southern continents were cut off at various times. Africa rifted off first. At Hope Bay, at the north end of the Antarctic Peninsula, Mount

Pacific Plate
The origins of the Antarctic Peninsula and its offshore islands such as Petermann, lie in the ancestral Pacific plate, which collided along the edge of "paleo-Antarctica" more than once.

Vulcan Relatives
The unusual alkaline chemistry of the active volcanoes in the Ross Sea region is shared by those in the Rift Valleys of East Africa. Evidence has shown that both continents are in the early stages of internal splitting of their tectonic plates.

Andean Arc
Pumice and ash from recent volcanic eruptions on Deception Island in the South Shetland group provide evidence of the volcanic arc, from the Antarctic Peninsula to South America, formed during the Andean Orogeny.

Fossil Evidence
Fossil leaves of the extinct *Glossopteris* (above) are found in all the fragments of the supercontinent Gondwana. These plant fossils were the key to proving Gondwana's former connections. When explorer Robert Scott was found dead in his tent, *Glossopteris* fossils were found with him: after abandoning unnecessary equipment, his group had retained the fossils they had collected.

Flora's spectacularly plant-rich beds of Jurassic age have no counterparts in southern Africa. Similarly, the vast early Cretaceous forests of *Nothofagus*, or southern beech, about 100 million years old—fossils of which are found in Antarctica and living forms in most of the Southern Hemsiphere—are not represented in Africa.

Antarctica was becoming isolated at a time when elsewhere land mammal species were diversifying and flourishing, populating all the other continents. It had long been thought that Antarctica was a migration path for marsupials moving between South America and Australia during earliest Cenozoic time, and proof of this was found in the 1982 discovery of a marsupial fossil on Seymour Island near the Antarctic Peninsula.

Around 50 to 60 million years ago, Australia was separated by rift-faulting. The final phase in Antarctica's isolation was its stretching apart from South America about 35 to 40 million years ago, during the formation of the Scotia Sea. Growth of that new sea, as well as the southeastern Indian Ocean, had important climatic and biological consequences in the ensuing development of one of the Earth's greatest currents, the Southern Ocean's Circumpolar Current, also known as the West Wind Drift. This current kept warm northern waters from reaching the coasts of Antarctica and, it is believed, it allowed the growth of the ice sheets. The migration routes of animals such as early marsupials were cut off. Fossils of a number of species of dinosaurs found in Antarctica—some near-cousins of Australian species— show that such migration routes had extended far back into Mesozoic times.

Collisions of the Pacific and other oceanic plates against the growing continental crust of West Antarctica and the Antarctic Peninsula continued from the mid-Mesozoic time through the Cenozoic epoch. Subduction of the oceanic crust produced basins in which sandstone and shale were formed, and these were accompanied by extensive intrusion of granite bodies and extrusion of

lavas of various types—materials widely exposed in the South Shetlands and the Antarctic Peninsula.

Volcanic activity indicates the continuation of the plate tectonic processes, as in the West Antarctic Rift System, in northern parts of the Antarctic Peninsula, and the highly active volcanoes of the South Sandwich Islands of the Scotia volcanic arc at the eastern end of the Scotia Sea. Continuing eruptions of the caldera of Deception Island, with major activity in 1967–70 that destroyed two research stations, mark the development of a new ocean-ridge spreading system: at the present time, the South Shetland Islands are moving westward from the Antarctic Peninsula and widening Bransfield Strait. Global positioning system instruments are now directly measuring crustal movements and movements of continents to test the various past speculations by geologists and geophysicists.   AF

Out of Africa
The Dufek Ventifact (below) is a large body of igneous rock in the northern Pensacola Mountains near the Weddell Sea. Geologists believe that it may mark a mantle plume associated with the rifting of Africa from Antarctica early in the break-up of Gondwana.

Evolution in Antarctica

Finding a Roothold
The tiny tussocks of the *Deschampsia antarctica* hair grass grow among rocks and in soil-filled cracks in the bedrock. Right to the southern limits of its distribution, *D. antarctica* can grow densely enough to form closed swards.

Sparse Coverage
Once mosses were widespread and diverse in Antarctica. Now they grow only sparsely along the continent's coast and on the Antarctic Peninsula. Vigorous moss beds sometimes form, often along drainage lines, where conditions are favorable.

Antarctic wildlife is usually perceived in terms of the animals that live in the region never visiting land or only using it as a temporary base for breeding. Land plants are mainly mosses and lichens, and only two flowering plants—a grass and a cushion plant in the Antarctic Peninsula. Probably as recently as 5 million years ago, however, woody flowering plants grew well inland, suggesting diverse and complex vegetation. There is evidence of more complex plant and animal communities even further back in time.

The Antarctic continent has occupied its present polar position for tens of millions of years, but long ago it was nowhere near the South Pole, and even when it was, it supported life forms as complex as those of any landmass in low latitudes: for example, the *Glossopteris* flora, which dominated Gondwana 248 to 290 million years ago. Scientists once believed that Antarctica's high latitude prevented the development of complex vegetation, because plants could not survive the long and extremely harsh polar winters. Experiments on living plants related to those that thrived in Antarctica in the remote past, however, have proved that plants can survive such conditions, especially if temperatures are not too extreme (either high or low). Now the very high latitudes of Antarctica are no longer seen as an impediment to the development of complex plant and animal communities.

The reduction of Antarctica's plant life to extremely sparse and simple vegetation is one of the great natural extinction events in the history of the Earth.

Climate Change

The cause of the extinction is clear—extreme climate change. The history of Antarctic life over the past 60 million years clearly shows the impact of climate change, and the potential impact of human activity on life on the Earth.

The Southern Hemisphere's climate is dominated by massive oceans, their currents spanning large latitudinal belts and exerting a profound influence on the weather. Water warmed by the sun in equatorial regions can transfer energy to high latitudes; however, the Antarctic Circumpolar Current, which circulates vast amounts of water around Antarctica, never leaves very high latitudes, and so draws very little energy from incoming sunlight. This body of extremely cold water is the main reason for the freezing conditions that exist in Antarctica today.

When the southern continents were part of Gondwana, no circum-Antarctic current could form because there was no separate Antarctic continent; the major southern currents flowed from equatorial to polar latitudes and back again, so that the water that reached the high latitudes

remained relatively warm and produced much milder conditions there; indeed, it was sometimes so warm that there was no polar-ice covering. Warm seawater leads to high evaporation, and, consequently, high rainfall. In those times, the land at very high latitudes was both warm and wet—perfect conditions for the development of complex forests.

The rifting of Gondwana, the separation of Australia, and the subsequent opening of the Drake Passage at the tip of South America allowed full development of the Circumpolar Current. The circulation of this cool water forced the formation of the Antarctic ice sheets (although prior to this, higher areas may have had ice caps), and these reflected much of the incoming solar radiation. This made the Antarctic even cooler, so that the ice sheets extended farther still, thus cooling the surrounding ocean. This feedback system is complex and often unpredictable, but its overall outcome was the modern Antarctic environment, where conditions suitable for complex forest growth are long gone.

Evidence for Past Antarctic Life

Fossils are usually quite common, but in Antarctica the fossil record is particularly difficult to assemble. Today, the continent lies mostly beneath two thick ice sheets that either conceal fossils or have scoured them from the surface and deposited them in the Southern Ocean. Fortunately, the fossil records of other Gondwanan

landmasses that were connected to Antarctica once—for example, South America, Australia, and New Zealand—are much more complete, and they provide many ideas about Antarctic species during times when the ice cover was greatly reduced, or even absent.

When the Antarctic climate was able to support complex ecosystems, the day length was a major factor affecting life. Generally, observation of similar living organisms casts light upon plant and animal communities of the past. However, all present-day high-latitude areas, with their very long days in summer and continuous darkness during at least part of the winter, have extremely cold climates that prevent complex forest vegetation from developing. There are no modern forest communities growing under these light conditions, which makes the reconstruction of the past more difficult.

One lucky fossil find in Antarctica was a forest of conifer trunks in their growing positions. The trunks are 10 to 16 ft (3 to 5 m) apart, and the tallest one is 23 ft (7 m) high. These relatively widely spaced trees would have captured light for photosynthesis very efficiently. Trees in high latitudes must not shade each other too much because during the day the sun moves around close to the horizon, so that incoming solar radiation is at a very low angle and comes from different directions throughout the day. To capture the optimum amount of light, the Antarctic conifers would probably have had their foliage cascading down the sides of the trunks, rather than spreading overhead like canopies of tropical forests, where the sun is predominantly high in the sky. Ancient Antarctic forests had open canopies to allow

Clinging to Life
Plant life in Antarctica today has a tenuous existence. This whale skull is encrusted with lichen, one of the major plant forms that is now present in Antarctica—a far cry from the polar forests that once covered much of the continent.

Antarctic Dinosaur
Cretaceous dinosaur remains occur on the southern Australian shoreline, which was once adjacent to Antarctica. The dinosaur *Leaellynasaura* was a small animal with relatively large eyes, possibly an adaptation to low winter-light levels. These dinosaurs may have been bare-skinned, or perhaps had feathers for insulation.

Still Standing
This 100-million-year-old petrified tree is still in its growth position. The roots were preserved in the fossil soil, and the trunk was entombed in flood-borne volcanic-rich sands.

adequate light uptake. The fossil tree rings are very clear and large, suggesting seasonal growth under good growing conditions.

These open forests provided habitats for many animals. One of the most interesting, from the early Cretaceous period (144–99 mya), was the dinosaur *Leaellynasaura*. Relatively small, these dinosaurs had unusually prominent optic lobes, which suggest they had very large eyes that allowed them to remain active during prolonged winter darkness. If so, they may have been warm-blooded to cope with the cold of the dark winters. These dinosaurs may have been feathered to provide better insulation. Several different carnivorous dinosaurs were also present at this time.

The forests were dominated by conifers and they were undergrown with now-extinct seed plants, ferns, mosses, liverworts, and lichens. The seed-bearing tree *Ginkgo* sometimes occurred: it is very probable that its deciduousness allowed it to survive the long winters.

Flowering plants evolved about 130 to 160 million years ago, probably in northern South America and Africa. The pioneering flowering plants had generalized pollination and seed dispersal, and this enabled them to spread over a wide range, and they soon began to disperse along coastlines of the rift valleys that formed as the supercontinent of Gondwana began to break up.

Fossil evidence for Antarctic animals is sparse, but some from nearby landmasses are helpful. Platypuslike bones around 61 to 63 million years old from Patagonia suggest that there were monotremes in South America and later across Antarctica. Similar evidence suggests that marsupials lived in Antarctica at least 60 million years ago. Other land animals have been recorded as rare fossils, including a giant—probably flightless—bird, 40 million years old, on the Antarctic Peninsula.

After the Break-up

Some prominent plants in the early Antarctic flora still exist elsewhere, and their ecology is well understood. A good example is *Nothofagus*, the southern beech, which has been studied extensively across its modern range. In South America, New Zealand, and the island of New Guinea, these species tend not to continuously regenerate where climatic conditions favor complex forests. In these environments, seeds germinate and establish in freshly cleared subsoil after catastrophic disturbances, which are common events in the unstable mountain chains. Such behavior is probably similar to the regeneration early in the history of *Nothofagus* in Antarctica, when developing rift valleys would have provided unstable habitats.

Until about 40 million years ago, these unstable habitats at high southern latitudes provided a migration pathway for early flowering plants. They also provided an opportunity for incidental populations of widespread plant species to become isolated, and this may have been a critical factor for evolution. There is evidence that most of these flowering plants migrated eastward across Antarctica, moving from South America toward Australia and New Zealand.

In the late Cretaceous period (99–65 mya), there were strong similarities between vegetation within the same latitudes of Australa, New Zealand, Antarctica, and the southern part of South America. Conifers and *Nothofagus* were the major species of the tall, open forests that grew in this climatic zone. There were also some Proteaceae and Myrtaceae, although both were less diverse than they were in Australia, where they remain prominent today. By 65 million years ago, vegetation dominated by woody flowering plants was well established in Antarctica.

Growth Rings
This fossil of a flowering plant wood (above) shows well-defined growth rings that are indicative of a strongly seasonal climate. The cells in the wood are characteristic of flowering plants and are for transporting water through the plant.

Perfect Form
A single, beautifully preserved leaf of *Nothofagus* from 40 million-year-old sediments found on King George Island, off the northern Antarctic Peninsula.

Long-time Survivor
The waratah (below) belongs to the ancient Gondwanan plant family, Proteaceae. This family occurred in Antarctica for tens of millions of years. Today, the species occurs only in Australia.

Major Extinctions
The end of the Cretaceous, about 65 million years ago, is noted as a time of major extinction, probably resulting from the impact of a massive extraterrestrial body on the Earth. Scant evidence of this event has been found in Antarctica, however. Scientists have suggested that the collision occurred around June (midwinter in the high southern latitudes), when the animals and plants—for example, winter-deciduous trees—would have been dormant. This scenario explains the paucity of evidence for the effect of the impact on Antarctic life. Because of the lack of information, it is difficult to conclude too much from current data.

Australia's separation from Antarctica began in the late Cretaceous period, although opportunities for plants and animals to cross water gaps might have persisted for a long time after that. The Drake Passage probably began to open about 23 to 25 million years ago, but there is little information about how much dry land there was between the Antarctic Peninsula and South America before then.

Few Antarctic fossils have been discovered from the past 65 million years, but some later ones are very instructive. Cenozoic plant macrofossils, which are less than 65 million years old, are known only from a drill hole in McMurdo Sound, the Antarctic Peninsula, and the Sirius Formation in the Transantarctic Mountains. Controversy surrounding these fossils centers on the accuracy of the identifications, and on the age of the sediments containing them. Most are poorly preserved, which makes classifying them difficult, especially for *Nothofagus* leaves, which are relatively common.

Pollen grains fossilize more readily than other plant parts. The pollen record, however, is often difficult to interpret because much of it is not preserved where it was deposited, but lies in offshore sediments where it has been carried by ice scouring. Nevertheless, the evidence of quite diverse endemic species of flowering plants in the early Cenozoic era exists. This vegetation persisted without much change until about 35 million years ago, probably into early phases of chilling when glaciers had reached sea level.

The pollen record does not reveal when increasing cold and ice cover eliminated the Cenozoic vegetation, largely because of confusion caused by the movement of the pollen once the glacial erosion and sedimentation had begun. This was complicated by the discovery of probable Pliocene fossils from the Sirius Formation.

There is evidence for several glaciations on the Antarctic Peninsula between about 20 and 100 million years ago. The glaciations are separated by non-glacial periods when macrofossils were deposited. During the late Cretaceous period and the early Cenozoic era, there were enough refugia (havens during climatic change) in Antarctica or nearby to allow woody plants to retreat to them during a glaciation period, then subsequently to recolonize West Antarctica during the climatic improvements that followed. However, the formation of Antarctic-wide ice sheets about 32 to 33 million years ago may have brought about a major change in plant communities, and may have removed all woody plants from the Antarctic Peninsula for several million years.

Evolutionary Novelties
Evolutionary novelties are the result of major evolutionary changes in plants and animals that occur over relatively short periods of time. These changes often produce quite distinct life forms, distinguished as genera or even families of species that have many characteristics in common. High latitude areas were, and still are, an important source of evolutionary novelties. The reason for this is not well understood, although several theories have been proposed. However, these theories are difficult to test.

The Weddellian Biogeographic Province, which extends from southern South America along the Antarctic Peninsula and West Antarctica to New Zealand, Tasmania, and southeastern Australia, was the center of origin and diversification for many organisms during the past 100 million years. The development of marsupials is an obvious example. Many new plant species arose and survived in this region, including important groups like *Nothofagus*, the casuarinas, and the Proteaceae.

Much of Antarctica's endemic plant and animal life remained isolated because of geographical, climatic, or biological barriers. However, many of the plants and animals spread northward from the Weddellian Province to mid- and low-latitude regions, where their descendants still occur today, long after their place of origin has become too cold to support such life forms.

Fire oak (*Casuarina cunninghamiana*).

Similar Situation

Mountain beech (*Nothofagus solandri*) occurs in New Zealand and forms part of a cool temperate ecosystem, which is probably similar in structure to forests that occurred in Antarctica prior to the separation of Australia and South America. The very wet conditions are conducive to fern and moss growth on the trees' trunks.

On Snow Hill

Sedimentary rocks on Snow Hill Island, and the adjacent Seymour, James Ross, and Vega islands, in the northwestern Weddell Sea, contain an unusual abundance of fossils and a unique record of the Earth's history across what scientists call the "K–T" boundary, which marks the end of the Mesozoic era and the extinction of dinosaurs.

Antarctica's growing isolation and lengthening plant migration routes meant that many species could not return when the glaciation waned.

Thus, the Antarctic Peninsula's macrofossil record suggests a flora in decline through time. Presumably, land fauna also declined, although there is no direct evidence for this. A single *Nothofagus* leaf fragment, 24 to 29 million years old, from the CIROS-1 borehole demonstrates the presence of at least that genus at a much higher latitude, and well away from the Antarctic Peninsula. The record from the Antarctic Peninsula ends about 24 million years ago, but it can be placed in the context of the Sirius Formation flora, which is probably Pliocene, 5.3 to 1.8 million years old.

Sirius Formation Flora

The Sirius Formation sediments from the Transantarctic Mountains contain beautifully preserved pollen, in situ roots, leaves, and wood; the plants that produced these fossils must have been growing on site when the fossils were deposited. *Nothofagus* dominates the fossil pollen, although pollen from other flowering plants and conifers

is recorded. In addition, the fossil wood and leaves are also *Nothofagus*, and it can be assumed that they all came from the same species.

The stems are up to ½ in (12 mm) in diameter, and several are gnarled, contorted, and contain branching junctions. The extremely narrow growth rings indicate extremely slow growth. More than 60 rings in one specimen were measured along a radius of less than ¼ in (6 mm). The stems are distinctly asymmetrical, in a pattern known as "reaction wood." In flowering plants, this forms on the upper side of branches and in the horizontal trunk wood. The dominance of reaction wood among Sirius samples suggests that these stems grew horizontally, like living dwarf Arctic willows, which grow close to the ground, and are thus protected from the freezing cold winds. Asymmetric growth rings are also present in prostrate forms of living *Nothofagus gunnii* from alpine Tasmania. Some of the fossil branches show evidence of traumatic events and scarring, which is also a common feature in living dwarf Arctic willows.

Many hundreds of fossil leaves have been found, each with a very strong ribbed and creased pattern where it was folded like a fan in the bud. In the living *Nothofagus* species, this indicates deciduousness and, along with the dense accumulation of leaves within a single thin layer, suggests that the fossil leaves are the result of a single, seasonal leaf fall.

Since the fossil wood and leaves from the Sirius group are found among glacial sediments, it is believed the Antarctic Pliocene epoch environment was similar to that of the present high Arctic. Pulses of glacial melt and rapid erosion, like those that occur in the Arctic spring and summer, probably damaged the tree stems and periodically retarded their growth. It is likely that the vegetation resembled living alpine communities in Tasmania, but was less diverse, restricted to winter deciduous *Nothofagus*, other flowering plants, and to a variety of conifers.

Since *Nothofagus* was identified from these Sirius sediments, other flowering plants and mosses have been found. They are consistent with sparse vegetation of limited diversity, growing in extremely cold conditions.

Controversy surrounding the Sirius sediments has centered on the interpretations of the prevailing climate, particularly temperatures, based on the presence of the *Nothofagus* fossils and their nearest living relatives, and the implications for the Pliocene climate. The dwarfed growth forms of the fossil *Nothofagus* suggest that, although summer temperatures may have been around 41°F (5°C), these conditions would have lasted for only about three months during the summer. For the rest of the year, temperatures would have been below freezing, and the plants would have remained dormant. The lower temperature limit would probably have been at least 5°F (–15°C) to –8°F (–22°C), and possibly much colder, if snow cover had protected the dormant plants. This gives a mean annual temperature that is well below freezing, and perhaps in the region of 3°F (–16°C).

Nothofagus-dominated vegetation must have been present in Antarctica for around 80 million years, until the time the Sirius *Nothofagus* was deposited. This is consistent with a progressive simplification of Antarctic vegetation, possibly with some form of tundra vegetation toward the end of the process. It should be noted that the Sirius Formation fossils are well inland in Antarctica, suggesting there may have been other, more diverse vegetation at lower sites closer to the coast.

No Simple Conclusions

Scarceness is the most obvious feature of the vegetation found in Antarctica today. Many scientists do not believe it possible that the Antarctic continent was once thickly vegetated. However, a combination of plate tectonics and other factors can be used to explain a very different climate at high latitudes in the distant past, and it is now known that diverse life was possible in Antarctica without other physical changes. What is particularly interesting is that, although it is now almost barren, the rest of the Southern Hemisphere is covered in wildlife that owes much to the most enigmatic continent of Antarctica.

The Cenozoic decline in Antarctic plants and animals was undoubtedly climatically induced, but there is little scientific agreement about other details. Evidence from the Antarctic Peninsula has demonstrated that a gradual impoverishment of the flora did occur in response to glacial cycles, and this, coupled with the continent's increasing isolation, strengthened migration barriers during the succeeding warmer phases. The probable Pliocene Sirius Formation fossils suggest, however, that woody vegetation remained in Antarctica long after most palaeoclimatologists had previously thought it was far too cold. The fact that one fossil can cause so much uncertainty shows how much there is yet to learn about the history of life on this most mysterious continent. RH

Science in Progress

Evidence of animal life in Antarctica is sparse. Here, a palaeontologist sorts through fossil teeth. The first dinosaur find was in the mid-1980s on James Ross Island. Since then, fossils of other dinosaurs and mammals have been collected from deposits in the same area of the northwestern Weddell Sea.

Fossil Sites

Cretaceous and Cenozoic fossils are primarily found in offshore sediments. The fossils are mainly pollen and, although originally laid into sediments on the continent, have been relocated (recycled) by ice scouring. The fossil record is often difficult to interpret for this reason.

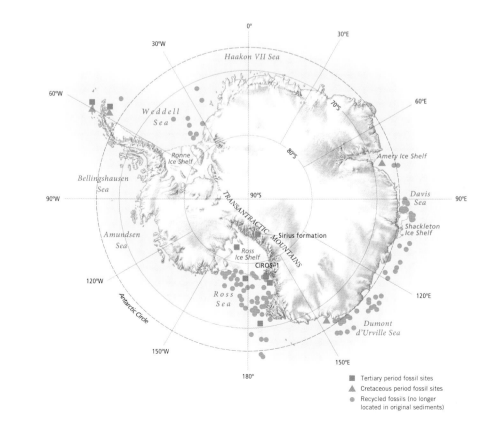

■ Tertiary period fossil sites
▲ Cretaceous period fossil sites
● Recycled fossils (no longer located in original sediments)

The Polar Front

Early explorers seeking a great southern continent found that both the air and sea became cooler as they sailed south across the Southern Ocean. The whalers and sealers who followed them in quest of profits noticed this too. These intrepid travelers also observed that strong currents set their ships to the east where the temperature change occurred. This transition between warm subtropical waters and cold polar waters became known as the Antarctic Convergence. Today, it is called the Polar Front.

Defining the Polar Front

The Polar Front encircles Antarctica between latitudes 40°S and 60°S. Changes in sea temperature across the front occur in a few sharp jumps—fronts—rather than as a gradual decrease across the width of the Southern Ocean. The most important fronts are the Sub-Antarctic and Polar fronts. The rapid temperature changes across the fronts are linked to the strong eastward flow of the Antarctic Circumpolar Current.

These fronts act as boundaries that define zones with different temperatures, salinities, and nutrient concentrations. The different zones also tend to be populated by distinct communities of animals and plants; early oceanographers could determine which side of the Polar Front they were on by the presence or absence of particular species of krill.

The Circumpolar Current stretches for more than 12,400 miles (20,000 km) around Antarctica. Although the surface speed of the current is modest, its great depth and width make it the largest of all currents in the world's oceans. It carries about 4,800 million cub ft (135 million cub m) of water per second from west to east around the Antarctic continent; this is equivalent to about 135 times the flow of water from all the rivers of the world combined.

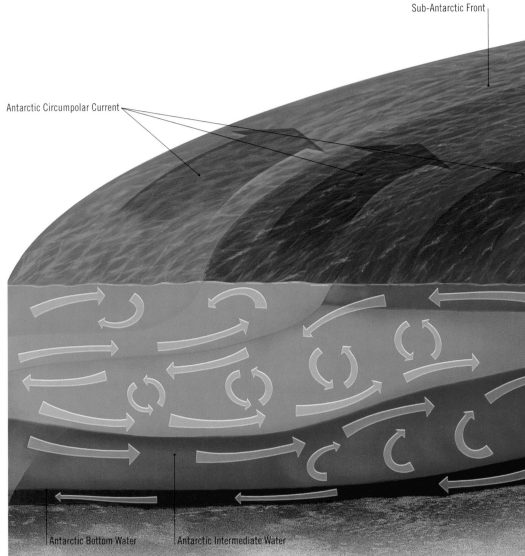

Sub-Antarctic Front

Antarctic Circumpolar Current

Antarctic Bottom Water

Antarctic Intermediate Water

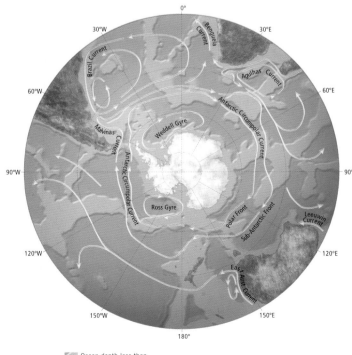

Ocean depth less than
11,480 ft (3,500 m)

Southern Ocean Currents
The currents of the Southern Ocean circulate west to east, unimpeded by land. This is the only place on the planet where the oceans can circulate around the globe, uninterrupted by any continental landmass. These currents allow water transfer between oceans, providing a key link in a global ocean circulation pattern that maintains the Earth's climate.

The massive flow of the Circumpolar Current is driven by some of the strongest winds on the Earth. The persistent westerly winds, which are punctuated by frequent violent storms, prompted sailors to dub the southern latitudes the Roaring Forties and the Furious Fifties. The strong winds acting over the circumpolar extent of the Southern Ocean also create the largest waves on the planet.

Ocean Conveyor Belt

The Antarctic Circumpolar Current plays a unique role in the Earth's climate system. Each of the world's major ocean basins is enclosed by land except at its southern boundary. The Circumpolar Current functions as a pipe that connects these basins, smoothing out variations in water properties between the basins, and—more importantly—permitting a truly global ocean circulation pattern. Water, which is made cold and saline at high latitudes, becomes heavy enough to sink into the deep ocean; warm water flows into the high-latitude regions to replace the sinking dense water. This exchange of warm and cold water carries heat from low latitudes to high latitudes, cooling the former and warming the latter, and so maintaining the Earth's climate. The Circumpolar Current, by allowing the free exchange

Polar Front

Antarctic Divergence

Ice sheet

Sea ice

Ice shelf

Continental shelf

When winter sea ice forms, cold, dense water is produced underneath it. This heavy water, called Antarctic Bottom Water, sinks and flows slowly northward to the Equator.

The Southern Ocean

The Polar Front encircles Antarctica between 40°S and 60°S, and is the region where cold Antarctic waters (dark blue) converge with the warmer waters of the northern oceans (green). The rapid temperature changes across the Polar Front are linked to the strong eastward flow of the Circumpolar Current. Near the Antarctic continent, the coastal current flows westward.

of water between all the ocean basins, is a key link in this "overturning circulation," or "ocean conveyor belt."

Polar Differences

The southern polar region differs in some important ways from the northern region. Antarctica is a continent surrounded by the Southern Ocean, whereas the Arctic is virtually a landlocked sea—and only relatively small amounts of water are exchanged through the narrow straits between the Arctic seas and the northern Pacific and Atlantic oceans.

The two polar regions also differ in the character of the sea ice. That in the Antarctic region tends to be carried away from the continent by winds and currents, whereas Arctic sea ice is trapped by landmasses and has a tendency to pile up, becoming much thicker than the sea ice of Antarctica. The freezing of the ocean surface in the Antarctic effectively doubles the size of the continent in winter, but in summer most of this ice melts, whereas the Arctic is ice-covered all year round.

These differences mean the Antarctic Polar Front has no Arctic equivalent. In the Northern Hemisphere the transition between polar and subtropical waters occurs much farther south, outside the Arctic, and is not circumpolar due to the presence of landmasses. SR

Force 11 Gale

The gale-force winds and huge seas battled by the early explorers are experienced by visitors to the Southern Ocean today. The vessels may have been strengthened, but the Southern Ocean's power to generate the wildest conditions on the Earth still poses a great challenge to modern sailors.

Cold, Dry, and Windy

The polar regions are "heat sinks" influencing the world's climate. A complex meteorological system is created by the different levels of energy that are received by the poles and the tropics, and is greatly affected by the spinning of the globe. Air heated at the Equator rises and flows toward the poles, where it cools and sinks; this dense, cool air then flows back to the low-pressure area created at the Equator by rising warm air.

Although the Antarctic polar ice sheets are crowned by constant high pressure, they are surrounded by a region of low pressure. Since the South American and Antarctic landmasses separated, about 25 million years ago, Antarctica has been completely encircled by the Southern Ocean, allowing the winds to flow unimpeded. Eastward-bound low-pressure systems are generated in never-ending succession here, circling Antarctica like a procession of spinning tops.

Shaped by the Wind
Sculpted by sunlight and wind, a piece of broken sea ice rests on the frozen surface of the Ross Sea. By summer, it will be gone. Wind shapes everything in Antarctica: snow, ice, rocks—and people's lives. Antarctica abounds in stories of being trapped inside tents for days by raging blizzards.

Weather observations, first taken by explorers and then by scientists, are invaluable sources of meteorological information about the Antarctic. The International Geophysical Year (IGY) of 1957–58 led to the establishment of many Antarctic research stations, where further records were kept. The later development of weather satellite technology has provided a wealth of additional data. Today, about 100 ground stations, attended (mainly coastal) and automatic (mainly in the interior), provide global weather information.

Over 97 percent of Antarctica is covered by snow, and this has a significant impact on climate all over the world. Solar energy is reflected from the Earth in an effect known as albedo. The Antarctic surface absorbs radiation for a short period only in mid-summer—for the rest of the year it re-radiates more energy than it receives, the balance being restored by heat transfer from the tropics.

World's Coldest
The lowest temperature recorded was by A. Budretski on July 21, 1983, near Russia's Vostok Station in East Antarctica: it was –128.6°F (–89.2°C), lower than anything experienced at the South Pole because of the higher altitude of Vostok—11,475 ft (3,500 m), compared with the South Pole's 9,300 ft (2,836 m). Vostok's record is likely to be broken only if a base is established at the top of the East Antarctic Ice Sheet—an altitude of 13,100 ft (4,000 m). Antarctica rapidly becomes colder in autumn, reaching its extreme in the depths of winter. Inland, the winter temperatures can range from –40°F to –94°F (–40°C to –70°C); winter coastal temperatures range from 5°F to –31°F (–15°C to –35°C). The Antarctic Peninsula is by far the warmest part of Antarctica; midsummer temperatures there can reach 59°F (15°C), while in East Antarctica they range from 32°F (0°C) on the coast to –13°F (–25°C) inland.

Antarctic temperatures are lower than those in the Arctic, which range from about 32°F (0°C) in summer to –31°F (–35°C) in the winter. The Antarctic latitudes experience about the same temperatures as the Arctic latitudes 310 miles (500 km) closer to the North Pole, because of the height of the Polar Plateau, and because Arctic temperatures are moderated by the ocean: the Southern Ocean generates little warmth because it is much colder than the Arctic Ocean and does not receive the warm currents of the Northern Hemisphere. As in Antarctica, the lowest Arctic temperatures occur far from the ocean. Verkhoyansk, on the Yana River in Siberia, has recorded –90°F (–67.8°C), but has peaked in summer at almost 100°F (38°C). The temperatures across the Arctic range widely, depending on proximity to warm or cold ocean currents.

"The Merciless Blast"

At Cape Denison, Mawson measured wind speeds over a year on a 24-hour average. The wind nearly always came from the south–southeast. On only one day was the wind speed less than 15 ft (4.5 m) per second, and for only about a third of the year was the average wind speed less than 59 fps (18 mps). There were six days when the average wind speed was greater than 88 fps (27 mps). The average wind speed over the year was 63½ fps (19.4 mps), or about 44 miles (70 km) per hour. In the stormiest hour of that year, on July 6, 1913, the wind speed was 141 fps (43 mps), or 96 mph (154 kph). The mean wind speed for the quietest month was 38½ fps (11.7 mps), or 26 mph (42 kph).

A French station that was later established nearby discovered that Mawson's anemometer may have been registering under-readings. This area certainly records the world's strongest winds at sea level— higher wind speeds may have been recorded at only a few mountain peaks.

Down-flowing Winds

Freezing katabatic winds sweep down the slopes of Terra Nova Bay in the Ross Sea. Unlike winds in other parts of the world, Antarctica's katabatic winds are caused by the shape of the land: the cold, dense air on the high ice sheet flows down the coastal slopes under the influence of gravity.

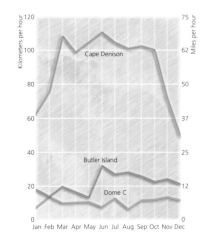

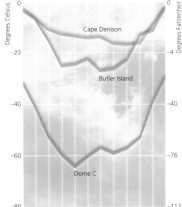

World's Driest

The air over the Antarctic is generally too cold to hold water vapor, so there is very little precipitation: snowfall on the Polar Plateau equates to less than 2 in (5 cm) of rain per year. Antarctica, the world's driest desert, ranks with the Sahara as the world's largest desert. However, Antarctica retains whatever moisture it receives—about three-quarters of the world's freshwater is stored as ice, and 90 percent of that ice is in Antarctica.

The Arctic is a desert too: its annual precipitation ranges from 4 to 16 in (10 to 40 cm). As it is in the Antarctic, the Arctic polar air is too cold to hold much moisture, but what there is falls on ocean—the only ice sheets are on Greenland and on nearby islands. The Greenland ice sheet is about one-tenth the size of the Antarctic ice sheets.

World's Windiest

Antarctica is the windiest place on the Earth. A variety of different winds blow there, from the inversion winds at the Pole to winds funneled between islands along the coast, and violent blizzards can develop with incredible speed. The wind creates its own landforms: sastrugi, irregular ridges up to 3 ft (1 m) high, carved by blowing snow. Their undulating surfaces indicate the prevailing wind direction, like sand dunes. The strongest winds, however, are on the coast, where wind speeds of up to 185 mph (300 kph)—twice the velocity of hurricane-force winds—have been recorded.

On January 8, 1911, the Australasian Antarctic Expedition arrived at Cape Denison in East Antarctica. That afternoon the wind rose as they unloaded their supplies. They waited for it to drop, but it never really did. Douglas Mawson called his published account of the 1911–14 expedition *The Home of the Blizzard*— not without good reason. He had built his hut where blasts of katabatic wind flow down to the coast from the Polar Plateau. Simply put, in Antarctica, katabatic wind is cold and dense air that pours down the ice slope to the sea, becoming denser and picking up speed as it goes. On average, the wind speed at Cape Denison reaches hurricane force every three days. DM

Cold and Windy

Temperature and wind speed vary greatly over Antarctica. In general, the heart of the continent (Dome C) is very cold, especially during the dark of winter. Antarctica's coastal regions (Cape Denison, Butler Island) are milder, warmed by the oceans, but are often the windiest areas, buffeted by freezing blasts of katabatic winds from the icy slopes.

The Antarctic Ice Sheets

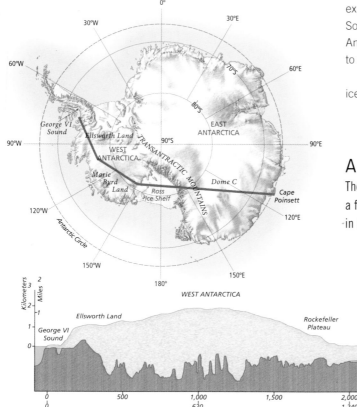

The Antarctica ice sheets are immense. They cover about 5,400,000 sq miles (14 million sq km) and average about 7,500 ft (2,300 m) thickness (from bedrock), with local thicknesses up to nearly 16,400 ft (5,000 m). The ice in the Antarctic potentially locks up about 75 percent of the world's freshwater. Surprisingly, recent seismic soundings have confirmed the presence of many large lakes under East Antarctica, but very little is known about them, especially about their age. Just above one of the largest of these lakes, Lake Vostok, the drilling of an ice core that was more than 1¼ miles (2 km) long was discontinued until techniques could be developed for sampling its waters without running the risk of contaminating them. It is believed that the lake's waters may contain some extremely ancient organisms (perhaps now extinct elsewhere). Now studies find that many of over 100 subglacial lakes are interconnected by large rivers running under the ice, so contamination of one could affect others.

Antarctica's Ice Sheets

The East Antarctic Ice Sheet contains a great volume of ice, and much, although not all, of it is grounded on bedrock near or well above sea level. In contrast, much of the far smaller West Antarctic Ice Sheet lies on rock below sea level, which will be an important factor if the global climate warms sufficiently for the polar ice sheets to melt. Glaciologists debate whether the ice sheets in Antarctica are expanding or decaying; however, most believe that the two continental ice sheets, especially that of East Antarctica, are more or less in a state of equilibrium. In contrast, some of the ice shelves seem to be breaking up. The Antarctic ice sheets are like settling tanks, trapping and accumulating materials from the atmosphere. They contain records of ancient climates,

and of global volcanic activity extending back for thousands of years, as well as trapped meteorite fragments.

Storm tracks seldom reach the interior of East Antarctica, and much of this region is true desert, where snow accumulates at very low rates, generally about ¾ in (18 mm) of equivalent water per year. As snow layers become buried, they transform into ice, brittle at upper levels, but becoming more plastic under pressure at the deeper levels. Under the thickest areas, the ice can reach its pressure melting point and water can form, allowing basal sliding—the rapid movement of ice over bedrock under the pull of gravity. The highest part of the ice sheet, around 13,100 ft (4,000 m) above sea level, lies not at the South Pole but in the center of the East Antarctic Ice Sheet at about 80°S latitude and 75°E longitude, south of the Amery Ice Shelf. From near there, ice flows outward more or less radially by flow under gravity. Where it reaches the sea, much of the ice calves off as icebergs, and flows onto and adds to nearby masses of floating ice shelves that fringe the coasts. In some places, ice movement is channeled into rapidly flowing ice streams. Regional ice "domes" occur at various places.

The East Antarctic Ice Sheet also flows toward and becomes largely dammed by the great Transantarctic Mountain range. In many places the dam is breached by outlet glaciers that squeeze through the mountains and plunge down to feed the Ross Ice Shelf or they flow into the Ross Sea, farther to the north. The giant, highly crevassed Beardmore Glacier was crossed in the early 1900s by Ernest Shackleton's and Robert Falcon Scott's expeditions up to the Polar Plateau on their way to the South Pole. The Axel Heiberg Glacier provided Roald Amundsen with a more direct route, and allowed him to reach the Pole first, in November 1911.

The West Antarctic Ice Sheet forms several major ice streams that feed the head area of the great Ross

Plateau to Shelf

Ice from the East Antarctic Ice Sheet covering the Polar Plateau reaches the Dufek Massif in the Pensacola Mountains, a long mountain chain running between the Foundation Ice Stream and a glacier shelf in the Ronne Ice Shelf.

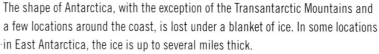

A Cross-section of Antarctica

The shape of Antarctica, with the exception of the Transantarctic Mountains and a few locations around the coast, is lost under a blanket of ice. In some locations in East Antarctica, the ice is up to several miles thick.

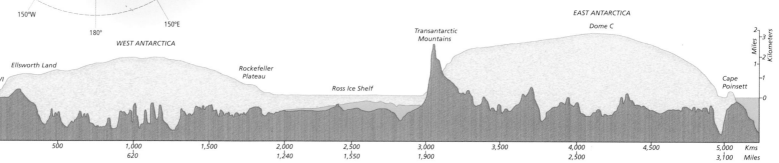

Ice Shelf, and are its main source. Its center is at only half the elevation of that of the East Antarctic Ice Sheet. The accumulation rate is much higher than that of East Antarctica because many maritime storm tracks cross the West Antarctic region.

The Antarctic Peninsula lies outside the area of the ice sheets, and it contains mostly glaciers of alpine type, as well as larger ice fields.

Theories of Origin

Today we live in an ice-age world (often referred to as an "interglacial" stage), with large ice sheets covering Antarctica and Greenland. Antarctica's ice sheets give the best picture of how northern North America must have looked 20,000 years ago, during the last glacial maximum, under the Laurentide Ice Sheet, which had largely disappeared by 10,000 years ago. In Antarctica, ice seems to have begun forming 30 to 35 million years ago and the ice sheets probably remain in much the same form as they have throughout the ice age. Major glaciations do not seem to be synchronous between the hemispheres (although there is some debate about this); in contrast, the advances and retreats of smaller valley glaciers tend to be more sensitive to both global and regional climate changes.

Ice started forming in Antarctica apparently soon after the continent became isolated from the lands to the north, by the consequental development of the Circumpolar Current. Refrigeration began with alpine glaciers in highlands, such as the Transantarctic and Ellsworth mountains. Valley glaciers coalesced to form piedmont glaciers and ice sheets. The record of ice build-up is found in the surrounding ocean sediments. Cores drilled from the Ross Sea in the 1970s recovered boulders and pebbles that must have been carried along by icebergs and dropped into sea-floor mud about 30 to 40 million years ago. Such rocks, deposited far from their place of origin, are called glacial erratics. The volume of south polar ice must have fluctuated greatly since the birth of the Antarctic ice sheets: the southern continent has probably been continuously glaciated since the sheets' formation, but there is some evidence of extensive deglaciation about 5 million years ago.

Evidence and opinions that attempt to explain the formation of glaciers and their oscillations often conflict. In 1920, Serbian mathematician Milutin Milankovitch found evidence for cyclic variations (of 23,000, 41,000, and 100,000 years) in the Earth's orbit around the sun and in movements of the Earth's axis, which might help explain variations in solar radiation absorbed to warm or cool the Earth. Scientists of many specialties have been refining the Milankovitch cycles in ever-increasing detail in order to improve correlations with the evidence from geological field work. Others regard the sun as the prime mover of climatic change, and search for variations in solar cycles (such as the 11-year sunspot cycle), and in radiation from a "turbulent sun." Such explanations for continental glaciations are not yet well understood.   AF

An Awful Lot of Ice
Ninety percent of the world's ice can be found in the Antarctic. Scientists have estimated that, if all the ice in Antarctica were to melt, sea levels could rise by as much as 260 ft (80 m).

Ice Shelves and Glaciers

Perpendicular Cliff
When Robert Scott arrived at the Ross Ice Shelf in 1902 he wrote: "such a phenomenon was unique, and for sixty years … many a theory had been built on the slender foundation of fact … It was an impressive sight, and the very vastness of what lay at our feet seemed to add to our sense of mystery."

Underwater Glacier
Barne Glacier flows down from Mount Erebus to the Ross Sea, where it forms an ice cliff grounded on the sea floor. The cliff rises about 150 ft (45 m) above sea level and extends far below the waterline.

Australian geologist Douglas Mawson may have been the first to use the term "ice shelves" to describe sheets of very thick, mostly floating ice (in 1912). Ice shelves make up nearly half of the Antarctic coastline. They are fed by glaciers or by ice streams from Antarctica's continental ice sheets, and are additionally nourished by snow that accumulates at their surface and probably, in some cases, by seawater freezing at their base. Some areas may be aground. The Ross Ice Shelf is the world's largest ice shelf; it averages about 1,100 ft (330 m) thickness and it increases to about 2,300 ft (700 m) toward its southern boundary.

Ice Shelves

Ice shelves generally fill embayments, and therefore they are landlocked on three sides. Their surfaces are level or gently undulating, and the seaward side spawns the tabular bergs characteristic of the Southern Ocean. Tabular icebergs are usually easily distinguished from bergs formed by a calving glacier: the typical shape of a newly calved glacier berg is irregular.

Where ice shelves are joined to the land, immense cracks or crevasses can develop due to ocean tides and currents. The Grand Chasm, at the head of the Filchner Ice Shelf, for example, is a zone of almost impenetrable crevasses. It was the first challenge facing the English geologist Vivian Fuchs and his companions on their way to the continent itself in the first successful continental crossing, from the Weddell to the Ross Sea in 1957–58. There had been two previous transcontinental attempts,

by Wilhelm Filchner in 1911–12, and Ernest Shackleton in 1914–16, but both expeditions were imprisoned with their ships in the pack ice.

Most of the Antarctic ice shelves are relatively small. The three principal ones, in decreasing order of size, are the Ross, the Ronne, and the Amery. The Ross Ice Shelf lies at the head of the Ross Sea and covers an area about the size of France. Another smaller shelf, the Larsen Ice Shelf on the Weddell Sea coast of the Antarctic Peninsula, has decreased greatly in recent years and seems to be breaking up. The Larsen Ice Shelf is the probable source of many of the tabular bergs seen by visitors on cruise ships.

The first humans to reach the Ross Sea were awed by the impressive cliff front of the Ross Ice Shelf, which rises up to 165 ft (50 m) above the sea to the south—and it seemed such an impediment to travel inland that it became known as the Ross Ice (or Great) Barrier.

Visitors to the Ross Sea and to McMurdo and Scott bases are very likely to see some of the Antarctic's most developed pressure ridges. This is the point where the Ross Ice Shelf, moving northward, collides with Ross Island and is thrown into spectacular giant folds.

Glaciers

In the Antarctic, rivers of ice, known as valley glaciers, form at the outlets of the polar ice sheets that flow down through the mountains to feed the ice shelves. The Lambert Glacier, 25 miles (40 km) wide, and probably the world's largest glacier by volume, drains a major

Glacier Tongue

Vast floating glacier tongues, such as that of the Vanderford Glacier (seen here), can spawn large tabular icebergs not easily distinguished from true ice-shelf bergs. Glacier tongues project out into the ocean and are landlocked at the rear, where the feeder glacier comes from; ice shelves are usually landlocked on three sides.

area of the East Antarctic Ice Sheet to feed the Amery Ice Shelf. The fastest-flowing Antarctic glacier known, the Shirase in Dronning Maud Land, flows at a rate of 1¼ miles (2 km) per year.

On the Antarctic Peninsula there is no ice sheet, so the glaciers there are defined by the area's mountain topography. Only glaciers of the alpine type are seen, including cirque glaciers that occupy high mountain amphitheaters, and valley glaciers, which flow down mountain chasms. Many of these glaciers show evidence of retreat in recent years.

Glaciers can carry rocks on their surfaces and deposit them along their path, where they end up as ridges of bouldery material called moraines. Such ridges show the former extents of glaciers long after the ice has melted back. AF

Cracked Shelf

The Brunt Ice Shelf (in the distance) is fed mainly by ice flowing from Dronning Maud Land. The shelf regularly cracks and breaks apart, sending icebergs into the eastern Weddell Sea.

ANTARCTICA'S ICE SHELVES

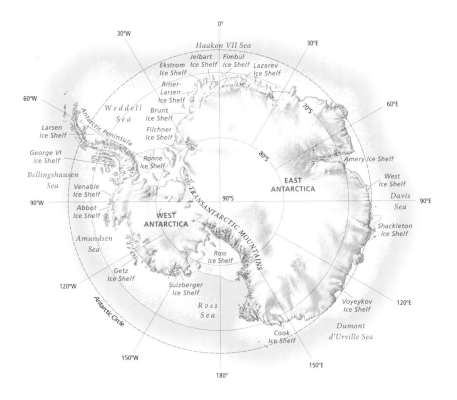

Adrift in Ice
These small bergs have run aground in shallow waters and are now locked in place by sea ice. Icebergs can remain grounded for many years. Jade bergs (in the foreground) form under special conditions at the base of floating ice shelves. The impact of large icebergs hitting the ocean floor may be nothing less than catastrophic for the dwellers on the ocean floor.

Icebergs

And now there came both mist and snow,
And it grew wondrous cold:
And ice, mast-high, came floating by,
As green as emerald.

And through the drifts the snowy clifts
Did send a dismal sheen:
Nor shapes of men nor beasts we ken—
The ice was all between.

—Samuel Taylor Coleridge
The Rime of the Ancient Mariner

There are now websites analyzing Samuel Taylor Coleridge's *Rime of the Ancient Mariner* for the geographical information it contains. Coleridge (1772–1834) never saw an albatross, let alone an iceberg, but his description is remarkably accurate. Although an iceberg is not likely to be "As green as emerald"—for green bergs do occur, but only when the ice holds a large quantity of organic matter—the relationship Coleridge describes between icebergs and the "snowy clifts" of an ice shelf rings very true.

Birth of an Iceberg
An iceberg adrift is the archetypal image of ice in the waters of Antarctica. But icebergs are not composed of frozen saltwater; they are born on land and contain freshwater. They begin life in the form of snow falling on the continental ice sheets; this accumulates and is compressed over many years as the ice sheets flow toward the sea. Upon reaching the sea, the ice spills into the water when a glacier calves, or floats on the ocean surface as an ice tongue or as a massive ice shelf. The Ross Ice Shelf is just a floating ice block, abutting the fast ice along its seaward front so that it rises and falls with the tides.

Tabular Berg
Icebergs are categorized according to their shapes and sizes. This iceberg is known as a tabular berg, with its table-like top and sheer sides. Tabular bergs can be enormous, sometimes hundreds of square miles in surface area. What appears above the water is just a fraction of the overall size of the iceberg. About four-fifths of an average iceberg is submerged.

Every year, tens of thousands of icebergs break off Antarctica and then scatter across the Southern Ocean, mainly below the Polar Front, where the water is cold enough to slow their melting. Far fewer icebergs occur in the northern polar region because there is much less glaciated land—only Greenland provides a comparable source of icebergs and it is only a fraction of the size of the Antarctic continent.

Huge tabular bergs sometimes break off ice shelves, and drift northward under the influence of winds and currents, eventually to melt in the warmer waters north of the Polar Front. On rare occasions, they have been found as far north as 45°S in the southern Pacific, and at 35°S in the Indian and south Atlantic oceans.

The giant Ronne–Filchner Ice Shelf of the Weddell Sea spawns giant flat-topped or tabular bergs. These are trapped in the field of sea ice that rotates clockwise and almost fills that sea. It was such a rotation that eventually carried Ernest Shackleton and his party to open water long after their ship, the *Endurance*, was crushed in the pack ice in 1916.

As well as the huge icebergs that regularly make the news headlines, there are innumerable smaller ones. One survey carried out in 1985 found 30,000 bergs in an area of 1,500 sq miles (4,000 sq km) of ocean. Today, satellite imaging and satellite tracking gather a more accurate record. The United States National Ice Center maintains a database of the world's large icebergs (those more than 6 miles/10 km long). It can be accessed at the website www.natice.noaa.gov, along with satellite images of the larger, tabular bergs.

Profiles of Bergs

Icebergs are named by their quadrant of origin. Those designated "A" are from 0° to 90°W (the Bellingshausen Sea and the Weddell Sea); "B" bergs are from 90°W to

Fantastic Shapes

The tall ice sculpture at the center of the picture is a weathered berg, which has eroded (probably from an old cave) to form a pillar on the left, and tilted so that an old water line is exposed at an angle across the berg. Bergy bits bob up and down in the foreground.

Blue Ice

As snow falls, its weight compresses the snow beneath, and it turns to ice. At first, the ice is air-filled and granular, but by the time it sinks to 165 ft (50 m), it has become solid ice clouded with bubbles of air. As it descends deeper, the bubbles are compressed out, and the ice becomes clear blue. When a glacier containing this blue ice breaks off and floats away, it becomes a "blue berg"—one of Antarctica's most attractive features.

180°W (the Amundsen Sea and eastern Ross Sea); "C" bergs are from 180°E to 90°E (the western Ross Sea and Wilkes Land), and "D" bergs are from 90°E to 0° (the Amery Ice Shelf and the eastern Weddell Sea). A large tabular iceberg can be 980 ft (300 m) thick, may weigh hundreds of million of tons, and may contain enough freshwater to supply a city of a million people for several years.

A vast hunk of the Amery Ice Shelf broke off in 1963, and, four years later, it collided with the Fimbul Ice Shelf to produce two tabular bergs, 68 by 47 miles (110 by 75 km) and 65 by 33 miles (104 by 53 km). The last recognizable chunk of these was seen off the Antarctic Peninsula in 1976. Toward the end of the twentieth century, there was considerable interest in the berg B-10A, which had broken off in 1992, and entered shipping lanes between the Antarctic Peninsula and South America in 1998. It measured about 62 by 31 miles (100 by 50 km).

But these giants were dwarfed in 2000, when the Ross Ice Shelf calved the biggest iceberg ever seen. The larger of the two segments that broke away from the shelf in March 2000 was about 185 miles long and

25 miles wide (300 by 40 km) and had an area of more than 4,215 sq miles (10,915 sq km). Soon afterward, on May 4 and May 6, two massive icebergs calved from the Ronne Ice Shelf. The first soon broke in two: these icebergs become known as A-43A, measuring 104 by 21 miles (168 by 33 km) and A-43B, which was 52 by 22 miles (84 by 35 km). The other was named A-44, and measured 37 by 20 miles (60 by 32 km).

Whereas ice shelves give birth to tabular icebergs, most glacial tongues (unless they are enormous) create icebergs of irregular shapes that can inspire flights of fantasy. Frank Worsley, who was captain of Shackleton's *Endurance*, described the ice threatening their lifeboat, the *James Caird*, as they struggled across the Southern Ocean to South Georgia: "Swans of weird shape pecked at our planks, a gondola steered by a giraffe ran foul of us, which much amused a duck sitting on a crocodile's head … All the strange, fantastic shapes rose and fell in stately cadence with a rustling, whispering sound and hollow echoes to the thudding seas."

Icebergs range from crystal clear and pure white to green, brown, and blue—even pink (from algae). Most bergs have fissures of bright electric blue and many are

Finding a Balance

Eroded by wind and waves, sections of tabular bergs break away. The iceberg slumps into the ocean, but finds its balance moments later, revealing a magnificent ice cave. The broken chunks of ice float up to the surface.

fringed by icicles. The shapes are unimaginable—from the hewn straight sides of a recently calved tabular berg to the grottoes, pinnacles, and arches of a berg that has rolled to reveal its partially melted underside.

Dangers and Difficulties

The expression—"just the tip of the iceberg"—takes on new significance in polar regions. Less than 20 percent of an iceberg is visible above the water, but the actual percentage can change depending on such factors as impurities in the ice and the salinity and temperature of the water. Although they look huge and unchanging, icebergs are inherently unstable, as they are constantly melting from both the top and the bottom.

Smaller icebergs (relative terminology if they are seen from onboard a ship) become top-heavy as they melt underwater, and they can roll quickly when their equilibrium shifts. Large bergs may seem to explode violently, as they collapse into many pieces, creating large waves and dangerous vortices.

Ice shelves spawn tabular bergs that may tip over to become capsized bergs, or they may create castellated bergs after a lot of the ice has melted from underneath. Glacial ice breaks into the ocean as smaller icebergs that decay to become "bergy bits," and then "growlers" when they have melted enough so that little can be seen above the water. Growlers are especially dangerous because they may float unnoticed by a helmsman or radar until a ship collides with one. The most notable shipping disaster involving an iceberg was the *Titanic*, which sank at a latitude of 41°46′N, on April 15, 1912, on its maiden voyage from Southampton to New York—with the loss of 1,503 lives. But the *Titanic* did not hit a growler but came to grief on the extended underwater "foot" of an iceberg.

Most icebergs simply drift toward warmer, temperate waters, where they quickly melt. However, some run aground on shallow shoals and stay intact for years. In the 1970s, there was considerable interest in the idea of towing icebergs toward the desert coastlines of South America, Australia, and Africa, where their freshwater could be used to irrigate barren land. This bold plan was not completely impractical, although the expense would have been enormous, but it was abandoned after consideration of all the details. One difficulty was that if an iceberg broke up while under tow in warmer water it would have created safety problems for shipping. But the main problem was the underwater bulk of the berg, which would almost certainly have become grounded on the continental shelf at a distance too far from land to be able to pipe the water ashore. DM

Eroded Berg
In the summer months, this iceberg has cracked and its edges have melted into the sea. Once calved, the iceberg may remain grounded on the ocean floor or it may drift north, melting and breaking as it goes.

Jelly Mold
The bulbous mound of an overturned iceberg is scarred by parallel bubble rills (or channels), which formed while the berg sat upright in the water. Smooth scallop features, the size of a cupped hand, pockmark the once submerged surface. The old crown of the iceberg juts skyward at the back.

The Frozen Seas

New Ice for Old

Beyond the edge of last winter's sea ice in McMurdo Sound, new sea ice is forming. Its appearance depends on conditions at the time. On calm seas a thin slick forms, which gradually thickens into glossy sheets just a few inches thick. Typically, disturbances from sea or wind break these fragile sheets up and they ride over each other, creating amazing patterns.

Strong Ice

By midsummer, there will be open water around the Cape Royds coastline with orcas and minke whales feeding (right), but in the spring the fast ice is strong enough to travel safely on. Broken floes are trapped in the sea ice, whose uneven surface and cracks bear testimony to some of the pressures acting on it.

As winter approaches, the shores of Antarctica go through a transformation. By March, after weeks of 24-hour sunlight, most of the sea ice that normally surrounds the continent has melted. All that is left is a rim of ice in some places, most notably in the Weddell and Bellingshausen seas.

The Formation of Sea Ice

Each evening, as the winter draws closer, the sun dips farther below the horizon and the air temperature falls enough to create a thin sheet of ice on the surface of the water, seen first as ice needles and tiny ice plates, known as frazil. It develops into a thin sludge that sailors call grease ice. At first, the ice crystals are about 1 in (25 mm) wide and less than ⅛ in (3 mm) thick, but when the sea is calm the crystals grow quickly to form sheets, especially when snow falls on top of this skin of ice. This plastic ice, up to 4 in (10 cm) thick, which bends readily with the waves, is called nilas. Wind and wave motion, however, may break the ice into plates, and then bang them together to create pancake ice—a clear sign of substantial freezing, and a warning to polar mariners that winter is certainly on the way.

If the temperatures stay low for several days, the pancakes meld and thicken to form floes of new ice. In a month, the ice may thaw, freeze, and be covered by snow, forming a sheet of sea ice that is 6 to 24 in (15 to 60 cm) thick. Sea ice that is attached to the edge of the continent is called fast ice; ice that drifts with the currents offshore is called pack ice. Fast ice thickens mostly as the result of new ice forming on the underside of the surface ice (this is also particularly true of the way

Arctic sea ice forms), whereas 50 percent or more of the thickening of pack ice is caused by the pancakes or small floes of ice piling one on top of another (known as rafting) during storms.

This expanse of new ice grows at an incredible rate: up to 23 sq miles (60 sq km) per minute. Throughout

winter, it advances about 2½ miles (4 km) each day, adding almost 40,000 sq miles (100,000 sq km) of new ice. In October, at the end of an average winter, it has effectively more than doubled the size of Antarctica. The 5,400,000 sq miles (14 million sq km) of land abut 7,700,000 sq miles (20 million sq km) of frozen ocean, reaching out more than 1,240 miles (2,000 km) from the coast. The ice is generally about 3 ft (1 m) thick, but can be up to 33 ft (10 m) in places. By October, the Pacific pack ice extends northward to about 62°S, while the Atlantic pack extends much farther north, to approximately 52°S. The northern range of icebergs is from 6 to 10 degrees beyond the pack ice.

New ice does not accumulate uniformly around the whole Antarctic continent. It starts to form in the Weddell Sea, and then in the Ross Sea and the Bellingshausen Sea. As Ernest Shackleton had ample time to observe, sea ice moves continually. In the Weddell Sea, where his *Endurance* was trapped and then finally crushed, a clockwise drift is more defined than the drift in either the Ross or Bellingshausen seas.

Neither is the spring melt a simple reversal of the freezing process. Ice that once formed near the coast can be found at great distances from the shoreline by December, and in late summer there can be a thick ring of pack ice far from land, while the coastal waters are relatively ice-free and navigable.

Polar Dissimilarities

Sea ice of the two polar regions differs in many ways. Arctic sea ice may have a lifespan of up to eight years, but most Antarctic sea ice is less than a year old—only

within the Weddell and Bellingshausen seas is sea ice found that is up to three years old. About 90 percent of Arctic sea ice is older than a year, more than 7 ft (2 m) thick, very strong, and low in salinity. With Antarctic sea ice, a similar proportion is younger than a year old, less than 7 ft (2 m) thick, structurally weak, and quite saline. Old sea ice is much stronger than first-year ice, and it is much more difficult for ships to break through.

The area of Antarctic pack ice is much larger than in the Arctic in winter: 7,700,000 sq miles (20 million sq km), compared with the Arctic's 5,400,000 sq miles (14 million sq km). In the summer, the sea ice in both regions is much reduced: 1,500,000 sq miles (4 million sq km) in Antarctica, as against the Arctic's 2,700,000 sq miles (7 million sq km).

Shattered Ice
Angular in shape, ice forms collide with one another. These floes may freeze together and break up several times before forming a solid cover. As the ice thickens, ocean swells break it into larger pieces.

WINTER–SUMMER ICE

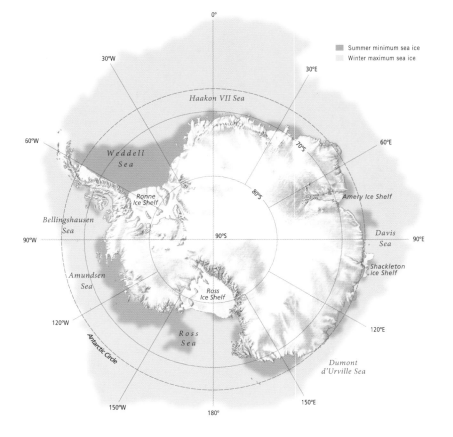

Summer minimum sea ice
Winter maximum sea ice

Beware of Cracks

A recurring crack in Erebus Bay shows that even the fast ice is subject to pressures, such as tides, currents, and weather. Fast ice can be a convenient traveling surface between winter and early summer—but care is necessary.

Sea Ice Nomenclature

Sea ice is named by size. Fragments of ice less than 7 ft (2 m) across—smaller than a grand piano—are called "brash ice." Ice floes are "small" if they are under 330 ft (100 m) wide, "medium" if less than 1,000 ft (300 m), and "large" if under 1¼ miles (2 km). A "vast floe" measures up to 6 miles (10 km) across; anything larger than that achieves "giant floe" status.

When wind and ocean currents push the ice apart, the open water is called a lead, and the heat released from the water below can create frost smoke. A very large area of open water is known by the Russian name: polynya. In shipping terms, "open pack" refers to water containing up to six-tenths ice, and "close pack" means seven- or eight-tenths ice, in which some water is still visible. An icebreaker or an ice-strengthened vessel is needed for movement in "very close pack"—meaning nine-tenths ice. "Compact" pack is all ice, no leads; no water at all is visible—even most icebreakers would have difficulty passing through it.

How Sea Ice Behaves

In some ways, it is miraculous that sea ice forms at all. Only a few substances are less dense as solids than they are as liquids; water is one of them. The density of pure ice is 91.7 percent that of pure water—so it floats with 8.3 percent of its mass above the water. Further, because of its salt content, seawater freezes at around 29°F (−1.8°C) and, unlike freshwater, it increases in density as it approaches freezing point. So, therefore, when it cools, it sinks. Only when the cold air above has cooled a top layer of water around 10 ft (3 m) thick to freezing point can sea ice form.

The underwater dynamics beneath sea ice are very different from those in a lake or river covered in ice. As freshwater cools, it increases in density to about 39°F (4°C), then it slightly decreases in density until 32°F (0°C). The density of the saltwater increases between 32°F (0°C) and its sub-zero freezing point. So, in an ice-covered lake, the coldest water is lighter than the warmer water, and the water forms bands: ice on top, colder water below it, and warmer water at the bottom. In the sea, the ice cools the water directly beneath it, some of which freezes to the bottom of the ice, while the rest of the water falls (when its density increases as it approaches freezing point), to be replaced by warmer, lighter water. So the ocean's water is circulating beneath the sea ice constantly.

The extent of the sea ice varies considerably from year to year: climatologists are searching for patterns that may reveal evidence of global warming. There is no doubt the presence of so much ice influences the climate of Antarctica. The sea-ice "skin" restricts the normal heat exchange between the atmosphere and the ocean. In spring, the sea ice reflects back solar energy that would otherwise warm the ocean and start summer earlier. At the start of winter, new sea ice reflects the last

Dirty Ice

Ice shelves are typically much "cleaner" than this area, which is called the Dirty Ice, lying at the edge of the ice shelf at McMurdo Sound in midsummer when the surrounding sea ice has broken out. Each of the ponds in the area has a unique chemistry and is full of cyano-bacterial microbial mats.

Pancake Ice

Pancake ice looks very much like pale waterlily leaves—each flat plate has a raised edge around it from collisions with its neighbors.

sunlight away, and isolates the (relatively) warm ocean from the air, so that winter comes much sooner.

Is Sea Ice Salty?

Sea ice always contains some salt, but generally its salinity is about one-tenth that of seawater. As seawater freezes, the salt is forced out but gets caught up in tiny brine pockets within the ice. The faster is the freezing, the more salt is captured. Over weeks and months, in response to gravity, the trapped salt migrates downward through the ice. So, gradually, the sea ice becomes less saline at the top than at the bottom. In the Arctic, the Inuit often use surface ice for their drinking water; sea ice that is more than one year old has a salt content of less than 0.1 percent.

The freezing and the melting of polar sea ice greatly affects the surface-water salinities of the sea. As sea ice forms, the exclusion of salts from the growing crystals of ice concentrates the salts into the seawater, and thus increases the sea's salinity. Throughout the spring and summer seasons, the surface seawater loses salinity as this sea ice melts. DM

Lights in the Sky

High Cloud

The delicate "mother-of-pearl" colors in these nacreous clouds are produced when fading light at sunrise or sunset passes through tiny ice crystals blown along on a strong jet of stratospheric air. The clouds are situated high in the stratosphere, some 12 miles (20 km) above the ground. The rare clouds only occur at high polar latitudes in winter months, when temperatures are less than about −115°F (−82°C).

A Gifted Doctor

Edward Wilson, a close confidant of Robert Scott, was a doctor, a naturalist, and artist. During Scott's expedition of 1901–04, Wilson sketched the *Discovery* beneath a stunning aurora australis.

Aurora, named for the Roman goddess of dawn, is a dramatic natural phenomenon that swirls or pulsates in the night sky, creating curtains and patches of colored light, predominantly in mixtures of green, red, and violet. An active display is an awe-inspiring sight.

More analytically, aurora is the light emitted by atoms, molecules, and ions that have been excited by charged particles, principally electrons, traveling along magnetic field lines into the Earth's upper atmosphere. Light is emitted when excited atmospheric constituents release energy while returning to a lower energy state. This is similar to the process which generates light in a neon tube. For neon, the dominant wavelength that is emitted corresponds to a red color.

What Causes Aurora?

The solar corona—the outer region of the sun—is hot enough to continually emit solar wind plasma (a gas consisting of charged particles composed mainly of protons and electrons), which streams into space at speeds of around 185 miles (300 km) per second. The Earth's magnetic field also confines plasma; this travels along the field lines and is bounced between the two hemispheres by the increasing field strength encountered in polar regions.

Electrons in the solar wind or trapped by the Earth's magnetic field do not typically have sufficient energy to generate aurora. The processes associated with the interaction of solar wind and the Earth's magnetic field accelerate these electrons. Solar plasma also carries a magnetic field. When the solar wind's magnetic field opposes the Earth's, energy is transferred by a process called magnetic reconnection. The extraction of energy from reconnecting magnetic fields can be demonstrated when the opposite poles of two bar magnets are brought together gradually. The Earth's magnetic field becomes linked to the solar wind by magnetic reconnection on the day-side and disconnected on the night-side, thus accelerating electrons on both occasions. Electrons may be accelerated by an interaction that resembles friction too—as the solar wind plasma which has not been magnetically reconnected to the Earth races past the flanks of its magnetic field, like wind passing over water and generating waves.

Global Occurrences

Aurora, in the form of an oval around the magnetic pole in both hemispheres, results from these interactions. Most commonly, aurora is located at magnetic latitudes of about 67 degrees on the night-side and of about 75 degrees on the day-side. When considering the chances of observing aurora at a specific location, it is important to take into account the separation of the magnetic and the geographic poles.

Aurora is known as aurora australis, or the southern lights, in the Southern Hemisphere; it is called aurora borealis, the northern lights, in the Northern Hemisphere. Auroral activity in the two hemispheres is strongly linked

Fiery Inferno

Folklore and legend has it that the aurora was the flickering of distant fires far over the horizon. Even Galileo attributed it to sunlight reflecting off the atmosphere. Blazing red displays like this at Mawson Station are, in fact, due to low-energy electrons crashing into oxygen atoms in the rarefied upper atmosphere above the base.

Split Corona

The expansive phase of an auroral substorm develops in minutes to the break-up stage. Individual folds in previously quiescent auroral arcs can split into spirals that splay outward from a single point.

in space and time, as electrons can travel freely in either direction along magnetic field lines. Brightenings in the two hemispheres coincide within seconds in magnetically linked regions. This was confirmed in the 1970s by measurements made on aircraft following flight paths linked by the Earth's magnetic field out of Christchurch, New Zealand, and Fairbanks, Alaska.

Sunspots (cooler regions of intense magnetic field) are signs of enhanced solar wind activity. They have a cycle of about 11 years, which recently peaked in 2000. Aurora is commoner at lower latitudes at peak sunspot times and during the subsequent two years; it slightly intensifies in spring and autumn.

Observing Aurora

It is possible to observe about 440 miles (700 km) of the approximately 5,000-mile (8,000-km) wide auroral oval from the ground. In quiet times, this appears as a shimmering curtain extending in an east–west direction.

More active displays, known as auroral substorms, occur when stored energy is rapidly released. During an explosive onset, rays and arcs may twist and move and break into small segments as enhanced electric currents flowing near auroral heights modify the field lines along which the charged particles are traveling. Pulsating patches of auroral glow may persist during the gradual recovery phase, and eventually quiet arcs re-form. A substorm may last an hour, and may recur up to three or four times each night.

The aurora produces many colors, each of them corresponding to specific energy level transitions of excited atoms, molecules, or ions. Three, however, are most significant. A green line emitted by an oxygen atom is dominant when a sharp lower border can be discerned, usually at about 60 miles (100 km). Red is generated by another atomic oxygen transition, and is prominent at altitudes around 155 miles (250 km). A violet band emitted by a nitrogen molecular ion is about five times less intense than the green line, but can be observed on the leading edge of active aurora because of the delay of about a second in the average emission time of auroral green light.   GB

Vanishing Point

The coronal point is where overhead auroral rays seem to meet. Along the line of the Earth's local magnetic field in this direction, individual rays delineate neighboring field lines that appear to converge like skyscraper spires viewed from immediately below.

The Antarctic Ozone Hole

Ozone, a form of oxygen, is a comparatively sparse constituent of the Earth's atmosphere—there are only about three molecules of it in every 10 million molecules of air. Ozone is extremely important to the atmosphere, however, and, for the past 600 million years or so, it has been the atmospheric sentinel protecting life on the planet from the biological damage caused by the sun's ultraviolet radiation. Many experimental studies of plants, animals, and humans have demonstrated that there are harmful effects from excessive exposure to ultraviolet-B radiation. This is the reason the maintenance of the ozone layer is of such vital concern, particularly in view of recent evidence that synthetic chemicals have caused significant loss of ozone from the atmosphere.

Damaging the Atmosphere

During the early 1970s, atmospheric scientists who were engaged in theoretical research first became aware that possible damage to the ozone layer might be occurring as a result of human activities. However, it was not until 1985 that British Antarctic Survey scientists identified the Antarctic ozone hole and solid evidence was found that the ozone layer might be under threat.

In 1987, the United States carried out a research program in which high-altitude NASA aircraft, fitted with sophisticated scientific equipment, were flown through the stratosphere from South America into the Antarctic ozone hole. These experiments produced irrefutable evidence that the ozone hole was mainly caused by the chemical destruction of ozone by atmospheric chlorine. Because the quantity of chlorine required to destroy the

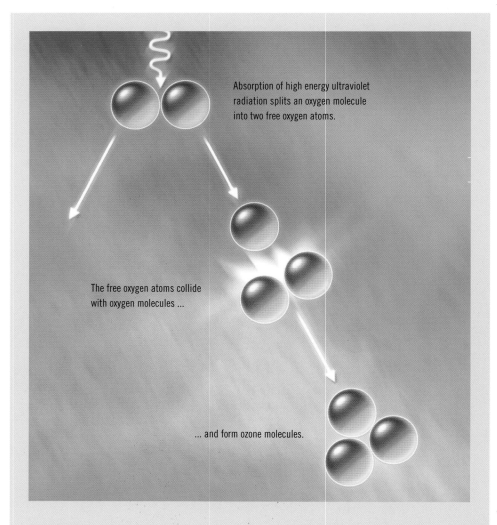

Absorption of high energy ultraviolet radiation splits an oxygen molecule into two free oxygen atoms.

The free oxygen atoms collide with oxygen molecules ...

... and form ozone molecules.

Making Ozone

The sun's ultraviolet radiation is more intense in the upper atmosphere, where it is not absorbed or reflected by atmospheric gases and particles to the extent that occurs lower down. This means that it can separate the atoms in oxygen molecules more often than it can lower down in the atmosphere, and the result is more free oxygen atoms. However, the joining of free oxygen atoms with oxygen molecules to produce ozone occurs more often lower down, where the atmosphere is denser, and these atoms and molecules collide more often.

Because there are more free oxygen atoms higher up, yet more ozone-producing collisions between oxygen atoms and molecules lower down, the creation of ozone is at its greatest at the intermediate altitudes—and it is this that forms an "ozone layer." The altitudinal range of the atmosphere in which the ozone layer resides is between about 6 and 31 miles (10 and 50 km), and is known as the "stratosphere."

Telling Instrument
A scientist measures how much ozone is overhead at New Zealand's Arrival Heights Laboratory in Antarctica, using a Dobson spectrophotometer. The ozone hole was first discovered from Dobson spectrophotometer data. The Dobson unit, named after British scientist Gordon Dobson, is the most common measurement of ozone.

amount of ozone lost could not originate from natural sources, another culprit had to be identified. Scientists reached the inescapable conclusion that pollution from synthetic chemicals was causing ozone depletion. The prime offenders were halocarbon compounds, such as chlorofluorocarbons (CFCs), containing chlorine atoms.

CFCs are non-toxic and non-flammable, and they can easily be converted from a liquid to a gas, and from gas to liquid. They have many useful commercial and industrial applications—for example, as cleaning and foam-blowing agents, as aerosol-spray propellants, and as the cooling fluid in refrigerators. Because CFCs were relatively chemically non-reactive with most substances

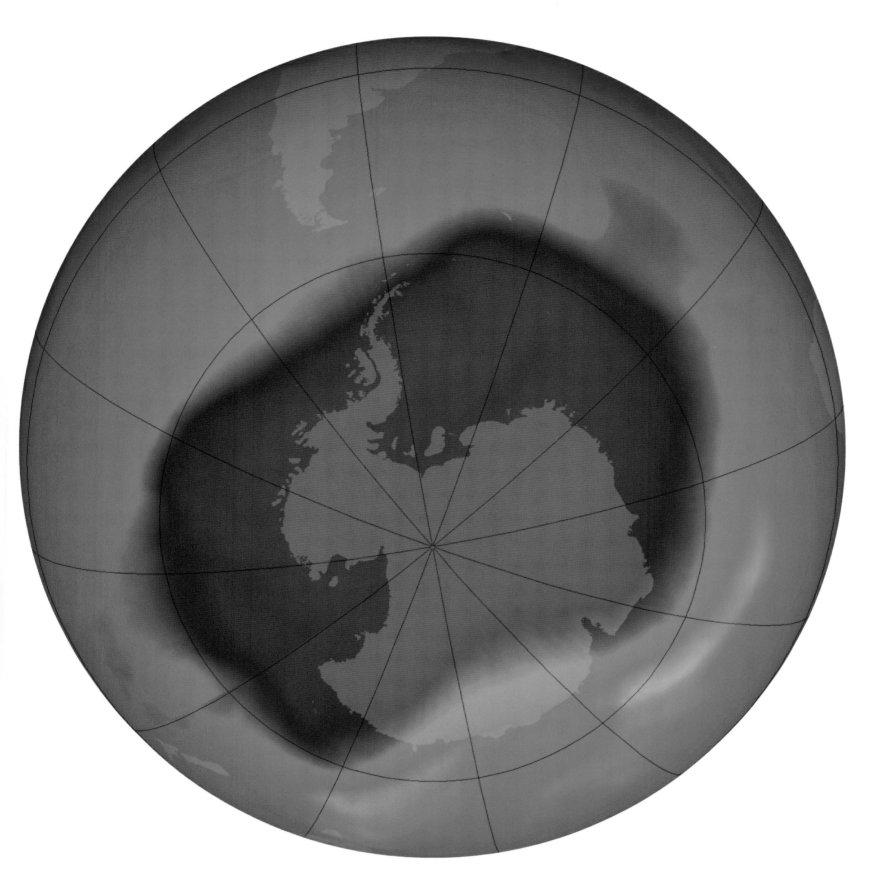

in the environment, this also meant that any of these substances would stay in the atmosphere a very long time if they escaped or were released.

Natural atmospheric circulation transports these chemicals high up into the stratosphere—there, the ultraviolet radiation is comparatively strong and can strip the chlorine atoms off the CFC molecules. Thus, chlorine atoms from synthetic chemicals are liberated into the stratosphere, and are able to "attack" ozone molecules immediately, and convert them to oxygen by pulling off single oxygen atoms. Through further chemical reactions in the atmosphere, these chlorine atoms are effectively recycled to destroy more ozone

molecules—one free chlorine atom can destroy tens of thousands of ozone molecules in the stratosphere before being removed from its destructive cycle by nitrogen compounds. This is the catalytic process that is the dominant and fundamental mechanism of ozone destruction. Through the process, large numbers of ozone molecules are converted into oxygen, and during this conversion their ability to absorb ultraviolet radiation is lost. To a lesser extent, ozone depletion can also be caused by pollution from chemical components other than chlorine, such as bromine. As a result, ultraviolet radiation at the Earth's surface increases to a level that has the potential to be detrimental to life.

NASA Records Largest Ozone Hole to Date

In late September 2006, the ozone hole over Antarctica reached a record size of around 11,275,000 sq miles (29.2 million sq km)—this is more than three times larger than the land area of the United States. The ozone hole (purple) can be seen over the entire continent and causes increases in ultraviolet radiation. Satellite sensors play an important role in the detection and monitoring of the ozone hole.

Patterns of Distribution

Ozone depletion in the stratosphere depends on latitude and season. For example, between latitudes 30°S and 60°S, the total ozone reductions year-round were about 5 percent between the late 1970s and the mid-1990s.

The southern ozone hole forms in the early spring (August) over the Antarctic continent, when the polar dawn occurs and sunlight splits chlorine molecules into the ozone-destroying chlorine atoms, which drastically deplete ozone in the stratosphere. In the years leading up to, and including, 2000, the ozone hole suffered an ozone loss of up to 70 percent, and has sometimes covered 11,275,000 sq miles (29.2 million sq km)—which is more than three times the area of the United States. The ozone hole slowly reaches its greatest depth and area in late September to early October, and then gradually recovers as fresh ozone from lower latitudes slowly enters the ozone hole region. On occasions, the ozone hole passes over Chile in South America, where people are alerted through the media to the concern of increased ultraviolet radiation. Typically, the ozone hole finally breaks up after mid-November, when the Polar Vortex—the wind system that contains it—weakens and breaks down. It is at this time that ozone-depleted air can move to lower latitudes and reduce ozone levels over Australia, New Zealand, and the other Southern Hemisphere countries. The effect of this decrease in ozone appears to persist into the following year.

In the Northern Hemisphere, ozone losses over the Arctic region during spring have reached up to about 30 percent, and therefore are considerably less than the losses that occur over Antarctica. The Arctic stratosphere suffers less ozone loss because it is generally warmer (approximately 50°F/10°C) than the stratosphere of the Antarctic, particularly in March, when sufficient sunlight is available to cause large ozone depletion.

International Problem Solving

Evidence that synthetic chemicals are responsible for the ozone depletion observed since the mid-1970s is overwhelming. In 1985, the Vienna Convention for the Protection of the Ozone Layer was established as a general treaty to facilitate international cooperation in progressively eliminating all known ozone-depleting substances. Twenty nations signed an agreement to take "appropriate measures … to protect human health

French Measure
Since the ozone hole was identified, there has been increased international research on the condition of the Earth's atmosphere, much of it based in the Antarctic. These scientists (above) are working for the French research organizations CNES and CNRS.

Monitoring Ozone

Progression of the Antarctic ozone hole from 1979 to 2006 (the first three measured by the Total Ozone Mapping Spectrometer on NASA's Nimbus-7 satellite, the fourth by the Ozone Monitoring Instrument on the Aura spacecraft), showing the averages for October. The blue and purple colors are where there is the least ozone.

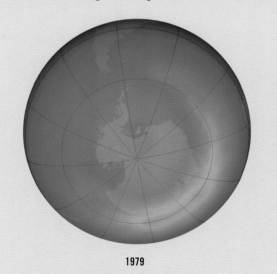

1979

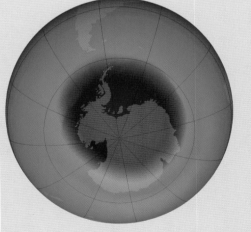

1989

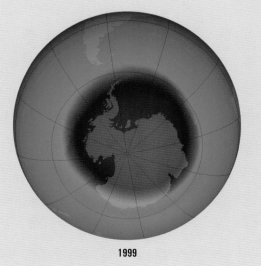

1999

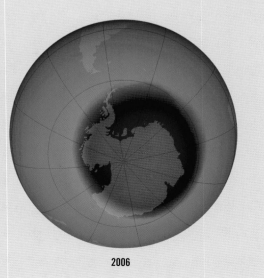

2006

and the environment against adverse effects resulting or likely to result from human activities which modify or are likely to modify the Ozone Layer." The measures, however, were not specified.

A protocol to this treaty, known as the Montreal Protocol on Substances that Deplete the Ozone Layer, was established in 1987 to identify and to implement specific procedures to regulate the use and production of ozone-depleting substances worldwide. Parties to the convention and protocol now number 191, of which more than 100 are developing countries. This is the first time that nations have agreed, in principle, to cooperate to solve a global environmental problem.

The combined stratospheric abundance of ozone-depleting chemicals has shown a downward trend since

Up, Up, and Away
Researchers preparing to launch a helium-filled balloon that carries instruments to measure ozone and properties of cloud particles involved in ozone destruction.

their peak values in the late 1990s. Globally, ozone levels appear to have stabilized at mid-1990 values in response to the Montreal Protocol, although it is debatable whether ozone recovery is yet significant.

Evidence is mounting that climate change may be influencing the recovery of stratospheric ozone by cooling the stratosphere and altering the stratospheric air circulation. Climate change may also cause more water vapor to enter the stratosphere over the tropics, and this will further exacerbate the chemical processes that destroy ozone.

These effects, coupled with the longevity of many of the ozone-depleting substances in the atmosphere already and the burden of future chemical emissions, mean that the length of time required for stratospheric ozone to recover to pre-1980 levels depends on many factors, some of them unknown. However, discoveries made since the early 1990s have significantly altered our understanding and forecasting of ozone depletion.

Whatever the future outcomes of research, it is vital that no further ozone-destroying chemicals are released into the stratosphere so that there is an opportunity for the ozone layer to repair itself. PL

Changing Weather Patterns

The Climate Museum

The bleakest and most hostile environments in the world, the two polar regions are the touchstone for measuring the changing weather patterns and reservoirs for much data about the Earth's climatic history.

Antarctica plays an integral role in the world's climate: it is a heat sink, it powers currents in the world's vast oceans, and it influences much of the weather in the mid-latitudes (from 30°S to 50°S) of the Southern Hemisphere. The Antarctic continent is surrounded and isolated by the most productive marine environment, the Southern Ocean, which also absorbs heat and carbon dioxide. Through the actions of wind, evaporation, and winter sea-ice formation, salty water accumulates at the ice sheet–sea margin and increases the density of surface waters. These waters, which are now heavy, descend to the ocean floor and form deep, nutrient-rich currents that flow around the world on the Great Ocean Conveyer Belt, or Thermohaline Current.

The Antarctic ice sheets are a "climate museum," providing the world with a legacy in the ice: tiny bubbles of air from past times. As snow and ice accumulate, the bubbles become trapped in the layers of annual snowfall, which are analogous to tree rings: the deeper the ice, the older the gas. Scientists analyze air bubbles from ice cores and these data contribute to the placing of the current global warming in context with recent and not-so recent climatic conditions. Atmospheric concentrations of carbon dioxide and other greenhouse gases, such as methane and nitrous oxide, have increased significantly since the mid-eighteenth century and far exceed pre-industrial values. For most of the last 800,000 years, the Earth's norm has been "icehouse," not "greenhouse."

The increase in carbon dioxide is mainly caused by the burning of fossil fuels and changes in land use, such as the clearing of forests. Increases in methane and nitrous oxide are due to increased agriculture. These changes in greenhouse gases in the atmosphere modify the energy balance of the Earth's climate systems—and the result

Halley Observers
A meteorologist services the equipment at Halley V Station. It was here, in 1985, that British scientists first measured the ozone depletion of the stratosphere.

Stories on Ice
The ice that covers Antarctica is a repository of the world's history, the ancient layers of ice revealing past temperatures, precipitation, and atmospheric conditions, and it tracks the many changes in ecosystems and species over time.

has been global warming, with rises in air and ocean temperatures, the widespread melting of snow and ice (particularly in the Antarctic Peninsula region, Alaska, the Arctic, and Asian high mountains), and increasing average global sea levels.

Is Antarctica Warming?
In the 1950s and 1960s, permanent scientific bases were established in Antarctica by Argentina, Australia, Britain, Chile, France, Japan, New Zealand, the Soviet Union, and the United States. Much of this activity was based around the 1957–58 International Geophysical Year (IGY). Weather stations at 18 of these stations have yielded continuous climate records for air temperature, pressure, and wind speed for around five decades. The message that scientists have extracted from these records is one of regional variation. Eleven stations had slight warming trends over the period; the remaining seven stations demonstrated slight cooling trends. The recent changing Antarctic weather patterns appear to be due to natural variability and human-induced climate change.

One of the stations with a cooling trend was the Amundsen–Scott Base at the South Pole. These data correspond with other signals to indicate that in recent times the interior of Antarctica has cooled slightly. The strong westerly winds that surround Antarctica generally restrict warm maritime air from penetrating into much of the interior of the continent. It is thought that most of East Antarctica is neither warming nor cooling.

There is some evidence of local summer cooling, on a decade-long scale, in the Dry Valleys region, however. Scientists from the United States established a Long-Term Ecological Research site (LTER) there in the early 1990s, at the beginning of this cooling phase, and have observed significantly reduced stream flow, the lowering of lake levels, and increased lake-ice thickness, which, in turn, has impacted on the ecosystems of these unique water bodies. However, in the summer of 2001–02, when a "relative" heatwave struck the region and melted local glaciers, the lake levels were rapidly restored to the levels observed before 1990. This heatwave also resulted in McMurdo Station recording a record-high air temperature of 50.9°F (10.5°C); this was recorded on December 30, 2001, and was 1.98°F (1.1°C) higher than the previous record set in the mid-1970s and well above the average December temperature of 26.2°F (–3.2°C). Around the same time, rain (as opposed to snow) fell over nine days at Dumont d'Urville on the Terre Adélie coast.

The most substantial and remarkable climate change signal is coming from West Antarctica. The Vernadsky weather station on the Antarctic Peninsula has recorded the substantial warming that is evident over most of the western side of the peninsula. Over the past 50 years, the mean air temperatures there have risen by nearly 5.4°F (3°C), with the greatest warming of around 9°F (5°C) occurring in winter months. This makes it one of the most rapidly warming parts of the world. The east coast of the peninsula is also warming, but not at the

Continuous Records
Glaciologists drill in sea ice for information about the ice growth and melt. Meanwhile, meteorologists are analyzing over half a century of data from weather stations in the Antarctic as they chart global changes in climate.

same astonishing rate. The warming has been linked to winds carrying warmer, moist air masses across the peninsula and onto the western ice shelves. Most of the glaciers on the peninsula's west coast have retreated in the last 50 years, as have the ice shelves around the peninsula, including, and most dramatically, the Larsen Ice Shelf in the Weddell Sea. A major recession has also occurred in the winter sea ice in the Bellingshausen and Amundsen seas in recent decades.

Two other major climate parameters—air pressure and wind speed—have also changed recently in the region. Variation in climate in the Southern Hemisphere is strongly dependant on the difference in such climate parameters between the Antarctic region and the mid-latitudes (approximately 40°S). Since the 1960s, there has been a clear trend in which sea-level air pressure in the Antarctic has been dropping (more low-pressure systems) and, conversely, increasing in mid-latitudes (more high-pressure systems). This has resulted in an increase in westerly winds around the continent and a poleward movement of the Polar Front jet stream and storm systems (non-tropical cyclones), which now track 1.5 to 2 degrees farther south on average.

In addition, a deep low-pressure system commonly forms off the Amundsen Sea. Changes in atmospheric pressure have led to more cyclones in the polar trough. These cyclones are believed to contribute to an increase in the frequency of maritime air masses passing over the Antarctic Peninsula and are responsible for "dragging" cold continental air masses from the interior, causing cooling around the continental edge.

Much of this change at the surface of the Earth is linked to changes occurring in the upper atmosphere over the last 30 years. The layers below the troposphere (5 miles/8 km in height) have warmed. Similar patterns have been observed around the world and are thought to be linked to increasing greenhouse gases, although warming in the troposphere above the Antarctic region is at a far greater rate than it is elsewhere. In contrast, temperatures above the tropopause—in the stratosphere—have cooled by up to 18°F (10°C). One of the features that is linked with this stratospheric cooling is the more frequent appearance of noctilucent clouds—"night-shining" clouds—very thin clouds of ice which form at around 30 to 50 miles (50 to 85 km) above the Earth and are indicative of extremely cold temperatures (lower than –202°F/–130°C).

The Circumpolar Vortex
Some cooling in the stratosphere is a natural annual event. During the darker months of winter, when limited energy is coming from the sun, the air mass at high altitudes above Antarctica naturally cools and moves in a westerly circular motion, which is in accordance with the Earth's rotation. On an annual basis, this air mass develops into an enormous persistent cyclone, called the Circumpolar Vortex, which circles around a near-stationary column of air above the South Pole. It is strongest in midwinter and breaks down in summer. But the vortex is spinning around Antarctica in a tighter circle, due to the recent additional cooling in the stratosphere, and it is breaking up later too. Thirty years

ago the vortex tended to break up in early November, but now it breaks up in late December. Some scientists argue that this cooling is linked to the ozone hole that forms in spring: without the protective layer of ozone, heat escapes into space and the stratosphere cools.

It has been suggested that the Circumpolar Vortex then has a downward effect into the lower atmosphere, which causes the lowering of sea-level pressure at the continental margin—and the result is stronger westerly winds between latitudes 55°S to 60°S.

Increases in carbon dioxide and other greenhouse gases are also linked with the cooling of the stratosphere. However, it is still not fully understood how the upper and the lower atmospheres are coupled; this is an area of science receiving significant attention at present. In addition, much attention is being given to working out what is natural variation and what is being forced by human-induced climate change.

Changes in the Southern Ocean

The Southern Ocean is vast and inhospitable, and it is dominated by the Antarctic Circumpolar Current, which moves from west to east and flows mainly between 40°S and 65°S. This current carries relatively cold water and is the reason why tundra and ice-covered sub-Antarctic islands, with glaciers flowing into the sea, exist at the same latitude in the Southern Hemisphere as the mild, temperate English countryside does in the north.

Because of the extreme conditions in the Southern Ocean, including the huge seas and ferocious winds, it is understandable that, historically, it has been poorly sampled scientifically, particularly prior to the 1950s. Despite this, there are some signals indicating that there have been significant large-scale changes in the ocean environment in recent years. Using data from ships and autonomous floating temperature sensors, evidence shows the Southern Ocean has warmed substantially since the 1950s. The warming has occurred at all depths in the ocean and it seems to be caused by a poleward shift of the Antarctic Circumpolar Current. Warmer waters are

now flowing a few hundred miles further south—this movement of the Circumpolar Current is possibly driven by the strengthening of westerly winds.

Changes on Sub-Antarctic Islands

The ring of sub-Antarctic islands lie in the path of the Antarctic Circumpolar Current. These islands have extreme maritime climates and, therefore, the climate-change signals observed in the Southern Ocean are also manifested in some way there. Most of the sub-Antarctic islands have experienced warming in some form over the last 50 years.

On Marion Island, one of the most northerly, for example, mean air temperatures have increased by about 2.7°F (1.5°C) and there has been a dramatic decrease in precipitation (by about one-third) during this time. In contrast, the more southerly Macquarie Island has had an increase in precipitation. The reason for this variation appears to be due to their positions in relation to rain-bearing storm tracks and strengthening of westerly winds across the Southern Ocean—both of which are currently tracking farther south than they did in the recent past. DB

The World's Climate Watchers

The Intergovernmental Panel on Climate Change (IPCC) was established in 1988 by the World Meteorological Organization and the United Nations Environment Programme. Its role is to provide a balanced view of climate change. The IPCC does not conduct any research or monitor climate-related data or parameters, but it assesses the latest scientific, technical, and socio-economic information on climate change—in a comprehensive, objective, and transparent way. The IPCC reports about every four years. Over 450 lead authors and 800 contributing authors were involved in producing the *Fourth Assessment Report*, released in 2007; over 2,500 expert reviewers, as well as governments, assessed it at various stages. Once an IPCC report is released, the cycle of reviewing scientific literature begins again for the next report. In 2007, the IPCC and former American vice-president Al Gore were jointly awarded the Nobel Peace Prize: "for their efforts to build up and disseminate greater knowledge about man-made climate change and to lay the foundations for the measures that are needed to counteract such change."

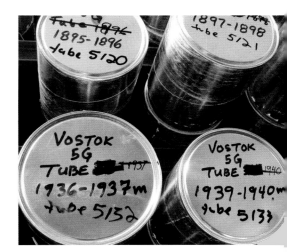

Core Library
Tubes of ice core drilled from Vostok are stored at the National Ice Core Laboratory in Littleton, Colorado, which holds over 13,000 tubes of ice, each just over 3 ft (1 m) long.

Ice in Context
Core samples of sea ice, like this one, are full of small channels that allow the brine in the salty water to escape from the ice. The bottom layer of sea ice is rough and colored by phytoplankton. Changes in concentration and duration of the sea ice affects phytoplankton production, which in turn impacts on the entire Antarctic ecosystem.

Climate Change and Melting Ice

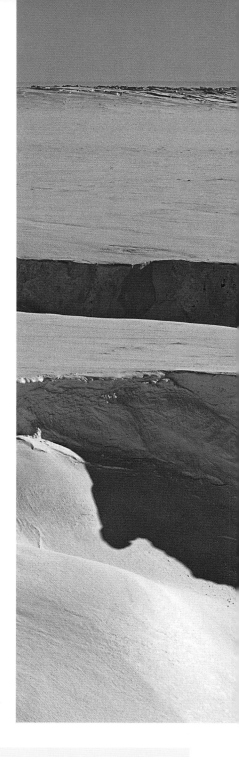

Today, the Earth supports six times more people than it did 200 years ago. In only the six decades between 1760 and 1820, the population of Britain jumped from 6 million to 14 million and the country's agrarian, handicraft economy became one that was dominated by industry and machines. The Industrial Revolution rapidly spread across the world, increasing the demand for energy. The burning of wood, coal, oil, and natural gas generates carbon dioxide.

Carbon dioxide is one of the greenhouse gases that occurs naturally in the atmosphere, absorbing the heat released by land and oceans in a process known as the greenhouse effect. Without the heat-trapping properties of these greenhouse gases, the Earth would be too cold for human habitation.

However, the concentrations of carbon dioxide in the atmosphere are more than 30 percent higher than they were 200 years ago. A typical car releases tons of carbon dioxide each year. Concentrations of methane, another greenhouse gas, are 140 percent greater: cows burp as much as 220 lb (100 kg) of methane per year. New greenhouse gases, such as the chlorofluorocarbons (CFCs), have been added to the air. This extra greenhouse effect is warming the lower atmosphere and changing the Earth's climate.

The average surface temperature of the Earth has already risen by between 0.5°F (0.3°C) and 1°F (0.6°C) during the past 100 years; the twentieth century was the warmest of the past millennium. The world rainfall patterns are changing, the sea level is rising, and glaciers are retreating. By the end of the present century, global temperatures are estimated to be between 2.5°F (1.4°C) and 10.4°F (5.8°C) higher than they are at the present time. This anticipated warming will be greater than any recent natural fluctuations, and it is estimated that it will happen faster than any other known changes to the Earth since the last ice age.

The Rising Seas

Current global climate change is having a major though variable impact on the polar ice sheets. It is now clear that losses from ice sheets in both the Antarctic and Greenland in recent times have contributed to the rise of mean global sea levels. The amount of ice in Antarctica or elsewhere, termed "ice mass," is a trade off between the amount of ice gained through the accumulation of ice and snow and the amount lost through glacial flow, calving, and melt. In effect, glaciers—or rivers of ice— are very large and slow drainage outlets, but the speed at which the ice flows into the ocean from some glaciers in the Antarctic has been increasing recently. The result is a thinning of parts of the ice sheets and the overall reduction of the ice mass.

The estimated rate the sea level rose globally from 1993 to 2003, reported by the Intergovernmental Panel on Climate Change, was around ⅛ in (3 mm) per year— and just under 10 percent of this rise was estimated to be from the Antarctic ice sheets. However, if fully melted,

The Disintegration of Larsen B

The Larsen Ice Shelf, named after a Norwegian whaler, fringes the eastern coast of the Antarctic Peninsula. It originally consisted of three separately embayed segments: Larsen A, B, and C. The northernmost segment retreated steadily from the mid-1980s until January 1995, when about 770 sq miles (2,000 sq km) of Larsen A and the part of the shelf in Prince Gustav Channel disintegrated within a few days. The northern section of Larsen B started to retreat in 1995. In early 2002, a section 1,255 sq miles (3,250 sq km), more than 660 ft (200 m) thick and weighing about 720 billions tons (653 billion tonnes), disintegrated, setting a swarm of icebergs adrift in the Weddell Sea. The disintegration took only 35 days; the photographs below show progress of the collapse, from January 31 (left) to March 7 (right). Although it had been predicted, scientists were stunned by the size and rapidity of the collapse.

One theory to explain these unexpectedly rapid collapses is linked to strong climate warming observed in this region since the late 1940s. The austral summer of 2001–02 was exceptionally warm on the Antarctic Peninsula. Warm winds had been crossing the peninsula and melting the surface ice, and in late January 2002 pools of water began to appear on the ice shelf's surface. The weight of water wedged open small crevasses, forcing them through the ice thickness and breaking the shelf apart.

The remaining large section of the ice shelf, now designated Larsen C and D, stretches farther south along the coast of Palmer Land, the southern portion of the Antarctic Peninsula. It is fairly stable at present but is showing signs of thinning. It is predicted that this section will also recede in the coming decade if the warming trend continues. IA

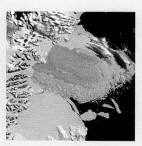
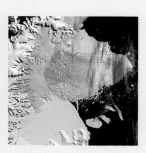

the total volume of ice in Antarctica would be equivalent to the sea rising around 190 ft (57 m); the IPCC points out that building and melting of the immense Antarctic ice sheets operates in timescales measured by century, if not by millennium. A recent NASA-sponsored study of satellite data up to 2006 suggests the rate of ice-mass loss in Antarctica may have increased lately.

The pattern of ice loss is regional: not all parts of the Antarctic region are losing ice. Indeed, it has been calculated that much of the East Antarctic Ice Sheet has been gaining ice mass (through fairly small increases in precipitation) or has been stable in recent times. Two areas in the east, however, do show ice-mass loss: the enhanced ice flow from the Totten Glacier and the Cook Ice Shelf may be related to the loss of buttressing ice shelves some time in the past.

The smaller West Antarctic Ice Sheet is losing ice mass, mainly due to ice draining into the Amundsen and Bellingshausen seas. Both the Thwaites and Pine Island glaciers, which terminate in the Amundsen Sea, have retreated substantially, and have become thinner and faster flowing. Glaciers on the western coast of the Antarctic Peninsula are also losing ice, particularly the glaciers that flow along narrow channels. Scientists from the British Antarctic Survey studied the behavior of 244

glaciers over a 60-year period and found that almost 90 percent of them were retreating. Comparison of satellite imagery of 300 glaciers in the Antarctic Peninsula over ten years (1993 to 2003) demonstrated their flow rate has increased by 12 percent.

The increased ice loss in West Antarctica has been attributed to warmer sea temperatures where glaciers meet the ocean, causing increased glacial flow rates and ice shelf-based melt. This reflects the southward migration of the Antarctic Circumpolar Current. Warmer waters within the current are flowing several hundred miles farther south now. This is significant in the Drake Passage region—the narrowest section of the Southern Ocean, where the Circumpolar Current flows through the gap between the tip of South America and the Antarctic Peninsula—because these warmer waters now come into contact with sea ice, ice shelves, and tide-water glaciers. Furthermore, as glaciers thin, the buoyancy of the ice can lift the glaciers off their rock beds, allowing them to slide even faster.

Increased northerly winds have also brought more precipitation to the Antarctic Peninsula, and because most of this precipitation falls as snow, in recent years there has been some ice-mass gain on the mountains in the peninsula region.

Natural Melt?
According to glacialogists, there have been four or more major advances and then retreats of continental ice sheets over the past 2 million years. The major question scientists face is how the human-made "greenhouse" effect modifies these natural cycles.

Ice-shelf Probe
On the Brunt Ice Shelf, scientists at Britain's Halley V Station have been conducting programs on glaciology, studies that are crucial in gaining a global perspective on climate change and sea-level rise.

Ice-shelf Collapse
Many Antarctic ice shelves are retreating. They retreat mainly by the calving of many small icebergs from their ocean margins and behind. The collapse of floating ice shelves has no direct impact on sea levels. However, ice shelves buttress the flow of grounded, glacial ice from the land behind them, and recent increases in glacier speed have been observed in places where Antarctic Peninsula ice shelves have disintegrated.

Changes in Sea Ice
The current plight of polar bears in the Arctic caused by the decreased sea ice, which has been receding since the 1970s, is well known. In Antarctica the picture is not as clear.

Like changes in air temperature, changes in the extent of sea ice vary regionally in Antarctica. But sea ice is not that easy to study: lack of data hampers attempts to establish if it has changed over recent times. Historical data are found in the logs kept by the early explorers and scientific voyagers, which noted when and where they encountered sea ice. Whaling records are another historical source: the locations where the whales were caught indicate the absence of sea ice in those areas. Extrapolating from these whale proxy records, researchers have suggested that the summer sea-ice extent in the East Antarctic region decreased by 25 percent between the mid-1950s and the early 1970s.

Shipboard observations provide localized data only. Continental-scale pictures of the extent of sea ice are obtained from satellite imagery, a dataset dating from 1973. Research that has compared observations from vessels with satellite information indicates that satellite imagery gives fairly accurate information on sea-ice edge locations, except in summertime when there is a consistent error of around 1 to 2 degrees in latitude due to the inability to distinguish slushy ice from solid ice. Using this knowledge and examining the historic ship records, scientists have not been able to confirm the decrease in East Antarctic sea ice that was estimated from the whaling records.

Strong evidence of sea-ice decline in recent years is apparent for the Weddell Sea and Antarctic Peninsula area, related to increased temperatures in the region. There is some evidence to suggest the sea-ice season

Sea Caves

When the ocean washes up against an iceberg, it exploits any area of weakness. Eventually the weakest section of the ice gives way, leaving a cavelike hollow.

Broken Tongue

When a 1,160 sq mile (3,000 sq km) iceberg collided with the Drygalski Ice Tongue (left)—the vast, floating portion of the land-based Davis Glacier, which stretches 45 miles (70 km) out into the Ross Sea—in March 2005, it broke two large pieces off the ice tongue, each about 27 sq miles (70 sq km) in area. For several months before, the berg had blocked ocean currents from sweeping the winter sea ice from McMurdo Sound, thus creating a serious problem for penguins, which had to travel farther to reach open waters and food. Although scientists believe the massive iceberg was part of a natural cycle, not the result of climate change, it highlighted some consequences of an increase in icebergs as the world warms.

has also been reduced by about a week over the last 30 years. Increases in the extent of sea ice, however, have been observed recently in the Ross Sea.

Changes in the Sub-Antarctic

Ice caps occur on a handful of sub-Antarctic islands: Marion Island, Grande Terre in Iles Kerguelen, South Georgia, and Heard Island.

Marion Island and Iles Kerguelen have relatively milder climates because of their position in relation to the Antarctic Circumpolar Current and the Polar Front. The mean temperature range at sea level on Marion Island is 39°F to 46°F (4°C to 8°C) and on Iles Kerguelen it is 36°F to 48°F (2°C to 9°C). As such, these sub-Antarctic islands have ice caps on their high interior mountains only. Marion Island has a small ice plateau that lies in between peaks, and it covers less than 1 percent of the island's 112 sq mile (290 sq km) land area. Grande Terre in Iles Kergeulen has an ice cap over 8 percent, or 210 sq miles (550 sq km), of the island. On both of these islands, in the last 50 years temperatures have increased and precipitation has decreased, and their glaciers and ice caps are retreating rapidly.

South Georgia and Heard Island, which both lie to the south of the Polar Front, have over 50 percent of their land surfaces covered in permanent ice and snow, with many glaciers running into the sea. The mean temperature range on South Georgia is 28.4°F to 41°F (−2°C to 5°C) and on Heard 32°F to 39°F (0°C to 4°C). Glaciers are retreating rapidly on these islands. In 1947, the total glacial cover on Heard Island was 79 percent, and this has now reduced to less than 69 percent of the island. Many of the glaciers that previously flowed into the sea are now grounded on land, resulting in major changes to the island's coastline. In places, extensive pro-glacial lakes are now established between the front of glaciers and a thin coastal strip: on Brown Glacier on the island's northeastern side, for example, the glacial

front has retreated over ⅔ mile (1 km) and the ice mass decreased in total volume by 38 percent, or up to 36 ft (11 m) in thickness.

When the Ice Melts

The huge weight of the polar ice sheets depresses the Earth's crust—if the ice were removed, the crust would rise again to its original level. The shoreline terraces of eastern Canada's Great Lakes were formed by this process of "glacial rebound." From near Chicago, they tilt gradually northward to a maximum of about 1,000 ft (300 m) near Hudson Bay, reflecting the disappearance of the 10,000-year-old continental ice sheets of North America. This model indicates that the rock surface of Antarctica would rise substantially by glacial rebound if the ice were to melt. The uplifted crust would displace seawater and the sea-level rise, but in small amounts and extremely slowly. Under East Antarctica's thickest ice, the crust is estimated to be depressed as much as 3,100 ft (945 m). How much sea levels would rise from that amount of uplift is unknown.

Melting ice would return water previously held on land to the world's oceans and, again, the sea levels would rise. Doubtless, this process would begin slowly, as climates warmed sufficiently to begin melting the polar ice sheets. Warming and thermal expansion of the oceans alone might result in a rise of some 1½ to 13 ft (0.5 to 4 m), but it might take some 1,000 years for atmospheric warming to reach the oceanic bottom waters. Already, however, sea levels are rising almost imperceptibly. A quickening rate would have serious consequences for the world's coastal settlements.

Antarctica contains 90 percent of the world's ice, enough to affect sea levels dramatically. Calculations suggest that sea levels rose about 425 ft (130 m) from the melting of 21,000-year-old Northern Hemisphere ice of the last ice age. The melting of all Antarctic ice might raise the sea level by about 260 ft (80 m).

The East Antarctic Ice Sheet lies on bedrock that is mostly at sea level or above, whereas much of West Antarctica's ice is based on bedrock well below sea level, and in places far below it. The decay of the ice in East Antarctica might result in a 200-ft (60-m) rise in the sea level, but at a very slow rate, perhaps over 10,000 years, because the ice would remain mostly grounded. Glaciologists worry that West Antarctica's ice could melt at a catastrophic speed, however, due to flotation from its bedrock below sea level. A rise in sea level of approximately 50 ft (15 m) might occur in a few tens to 100 years. Melting of floating ice, such as the immense Ross Ice Shelf, would not affect sea levels as the ice is already in flotational equilibrium.

If its ice were to melt, Antarctica would be a much smaller continent. East Antarctica would be the main landmass, joined to a peninsula of today's Ellsworth Mountains. Present-day Antarctic Peninsula would be a separate island mass, with Marie Byrd Land being a sea dotted with volcanic islands separated from the "Antarctic Peninsula" by a deep marine trough, even after glacial rebound. Many mountain ranges now under ice, such as the sub-glacial Gamburtsev Mountains, would rise above the plains of the continent. DB

Climate Change and the Antarctic Ecosystem

It is anticipated that global temperatures will rise significantly by the end of the twenty-first century. Land will warm more than the sea, and the greatest warming is likely to occur near the poles. Changes in the terrestrial, freshwater and marine environments over the last 50 years have been found across Antarctica, but patterns of change are not universal, and instead reflect variations in the impact of climate change.

Changes on Land

Antarctica is at the end of the planetary spectrum of life. The treeless, sub-Antarctic islands are dominated by tundralike flowering plants, lush carpets of moss, and generally the largest native terrestrial animals (as opposed to sea birds and seals) are insects. Farther south, the numbers of flowering plant species diminish, the insects disappear, and life is mainly represented by algae, moss cushions, lichens, and mites. In freshwater bodies, life is dominated by small crustaceans. The amazing feature of Antarctica is that where one finds liquid water, one finds life.

On Marion Island, one of the more northerly sub-Antarctic islands, air temperature has increased and precipitation has decreased. As a result, plants in some mires—the soggy vegetation areas in drainage basins—have died out and been replaced by local species that thrive in drier conditions. On Heard Island and South Georgia, warming has resulted in glacier retreat. This has created ice-free areas for colonization and the new areas are rapidly being covered in vegetation. Also, in 1986, a new plant species, a buttercup with a more northerly distribution, arrived on Heard Island and has established a new population in an area that was glacier-covered as late as 1950s.

Major changes have been noted in many lakes. On Signy Island, for example, lake temperatures have been increasing dramatically since 1980. Lake ice and snow cover have also reduced: the open-water time period in the mid-1990s was more than two months longer than it was 15 years earlier. With the melting of the ice from icefields, nutrients, in the form of sediments, have also flowed into the lake. These changes have rapidly altered the ecosystems in the lakes.

On the Antarctic Peninsula, warmer conditions have allowed the expansion of the continent's only two flowering plant species, a hair grass and a little cushion plant. At most localities these species have increased their extent and abundance. They have not extended their distribution farther southward, however.

In a number of different places, one of the major changes has been an increase in rainfall, as opposed to snow. Rain increases the availability of water to life, and, conversely, increases in temperature can reduce the spring and summer melt run-off, resulting in local drought. In extensive moss beds close to Casey Station in East Antarctica, for example, lichens appear to be growing over dying mosses in locations that seem to be drying out. Also at Casey, there is evidence that the ozone hole may be having an impact on some Antarctic organisms: botanists report stunted growth in the shoots of an Antarctic moss species, whereas plants shielded from ultraviolet light generally had normal leaves.

Less Ice, More Plants

It is anticipated global temperatures will rise significantly by the end of the twenty-first century. Land will warm more than the sea, and the greatest warming is likely to occur near the poles. Luxuriant vegetation, such as that below, is now found on only a few sub-Antarctic islands, but may well become more widespread.

An additional major threat that is associated with climate change in the region is the potential impact of non-native species which have been introduced by humans. Under warmer conditions such species may out-compete native species, whereas in the past they would not have survived.

Changes in Marine Ecosystems

Trying to identify adverse effects on ecosystems from climate change in the Southern Ocean is one of the great challenges facing marine biologists. The Southern Ocean is vast (covering approximately 10 percent of the planet's surface), difficult to work in, and very hostile to humans. Human activity in the ocean over the last 200 years makes it very difficult to disentangle the past and present changes. Top predators, including whales,

seals, and penguins, were hunted to extinction or near-extinction in many localities: in the early 1800s on sub-Antarctic Macquarie Island, for example, the entire fur seal population was harvested in only ten years—and it is still not known which species it was. The past over-exploitation of several fish species has resulted in their stocks collapsing and, despite stringent management systems being in place since the 1980s through the Convention on the Conservation of Antarctic Marine Living Resources (CCAMLR), the over-exploitation of the Patagonian toothfish is still occurring through illegal fishing activity. Controlled fishing, particularly of krill, is also occurring, although the exact size of the stock is currently unknown.

Removal of either prey or predator species causes major disruptions in any ecosystem. Some changes observed in the Southern Ocean ecosystem are due to these past or present human activities, and may include recovery from previously low populations and the flow-on effect of recovery on another species. Change will also be caused by natural decadal-scale variation and long-term climate change. But sorting out the various signals and attaching or partitioning causes can be an extremely difficult process, particularly because there are insufficient long-term data that have been collected in a consistent manner. Another important issue is that much of the data recorded or the trends noted are local responses—that is, they cannot be used to generalize trends across Antarctica.

Changes Close to Shore

Major changes in the local ecosystem have occurred off the northeastern coast of the Antarctic Peninsula. The collapse of large sections of the Larsen Ice Shelf in 1995 and 2002 suddenly provided a new sea floor of colonization. In 2007, as part of the Census of Antarctic Marine Life (CAML), a German research team explored these new habitats and found extremely varied life in the sediments, although the number of animals was

Global Warning
Monitoring the changes in the delicate Antarctic ecosystem provides important information about changes in the global climate and how they may impact on other parts of the world.

Measuring Growth
A researcher monitors moss growth close to a volcano on one of the sub-Antarctic islands. Farther south, in East Antarctica, mosses are dying or their growth is being stunted.

Moss Beds
In the Dry Valleys, researchers examine ancient moss deposits believed to be 13 million years old. By studying these deposits, a freeze-dried bed of ancient life from a time when Antarctica was warmer, they are trying to understand what processes caused the climate to change in the past and thus what may happen in the future.

New Species

The collapse of the Larsen Ice Shelf exposed new seabed species, including (clockwise from top): an *Epimeria*, a species of amphipod crustacean; a cirriped crustacean, a giant Antarctic barnacle; an unidentified sea star; a giant amphipod crustacean of the genus *Eusirus*; and a sea spider that is bearing its eggs.

Marine Change

Glacial sediments (captured here in ice) carried in meltwaters can increase the amount of nutrients available in ocean waters. This, in turn, affects all the levels of the marine ecosytem, from phytoplankton to sea mammals.

nowhere near as abundant as on the outer seabed. New species colonizing the Larsen Zone include fast-growing, gelatinous sea squirts and the slow-growing animals called glass sponges. Sea cucumbers, deep-sea lilies (members of a group called crinoids), and sea urchins were also found in the relatively shallow waters. These species are usually found in the greater depths of the ocean where the nutrient resources are scarce, but similar conditions are also found under ice shelves, so these species may have been under the Larsen Ice Shelf before sections disintegrated.

In the course of studying the marine ecosystem around the Antarctic Peninsula, where substantial warming has been occurring over the last 50 years, scientists from the University of California have found that the increased meltwater from glaciers during the summer months has significantly altered not only the immediate coastal water but also the seawater up to 60 miles (100 km) offshore.

The most significant change with glacial melt has been a reduction in salinity because of an influx of freshwater. Furthermore, the freshwater floats on the surface and speeds up the process of surface-water warming, resulting in a substantial increase in ice-edge retreat. Meltwaters also carry glacial sediments, which increase the turbidity of the near-shore waters and the amount of nutrients available to the ecosystem. Over a ten-year study, researchers found that meltwaters were triggering phytoplankton blooms (planktonic plant life), both in near-shore and offshore areas, and this has also altered the species composition of phytoplankton.

As phytoplankton are the basis of the marine food chain, it is anticipated that such changes will have an impact further up the food web. One of the significant changes observed in the Antarctic Peninsula and Scotia Sea areas has been a reduction in the numbers of krill and an increase in salps. Salps are jellylike animals that prefer warmer waters; they appear to be extending their range southward and they may be replacing krill—and, whereas krill are a major food source for many animals, few eat salps. British Antarctic Survey scientists report that over the last 30 years the populations of krill in the area have declined by between 50 and 80 percent. But such a large reduction is difficult to verify—it may be related to the netting methods used, because parallel acoustic surveys using sonar in the same region do not report the same patterns. Regardless of these problems in estimating populations, any negative trends in krill abundance are a cause for concern. Declines are linked to the reduction in the extent of sea ice, particularly in winter. A poor covering of winter sea ice corresponds with a decrease in the abundance of krill the following summer. Sea-ice extent is important to krill, particularly to the juveniles, as they spend winter under the sea ice, feeding on algae and plankton.

Changes in Predator Numbers

There have been notable local changes in the predator populations within the Antarctic region but, once again, the major issue for scientists is trying to sort out what components are due to climate change compared with other potential reasons.

Some of the changes in predator numbers observed include a loss of around 10,000 breeding pairs of Adélie penguins on Anvers Island in the Antarctic Peninsula over the past 30 years. This amounts to about 70 percent of the local population. Scientists believe this reduction is associated with loss of winter sea ice and an increase in local snowfall: the reduction in sea-ice extent disrupts food gathering, and snow melt in spring causes flooding of the nests. In contrast, the penguin species that are generally "ice avoiders"—chinstraps and gentoos—are extending their distribution southward and are breeding on the Antarctic Peninsula now, something that has not occurred for 800 years. Adélie penguins seem to be very dependent on krill, whereas the chinstraps and gentoos have more flexible diets, and are eating more fish and squid. A colony of emperor penguins (also krill-eaters) in the region has declined as well.

During the 1970s, the emperor penguin population at Terre Adélie also declined by 50 percent. Scientists believed this decline was due to adult penguins dying during a prolonged, abnormally warm period, which reduced the amount of sea ice and warmed the sea-surface temperatures. Such conditions appeared to have either reduced available food stocks or ice for molting. A molting emperor penguin needs solid sea ice for three to four weeks in January and February and will perish if it has to swim any distance. In contrast, scientists have observed that the survival rate of adult emperor penguins was high when the winter sea ice was extensive, but that they hatched fewer eggs. Why the emperor penguin numbers in this region have not returned to their pre-1970s levels is unknown and may be linked to overall climate change during the period; numbers elsewhere in East Antarctica do not appear to have changed significantly over the last 20 years.

On the sub-Antarctic islands, populations of some species, such as rockhopper penguins and elephant seals, have declined over the last 50 years (particularly during the late 1960s and early 1970s). However, some other species, such as fur seals and king penguins, have increased their populations. Once again, identifying the drivers of these changes is the focus of many scientific studies. Fur seals, for example, may still be recovering from seal harvesting days.

Changes in the Deep Sea

One of the most significant impacts of climate change in the Antarctic marine environment is ocean acidification. About half of the carbon dioxide emitted from human activities has dissolved in the ocean. This has changed the chemistry of ocean water, making it more acidic due to the formation of carbonic acid, the same weak acid used in carbonated drinks. The world's oceans, including the Southern Ocean, have suffered about a 30 percent increase in acidity. Increasing levels of carbonic acid are interfering with the formation of calcium carbonate, a structural component of the shells of many plankton species, and are affecting the availability of nutrients.

As it becomes more difficult for calcium carbonate to form, it will become more difficult for some planktonic organisms to form shells. If their shells are thinner and/ or deformed, the organisms may be unable to function properly. Many of these organisms are key components of the food chain, being important in the diets of krill, fish, squid, penguins, seals, and whales. In addition, they play a key climate role by assisting the removal of carbon from surface waters to the deep ocean and the release of oxygen into the air. Important metabolic processes, such as respiration in fish, may be impaired by the acidity as well. Scientists are watching for some of the first indications of such impacts in the Southern Ocean, because it contains more carbon dioxide than other oceans—and because cooler water absorbs more than warmer water. DB

Feast or Famine?
A small decrease in the abundant supplies of fish and squid near these king penguins at Salisbury Plain in South Georgia, could inflict great harm on such a large, closely packed colony in a very short time.

Lovers of Stability
The tagging of Weddell seals provides information on the distances they travel and how deep they can dive. Weddells, the most southerly breeding of all the mammals, prefer to give birth in the very stable, coastal fast-ice areas, which are at risk if the world heats up.

The International Polar Year, 2007–08

The International Polar Year, 2007–08 (IPY) was launched on March 1, 2007, half a century after the International Geophysical Year of 1957–58, which had led to major advances in knowledge about Antarctica. Co-sponsored by the International Council for Science (ICSU) with the World Meteorological Organization (WMO), the IPY was an internationally coordinated campaign of intensive scientific research and observations in the Antarctic and the Arctic regions. There was a strong interdisciplinary emphasis, and included geophysical, ecological, and, in the Arctic, the social sciences.

The concept of an IPY in 2007–08 was suggested by a number of scientists and organisations, and it was fostered when an IPY planning group was established by ICSU in June 2003. In consultation with researchers, the group developed six themes for IPY: to determine the present environmental status of the polar regions; to quantify and understand changes occurring there; to better understand links between them and the rest of the Earth; to investigate frontiers of science in polar regions; to use the polar regions as unique vantage points; and to investigate the cultural, historical, and social processes of circumpolar human societies.

Range of Research

IPY science covered an enormous range of topics, with a strong focus being on the impacts of changing climate on the polar regions. Among the projects were studies of reductions: in the extent and mass of glaciers and ice sheets; in the area and duration of snow cover; in permafrost; and in the extent and thickness of sea ice. Surface air temperatures over large areas of the Arctic and on the Antarctic Peninsula have risen considerably faster than the global average, partly because of the ice–albedo feedback that amplifies climate change in polar regions. Many of the changes to polar ice and snow have consequent impacts on other parts of the Earth, and IPY projects also investigated these: for example, permafrost degradation affects local ecology, hydrology, and coastal and soil stability; and changes in the large ice sheets have global impact on sea levels, affecting human populations in low-lying coastal areas.

Altogether, over 50,000 scientists from 62 countries were involved in projects for the IPY. New funding of more than US$400 million was allocated for IPY activities from 19 national and regional programs. In addition to this new funding, some US$800 million of perennial polar funding was directed specifically to IPY science activities, and substantial new infrastructure funding was provided from many nations for resources, such as research stations, to support this science.

The official IPY observing period ran from March 1, 2007, until March 1, 2009: this allowed the inclusion of a complete annual cycle of observations in both polar regions. There was an emphasis on Arctic projects, with nearly 55 percent of the science projects being based in the north. Of the rest, 25 percent dealt with the two polar regions, and 20 percent with the Antarctic alone. In the Antarctic projects, there was a strong emphasis on studies of the ocean and of ice. The analyses of the data from these projects will continue for many years.

Undersea Profiling

Much of the IPY Antarctic program focused on the marine environment, which is one of the most difficult parts of the Earth to explore. This airgun array sends soundwaves to the sea floor. The soundwaves penetrate the sediment and allow a profile to be built up of the sea floor.

Trawling for Life
Researchers prepare to lower a "sledge" into the ocean. It is used for benthic sampling of the marine life that lives on and above the ocean floor. This is a key part of the IPY's Census of Antarctic Marine Life (CAML).

Antarctic Projects

One of the major Southern Hemisphere oceanographic projects conducted during the IPY was the Census of Antarctic Marine Life. This program investigated the distribution and abundance of Antarctica's vast marine biodiversity in order to establish a benchmark, and to establish how this ecosystem might be affected by climate change. Ships from seven nations—Australia, Britain, France, Germany, Japan, New Zealand, and the United States—undertook dedicated marine ecosystem surveys. Ships from several other nations contributed to this study of both the deep waters of the Southern Ocean and of the Antarctic continental shelf. A large number of new species were discovered during these groundbreaking marine surveys.

A large-scale international study of the Southern Ocean and its role in the global climate system involved six north–south ship transects to measure the physical environment of the full ocean depth. In addition, more detailed but shorter surveys radiated outward across the Antarctic continental shelf and slope. A large number of robotic buoys were deployed to measure ocean temperature, drift, and salinity, relaying their data back via satellite links. Satellite-linked sensors were fitted to deep-diving seals to provide data on the location of the animals, their dive patterns, and the temperature and salinity profiles of the Southern Ocean, including those areas that are covered by sea ice and thus not accessible to robotic floats.

The Antarctic Continental Margin Drilling Program, a large multinational collaboration, used new drilling technology to recover ocean sediment cores beneath floating sea ice and ice shelves. When fully analysed, these ice cores will provide information on Antarctic environmental history over the past 60 million years, including the history of climate and ice sheets, and the development of polar ecosystems.

A number of teams traveled overland into the high interior of the Antarctic ice sheets. A US–Norwegian project surveyed the region surrounding the Pole of Inaccessibility, the point farthest from the ocean. On a Chinese expedition to Dome A, the 13,428-ft (4,093-m) high summit of the East Antarctic Ice Sheet, Chinese scientists collaborated with Australians in deploying an array of automatic weather stations, and with French scientists in recovering an ice core for use in preliminary investigations of past climates.

The Chinese have plans to build a permanent station at Dome A. This might prove to be a suitable place to recover a deep ice core, which would provide a detailed climate record of more than 1 million years.

Beneath Dome A are the sub-glacial Gamburtsev Mountains, a mountain range larger than the European Alps. These were investigated by another IPY project using instrumented aircraft and ground measurements in order to gather information on the mountains' structure, and of the overlying ice.

Data from a range of sophisticated sensors onboard a variety of satellites were essential for many of the IPY projects. Satellite agencies of many of the participating nations made substantial contributions to the scientific program by targeting data collection from specific areas, and making satellite data and products freely available to the IPY researchers. IA

Official Visit
Marking the International Polar Year, the Secretary General of the United Nations, Ban Ki-moon (second from right) visited King George Island, off the Antarctic Peninsula, in November 2007. He is the first UN leader to make an official visit to Antarctica.

Practical Cooperation
IPY science is not the only area of cooperation among nations in Antarctica. Here, an Italian resupply cargo vessel and a United States fuel tanker use a path cut through the pack ice by the United States Coastguard icebreaker, the *Polar Sea*.

Antarctic Regions

Antarctica

The polar regions are the most inhospitable on the Earth for humans yet, in the first decade of the twenty-first century, the Antarctic region houses more than 40 permanent stations where people live and work all year. From the desolate and scattered islands south of the Antarctic Circle to the Antarctic Peninsula, the Antarctic continent and the South Pole itself, explorers have left their mark. They have been followed by scientists, in the hope of unlocking the secrets of Antarctica, and tourists can now visit this most challenging of environments.

The Antarctic Circle is really only a theoretical line on a map, or a reading on a global positioning system (GPS), unless it is seen at the solstice, when the sun circles the horizon.

The Antarctic Continent

At the South Pole, there is no east or west—only north. However, geographers divide the world into eastern and western hemispheres, and these divisions are matched in Antarctica. East Antarctica—also known as Greater Antarctica because it contains most of the landmass of Antarctica—lies below Africa, Asia, and Australia. It is characterized by an ancient, stable Precambrian rock shield covered by an ice sheet about 1¼ miles (2 km) thick, and has a roughly semicircular shape that quite closely follows the Antarctic Circle. West, or Lesser, Antarctica lies roughly between the Americas and New Zealand—east of the International Date Line, toward the Greenwich Meridian—and it includes the Antarctic Peninsula, which is marked by the deep indentations of the Weddell and Ross seas and the extended arm of the Antarctic Peninsula. The Transantarctic Mountains form a natural barrier between East and West Antarctica.

There are some significant differences between the underlying landforms of East and West Antarctica, although these are not immediately apparent because of the deep covering of ice across the continent. If all the ice were to disappear, East Antarctica would be a continent that is similar in size and nature to Australia, whereas West Antarctica would be a much smaller landmass, tapering away to a series of small islands and one large one, with a long, narrow spine running along what is now the Antarctic Peninsula. In an ice-free East Antarctica, large mountain ranges would stand revealed and there would also be a few large basins that would become virtually inland seas.

East Antarctica contains about 90 percent of the ice in Antarctica, much of it in the vast Polar Plateau. It has very little exposed land, virtually all of it along the coast, and that is generally occupied by wildlife and scientific bases in uneasy conjunction. Not many people have ever seen this remote region, and there are only a few weeks each year when even icebreakers can approach its coasts. Until quite recently, exploration was largely along the coast, and the place names reflect the nationalities of early Antarctic explorers and expeditions. From west to east, the 11 major "lands" on this side of

Antarctica are Dronning Maud Land, Enderby Land, Kemp Land, Mac.Robertson Land, Princess Elizabeth Land, Wilhelm II Land, Queen Mary Land, Wilkes Land, Terre Adélie, George V Land, and Oates Land.

West Antarctica is the name given to the part of Antarctica that extends into the Western Hemisphere from the Transantarctic Mountains, and it comprises Ellsworth Land and Marie Byrd Land, bordered by the Bellingshausen and Amundsen seas.

West Antarctica is considered a distinctive region, based upon its different climate (generally warmer), bedrock topography (largely below sea level), geology (younger rock formations bearing a close resemblance to those in South America), and its shorter glacial history. The juxtaposition of East and West Antarctica is the result of continental drift in the relatively recent geological past. Several active volcanoes, including Erebus on Ross Island, serve as reminders of the more turbulent geological history of West Antarctica, which remains tectonically unstable. Considerably more mountainous than East Antarctica, the region contains Antarctica's highest peak: Vinson Massif (16,066 ft/ 4,897 m) in the northern Ellsworth Mountains.

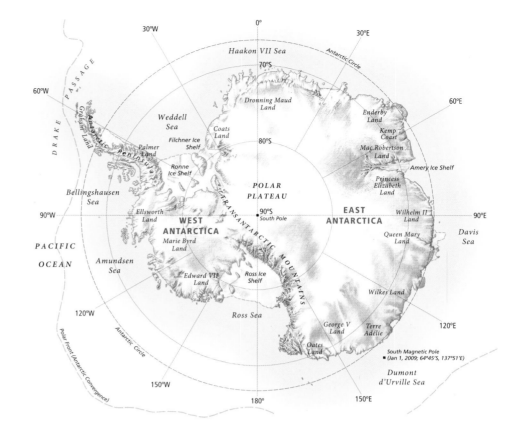

Antarctic Regions
Continental Antarctica is divided into east and west—two geologically distinctive regions. The spine of the Antarctic Peninsula and some of the larger outlying islands support several small, thin plateau ice caps.

Polar Oasis
The Taylor Valley (left) is part of an ancient polar desert, the Dry Valleys, in the Transantarctic Mountains. The small, steep Rhone Glacier (right) is typical of the Dry Valleys' "alpine" glaciers: because they are frozen to the glacier bed, they are slow moving and cannot readily slide over it.

Sheer from the Sea

The stark contrast between sharp, dark rocks and flowing glaciers on the peaks of Graham Land, in the north of the Antarctic Peninsula, creates some of the world's most spectacular scenery.

Sculptures in Snow

Light winds deposit snow in the lee of obstacles and undulations across the landscape of the Polar Plateau, and then blizzards erode the snow into sculptured shapes known as sastrugi. These polished forms can be taller than people and hard as concrete.

The West Antarctic Ice Sheet covering the region is markedly younger, less extensive, and lower than its East Antarctic counterpart, and it contains only 10 percent of the Antarctic ice sheets' entire volume. Underneath the ice sheet is a chain of mountainous islands separated by deep ocean trenches and basins: the average bedrock elevation of West Antarctica is 1,445 ft (440 m) below sea level—unlike the East Antarctic Ice Sheet, which rests on a vast and ancient continental landmass with a surface above sea level.

The Antarctic Peninsula

The Antarctic Peninsula (known as Tierra O'Higgins to Chileans and Tierra San Martin to Argentinians) is a range of rugged ice-covered mountains that rise dramatically out of the sea. This spectacular mountain range, which has its highest peak, Mount Jackson, at 10,446 ft (3,184 m), extends northward, 775 miles (1250 km), from West Antarctica toward the southern tip of South America, from which it is separated by about 620 miles (1,000 km) across the Drake Passage and Bransfield Strait. The Antarctic Peninsula ranges in width from only about 15 miles (24 km) in the north to 180 miles (290 km) in the south, where it separates the Weddell and Bellingshausen seas. Although narrow, the peninsula acts as a major climate divide: its eastern side is much colder and harsher than the west, where a warmer maritime climate prevails. The northern part of the peninsula is Graham Land. The southern section is called Palmer Land, which is higher and broader, with a single large island offshore to the west, namely Alexander Island. DM & RM

Bared by the Sun

The maximum amount of underlying rock is exposed at the end of the snow melt as summer is ending. This is the 4-mile (6.5-km) long Browning Peninsula in East Antarctica, at the southern part of the Windmill Islands, a biologically rich region that is close to Australia's Casey Station.

Antarctic Peninsula: The North

The northern peninsula and nearby islands have the highest concentration of research stations in Antarctica, plus a number of Specially Protected Areas (designated under the Antarctic Treaty System) and Historic Sites and Monuments.

Hope Bay

This notch in the tip of the Antarctic Peninsula, 1¾ by 3 miles (3 by 5 km), was first visited in January 1902 by Otto Nordenskjöld's Swedish South Polar Expedition. They named it Hope Bay, after three members of their party inadvertently wintered there during 1903; they had been dropped off not long before their ship, the *Antarctic*, sank. The sound beyond the bay bears the name of their doomed ship. Their makeshift stone hut (where hope was about all they had) stands near the dock—it was largely rebuilt over the summer of 1966–67.

The Argentinian Esperanza (Spanish for "hope") Station was founded on Hope Bay in December 1952, and has operated continuously ever since. The first Catholic chapel in Antarctica was consecrated there in 1976. Two years later, a substantial expansion program brought whole families to Esperanza. A school was built by Tierra del Fuego's Education Ministry; it burned down in 2007. The population ranges from about 55 in winter to 90 in summer, of which about a third are children or spouses.

In a move to strengthen Argentina's territorial claim to the Antarctic Peninsula, Silvia Morello de Palma, wife of the base commander, came to Esperanza when she was seven months' pregnant. She gave birth to the first

"Antarctican," Emilio Marcos de Palma, on January 7, 1978. Other babies have been born at Esperanza since then, and the presence of families gives this base the atmosphere of a tiny town. It has about 1 mile (1.6 km) of gravel roads and 43 buildings, including two scientific laboratories and an infirmary staffed by a doctor and a paramedic. LRA 36 (Radio Nacional Arcangel San Gabriel), the base's radio station, began broadcasting on October 20, 1979.

Esperanza Station is supplied by ship twice a year, in December and January, and it is also served year-round by Twin Otter ski planes, which land on a glacier airstrip 1 mile (1.6 km) from the base. Hope Bay often experiences northeasterly winds that may exceed 125 mph (200 kph), so landings are not always possible.

Uruguay's Lieutenant Juan Ruperto Elichiribehety Station is about 1 mile (1.6 km) from Esperanza. This summer base, named after the Uruguayan captain who tried to rescue Shackleton's men from Elephant Island, houses up to 12 people.

The building was originally called Trinity House and was built in 1952 by the British to replace an Operation Tabarin hut that burned down in 1949. The British left here in 1964, and their scientific base was transferred to Uruguay in 1996.

Brown Bluff

Situated in the Antarctic Sound, some 9 miles (15 km) southeast of Hope Bay, Brown Bluff is the point where the eastern edge of the Tabarin Peninsula drops almost sheer to the water from an ice-capped summit. From a rocky beach, a steep scree slope rises to a towering, rust-colored cliff of volcanic rock. This could be central Australia or the Badlands in the United States were it not for the penguins and Weddell seals along the shore and the ice cap high above.

Adélie penguins number the tens of thousands on Brown Bluff, and several hundred gentoo penguins also live there. The slope is covered with loose rubble and rock slips are common. Many scientists believe that landslides and falling boulders may have obliterated some of the penguin colonies.

Popular Attraction
The Antarctic Peninsula is a major tourist destination by virtue of its wonderful scenery, abundant wildlife, and proximity to South America.

Sold for a Song
The break-up of the Soviet Union gave rise to some interesting complications in Antarctica. Russia kept all the bases that were previously Soviet, but of the other Soviet states only the Ukraine now has an Antarctic presence. The Ukrainian government purchased the Vernadsky Base (shown here), in the Argentine Islands near the Antarctic Peninsula, from the British in 1996 for a nominal sum.

70°W

66°S

72°W

68°S

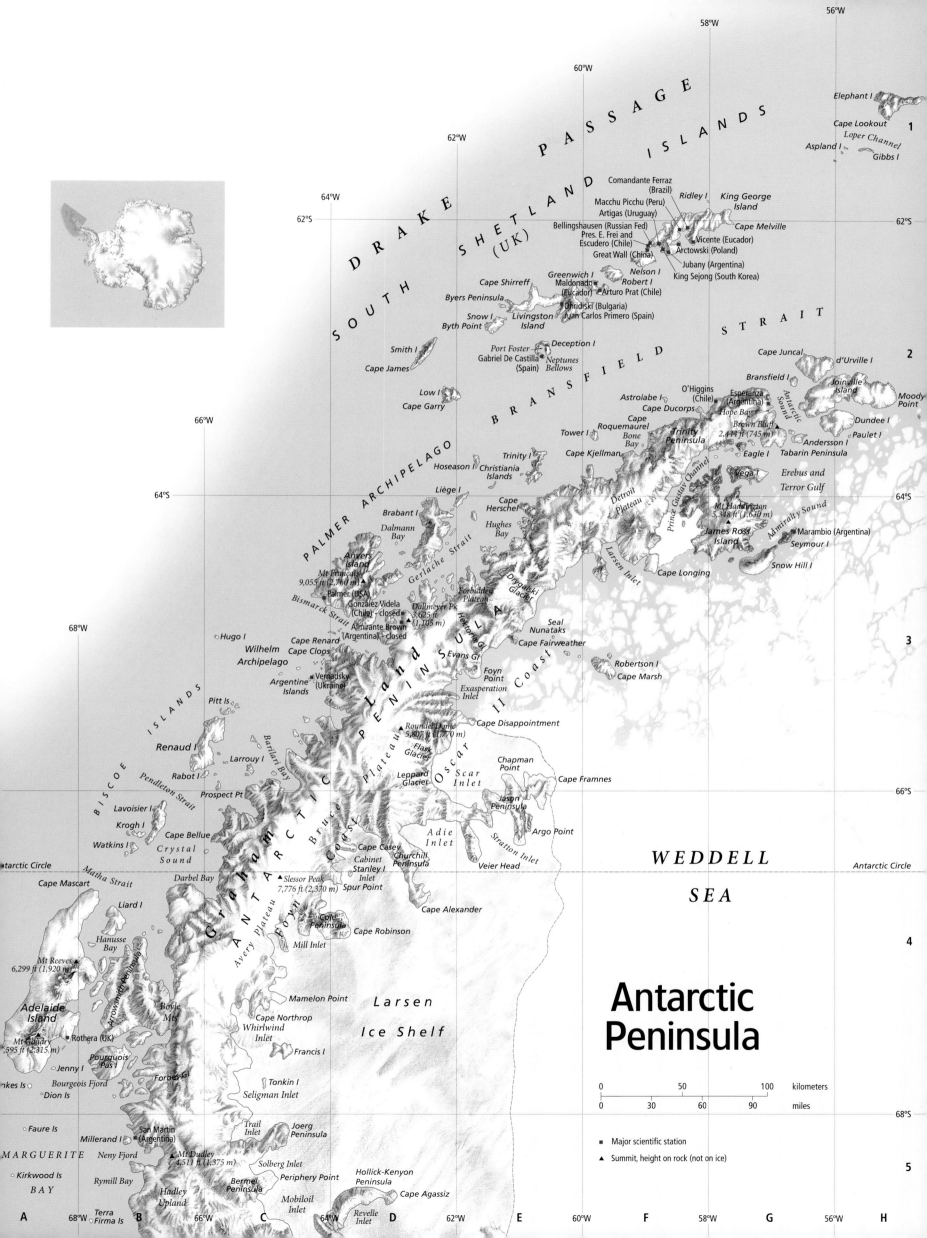

Antarctic Peninsula

DRAKE PASSAGE

SOUTH SHETLAND ISLANDS (UK)

56°W
58°W
60°W
62°W
64°W
66°W
68°W

62°S
64°S
66°S
68°S

Elephant I
Cape Lookout
Loper Channel
Aspland I
Gibbs I

Comandante Ferraz (Brazil)
Macchu Picchu (Peru)
Artigas (Uruguay)
Ridley I
King George Island
Bellingshausen (Russian Fed)
Pres. E. Frei and Escudero (Chile)
Great Wall (China)
Vicente (Ecuador)
Arctowski (Poland)
Jubany (Argentina)
King Sejong (South Korea)
Cape Melville

Nelson I
Robert I
Greenwich I
Maldonado (Ecuador)
Arturo Prat (Chile)
Ohridski (Bulgaria)
Juan Carlos Primero (Spain)

Cape Shirreff
Byers Peninsula
Snow I
Byth Point
Livingston Island

Smith I
Cape James

Port Foster
Gabriel De Castilla (Spain)
Deception I
Neptunes Bellows

Low I
Cape Garry

Cape Juncal
d'Urville I
Bransfield I
Joinville Island
Dundee I
Moody Point

BRANSFIELD STRAIT

PALMER ARCHIPELAGO

Astrolabe I
O'Higgins (Chile)
Esperanza (Argentina)
Hope Bay
Brown Bluff 2,444 ft (745 m)
Andersson I
Paulet I
Tabarin Peninsula

Tower I
Cape Ducorps
Cape Roquemaurel
Bone Bay
Cape Kjellman
Trinity Peninsula

Trinity I
Cape Herschel
Eagle I
Vega I
Erebus and Terror Gulf

Hoseason I
Christiania Islands
Liège I

Hughes Bay
Detroit Plateau
Mt Haddington 5,348 ft (1,630 m)
Admiralty Sound

Brabant I
Dalmann Bay
Anvers Island
Mt Français 9,055 ft (2,760 m)

James Ross Island
Marambio (Argentina)
Seymour I
Snow Hill I

Prince Gustav Channel

Palmer (USA)
González Videla (Chile) - closed
Dallmeyer Pk 3,625 ft (1,105 m)
Almirante Brown (Argentina) - closed
Drygalski Glacier
Larsen Inlet
Cape Longing

Hugo I
Wilhelm Archipelago
Cape Renard
Cape Cloos
Forbidden Plateau
Cape Fairweather
Seal Nunataks
Robertson I
Cape Marsh

Bismarck Strait
Evans Gl
Foyn Point
Exasperation Inlet

Pitt Is
Vernadsky (Ukraine)
Argentine Islands
Cape Disappointment

BISCOE ISLANDS

Renaud I
Larrouy I
Rabot I
Prospect Pt
Barilari Bay
Roundel Dome 5,807 ft (1,770 m)
Flask Glacier
Chapman Point
Cape Framnes

Leppard Glacier
Oscar II Coast
Scar Inlet
Jason Peninsula
Argo Point

Lavoisier I
Krogh I
Watkins I
Crystal Sound
Cape Bellue
Pendleton Strait
Adie Inlet
Stratton Inlet
Veier Head

WEDDELL SEA

Matha Strait
Cape Mascart
Darbel Bay
Cape Casey
Cabinet
Churchill Peninsula
Cape Alexander

Liard I
Stanley I Inlet
Spur Point
Slessor Peak 7,776 ft (2,370 m)
Hanusse Bay
Cole Peninsula
Cape Robinson
Mill Inlet

Mt Reeves 6,299 ft (1,920 m)
Boyle Mts
Avery Plateau
Mamelon Point
Cape Northrop
Whirlwind Inlet

Adelaide Island
Mt Gaudry 7,595 ft (2,315 m)
Rothera (UK)
Francis I

Jenny I
Pourquoi Pas I
Forbes Gl
Tonkin I
Seligman Inlet

Faure Is
San Martin (Argentina)
Trail Inlet
Joerg Peninsula
Hollick-Kenyon Peninsula

Kirkwood Is
Millerand I
Neny Fjord
Mt Dudley 4,511 ft (1,375 m)
Solberg Inlet
Cape Agassiz

MARGUERITE BAY

Rymill Bay
Bermel Peninsula
Periphery Point
Revelle Inlet

Terra Firma Is
Hadley Upland
Mobiloil Inlet

Larsen Ice Shelf

Antarctic Circle

GRAHAM LAND

Arrowsmith Peninsula
Foyn Coast
Bruce Coast

Bourgeois Fjord
Dion Is

Antarctic Peninsula

0   50   100   kilometers
0   30   60   90   miles

■ Major scientific station
▲ Summit, height on rock (not on ice)

A   B   C   D   E   F   G   H

1   2   3   4   5

Volcanic Colors
Brown Bluff is one of the most notable features of the Tabarin Peninsula. The sheer bluff, named for its characteristic reddish-brown volcanic rock, is the remnant of an ancient volcano that was subsequently shaped by glaciation.

Paulet Island
James Clark Ross, aboard the *Erebus* in search of the South Magnetic Pole, named Paulet Island after Lord George Paulet of the British Royal Navy. A circular volcanic cone, the island rises from the sea 1¾ miles wide and 1,158 ft high (3 km and 353 m).

This is as far as many Antarctic travelers will be able to venture into the Weddell Sea, because the clockwise current ensures that even this northern region has its share of spectacularly large icebergs. Huge tabular bergs encountered on the way through the Antarctic Sound give some idea of the difficulties faced by the early explorers, most notably the expeditions of Otto Nordenskjöld and Ernest Shackleton.

One of the largest colonies of Adélie penguins in Antarctica occupies Paulet Island—a fact that is only too apparent from the smell downwind, up to ⅔ mile (1 km) offshore. At least 100,000 breeding pairs crowd the beach and the nearby steep slopes for the short summer—a fine example of avian tenement living. Snowy sheathbills, skuas, and giant petrels live off the rookery, as do the opportunistic leopard seals that patrol the shores in the certainty of easy prey.

When the *Antarctic* sank 25 miles (40 km) away in 1903, Captain Carl Larsen and his crew rowed to Paulet Island. A small hut, only 65 by 23 ft (20 by 7 m), which the 20 marooned men constructed out of the island's regularly shaped volcanic rocks, became their winter quarters. The grave of Ole Wennersgaard, who died on June 7, 1903, is still visible on the shore about 1,000 ft

Island Haven
Determined to retrieve Nordenskjöld's men, Captain Larsen tried to sail the *Antarctic* to Paulet Island, and when the ship finally sank, the men made a 14-day crossing of ice and water to reach the island.

Chilling Out
Out of the water, Weddell seals, like this dozing pup, appear to do little more than rest and sleep. They have been observed on the beach below Brown Bluff, and are common in most locations along the northeastern end of the Antarctic Peninsula.

(300 m) from the ruined building. In summertime, the nesting penguins can make it difficult to approach the hut or walk past it to an attractive lake.

Seymour and Snow Hill Islands
Visiting Seymour Island and Snow Hill Island is possible only when benign sea-ice conditions in the Weddell Sea allow access to the southern end of Erebus and Terror Gulf. When James Clark Ross first saw the northeastern part of Seymour Island in January 1843, he named it Cape Seymour after a British rear admiral. In 1903, when Carl Larsen defined the whole island, 10 miles long by 5 miles wide (16 by 8 km), the nomenclature Ross had given it was adopted.

The rocky outcrops of Seymour Island contain fossil remains of penguins—giant birds compared with the size of today's species. Nordenskjöld found the first of these in December 1902, and palaeontologists continue to make exciting discoveries here.

Argentina operates Vicecomodoro Marambio Station high on the northern side of the island, which includes a small airstrip accessible by C-130 Hercules aircraft from Rio Gallegos. Some 40,000 Adélie penguins nest on Penguin Point at the island's southeastern corner.

While Seymour Island is largely ice-free and muddy, Snow Hill Island is almost completely covered in snow and supports no wildlife. This island, lying southwest across Admiralty Sound, is 20 miles long by 6 miles wide (32 by 10 km); uncertain whether it was part of the mainland, Ross named it Snow Hill. In 1902 Otto Nordenskjöld based his Swedish South Polar Expedition there. Now, the Argentinians care for their hut, and a few artefacts from that expedition, including a primus stove and coffee grinder, are kept at the Antarctic Office on the dock in Ushuaia. Aviator Lincoln Ellsworth used Snow Hill Island in 1935 as the starting point for one of his attempts to fly across the Antarctic continent.

Astrolabe Island
French explorer Dumont d'Urville sighted this island in the Bransfield Strait during his 1837–40 voyage, and named it after his expedition ship, the *Astrolabe*. The rocky island is 3 miles (5 km) long, with cliffs plunging more than 330 ft (100 m) directly to the sea. Chinstrap penguins, fur seals, and Weddell seals breed on the low ground, and skuas and fulmars nest along the craggy heights. Shore landings are possible here, and the surrounding ice provides a spectacular Zodiac cruise.

Building Blocks
In this small hut on Paulet Island, the 20 shipwrecked sailors (and cat) from the *Antarctic* lived for nine months over a long polar winter. Fortunately for the stranded men, building the hut was quite easy: the regular freezing and thawing of the island's basalt caused the rock to break into even-sized flat stones—ready-made building blocks.

Mikkelsen Harbor, Trinity Island
Mikkelsen Harbor, on the south side of Trinity Island, was named by Nordenskjöld's 1901–04 expedition, but the source of the name has been lost. The term "harbor" is deceptive as the landing site is in fact a small island in a bay that has no harbor at all. Rocky shoals lead to a rocky beach, with glaciers forming a magnificent backdrop on a fine day.

There was a hut on the island, and a radio mast at its low summit. The chief residents are gentoo penguins on its slopes, and slumbering Weddell and elephant seals on the beach. DM

Antarctic Peninsula: The Gerlache Strait

One of the major waterways of Antarctica, the Gerlache Strait stretches between the coast of the Antarctic Peninsula and the string of islands to the west. It is enclosed by mountains, rocky cliffs, and glaciers.

Curtiss Bay

On Argentinian charts, Curtiss Bay is rather harshly called Bahia Inutil ("Useless Bay"). Its English name was coined in 1960 for Glenn Curtiss, the first American builder of seaplanes, hence Seaplane Point within the bay. In the small glacier-lined cove behind, ice cliffs 1¾ miles (3 km) long are active, and large blocks of ice frequently break off and crash into the water. The bay is often visited by whales, and leopard and other seals.

Alcock Island

This island, which is less than ⅔ mile (1 km) across, commemorates another aviator: Sir John Alcock. Whalers called it Penguin Island, but the name was changed to avoid confusion with an island of the same name in the South Shetland Islands. Along most of Alcock Island's coastline are towering cliffs that chinstrap penguins seem able to scale with ease.

Portal Point

Portal Point, the entrance to Charlotte Bay on the west coast of Graham Land, received its name from a tiny refuge hut built there by the British in 1956 and used as a portal up onto the high plateau. Icebergs, glaciers, and mountains make Charlotte Bay one of the most attractive inlets along the Antarctic Peninsula. The bay indents for about 12 miles (20 km) into the coast. Because there is little resident wildlife, Portal Point is sometimes used by expeditioners as a camp site.

Wilhelmina Bay

South of the Reclus Peninsula, and bounded on the south by the Arctowski Peninsula, Wilhelmina Bay is frequented by whales and littered with islands which provide some shelter in almost any weather.

The exception is when ferocious katabatic winds howl down from the Forbidden Plateau, more than 7,000 ft (2,100 m) above the bay, and whip the water into foam. Adrien de Gerlache visited in 1898 and named the bay for the 18-year-old Queen Wilhelmina of the Netherlands.

Spigot Peak and Orne Harbor

On the northwestern side of the Arctowski Peninsula, the towering black spire of Spigot Peak marks the southern entrance of Orne Harbor. The 938-ft (286-m) peak takes its name from its resemblance to a spigot or cask plug. Orne Harbor, discovered by Adrien de Gerlache in 1898, is a mere indentation on the west coast of Graham Land.

Frozen Memorial
Arriving in late summer 1898, the expedition led by Adrien de Gerlache explored and mapped the islands and coast bordering the strait that is named for the explorer. The mountainous coast commemorates geophysicist Emile Danco, who died later on the voyage, and Wiencke Island is named for a young seaman who fell overboard.

65°S

Mountain Sentinels
Mountain spires of massive black rocks rise above the scenic Gerlache Strait, the channel that separates the island-dotted Palmer Archipelago from the west side of the Antarctic Peninsula.

South from Gerlache Strait

62°W
61°W

Low Island
Cape Garry

1

0   25   kilometers
0   10   miles

■ Major scientific station
▲ Summit, height on rock (not on ice)

Cape Wollaston
Tower Hill
3,691 ft (1,125 m)
Milburn Bay

63°W

Hoseason Island
▲ Stopford Peak
1,624 ft (495 m)
Angot Point

Spert I
Trinity Island

Christiania Islands ～ Intercurrence I

Mikkelsen Harbor
Orléans Strait

Liège Island   Neyt Point

Cape Andreas
Curtiss Bay
Seaplane Pt

64°S

64°W

64°S

Cape Roux   Cape Cockburn
Pasteur Peninsula
Guyou Bay
Claude Point
Bart Bay

Bouquet Bay
▲ Pavlov Peak
2,795 ft (852 m)
Macleod Point
Davis I
Spallanzani Point
Lister Gl

Small I

Cape Herschel

Cierva Cove

Davis Coast

Two Hummock Island
Hydrurga Rocks

Laennec Gl   Hill Bay

Brabant Island

Mitchell Point

HUGHES BAY

Alcock I

Brialmont Cove

Melchior Islands

Lanusse Bay
Mackenzie Gl
Lecointe Island

Spring Point

▲ Tournachon Peak
2,818 ft (858 m)

P
A
L
M
E
R

A
R
C
H
I
P
E
L
A
G
O

Eta I
Omega I

Dallmann

Gand I

Bay

Hippocrates Glacier

Solvay Mountains

Hunt I
Pinel Point

Valdivia Point

Salvesen Cove

Bluff Island

Buls Bay

Eckener Point
Mt Zeppelin
4,150 ft (1,265 m)

Wellman Glacier

2

Perrier Bay

Gourdon Pen

Thompson Pen

Hulot Peninsula
▲ Mt Bulcke
3,386 ft (1,032 m)

Strath Point

Recess Cove

Portal Pt

Nobile Glacier

Hamburg Bay

Fournier Bay

Cape Anna

Enterprise I

Bancroft Bay

Reclus Peninsula

Charlotte Bay

G
e
r
l
a
c
h
e

S
t
r
a
i
t

D
a
n
c
o

C
o
a
s
t

Anvers Island

Delaite I
Emma I
Nansen Island

Brooklyn I

Pelseneer I

Danco Coast

Mt Français
9,055 ft (2,760 m)

Orne Harbor
Spigot Peak
938 ft (286 m)
Cuverville I

Wilhelmina Bay

Leonardo Glacier

Plateau

Hooper Glacier
William Glacier

Lion I

Mt Tennant
2,257 ft (688 m)

Arctowski Peninsula

Renard Glacier

Forbidden

Copper Peak
3,690 ft (1,125 m)

Ketley Point
Rongé Island
Danco I

Errera Channel

Piccard Cove

Mt Moberly
5,029 ft (1,533 m)

Börgen Bay

Beneden Head

Laussedat Heights

Mt Walker
7,740 ft (2,359 m)

A
N
T
A
R
C
T
I
C

Cape Monaco
Wylie Bay
Arthur Harbor

Mt William
5,249 ft (1,600 m)

Neumayer Channel

Waterboat Pt

Andvord Bay

Deville Glacier

A
r
a
g
o

G
l
a
c
i
e
r

Moser Glacier

P
E
N
I
N
S
U
L
A

Nordenskjöld Coast

Palmer (USA)

Biscoe Bay

Nemo Peak
2,841 ft (866 m)

Lemaire I

González Videla
(Chile) - closed

Neko Harbor

Rudolph Glacier

L
A
N
D

Access Point
Cape Lancaster
Py Point

Port Lockroy

Paradise Harbor

Almirante Brown
(Argentina) - closed

Dallmeyer Peak
3,625 ft (1,105 m)

aubin Islands

Doumer I
Wiencke Island

Fridtjof I

Bryde Island

G
r
u
b
b

G
l
a
c
i
e
r

Bagshawe Glacier

G
R
A
H
A
M

B
i
s
m
a
r
c
k

S
t
r
a
i
t

Knight I

Cape Errera

Truant I
Bob I

Cape Willems

Kershaw Peaks

Hektoria Glacier

Cape Fairweather

65°S

Wauwermans Islands

Butler Passage

Puzzle Is

F
l
a
n
d
r
e
s

B
a
y

Nimrod Passage

Dannebrog Islands

Cape Renard
False Cape Renard

Guyou Is

D
a
n
c
o

C
o
a
s
t

Evans Glacier

O
s
c
a
r

I
I

C
o
a
s
t

W
i
l
h
e
l
m

A
r
c
h
i
p
e
l
a
g
o

Booth I

Lemaire Channel

Pléneau I
Vedel Is   Cape Cloos

Danco Coast

Foyn Point

3

Hovgaard I
Petermann I

Leay Glacier

Vernadsky
(Ukraine)

▲ Mt Scott
2,894 ft (882 m)

Exasperation Inlet

gentine Islands
Yalour Is

▲ Mt Shackleton
4,806 ft (1,465 m)

Penola Strait

Cape Tuxen

Wiggins Glacier

Berthelot Is

Bussey Glacier

Collins Bay   Trooz Glacier

Cape Disappointment

A   64°W   B   63°W   C   62°W   D   61°W   E

At the eastern end of the beach, the most visited penguin rookery overlooks a tiny, picturesque rocky cove. The western rookery, once off-limits to tourists, is even more spectacular, as the mountains of Rongé Island loom close across the channel. Behind the rookeries, skuas guard their nests. Antarctic shags also raise young on the cliffs, and pintados and snow petrels nest at the highest points. Crabeater and leopard seals often bask on ice trapped on shallow rocks around the island. In late summer, patches of vegetation—mosses and lichens, as well as Antarctica's two flowering plants: Antarctic hair grass (*Deschampsia antarctica*) and the pearlwort (*Colobanthus quitensis*)—are visible.

Danco Island

This island, only about 1 mile (1.6 km) long and 590 ft (180 m) high, lies to the south of Cuverville Island in the Errera Channel. It was named for the geophysicist who died on board the *Belgica* while she was trapped in the ice of the Bellingshausen Sea. Nesting gentoo penguins surround hut foundations at the northern end of the island during the breeding season. Built in 1956 by a British Antarctic Survey team, and occupied until 1959, the hut was removed in 2002.

Andvord Bay and Neko Harbor

South from Errera Channel, the next major indentation is Andvord Bay. The quite deep expanse of water, which is surrounded by mountains and glaciers spilling down to the shoreline, is 3 miles wide and 9 miles long (5 by 14 km). Exasperation Inlet, off the Weddell Sea, lies 25 miles (40 km) away, across the mountains that soar to 5,300 ft (1,600 m) at the head of the bay.

On the eastern side of Andvord Bay is the sheltered cove of Neko Harbor, a place of beauty carrying a name associated with slaughter: the *Neko* was a whaler, a factory ship that operated in the region between 1911 and 1924 and frequently moored in the bay.

An unoccupied Argentinian hut stands near the water by the end of a small point; snowy sheathbills sometimes nest in the hut's foundations. Weddell seals frequent the pebbly beach below the hut, and a gentoo penguin rookery on the steep slope behind the point may total 1,000 birds at peak season. The shallows of Neko Harbor are popular with gentoos, and whales sometimes enter Andvord Bay. The glacier that fills the bay to the north of the landing site is very active, and huge waves can wash up from large iceberg calvings.

Waterboat Point

Coming from the north, the entry point into Paradise Harbor is Waterboat Point, now the site of González Videla, the base operated by the Chilean Air Force. Some regard Waterboat Point as an island because the spit joining it to the mainland is covered at high tide. Whichever, it is the end of the peninsula between Andvord Bay and Paradise Harbor, and was first surveyed by de Gerlache in 1898. The British Imperial Expedition of 1920–22 named it for the abandoned waterboat that housed two of them when the expedition collapsed. Thomas Bagshawe, a geologist, and Michael Lester, a surveyor, stayed on to do some scientific work.

A Rocky Refuge

The dark bulk of Spigot Peak marks the southern end of Orne Harbor. During the brief summer months, its cliffs and crags are home to several bird species, including a colony of shags that nest close to the waterline. Crabeater seals are common around the point, and toward the end of the season fur seals rest along the shore.

Birth of Icebergs

Andvord Bay, near Anvers Island, has the spectacularly rugged landscape of mountains and glaciers that is typical of the Antarctic Peninsula. These alpine glaciers occur in the high mountains and flow all the way down to the coast. When the ice flow reaches the coastline, the glaciers break off in cliffs and spawn myriad icebergs.

Errera Channel

Narrow Errera Channel lies between the coast of the Arctowski Peninsula and nearby Rongé Island—two largely matching shorelines. The channel, roughly L-shaped and about 5½ miles (9 km) long, contains Cuverville and Danco islands, and is frequented by pods of humpback whales. There is often ice in the water and on the mountains which tower more than 3,000 ft (900 m) on either side of the channel.

Rongé Island

Rongé Island was named for Madame de Rongé, a sponsor of the 1897–99 Belgian Antarctic Expedition. George's Point, at its northern tip, is home to a small gentoo penguin population and several little chinstrap penguin rookeries.

Cuverville Island

Cuverville Island stands at the northern end of Errera Channel. Five thousand pairs of gentoo penguins breed on the long, shingly beach and in coves at the northern end of the island. Cuverville has a single landing site; elsewhere the terrain rises steeply to the 827-ft (252-m) rocky dome that makes up most of the island.

They survived under the upturned boat in makeshift quarters extended with food boxes until January 1922, and recorded extensive scientific data while living on penguins and minced seal meat. The few remnants of their boathouse remain near the station as a Historic Site. A short distance away is a hut built at Coal Point in 1950 to commemorate the visit of Chilean president, Gabriel González Videla, the first head of state to visit.

Paradise Harbor

Tough whalers were the first to call this wide bay sited behind Bryde and Lemaire islands "Paradise Harbor." The bay is indeed spectacular. The name has been in use since at least 1920, and it has proved to be an irresistible lure to expedition ships following the western side of the peninsula.

Almirante Brown, an Argentinian base that closed in 1999 and then reopened in 2008, is in the middle of Paradise Harbor. This base along with Chile's González Videla at Waterboat Point together receive more tourists than almost anywhere else in Antarctica, partly because both offer the opportunity to set foot on the Antarctic continent. The Chileans, identifying themselves as the Paradise Harbor Port Authority, may radio approaching ships and give permission to enter Bahia Paraiso, even if the destination is the Argentinian base across the bay. This is probably to establish a record of occupation and management in case of a territorial claim.

Directly behind Almirante Brown, a steep 165-ft (50-m) snow-covered slope is easily climbed for a fine view of Paradise Harbor; sliding back down the snow takes a matter of seconds. The base was built in 1951 and named after a founder of the Argentinian Navy; the burned foundations near the existing buildings are a reminder of a fire in April 1984.

Anvers Island

It can be difficult at times to distinguish the mainland of the Antarctic Peninsula from the offshore islands of the Palmer Archipelago. In this world of black basalt and white glaciers, only sea charts may reveal the real difference. This is particularly true of Anvers Island, one of the largest islands north of the Antarctic Circle on the western side of the peninsula. Anvers measures about 37 miles (60 km) in length, and is separated from the Danco Coast by the Gerlache Strait. De Gerlache named the island after a province in Belgium.

Palmer Station

Palmer Station, operated by the United States, was named after Nathaniel Palmer, young skipper of the *Hero*, who was one of the earliest explorers to reach this part of Antarctica. Palmer Station is situated on southwest Anvers Island on the shores of Arthur Harbor. The harbor contains the remains of an ecological disaster. On January 28, 1989, the Argentinian naval vessel, the *Bahia Paraiso*, ran onto a reef, tearing a hole in its side through which a vast quantity of diesel oil leaked. The coastline was covered in oil and terrible damage was inflicted on the local wildlife. Today, all that is visible of the inverted ship is the glint off the exposed hull on the west of the channel.

Britain named Arthur Harbor in 1955 after Oswald Arthur, governor of the Falkland Islands. "Station N," a small British base there, was first occupied in 1955. The building was lent to nearby Palmer Station in 1963 for use as a laboratory. Fire destroyed it in 1971, and the United States removed the remains in 1991.

Palmer Station began in 1964 as a single hut, and was extended the following year. The current station, which has linking walkways above snow level, is quite small, with two main buildings and a population that ranges from ten in winter to up to 43 in summer. Palmer is a popular destination—so popular, in fact, that the station has initiated a roster system and applications to visit are lodged months in advance. Tourists are shown a well-run operation doing useful research and a fascinating aquarium of Antarctic marine life. Palmer also has perhaps the best souvenir shop in Antarctica.

Palmer is the base for a research program that is looking at marine life to see how changing sea-ice cover affects the ecosystem. The study area covers a region of 69,500 sq miles (180,000 sq km) that includes land and sea, and a wide range of ice cover.

Port Lockroy

Though Port Lockroy is the name for the whole harbor within the southwestern coast of Wiencke Island, it is now used for the old British station on Goudier Island, one of a string of rocks off Jougla Point. Port Lockroy was established on February 16, 1944, and operated until January 1962. The original hut remains the core of the station; it is operated during the summer months as a museum and visitor center. Damoy Point, on the northern end of Port Lockroy, offers a quieter landing option on Wiencke Island. From the beach, there is a gentle slope to some scattered gentoo rookeries. The British built a hut behind the beach in 1975 as an air facility. Though they can be covered in deep snow, and are more open to the wind than Port Lockroy, Damoy Point and nearby Dorian Bay are beautiful. DM

Cool Customers

Visitors to Port Lockroy, on Goudier Island, are constantly surprised at how little notice gentoo penguins take of the passing parade, and wonder if the birds are as relaxed as they appear. Studies so far suggest that human disturbance has little impact on the number of chicks fledged.

Flying South

Snowy sheathbills are the only birds without webbed feet that breed in the Antarctic—to get there, they must fly nonstop across the Drake Passage. They nest on the edge of penguin colonies, where rewards are plenty for scavengers: dropped food, unfortunate penguin chicks, but mainly guano.

Antarctic Peninsula: The South

On a clear, still day, the glaciers and peaks that line the Lemaire Channel are reflected perfectly in the deep, dark water. But on a foggy day, visibility closes in, and there is an eerie sensation of sailing off the end of the Earth—in fact, this is heading deep into the Antarctic region.

Lemaire Channel

Ships proceeding south from Palmer Station invariably sail eastward along the Bismarck Strait, named in 1874 by Edward Dallman's German expedition for Prince Otto von Bismarck. There they join those sailing due south from Port Lockroy on the northeastern coast of Wiencke Island and those coming from points farther north along the Gerlache Strait, to proceed down the Butler Passage toward Cape Renard on the Antarctic Peninsula's coast. The tall, twin peaks of Cape Renard mark the northern entrance to the Lemaire Channel, which is one of the highlights of any voyage along the western coastline of the Antarctic Peninsula.

Perversely named in 1898 by Adrien de Gerlache for Charles Lemaire, a Belgian who had explored the Congo—not for Jacob Le Maire who, with Willem and Jan Schouten, discovered the route through the Drake Passage and around the southern end of Tierra del Fuego in 1616—this narrow seaway between Booth Island and the Antarctic Peninsula is a narrow cleft between towering cliffs. That de Gerlache was prepared to risk sailing through the channel seems an impressive act of nineteenth-century seamanship and courage. Sometimes it is impassable—currents and wind can fill it with sea ice and large icebergs within hours—but, when the ice permits, the hour's voyage each way is an unforgettable experience. The sea ice often supports basking crabeater seals, and minke, humpback and orca whales navigate the passage.

Although the Lemaire Channel is defined as the whole passage between Cape Cloos in the south and False Cape Renard in the north, its most spectacular stretch is the 4 miles (7 km) where the sheer rock and

Towering Peaks

The spectacular northern entrance of the Lemaire Channel is marked by Cape Renard (in the middle of the photograph). Beyond the cape's twin soaring peaks, the channel narrows dramatically to become what is known as False Cape Renard.

Private Wonderland
A maze of islands, rocks, and grounded icebergs stretches south beyond Petermann Island. The area has several sheltered anchorages and many private expeditions venture south of the Lemaire Channel to explore this magical place.

Grounded Icebergs
Large icebergs, often driven in from the west by wind and currents, run aground in the shallow waters north and west of Pléneau Island. They can remain for years as they slowly melt, creating an impressive spectacle of form and color.

Enduring Contest
In view of the intense rivalry between Robert Scott and Ernest Shackleton, Scott must have turned in his icy grave when the soaring 4,806-ft (1,465-m) peak (pictured on the right) was named Mount Shackleton, while the smaller peak was named in his honor.

ice walls of Booth Island tower above it. Midway along this section, the channel opens up on the eastern side into Deloncie Bay, which cuts back into the peninsula to the Hotine Glacier. Immediately south of there, the channel narrows dramatically to less than 2,600 ft (800 m), with Mount Cloos on the mainland and Wandel Peak on Booth Island both looming overhead.

Pléneau Island
To the south, the Lemaire Channel opens into the Penola Strait, named by John Rymill after his schooner and his Australian sheep station, which were both called Penola. The island immediately to the right of the channel's entrance is Pléneau Island, a snow-capped rocky islet just over ⅔ mile (1 km) long, which was once

thought to be a peninsula of Hovgaard Island to the south. Jean-Baptiste Charcot named it in 1904 for Paul Pléneau, company director, supporter, and friend, who had sailed with him as his expedition's photographer. The island is home to numerous gentoo penguins, elephant seals form communal wallows slightly inland, and Weddell seals are often found alone or in pairs along the shoreline. Its highest point commands a stunning view across to Booth Island and the almost indiscernible entrance to the Lemaire Channel.

The most spectacular feature of Pléneau Island is not on the land at all but a field of grounded icebergs usually, but not always, found west of the Lemaire Channel between Pléneau and Booth islands. Because these bergs are grounded, they are more stable than

floating ones, and it is possible to navigate through them. They form a fairyland of arches and pinnacles, and range in color from pellucid turquoise, where shallow water laps at their feet, to brilliant blue in the cracks in the ice. The area is prone to strong winds that sweep off Booth Island and churn the smooth waters of the channel between the islands into a lather of foam.

Pléneau Island, is less than 100 miles (160 km) north of the Antarctic Circle, and is often the turning point for cruise ships. There are some good reasons for continuing south, but Hovgaard Island is not one of them. It is large and rises to 1,210 ft (370 m), but has neither the icebergs and seals of Pléneau Island to the north, nor the Adélie penguins and history of Petermann Island to the south.

Penguin Island
Adélie penguins come ashore onto Petermann Island to reproduce in spring. The parents take turns remaining with the eggs over winter and brooding the chicks.

Crystal Clear
Only toward the end of summer is Crystal Sound and the approach to the Antarctic Circle so ice-free. As winter approaches, ice crystals are quick to form. The sound was named by the British for the ice-crystal research that was carried out in the area.

Petermann Island

The German geographer, August Petermann, strongly supported his country's polar exploration in the mid-nineteenth century. His name lives on in the dry, dusty Petermann Range in Western Australia, and the contrasting Petermann Ranges of Dronning Maud Land in East Antarctica. Snowy, cold Petermann Island, 1 mile (1.6 km) long and 436 ft (133 m) high, discovered and named by the German Dallman expedition of 1873–74, lies close to the Antarctic Circle.

Most of Petermann Island's historical links, however, are French, not German. On the southeastern side of the island, the landing site is named Port Circumcision because the small cove was discovered by Charcot on January 1, 1909, the feast day of the Circumcision (when, tradition holds, Christ was circumcised). Charcot securely moored his ship, the *Pourquoi Pas?*, in the cove, ran iron hawsers across the mouth of the harbor to keep out icebergs, and settled in for the winter.

Charcot named a rocky hill in the south of the snow-covered island Megalestris Hill, after the Latin term for a skua. He built a stone cairn to commemorate his expedition—it was restored by a British Antarctic Survey (BAS) team in 1958, and it is now an Antarctic Historic Monument.

The hut by the cove is an abandoned Argentinian *refugio*; built in 1955, it now provides shelter to scores of gentoo penguins and their chicks, and to snowy sheathbills seeking scraps. For much of the summer, it is used as a field camp by scientists from nearby Vernadsky Station. The memorial cross close by is for three British scientists who died near here on August 14, 1982. Across Penola Strait, on the mainland, stands the U-shaped massif of Mount Scott, 2,926 ft (892 m) high, and behind it the sheer-walled peak of Mount

Shackleton, 4,806 ft (1,465 m). Charcot named both of the mountains for the British explorers.

While gentoo penguins have claimed the low rocks, Adélie penguins dominate the rocky summit up the slope to the north. Due east from the Adélie hordes, in a tiny rocky bay formed by a basalt dike, Antarctic shags claim almost every crag at peak season. Around the island's convoluted southern coast, large groups of crabeater seals rest on floes or can be seen streaking through the water, and the occasional leopard seal may be out cruising for wayward penguins.

Yalour Islands

The rocks and islets scattered over 1½ miles (2.4 km) to the south of Wilhelm Archipelago hardly merit the title of islands, but they are a good place to observe the antics of some 8,000 pairs of Adélie penguins and their chicks who live there each summer.

Charcot named the Yalour Islands after Lieutenant Jorge Yalour, an officer on the Argentinian corvette, the *Uruguay*, which rescued the Nordenskjöld expedition in November 1903. Charcot's first Antarctic expedition intended to go to the aid of the Swedes, and diverted to this side of the Antarctic Peninsula only after learning the expedition members had been evacuated.

Argentine Islands

Charcot named these islands to thank the Argentinian government for its varied assistance during his *Français* expedition of 1903–05.

Even in the middle of summer, reaching the islands through the ice is tricky. The main reason to visit here is to go to the Ukrainian base, Academician Vernadsky, which is one of the most comfortable and welcoming scientific research stations in Antarctica. Until 1996, this was Britain's Faraday Station, named after Michael Faraday, discoverer of electromagnetism; then it was sold to the Ukraine for the nominal sum of one pound. The Ukrainian government was keen to maintain an Antarctic presence following the break-up of the Soviet

Union, when all the bases previously operated by the Soviets were transferred to the Russians.

The first British building on the Argentine Islands was constructed on Winter Island by the British Graham Land Expedition in 1935–36, but it was destroyed in 1946, perhaps by a tidal wave. Undeterred, in 1947 Britain erected another hut in the same location, and named it Wordie House after Sir James Wordie, who was on Shackleton's *Endurance* expedition (1914–17), and served on the advisory committee for Operation Tabarin (1943–45). In 1954, the station was transferred to Galindez Island, and it is this base that has become Academician Vernadsky, named for Vladimir Vernadsky, first president of the Ukrainian Academy of Sciences. British teams have restored Wordie House to the way it would have been in the 1950s. Vernadsky's personnel now maintain it, and visiting ships bring passengers by Zodiac craft down a narrow, picturesque channel or by the shorter and more spectacular Cornice Channel. Faraday was occupied continuously by the British for 49 years and 31 days.

At Vernadsky, the bar is the most popular place for expeditioners and visitors alike—the Ukrainians have maintained its distinctly British atmosphere, right down to the billiard table and dart board. Hanging above the bar is an amazing photograph of a man and a minke whale in eye-to-eye contact. Although some question the photograph's authenticity, base members swear that it is real. The infamous collection of brassieres, including some of improbable dimensions, continues to grow as some visitors elect to contribute. When ships call, the staff open "the southernmost gift shop in the world," stocked with souvenir cloth badges, postcards, and attractive carved plaques.

Vernadsky continues Britain's long-term research programs, with some 12 staff in winter and 24 in the summer. The meteorological records here reveal that the local mean annual temperature has increased by 4.5°F (2.5°C) since 1947. Visitors to the station can see the equipment that is used for measuring the ozone in the

atmosphere, and they often learn about the trends that appear in the newspapers months later.

Antarctic Circle

Ships voyaging to the Antarctic Circle normally pass through Crystal Sound, which was named by Britain in 1960 because most nearby places were called after researchers studying ice crystals. It is a good region for ice—even in late summer, when north of the Lemaire Channel is comparatively warm, the Antarctic Circle there may be distinctly colder; late in the season some pancake ice persists, when there is not even the less dense frazil or grease ice to the north.

Between Petermann Island and the Circle the only potential landing site is the Fish Islands, at the foot of towering Prospect Point, a small group of rocky islets that are home to nesting Antarctic cormorants and Adélie penguins. The main attraction here is the ice, including impressive tabular icebergs that often drift north from the ice shelves of the Bellingshausen Sea.

Voyaging through or around the ice south of the Circle requires a lot of effort for relatively little reward. Detaille Island Station, at 66°52′S, exemplifies some of the problems. Established by Britain in February 1956, it was abandoned in March 1959 when it proved to be impossible to reach regularly.

Considerably farther south, at 67°34′S, is Britain's Rothera Station on Adelaide Island, its principal base in Antarctica. It was opened in October 1975 to replace the Adelaide Station (1961–77), when that station's ice runway deteriorated. Fire is a major fear in Antarctica and that fear was reinforced when Rothera's Bonney Laboratory burned down in 2001—it has since been rebuilt. From Rothera, excursions are possible to the Dion Islands, at the northern end of Marguerite Bay, the site of the peninsula's sole colony of emperor penguins.

Toward the Bellingshausen Sea, the unprecedented warming that has occurred in the Antarctic Peninsula over the last five decades is nowhere more evident than in the break-up of the Wordie Ice Shelf in the 1980s and the fast disintegration of the Wilkins Ice Shelf, near Alexander Island, in 2008. DM

Changeable Channel
The ice that often makes the short passage through the Lemaire Channel so memorable can move rapidly under the influence of wind and currents, and within hours block the way with icebergs. This ship is approaching the narrow northern end of the channel.

Environmentally Suited
Crabeater seals, unlike their nearest relatives, Weddell and leopard seals, avoid land, preferring to come out of the water onto ice. The crabeaters' musculature is designed for swimming not walking—movement on ice rather than on land is much easier for them because they can slide along, without having to lift their bodies at all.

West Antarctica

Because of the difficult ice conditions and its remoteness, this region west of the Antarctic Peninsula stretching to the Ross Sea has some of the least studied parts of the Antarctic. However, with the advent of satellite studies, the region has garnered interest in recent years as one in which the effects of climate change are becoming evident, with the retreat of local ice shelves and of the sea-ice edge.

Bellingshausen and Amundsen Seas

The Bellingshausen Sea lies to the west of the Antarctic Peninsula and is separated from the Amundsen Sea by the northward projection from Ellsworth Land of a group of islands dominated by Thurston Island, situated at about 100°W. It extends between longitudes 55°W and 100°W, and is bounded on the east by Alexander Island and the Antarctic Peninsula. Only one island is known in the Bellingshausen Sea: Peter I Øy. The northern boundary is vague because the sea deepens gently to the deep sea floor of the Bellingshausen Abyssal Plain.

The Amundsen Sea is farther west and its western boundary is variously taken as the eastern extremity of the Ross Sea (158°W) or somewhat farther east, at about 126°W. Its eastern boundary is Thurston Island. The landward edge of the Amundsen Sea is dotted with islands, but there are none out to sea, although there is a noteworthy seamount (Marie Byrd Seamount), which rises from water over 9,840 ft (3,000 m) deep to within 1,820 ft (555 m) of the surface.

Some of the world's greatest maritime explorers have been associated with the Amundsen Sea area. Cook had reached 71°10'S, 107°W, in 1773, his

farthest point south. Bellingshausen followed in January 1821 and discovered Peter I Øy. Amundsen spent the winter of 1898 in the Bellingshausen Sea as part of the first Antarctic wintering party on the *Belgica* expedition.

In contrast with most of the Antarctic coast, the continental shelves adjacent to these seas are relatively wide—mostly 135 to 200 miles (220 to 320 km) but extend to 400 miles (640 km). Like the rest of the coast, the shelf edge is deep, about 1,640 ft (500 m), and it varies considerably. As with the rest of the continental margin, the shelf is crossed by deep gullies, which are glacial in origin and of unknown age. At their seaward ends, they may be marked by raised accumulations of sediment, termed sediment drifts.

Until recently, sea-ice extent was more stable than elsewhere but the margin of the sea-ice zone has retreated some 1.5 degrees of latitude since the early 1980s as temperatures over the western Antarctic Peninsula have risen; the date of minimal sea-ice extent is now February rather than March. Sea-ice extent is less here than elsewhere and sea-ice growth and decay are at comparable rates; thus, there is no sudden springtime release of nutrients at the ice edge. The low variation in sea-ice extent leads to lower generation of Antarctic Bottom Water than occurs elsewhere around Antarctica—a particular contrast with the Weddell Sea.

Oceanographically, the region south of the Pacific Ocean differs from other areas around Antarctica. There is little upwelling of deep water along the continental shelf edge and thus the availability of nutrients is less, possibly a result, in part, of the width of the continental shelf. As a result of low iron concentrations and other

Cold, Cold Sea
The Amundsen Sea, off the coast of Marie Byrd Land, was first mapped by the 1939–41 United States Antarctic Services Expedition, which was led by Richard Byrd. Beneath the surface, the sea grades imperceptibly north into the Amundsen Abyssal Plain.

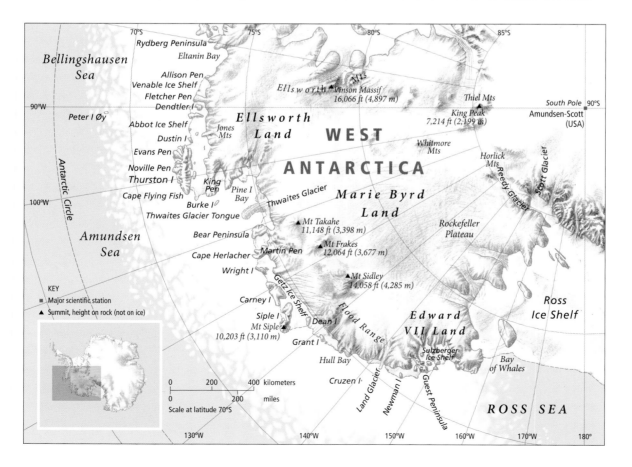

Thick Ice
Scientists measure the ice thickness of the Bellingshausen Sea. Because it remains frozen in pack ice, this is one of the more inaccessible and least studied parts of Antarctica. Until recently, the sea-ice extent was more stable here than elsewhere.

differences at the sea-ice edge, the mix of micro-organisms at the ice-edge is different, in turn leading to contrasts in dominant grazing organisms and higher predators, both in abundance and species mix.

In the former supercontinent of Gondwana, the Campbell Plateau, which now lies to the south of New Zealand, was adjacent to this coast, resulting in many commonalities in early geology between this region and parts of New Zealand. The modern sea floor formed as part of the separation of the two continental fragments over the last 100 million years.

Peter I Øy

Peter I Øy, the first land ever seen inside the Antarctic Circle, was discovered by Thaddeus Bellingshausen in January 1821 and named after the Russian czar, Peter the Great. The island, 11 miles long by 5 miles wide (18 by 8 km), is an extinct volcano with Lars

Christensen Peak rising 5,760 ft (1,755 m) from its center. Its shores are predominantly ice cliffs, making visits extremely difficult. Ola Olstad and his *Norvegia* expedition achieved the first landing in 1929, and they claimed the island for Norway.

Ellsworth Land

Ellsworth Land is a vast region that lies between 70°W and 100°W, at an average altitude of some 3,280 ft (1,000 m), along the coast of the Bellingshausen Sea. It links the southern Antarctic Peninsula to Marie Byrd Land. Likewise, it forms a bridge between the Ronne Ice Shelf, below the southern Atlantic Ocean, and the Bellingshausen Sea, south of the Pacific. It was named by Lincoln Ellsworth for his father, James W. Ellsworth, during his history-making flight of November 1935. During this flight, Ellsworth is reputed to have dropped an American flag and claimed the territory for his home country. Thurston Island, at the western end, forms a rough peninsula that marks the coastal separation of Ellsworth Land from Marie Byrd Land.

Most of the area is part of the West Antarctic Ice Sheet, with ice thickness commonly in the order of 5,000 to 6,560 ft (1,500 to 2,000 m). Ellsworth Land is the even surface expression of the ice sheet, but underneath the ice is another story. The land beneath Ellsworth Land is almost all well below sea level, by as much as 3,950 to 4,270 ft (1,200 to 1,300 m); only very locally is it above sea level. If the Antarctic ice were to melt, very little land would be visible until isostastic effects raised some of the land—even then, most would remain below sea level. The southern side of Ellsworth Land includes the north–south oriented Ellsworth Mountains in the Weddell Sea region.

Eastern Ellsworth Land, that portion at the southern end of the Antarctic Peninsula plus Alexander Island, is very different geologically from the part adjacent to Marie Byrd Land. The former is the complex remnant of a Jurassic–Cretaceous volcanic arc from the time when southern New Zealand was adjacent in Gondwana.

Abbott and Venable Ice Shelves

The larger of the two relatively "small" ice shelves in the region, Abbott Ice Shelf lies between Thurston Island to the north and Ellsworth Land to the south, and between the Bellingshausen and Amundsen seas. The ice shelf runs east–west, and measures some 155 by 40 miles (250 by 65 km). Its eastern boundary consists of both the Bellingshausen Sea and the north–south oriented Fletcher Peninsula. There are several islands within the ice shelf and these ensure its stability. It was first seen from the air by planes from the ship, the *Bear*, in 1940 on Richard Byrd's third expedition. It was named for Rear Admiral J. L. Abbott, who commanded the US naval presence in Antarctica between 1967 and 1969.

The Venable Ice Shelf is even smaller, at 37 by 15 miles (60 by 24 km), and lies to the east of Abbott Ice Shelf between the Fletcher and Allison peninsulas, with Ellsworth Land as its southern edge. It too is oriented east–west, and is bounded to the north by the Bellingshausen Sea. It is named to honor J.D. Venable, also with the US Navy in Antarctica in 1967 and 1968.

Remote and Inaccessible
In 1999, a two-year-old elephant seal hauled out on this sliver of beach on Peter I Øy. A tag identified the seal as one seen on Macquarie Island, over 2,500 miles (4,000 km) northwest.

Frozen Fastness
Peter I Øy, named by Bellingshausen for Russia's greatest czar, is one of the world's most inaccessible islands. The first landing was more than 100 years after discovery, and barely a dozen expeditions have since broached these forbidding shores. Only one has reached the interior—by helicopter.

Marie Byrd Land
The largest expanse in the West Antarctic region, Marie Byrd Land includes all the land and ice sheet between the eastern edge of the Ross Sea and Ellsworth Land. Its southern boundary is the Transantarctic Mountains.

The major part of Marie Byrd Land that is above sea level is to its north, where it is oriented east–west and abuts the Amundsen Sea; at this point it reaches elevations well over 13,000 ft (4,000 m).

Between the continental fragment at the north and the Transantarctic Mountains in the south, much of the ground surface is well below sea level, forming the Byrd Sub-glacial Basin; at its eastern end, this deep, narrow trough, which is more than 6,560 ft (2,000 m) below sea level, acts as the boundary between Marie Byrd Land and Ellsworth Land. The basin and its constituent elements can be seen as a continuation of the Ross Sea to the east from under the Ross Ice Shelf. This depression is the main feature that separates East and West Antarctica.

The region is a rift valley system that is underlain by thinned continental crust, most less than 18½ miles (30 km) thick—in contrast with East Antarctica, where it is normally about 25 miles (40 km). As a result, much of the surface is below sea level and is not depressed by ice load. The rift system formed as the northern fragment moved away from the rest of Antarctica to create the Ross Sea depression; this also allowed the Transantarctic Mountains to rise. The rift valley beneath the ice is marked by higher and lower fragments, the low land reaching more than 6,560 ft (2,000 m) below sea level in the Bentley and Byrd sub-glacial basins.

Many of the mountains of Marie Byrd Land lie at its north, on the higher land, and are dormant volcanoes. There are some active volcanoes below the ice sheet, and these are marked at the surface by depressions in the ice sheet. The volcanoes commonly rise to high above 13,000 ft (4,000 m).

Marie Byrd Land coincides with the unclaimed sector of Antarctica. Richard Byrd and Lincoln Ellsworth undertook acts to claim the region for the United States, although the claim was never enacted. Historically, it is famous for their pioneering aviation exploits during the 1930s, particularly those of Byrd. There are no permanent stations in Marie Byrd Land at present, but for many years the United States operated Byrd Station at exactly 80°S, 75°W, to study magnetic phenomena free of the influence of nearby land.

Transantarctic Mountains
The Transantarctic Mountains are one of the world's greatest, longest, and most spectacular mountain chains, and are marked by clean, treeless, stark peaks, commonly over 10,000 ft (3,000 m) high. The highest peak—14,856 ft (4,528 m)—is Mount Kirkpatrick at 84°20'S, 375 miles (600 km) from the South Pole. The mountains stretch from one side of Antarctica to the other—from the entrance to the Ross Sea across to the southeastern corner of the Weddell Sea, a distance of 1,990 miles (3,200 km). Many of the peaks have not been climbed yet because of the logistic problems of expense and access.

The Transantarctic Mountains were discovered by James Clark Ross in the summer of 1840–41 and much of the nomenclature we use today follows his efforts. To him, they were majestic but also a source of frustration because they prevented him from reaching the South Magnetic Pole, one of the chief aims of his expedition. Robert Scott also suffered difficulties caused by the mountains as he ascended the Beardmore Glacier on his way to the South Pole. Roald Amundsen had much less difficulty when he found the Axel Heiberg Glacier.

The mountains act as the margin between West and East Antarctica. They are made of ancient shield rocks that are very comparable to those of Australia. Over the basement, the mountains are marked by several rock groups (granites and sediments), of which the roughly horizontal sediments of the Beacon Group make up most of the major peaks. They also contain fossils that are similar to those of other Gondwanan fragments. Perhaps the most enlightening discovery was made by Scott, who brought back with him from the South Pole the plant leaf fossil *Glossopteris*—the term Gondwana was defined to explain the distribution of *Glossopteris* and its discovery in Antarctica was the final clue in the jigsaw that makes up Gondwana.

When Antarctica was the centerpiece of Gondwana, much of the geology of the Transantarctic Mountains was continuous with that of the eastern part of Australia, particularly Tasmania. It seems that the mountains began to rise some 45 million years ago, but many aspects of their age and evolution are still not well understood. There are many other mysteries, such as the age and environment when abundant fossil leaves of the southern beech (*Nothofagus*) accumulated at what is now the 3,950 to 5,900 ft (1,200 to 1,800 m) elevation at the head of the Beardmore Glacier.

The mountains act as a dam to contain the East Antarctic Ice Sheet. In many places, the mountains are breached by massive glaciers that drain from inland Antarctica to the Ross Sea, in part feeding the ice of

the Ross Ice Shelf. These glaciers carved valleys as the mountains rose, and continue to cut their way down, carrying ice and sediment to the Ross Sea. The southern Victoria Land section of the Transantarctic Mountains is home to the famous Dry Valleys, which is one of the driest, most inhospitable and breathtakingly beautiful localities on the Earth.

Ice from the eastern ice sheet rides up the inland side of the mountains and, as it does, much of the ice is blown away or ablated, leaving behind the residue of rocks contained in the ice. One legacy of this ablation is a concentration of meteorites: Antarctica, particularly near the Transantarctic Mountains, is the source of 65 to 70 percent of the Earth's meteorites. PQ

Significant Site

Allan Hills in the Transantarctic Mountains is a site rich in fossil remains. The nearby Allan Hills ice field is the location of a number of significant meteorite finds, including a 4½ billion-year-old rock believed to have once been part of Mars.

Rock Glaciers

A light dusting of snow highlights the rock glaciers at the head of the Beacon Valley, in the Dry Valleys in the Transantarctic Mountains. Composed of rock debris held together with interstitial ice, they move very slowly down the valley—at only 1½ to 7 ft (0.5 to 2 m) per year. The dark rocks are Jurassic basalt magma that was forced sideways, between layers of lighter sandstone, as thick horizontal sheets.

The Ross Sea Region

The Ross Sea lies at the southernmost limit of the Pacific Ocean sector of Antarctica beneath New Zealand. When James Clark Ross sailed into the expanse of sea that now bears his name, he hoped that it would allow him sailing access to the South Magnetic Pole. However, the impressive line of mountains that he saw to the west soon quelled this ambition. The western shore of the Ross Sea is delineated by the Victoria Land coast up to Cape Adare; its eastern boundary is Cape Colbeck, at the northwestern tip of Edward VII Land. Its total area is almost 385,000 sq miles (1 million sq km).

Most of the Ross Sea is quite shallow—between 1,000 and 3,000 ft (300 and 900 m)—but to the north it plunges sharply to depths of more than 13,000 ft (4,000 m). At its southernmost edge is the Ross Ice Shelf, one of Antarctica's most spectacular features.

At the northwestern corner of the Ross Ice Shelf is Ross Island, which is the most historically significant location in Antarctica. It was from here that Scott, then Shackleton, then Scott again, led parties out on the first great attempts to walk to the geographic South Pole— and it is here that Mawson, David, and Mackay returned after walking to the South Magnetic Pole. Today, Ross Island is the site of McMurdo Station, the largest of the Antarctic bases, and of New Zealand's Scott Base. All volcanic, the island is roughly shaped like a four-pointed star, about 43 miles (70 km) wide.

The sea ice around the island does not break up until late January and re-forms in early April. About six ships visit over the short summer season. Across its southern side, the island is attached to the Ross Ice Shelf from near the tip of Hut Point Peninsula to Cape Crozier. Glaciers spill onto the ice shelf and onto the western shore. The island is also linked by ice shelf to Victoria Land; the waterway in between, known as McMurdo Sound, is about 95 miles long and 45 miles across (150 and 75 km) at its widest point.

Mount Erebus looms over Ross Island. After first seeing the mountain on Robert Scott's last expedition, Apsley Cherry-Garrard wrote: "I have seen Fuji, the most dainty and graceful of all mountains; and also Kinchenjunga; only Michael Angelo among men could have conceived such grandeur. But give me Erebus for a friend. Whoever made Erebus knew all the charm of horizontal lines, and the lines of Erebus are for the most part nearer the horizontal than the vertical. And so he is the most restful mountain in the world, and I was glad when I knew that our hut would lie at his feet. And always there floated from his crater the lazy banner of his cloud of steam."

Erebus was seared into the world's consciousness on November 28, 1979, when an Air New Zealand DC10 on a sightseeing flight crashed into the mountain, killing all 257 people on board. At the time it was the world's fourth worst air disaster. It was a major tragedy which had a profound effect upon the nation that sent the flight, and it is still recalled with horror by those who were based at Scott and McMurdo stations at the time that the crash occurred. DM

Sleeping Volcano
A team of New Zealanders driving a Hagglund ATV (all-terrain vehicle) traverse the vast, flat Ross Ice Shelf toward Mount Discovery. The youngest vents of this dormant volcano, named by Captain Scott after his ship, are 1.8 million years old.

Mountains to the Sea
The lower slopes of Ross Island are shaped by heavily crevassed glacial ice, formed by the compaction of accumulated snow. Here, at the sea edge, flat sea ice meets the glacial ice. Unbroken sea ice attached to a coastline is called fast ice, and can be strong enough to support ice vehicles, and even to land large aircraft on.

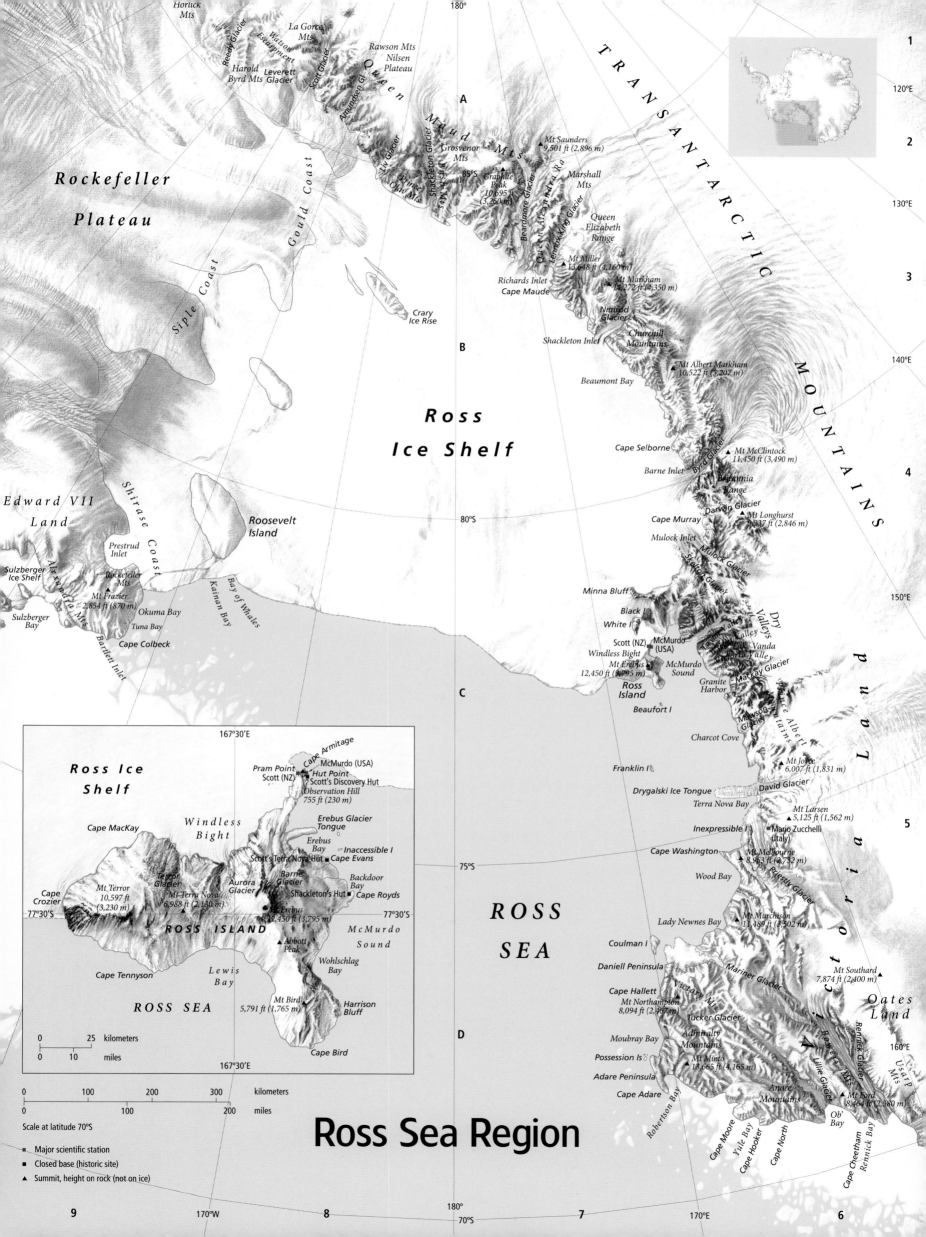

Ross Sea Region

Ross Ice Shelf

Rockefeller
Plateau

Edward VII
Land

Gould Coast

Siple Coast

Horlick
Mts

La Gorce
Mts

Reedy Glacier

Watson
Escarpment

Harold
Byrd Mts

Leverett
Glacier

Scott Glacier

Amundsen Gl

Liv Glacier

Shackleton Glacier

Queen
Maud
Mts

Nilsen
Plateau

Rawson Mts

Grosvenor
Mts

85°S

Graphite
Peak
10,695 ft
(3,260 m)

Beardmore Glacier

Queen Alexandra Ra

Mt Saunders
9,501 ft (2,896 m)

Marshall
Mts

Lennox-King Glacier

Mt Miller
13,648 ft (4,160 m)

Queen
Elizabeth
Range

Mt Markham
14,272 ft (4,350 m)

Richards Inlet
Cape Maude

Nimrod
Glacier

Churchill
Mountains

Shackleton Inlet

Mt Albert Markham
10,522 ft (3,207 m)

Beaumont Bay

Cape Selborne

Mt McClintock
11,450 ft (3,490 m)

Barne Inlet

Byrd Glacier

Britannia
Range

Cape Murray

Darwin Glacier

Mt Longhurst
9,337 ft (2,846 m)

Mulock Inlet

Mulock Glacier

Skelton Glacier

Minna Bluff

Black I

White I

Scott (NZ)
Windless Bight

McMurdo (USA)

Lake Vanda

Dry
Valleys

Mackay Glacier

Mt Erebus
12,450 ft (3,795 m)

Ross
Island

Granite
Harbor

McMurdo
Sound

Mawson Glacier

Prince Albert Mountains

Beaufort I

Charcot Cove

Mt Joyce
6,007 ft (1,831 m)

Franklin I

Drygalski Ice Tongue

David Glacier

Terra Nova Bay

Mt Larsen
5,125 ft (1,562 m)

Inexpressible I

Mario Zucchelli
(Italy)

Cape Washington

Mt Melbourne
8,963 ft (2,732 m)

Wood Bay

Lady Newnes Bay

Mt Murchison
11,489 ft (3,502 m)

Coulman I

Mariner Glacier

Daniell Peninsula

Mt Southard
7,874 ft (2,400 m)

Cape Hallett

Mt Northampton
8,094 ft (2,467 m)

Victory Mts

Tucker Glacier

Moubray Bay

Admiralty
Mountains

Possession Is

Mt Minto
13,665 ft (4,165 m)

Priestley Glacier

Adare Peninsula

Anare
Mountains

Cape Adare

Robertson Bay

Lillie Glacier

Bowers Mts

Rennick Glacier

Usarp Mts

Mt Ford
8,464 ft (2,580 m)

Cape Moore

Yule Bay

Cape Hooker

Cape North

Ob' Bay

Cape Cheetham

Rennick Bay

TRANSANTARCTIC

MOUNTAINS

Victoria Land

Oates Land

Ross
Ice Shelf

Crary Ice Rise

Roosevelt
Island

Bay of Whales

Kainan Bay

Roosevelt Island

Shirase Coast

Prestrud
Inlet

Rockefeller
Mts

Mt Frazier
2,854 ft (870 m)

Okuma Bay

Tuna Bay

Cape Colbeck

Bartlett Inlet

Sulzberger
Ice Shelf

Sulzberger
Bay

Alexandra Mts

Ross
Sea

ROSS

SEA

McMurdo
Sound

Inset map

**Ross Ice
Shelf**

167°30'E

Cape Armitage

Pram Point
Scott (NZ)

McMurdo (USA)

Hut Point
Scott's Discovery Hut

Observation Hill
755 ft (230 m)

Cape MacKay

Windless
Bight

Erebus Glacier
Tongue

Erebus
Bay

Inaccessible I

Cape Evans

Scott's Terra Nova Hut

Barne
Glacier

Backdoor
Bay

Terror
Glacier

Mt Terror
10,597 ft
(3,230 m)

Aurora Glacier

Shackleton's Hut

Cape Royds

Mt Terra Nova
6,988 ft (2,130 m)

Cape
Crozier

Mt Erebus
12,450 ft (3,795 m)

ROSS ISLAND

Abbott
Peak

Wohlschlag
Bay

77°30'S

77°30'S

Cape Tennyson

Lewis
Bay

McMurdo
Sound

ROSS SEA

Mt Bird
5,791 ft (1,765 m)

Harrison
Bluff

Cape Bird

167°30'E

			kilometers
0	25		
0	10		miles

				kilometers
0	100	200	300	
0		100		200 miles

Scale at latitude 70°S

■ Major scientific station
■ Closed base (historic site)
▲ Summit, height on rock (not on ice)

180°

170°W

170°E

160°E

150°E

140°E

130°E

120°E

1

2

3

4

5

6

7

8

9

70°S

75°S

80°S

A

B

C

D

The Ross Ice Shelf

Icy Poles
Smooth mounds appear where wind-blown snow has been plastered onto blocks of ice on the walls of an open Ross Ice Shelf crevasse. The warmth of the summer sun generates considerable melt, causing water drops to refreeze into long stalactite spears; stalagmite spikes jut skywards (top left) beneath dripping overhangs above.

Long Wall of Ice
The pale light of the setting sun does little to soften the chilling bulk of the Ross Ice Shelf. Looking for an opening, James Clark Ross sailed 390 nautical miles (720 km) along these ice cliffs. At no point was the unbroken wall lower than 50 ft (15 m), and all that could be seen from mast height was "an immense plain of frosted silver."

A glance at most world maps suggests that there is a bay in the Southern Ocean that extends almost to the South Pole. But closer inspection reveals a subtle change of shading. This is the Ross Ice Shelf—the feature that makes the South Pole such a challenging overland destination. It is roughly triangular in shape, and covers about 190,000 sq miles (500,000 sq km)—about the same area as France, and twice that of Britain. Most remarkably, virtually all of it is floating.

From the ocean, the Ross Ice Shelf looks like an endless white cliff towering 30 to 200 ft (10 to 60 m) above the sea. When James Clark Ross discovered it in 1841, he described it as a "perpendicular cliff of ice" that would be as difficult to penetrate as the cliffs of Dover, and named it the Victoria Barrier. (The name of the ice shelf was changed to honor Ross himself.) Some six decades later, Roald Amundsen was to express some of the wonder of the spectacle: "Slowly it rose up out of the sea in all its imposing majesty. It is difficult … to give any idea of the impression this mighty wall of ice makes on the observer who is confronted with it for the first time."

The shelf is about 500 miles (800 km) wide, and it extends back almost 620 miles (1,000 km) toward the Pole. Although it appears to be static, in fact it is fed by glaciers flowing off the polar ice cap, and can grow outwards by more than ⅔ mile (1 km) each year. There has been a somewhat morbid suggestion that the grave site of Scott and his companions may soon reach the sea toward which they were struggling.

Birth of an Iceberg
On average, the thickness of the Ross Ice Shelf is about 1,000 ft (300 m) at its seaward edge, and up to 2,300 ft (700 m) farther to the south. Although it is hinged along the coast, powerful forces operate on this relatively thin skin of ice, breaking off huge blocks that become the tabular bergs thronging the Ross Sea. Apart from a questionable sighting in 1956, the largest iceberg ever recorded calved from the eastern Ross Ice Shelf in late March 2000; it was 183 miles long and 23 miles wide (295 by 37 km), with an area of about 4,250 sq miles (11,000 sq km). Prosaically named B-15, the iceberg gradually broke up and knocked other bits off the ice shelf as it drifted westward toward Ross Island. The first

Strange Birds

Sailing through the Strait of Magellan with Francis Drake, Francis Fletcher described a "great store of strange birds which could not flie at all," with a body size "less than a goose," and which laid their eggs on the ground. The name "penguin" seems to have been first used to refer to the familiar black and white birds of Antarctica by Francis Petty during a circumnavigation of the globe by Thomas Cavendish in 1586–88; Petty noted that "we trimmed our saved pengwins with salt for victual."

Solid Waves

As the vast Ross Ice Shelf approaches immovable Hut Point Peninsula on Ross Island, it buckles, looking like huge waves about to break on land. The flat ice in the foreground is sea ice just a few feet thick; later in summer it will probably break out, leaving open sea before lower autumn temperatures freeze it again. The long, straight line in the distance is a service road.

cracks that gave rise to B-15 were observed by satellite as far back as 1990.

The ice is low enough to permit landing at the Bay of Whales, an indentation toward the eastern end of the ice shelf. It was discovered by Carsten Borchgrevink in 1900 and named eight years later by Ernest Shackleton, who wrote that "it was a veritable playground of these monsters." Encroaching ice drove Shackleton away, so he could not use the bay as his base, but Amundsen calculated that it was about 60 miles (100 km) closer to the South Pole than Ross Island, and used it to make his successful bid in 1911. It is also closely associated with Richard Byrd, who established his Little America bases there on his three expeditions between 1928 and 1941 and, later, during Operation Highjump.

In 2000, B-15 took the existing Bay of Whales with it, and the ice where Amundsen and Byrd camped is long gone. But the bay remains because it marks the junction of two ice systems: not far to the south, within the shelf, the ice must divide and flow around the large, ice-covered Roosevelt Island. With B-15 gone, the edge of the ice shelf lies farther south than previously and, on January 11, 2001, the Russian icebreaker *Kapitan Khlebnikov* set a new southernmost shipping record— 78°37′S in the reconfigured Bay of Whales. DM

Ross Island

Ross Island is a small outcrop of volcanic rock perched between the immensity of the Ross Ice Shelf and the vastness of the Ross Sea. It was a traditional base location for early expeditions. Its Hut Point Peninsula is now the site of Antarctica's largest research station, McMurdo.

Hut Point

This is the site of the *Discovery* hut—a prefabricated wooden hut erected by Scott's 1901–04 expedition. Records show that the almost-square building cost 360 pounds 13 shillings and 5 pence. The hut lacks the atmosphere that might be expected from so historic a structure, possibly because it was used by various expeditions over many years, or perhaps because it is dwarfed by the huge modern base—McMurdo Station—on its doorstep.

The original hut was designed after the classic Australian outback homestead (with verandas!), but with several modifications. The walls and floor were made of two layers of Douglas fir packed with felt and there were two stoves. Even so, it was very cold, and Scott decided that the expeditioners would spend the winter aboard the *Discovery*, immobilized in the ice

nearby. The hut provided accommodation for the Shackleton expedition of 1908 and Scott's 1911 expedition, as well as for the marooned members of Shackleton's Ross Sea shore party, who survived on seal meat and the meager supplies that Scott's party had abandoned more than a decade earlier.

After 1916 the hut was unvisited until 1947, but when McMurdo Station was established it became a favorite target for pilferers. Restoration was begun in 1964, and when the snow and ice were dug out many artefacts from the old expeditions were found. The present hut is as it was when Shackleton's party left in 1916, and quite unlike the original layout.

Scott Base

Now a modern scientific station, New Zealand's Scott Base was built by Edmund Hillary's Commonwealth Transantarctic Expedition party of 1955–58. One of the original buildings erected during that expedition still stands and houses memorabilia from the station's subsequent history. Scott Base opened on January 20, 1957, and has been in constant use ever since. It has been considerably enlarged over the years, notably during an extensive rebuilding in 1976. Its eight

High Point

Castle Rock on Hut Point Peninsula is a popular destination for off-duty staff from nearby Scott Base and McMurdo Station. It is one of the oldest of a line of small volcanic craters that extend southward from the lower slopes of Mount Erebus, and is composed of volcanic breccia that formed from eruptions below the ice.

Exploring a Crevasse
Crossing Hut Point Peninsula, a New Zealander lowers himself into a large crevasse where the ice roof has collapsed. Snow will collect on each lip of a crevasse, building outwards until the gap is bridged and the crevasse becomes invisible, and potentially very dangerous.

Mount Erebus
Its summit hidden by cloud, Mount Erebus is the youngest of several volcanoes that form Ross Island, and the southernmost active volcano. Castle Rock, the snow-free knob (left) is Hut Point Peninsula's tallest point. In the foreground, pressure ridges of jumbled sea ice push against the shoreline.

Vast and Featureless
Two figures near Hut Point Peninsula give some idea of the vastness of the Ross Ice Shelf. About the size of France, it is a flat-topped body of snow-covered glacial ice, mostly floating, except along coastlines and over other shallow areas.

interlinked green buildings can accommodate up to 86 people in summer and about ten in winter.

New Zealand uses the base for meteorological observations and for research into the environment, the aurora, the ionosphere, marine biology, and tides. Scott Base is also fundamental to New Zealand's claim to a part of Antarctica: it is the only New Zealand station on the Antarctic continent. Scott Base, and the whole of Ross Island, fall within New Zealand's Ross Dependency, and New Zealand has responsibility for the island's historic huts. Tourists can enter them only in the company of a supervisor from the New Zealand Antarctic Heritage Trust.

Scott Base operates in close cooperation with the US's McMurdo Station, to which it is linked by a gravel road 3 miles (5 km) long. Over summer, New Zealand aircraft make some 15 flights from Christchurch, where the air support base for the United States Antarctic Program is also located; numerous US flights carry personnel and supplies for Scott Base and McMurdo Station from Christchurch to McMurdo's airfields. McMurdo supplies the Amundsen–Scott South Pole Station by air.

McMurdo Station
McMurdo Station is the largest of the Antarctic research bases; at the height of summer there may be as many as 1,200 personnel here. The station is named after the adjoining McMurdo Sound. Near the station's dock area at Hut Point stands the *Discovery* hut from where Scott set out, with Wilson and Shackleton, on the first serious attempt to reach the South Pole in 1902.

Overlooking McMurdo Sound and the station is Observation Hill, 750 ft (230 m) above sea level and crowned with a cross commemorating Scott's polar party. The base of the hill was the site of Antarctica's only nuclear reactor, which was designed as a portable nuclear power plant, installed in December 1961 and

commissioned in March 1962. The reactor operated until 1972, when it was shipped back to the United States; since that time more than 130,000 cubic yds (10,000 cubic m) of soil contaminated by radioactive waste have been removed and freighted out.

McMurdo Station first took shape in December 1955 as Naval Air Facility, McMurdo Sound, one of six of the research stations built by the United States in preparation for the International Geophysical Year (1957–58). The National Science Foundation of the United States funded research work through its National Academy of Sciences, and the Department of Defense separately provided operational support. In recent years, there has been a shift away from military support to contractor services; scientific research is now the primary expression of the US presence in Antarctica. A recent government report stated "no armaments are currently stored or in use at McMurdo Station."

McMurdo Station, which sprawls in the summer mud on the bare volcanic rock of Hut Point Peninsula, is a cluttered, bustling settlement reminiscent of an Alaskan frontier town. It is littered with above-ground telephone and power cables, and pipes for water and sewerage. The station's 85 utilitarian buildings comprise laboratories, workshops, dormitories, administration blocks, and warehouses, as well as the Chapel of the Snows, a fire station with eight fire trucks, a coffee shop, a gymnasium, and a hydroponic greenhouse. There are also airfields, a helipad, and an impressive inventory of vehicles, including trucks, personnel and cargo carriers, front-end loaders and forklifts, tractors, and motor toboggans.

The marked annual population cycle of McMurdo begins in August (the Southern Hemisphere's spring), when several Winfly (winter fly-in) flights from New Zealand raise the station's population from its winter maximum of about 250 people. During summer there are arrivals and departures several times a week. In late December, the "Christmas notch" coincides with the melting of the sea ice, and scientists (particularly biologists) who need the ice as a working platform give way to those (such as geologists) who do not. With the approach of the Antarctic winter in late February, shortening days and plummeting temperatures make field research impractical, and the population of the station falls accordingly.

McMurdo Station has three airfields. Williams Field is a ski-way on the Ross Ice Shelf, 10 miles (16 km) from the station. It operates from December until late January and has a main runway of 10,000 ft (3,050 m) and a cross-runway of 8,000 ft (2,440 m). Pegasus, a permanent, hard-ice airstrip for wheeled aircraft, also on the Ross Ice Shelf, is much farther away. It is usable from mid-January until the end of summer. And from late September until early December, there is a sea-ice airfield suitable for use by heavy, wheeled aircraft.

The Albert P. Crary Science and Engineering Center at McMurdo, a major science laboratory that is as well equipped as one anywhere else in the world, was built in three phases and opened in November 1991. The vast laboratory supports studies of marine and terrestrial biology, biomedicine, glaciology and glacial geology,

A Feeding Expedition
Emperor penguins gather at the edge of the fast sea ice at Cape Crozier, about to set out onto the thinner ice for a feeding trip. It is springtime, and parents are taking it in turns to feed their rapidly growing chicks.

The Western Cape
Sharing your home with large colonies of Adélie penguins—as several early expeditions did on Cape Royds—means enduring the stench, as well as the noisy squawking that can be picked up tens of miles away.

geology and geophysics, meteorology, aeronomy, and upper atmosphere physics, as well as carrying out the obligatory environmental monitoring.

Cape Evans

The hut at Cape Evans, 7 miles (11 km) from Cape Royds, is the destination to which Scott and his polar party never returned. It is an evocative place; so close do visitors feel to the early inhabitants that some have been seen surreptitiously touching the old stove to see whether it is still warm.

The hut is on a promontory named after Scott's second-in-command, Lieutenant Edward ("Teddy") Evans. Nearby, a ridge 165 ft (50 m) high leads to the base of Mount Erebus. Measuring 50 by 25 ft (15 by 8 m) to house 25 men, and with multiple-insulated floor, walls, and ceiling, it was well designed and constructed. Davis, the ship's carpenter, declared that "hut" was a misnomer—it was more like a parish hall. Manufactured in England, the hut was re-erected in Lyttelton, New Zealand, and some problems rectified.

After the bodies of Scott's polar party were found, the *Terra Nova* left Cape Evans on January 19, 1913. "Teddy" Evans wrote: "We have left at Cape Evans an outfit and stores that would see a dozen resourceful men through one summer and winter at least." They were prophetic words: when a storm wrenched the *Aurora* off her anchors, those stranded relied on the supplies for 20 months. Three died before help arrived.

The restoration of the ice-filled hut began in 1960. Everywhere there are artefacts and links to the Heroic Age parties, even a mummified penguin on the chart table. A guest book on the oak wardroom table features in Herbert Ponting's photograph of Scott's last birthday party, held on June 6, 1911.

Cape Royds

The homely little hut from Ernest Shackleton's *Nimrod* expedition of 1907–09 is situated up a slight rise from Backdoor Bay, in a hollow largely filled with the icy Pony Lake. On the slope behind is the world's southernmost Adélie penguin rookery (about 4,000 breeding pairs).

In March 1908, it was from there that Edgeworth David led the first ascent of Mount Erebus, the foothills of which start to rise only 1¼ miles (2 km) from the hut, although the summit is 15 miles (24 km) away.

Cape Royds is the westernmost point of Ross Island. Shackleton had promised Scott that he would not base his 15-man expedition at Ross Island. However, the heavy sea ice and other factors made the island the only viable option. The hut, erected in February 1908, was modified when the men discovered how cold the unlined floor was. They improvised an airlock by sealing off the underfloor space with boxes around the outside of the hut, a technique that was partially successful. When the provident Shackleton abandoned the hut in 1909, he left "a supply of stores sufficient to last 15 men one year. The vicissitudes of life in the Antarctic are such that such a supply might prove of the greatest value to some future expedition." He was right: the hut was visited by Scott's *Terra Nova* expeditioners and by his own marooned Ross Sea party of 1915–17.

By 1947, when the next visitors called at Cape Royds, there were holes in the roof and the whole edifice was in a state of disrepair. The garage built to house Antarctica's first car (a Scottish 12-horsepower Arrol-Johnston) had collapsed, and the stables had filled with snow. The hut was restored and repaired in the summer of 1960–61. There were still chemicals in the darkroom and supplies in the biology laboratory, and many more artefacts were discovered nearby.

Because so many expeditioners used the hut, it is difficult to determine the provenance of the clothes and objects there. The seal blubber that lay thick around the stove before restoration was almost certainly left by the stranded men of the *Aurora*, who burned the blubber for fuel. The canned goods include Bird's Egg Powder, curried rabbit, stewed rump steak, canned cabbage, and whole tins of Colman's Mustard.

Scott, with his Royal Navy background, believed in segregating officers from crew, but Shackleton was from Merchant Marine stock and treated all men as equals. The difference is reflected in graffiti on the walls at Cape Royds, including about *Aurora Australis*, the first book printed in Antarctica: "Wild & Joyce, printers, book binders etc. Gentlemen only." Among the signatures is Shackleton's; he wrote it upside down on the casing that formed his bed end. Such personal glimpses are notably lacking in Scott's huts.

Cape Crozier

On its eastern side, Ross Island bulges around Mount Terror and rounds into Cape Crozier, where ice, sea, and island meet. This high cape was discovered by the Ross expedition in 1841 and named after Francis Crozier, captain of the *Terror*. Thousands of Adélie penguins nest on its gravel slope, and lower down a colony of emperor penguins occupies the ice shelf during the winter.

Cape Crozier is the subject of Apsley Cherry-Garrard's *The Worst Journey in the World*. In the Antarctic midwinter of 1911, Cherry-Garrard, with Bill Wilson and "Birdie" Bowers, visited the cape to collect emperor penguin eggs. Cherry-Garrard thought "the Emperor is probably the most primitive bird ... the embryo ... may prove the missing link between birds and the reptiles ..." The timing of "the weirdest bird's-nesting expedition" was dictated by Wilson's calculation that the birds would lay in early July. The three barely survived, and the remains of their igloo are still there. "I don't think anybody could have made a better igloo with the hard snow blocks and rocks which were all that we had," wrote Cherry-Garrard. DM

In Ice and Snow
The Cape Royds area looks over the frozen sea ice of McMurdo Sound toward the Transantarctic Mountains. Shackleton's hut, still in remarkably good condition, is wired to the ground in the foreground. Beyond the hut lies Pony Lake, also solidly frozen.

Frozen in Time
Still looking as if the men might walk back inside at any minute, Shackleton's hut at Cape Royds—like Scott's at Cape Evans—is frozen in both time and in reality. The cold temperatures and dry air of Antarctica have preserved the hut's contents well.

The Dry Valleys

Fieldwork Fashions
United States scientists camp on a frozen tarn in one of the small valleys southeast of the main Dry Valleys area. A helicopter brings supplies to the people working here. The design of polar tents has not changed greatly since Scott's time, a century ago.

The Dry Valleys
Located in the Transantarctic Mountain region in south Victoria Land, at approximately 78°S, the Dry Valleys region is an oasis of more than 1,550 sq miles (4,000 sq km).

Like all the world's deserts, Antarctica has oases. However, unlike hot, sandy deserts, where oases are places water is found, the oases of Antarctica are defined as ice-free areas. These are special places on a continent that is 98 percent ice-covered, and where much of the remainder is exposed coastline or rugged mountain ranges.

The most extensive Antarctic oasis areas are the Larseman and Vestfold hills, near Australia's Davis Base; the Bunger Hills, near Casey Station (also Australian); and the Dry Valleys, near McMurdo Station (United States) and Scott Base (New Zealand), on the western side of McMurdo Sound. The Dry Valleys is by far the largest of these oases, its ice-free area extending more than 1,550 sq miles (4,000 sq km).

The Dry Valleys exists because this section of the Transantarctic Mountains acts as a dam wall containing the polar ice cap. The polar ice sheets have fallen some 1,640 ft (500 m), so now only a few glaciers breach the threshold into the valleys.

Formed by Glaciers
Like the other Antarctic oases, the valleys are close to the coast, and are now bare because the glaciers that formed them have retreated and most of the continental ice they contain is trapped behind the rock walls at the head of the valleys. The region's three main valleys—Taylor, Wright, and Victoria—run parallel to each other on an east–west axis.

The Taylor Valley is the closest to the coast, and is the one normally visited by helicopter from tourist ships or from Ross Island. This valley and the Taylor Glacier at its western end were named by Robert Scott in 1903 for Australian geologist Griffith Taylor. Scott's report of the

discovery of the Dry Valleys reflects the awe everyone experiences here: "I was so fascinated by all these strange new sights that I strode forward without thought of hunger until Evans asked if it was any use carrying our lunch further … from our elevated position we could now get an excellent view of this extraordinary valley, and a wilder or in some respect more beautiful scene it would have been difficult to imagine … we have seen no living thing, not even a moss or a lichen; all that we did find, far inland amongst the morrain heaps, was the skeleton of a Weddell seal."

A Living Museum
The Dry Valleys region is aptly named: its microclimate is dominated by dry winds, and therefore evaporation far exceeds precipitation. In fact, the valleys are far from lifeless. In the 1970s scientists discovered that minute bacteria, algae, and fungi lived in tiny cracks in the

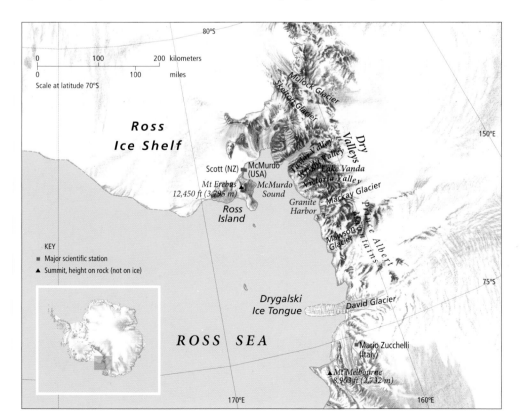

Moving Ice

The 56-mile (90-km) long Taylor Glacier is polythermal—partly wet-based rather than frozen, as are others in the region, and it has annual surface movements of up to 45 ft (14 m).

rock, and blue-green algae have been found at the bottom of the few lakes that are not frozen solid. At the upper end of this simple food chain are three species of microscopic nematode worms that live on bacteria and can survive being freeze-dried in this extraordinary environment. The few seals that stray into the valleys do not survive, and mummified seal remains that have been found may be hundreds of years old. In the absence of precipitation, the strong winds that blow in the valleys have sculptured the rocks into fantastic shapes and wafer-thin hollow shells.

The Dry Valleys is an environmentally fragile region, and human visits are carefully controlled; some parts of the valleys are so sensitive that tourist visits are not permitted, and even scientists need special approval for their studies in certain areas. For this reason few people have laid eyes on Antarctica's longest river, the Onyx, a meltwater stream flowing for 20 miles (32 km) down the Wright Valley from the Wright Lower Glacier to Lake Vanda. In the same valley is Don Juan Pond, a small lake that is so salty that it never freezes, although it is at most 2½ ft (0.75 m) deep. When a new mineral, a calcium salt, was discovered there it was dubbed antarcticite. The lake is not named for the archetypal seducer; the name is a conflation of the names of two United States lieutenants—Donald Roe and John Hickey—on a 1961 field party.

The Dry Valleys region constitutes an environmental extreme, an extraordinary place where plant and animal life can survive against incredible odds. In this cold and arid local climate, even the smallest climatic changes can have catastrophic consequences, and scientists continue to monitor weather conditions and conduct research in the Dry Valleys in the hope of recognizing the early effects of climatic changes that may be of worldwide significance. DM

Waterless Wastes

The lunar landscape of Bull Pass between McKelvey and Wright valleys is characteristic of the Dry Valleys terrain. The valleys receive less than ½ in (about 12 mm) of snowfall each year but evaporation greatly exceeds precipitation. The ground surface is dry, but there is permafrost beneath.

Lake of Contrasts

Lake Vanda is ice-covered year-round (although in this late summer view there is a narrow moat of ice-free water around its shoreline), but the bottom of the lake is warm—a remarkable 77°F (25°C). Although the surface water is fresh, the water at the bottom of the deepest part of the lake is three times as salty as seawater.

The Ross Sea Coast

Port in a Storm
From 1957 to 1964, Cape Hallett was the site of a year-round joint New Zealand/United States base, which then operated as a summer station until 1973. After its closure, many of the shelters were cleared away. Today, the main sign humans have ever been here is this dome that was left standing to offer sanctuary in an emergency.

Italian Presence
The Italian station at Terra Nova Bay is built on a small granite peninsula at the southern end of Gerlache Inlet. One of the world's largest emperor penguin colonies is found on the sea ice at Cape Washington, 25 miles (40 km) to the east of the station.

S everal important Antarctic sites lie along the western shore of the Ross Sea. Franklin Island, 8 miles (12 km) long, was named by James Clark Ross in honor of Sir John Franklin, governor of Tasmania. Franklin was later to seek (and perish in the attempt) the Northwest Passage with the *Erebus* and the *Terror*, the ships Ross had used in Antarctica.

Terra Nova Bay is the site of an Italian summer station that was built in 1986–87 and can house 90 people. Scott discovered the bay and named it after his relief ship, which he later used as the main ship for his last expedition. Scott's six-man northern party also named Inexpressible Island, just off the coast, when they were forced to winter in a cave there from March until September 1912. They had disembarked for six weeks of summer exploration, but ice conditions prevented their ship from returning, and at the end of winter they had to cross 220 miles (350 km) of sea ice to their expedition headquarters at Cape Evans. Mount Melbourne is a volcanic cone on the northern side of Terra Nova Bay. One of the few volcanoes on the Antarctic continent itself, it is 8,950 ft (2,730 m) high. Ross named it for Lord Melbourne, who was then the British prime minister.

Coulman Island, which is offshore from the Victory Mountains, is part of an extinct volcano, 19 by 8 miles (30 by 13 km) in extent and with a height of 6,560 ft (2,000 m). Today it is best known for its penguins: Adélies along the shore and emperors on the winter fast ice. Ross named the island after his father-in-law, Thomas Coulman. Cape Hallett and the Hallett Peninsula take their name from the purser on Ross's *Erebus*. An International Geophysical Year base was built jointly by New Zealand and the United States on Cape Hallett in 1957, but was closed in 1973 and largely dismantled, leaving the site to the thousands of Adélie penguins who had abandoned the area when the base was constructed.

A scattering of nine small islands and numerous volcanic rocks, the Possession Islands lie about 5 miles (8 km) off the coast of Victoria Land. On January 12, 1841, there was too much ice to permit a landing on the mainland, so instead Ross raised the British flag on the northernmost and largest of these islands and drank a toast to the queen. He noted the presence on the islands of a large population of Adélie penguins that seemed to resent the invaders. There are now more than 300,000 birds here, and they still object to human incursions on their territory.

Cape Adare
Cape Adare lies at the tip of a peninsula that extends northwards from Victoria Land at the top of the Ross Sea. Ross named it in 1841 after his friend Viscount Adare; the adjoining land was later called the Adare Peninsula. All around are ice and snow, but the cape itself is a barren basalt ridge with a small triangular patch of gravel at sea level on the western side.

This is penguin country; in the peak breeding season it swarms with more than half a million adult Adélies, plus their chicks. It is a malodorous rookery with a history. It was here that the men of the whaler *Antarctic* disembarked in 1895. They squabbled ever afterward about who was first ashore to achieve what they believed was the first landing on the southern continent. Carsten Borchgrevink, who was one of the landing party, later raised money for the *Southern Cross* expedition to winter here in 1898—the first expedition to winter on the continent. Their living quarters were designed and prefabricated in Norway of spruce, and have been restored by the New Zealand Antarctic Heritage Trust. Borchgrevink's was not a happy party, and the hut, just 19 by 18 ft (6 by 5.5 m), has little of the evocative atmosphere of the Ross Island huts.

Ridley Beach, named after Borchgrevink's mother, lies in front of a ridge upon which the continent's first burial site stands. When the *Southern Cross*'s biologist Nicolai Hansen was dying in October 1899, he asked to be buried on top of the 1,150-ft (350-m) ridge. His wish was granted, with a lot of hard work and the use of explosives to create the rocky grave. The brass plaque and iron cross that his fellow expeditioners lugged to the summit and erected are still there. DM

Bleak Ramparts
Cape Adare looms over desolate Ridley Beach, site of Borchgrevink's base. Although a good spot for the camp itself, the inaccessible terrain behind prevented the expedition from making any significant inland journeys of exploration. The body of Nicolai Hansen, who died during the long winter of 1899, was carried up the steep cliffs to a grave on the ridge above.

Peak Landscape
Spring sunshine bathes the ice-locked coastline and Transantarctic Mountains north of Terra Nova Bay on the western shoreline of the Ross Sea. Mount Melbourne, one of Antarctica's few "active" volcanoes, is near the coastline (right). The Deep Freeze Range occupies the middle distance, with the Eisenhower Range beyond it.

Dumont d'Urville Coast

The wide stretch of water along the coast of Terre Adélie in East Antarctica is the Dumont d'Urville Sea. The French explorer sailed these waters in January 1840. When d'Urville's party struggled onto an islet, they established that they had reached land, not part of an iceberg, when they found a few chips of granite. They unfurled a flag and claimed the land for France. Terre Adélie, named by d'Urville after his wife, is a small slice of French claim between Australia's extensive territorial claims.

Commonwealth Bay

On January 8, 1912, Douglas Mawson's Australasian Antarctic Expedition reached a bay in the Dumont d'Urville Sea. They called it Commonwealth Bay, and the cape to the east they named Cape Denison, after a Sydney sponsor of the expedition. Time was running out, Mawson was anxious to find a base site, and this landfall seemed perfect. Mawson wrote: "The sun shone gloriously in a blue sky as we stepped ashore … The rocky area at Cape Denison … was found to be about one and a third miles [2 km] in length and half a mile [800 m] in extreme width. Behind it rose the inland ice, ascending in a regular slope and apparently free of crevasses." The expedition began to bring their stores ashore. "The day had been perfect," wrote Mawson, "vibrant with summer and life, but towards evening a chill breeze off the land sprang up …" This "breeze" scarcely relented over the two years they were there: Cape Denison lies directly in the path of the katabatic winds that stream down from the polar ice cap, winds that can still make landings impossible.

The main attraction is Mawson's hut—actually two huts, one used mainly for accommodation and the other as a workshop. In Cape Denison's savage climate, these structures deteriorated badly before there was any thought of preserving them, but restoration began in the early 1980s and continues to this day. To see how tough the conditions are, one need only look at the much better condition of the magnetograph house nearby. It has survived well because the topography protects it from the southerly winds. It and the adjoining magnetic

At the Whim of the Wind
Katabatic winds regularly whip Commonwealth Bay into a lather. The gusts there can be very localized. As Mawson noted: "Laseron one day was skinning at one end of a seal and remained in perfect calm, while McLean, at the other extremity, was on the edge of a furious vortex."

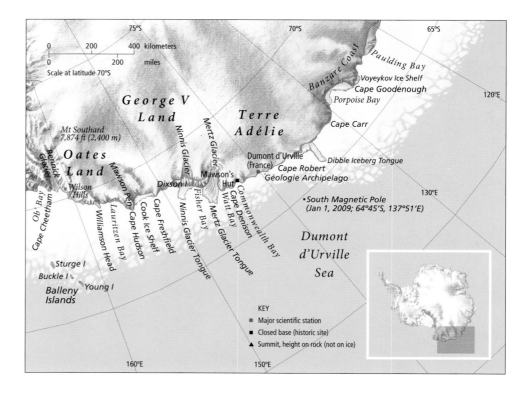

Ever-present Snow

In his book, *The Home of the Blizzard*, Mawson captioned this photograph taken by Frank Hurley: "An incident in March soon after the completion of the hut: Alfred Hodgman, the night watchman, returning from his rounds outside. Pushes his way into the veranda through the rapidly accumulating drift snow."

Preserving History

By the 1980s, the roof of Mawson's main hut (far right) was largely supported by the snow inside. After several restoration projects, in 2001 it was announced that the buildings had been preserved to a state that would endure for at least another decade.

absolute hut were built to hold the delicate equipment necessary to measure the changing magnetic fields so close to the South Magnetic Pole. The transit hut was an essential windblock while taking star sightings. In memory of B. E. S. Ninnis and Xavier Mertz, two members of Mawson's party who died on the ice, in November 1913, "on the highest point of Azimuth Hill, overlooking the sea, a Memorial Cross was raised to our two lost comrades."

Dumont d'Urville Station

The base lies between Pourquoi Pas Point (136°11′E) and Point Alden (142°02′E). The French have been active there since 1950. The present Dumont d'Urville Station was built in 1956 to replace an earlier one that burned down in 1952 at Port-Martin, 60 miles (100 km) to the east. The current station is on Petrel Island, at the southeastern end of the Géologie Archipelago and about 1¼ miles (2 km) from the continental coast. Forty-nine buildings are now scattered over an island that is less

than 1 mile (1.6 km) long. The station, which is almost universally known as "DuDu," accommodates about 30 people in winter and 100 in summer. It was first used as a winter base in 1958.

Its beautiful setting is marred by a gravel runway that was created, amidst considerable controversy, by dynamiting an island and some islets that held Adélie penguin rookeries. In January 1984, when the runway was complete but unused, a huge storm-driven iceberg damaged both the runway and hangar. The French government yielded to pressure and decided not to repair them, so the runway remains unused.

Ironically, Dumont d'Urville is an important center for the study of the rich local wildlife, including seals, petrels—and penguins. Between early December and early March, there are four resupply voyages by the French icebreaker *L'Astrolabe*. Three miles (5 km) away at Cape Prud'homme there is an array of over-snow vehicles jointly owned by France and Italy to service the inland Dome C Concordia project. DM

Edge of a Glacier

The Mertz Glacier is some 45 miles (72 km) long and 20 miles (32 km) wide, and where it reaches Cape Hurley it extends into the water as the Mertz Ice Tongue, seen here. These icy features were named by Mawson for Xavier Mertz who died on January 7, 1913, on the far-east sledge journey.

The Davis Sea Region

The East Antarctic coast edged by the Davis Sea stretches from the Vincennes Bay area, where Casey Station is located, to the Amery Ice Shelf. It includes Wilkes Land, Queen Mary Land, Wilhelm II Land and Princess Elizabeth Land. In February 1840, Charles Wilkes reached the Shackleton Ice Shelf, which projects almost 100 miles (160 km) into the sea. Erich von Drygalski's *Gauss* expedition spent the winter off the coast of Wilhelm II Land in 1902 and conducted the first scientific research in the area.

Casey Station

Australia's Casey Station is situated on Vincennes Bay, just outside the Antarctic Circle among low, rocky islands and peninsulas near the edge of the Antarctic ice cap. Law Dome, a circular ice cap that is 125 miles (200 km) in diameter and 4,577 ft (1,395 m) high, lies 70 miles (110 km) inland. It was named after Phillip Law, head of the Australian Antarctic Division from 1949 to 1966, who directed the exploration of much of Australia's Antarctic claim.

The first structure in the area was Wilkes Station, erected on Clark Peninsula by the United States in 1957 and operated as a joint base until it was handed over to Australia in 1963. By then, it was falling apart and largely buried under snow, A new base, built a short distance away on rocky Bailey Peninsula, was completed in February 1969 and named for Lord Casey, Australia's governor-general at the time. The station was built on stilts so that snow could blow under its 13 buildings, which were linked by a tunnel, but it vibrated in high winds and panels blew off in blizzards.

In 1979, work began on a new Casey station, not far to the west and also on Bailey Peninsula. The new station was opened in 1988. Its 16 multi-colored buildings, known affectionately as Legoland, house 17 to 20 people over winter and up to 70 in summer. With steel frames on concrete foundations and an external skin of steel-clad polystyrene panels, they are a far cry from the plywood of Wilkes, some of the roofs of which are still visible above the snow.

Mirny Station

The Soviet Union started regular Antarctic research in 1956, and opened Mirny Station in February of that year in preparation for International Geophysical Year. Mirny, named for one of Thaddeus von Bellingshausen's ships, was transferred to Russia after the break-up of the Soviet state. The average winter population in Mirny's 30 buildings is 60, rising to a maximum of 169 in summer. One supply ship visits each summer when the ice has broken out—otherwise supplies must be transported 25 miles (40 km) across the ice, since fast ice has re-formed by early April. In winter, the ice near the Haswell Islands in front of Mirny is the home to an important colony of emperor penguins, making the accumulation of 50 years of abandoned vehicles and debris at the station a cause for concern.

The Vestfold Hills

The Vestfold Hills on the coast of Princess Elizabeth Land were discovered in January 1935 by Norwegian whaling captain, Klarius Mikkelsen, and his wife, Karoline. When they landed here, Karoline became the first woman to set foot on the Antarctic continent. The largest ice-free area on the Antarctic coast—it covers 155 sq miles (400 sq km)—it is a beautiful place of lakes and fiords; the southernmost nesting site of giant petrels is on nearby Hawker Island.

An Australian party under Phillip Law landed in the Vestfold Hills in March 1954 and decided that establishing an Australian base there would pre-empt possible Russian plans to move in. Davis Station, on the edge of the hills, was built in January 1957 for the International Geophysical Year and—like the Davis Sea—named after John King Davis, ship's captain for Shackleton and Mawson. It is both the most southerly and the most temperate of the Australian bases on the Antarctic continent. The hills do provide some shelter, but Davis has experienced wind gusts of 130 mph (206 kph). The station has 29 buildings, supporting 22 in winter and up to 70 in summer. There are three ship's visits, one in mid-October when goods must be transported over the sea ice from the ship to the shore.

Aircraft Aiding Science

A blanket insulates the engine of a Russian AN-2 on the Amery Ice Shelf. Like the helicopters in the background, such craft have long been used for long-range surveillance and survey along the coast and deep into the interior, opening up a whole new range of scientific research opportunities.

Ebbs and Floes

On BANZARE's second voyage, easier ice conditions allowed the *Discovery* to sail closer to the shore and into the great indentation of Prydz Bay, among the tabular icebergs spawned by the Amery Ice Shelf. Many whalers also took advantage of the good season, several operating as far west as Prydz Bay.

Amery Ice Shelf

The Amery Ice Shelf (70°S, 70°E) is a 23,000 sq mile (60,000 sq km) floating body of ice in an embayment in Mac.Robertson Land. Ice from 580,000 sq miles (1.5 million sq km) of the interior of East Antarctica drains, via the Lambert Glacier and other large ice streams, through the Prince Charles Mountains into the Amery, and thence into Prydz Bay. BANZARE first mapped it in February 1931.

The ice shelf's glaciology was first investigated in the early 1960s. In 1968 a four-man Australian team over-wintered on the ice shelf, living in caravans that quickly became buried. They recovered a 1,035-ft (315-m) ice core from the shelf, and made detailed measurements of its movement and thickness.

The floating ice extends more than 300 miles (500 km) from the front in Prydz Bay to beyond 73°S. The thickness of the Amery ranges from 8,200 ft (2,500 m) where it first starts to float, to less than 1,000 ft (300 m) at the front. The front of the ice shelf is advancing at the rate of about 4,430 ft (1,350 m) per year. Presently, rifts near the front of the Amery are initiating formation of a 19 by 19 mile (30 by 30 km) iceberg nicknamed "the loose tooth."

Ice is also lost by melting from the ice shelf's base, which is in contact with seawater. But beneath the northwestern region of the Amery, ice refreezes onto the base. This refrozen layer is up to 660 ft (200 m) thick, though the lowest parts are sometimes porous. DM & IA

Glacial Erratics

Millions of years ago, these banded sandstone boulders were picked up and carried many miles in the ice before being deposited by the Vanderford Glacier on this smooth granite surface in East Antarctica.

Zhongshan Station

China acceded to the Antarctic Treaty on May 9, 1983, and was accepted as a consultative party on October 7, 1985. The Polar Research Institute of China had begun research in Antarctica in January 1980 when it sent two scientists to Casey Station. The first Chinese base established was the Great Wall Station on King George Island in 1985. Zhongshan Station was completed on the Larsemann Hills of Prydz Bay in February 1989. It can accommodate 15 people in winter and up to 30 during the summertime.

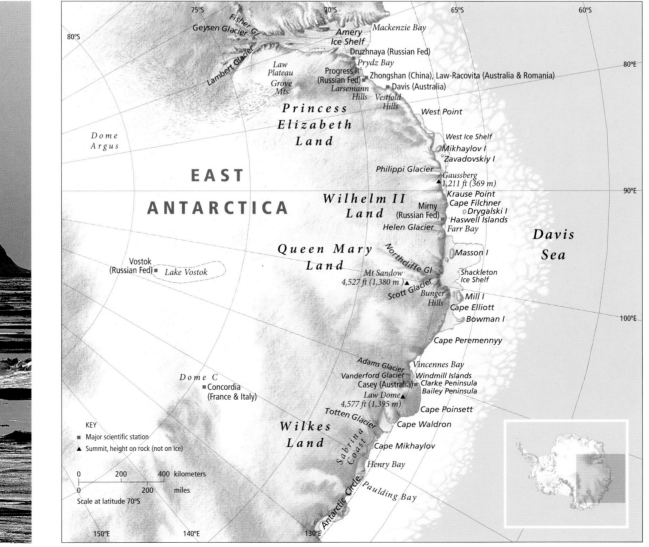

From Amery to Enderby Land

This section of East Antarctica in the Indian Ocean sector stretches west from the edge of the Amery Ice Shelf to Dronning Maud Land. Enderby Land was first explored by John Biscoe in 1831 and two years later the neighbouring coast, known as Kemp Coast, was visited by English whaling captain Peter Kemp.

On New Year's Day, 1930, during the BANZARE Expedition, Douglas Mawson took off in a seaplane and saw an area near the Amery Ice Shelf he called Mac.Robertson Land, which created a permanent punctuation mystery; it is named for Sir MacPherson Robertson, a supporter of the expedition.

Treading Lightly

Hagglunds are used for fieldwork. They can tow a trailer of equal weight, yet the footprint of their tracks is less than half the weight of a person's. Easy to drive, power is transmitted to each of the four tracks, so they can go virtually anywhere and tackle most gradients.

Mawson Station

Throughout his life, Mawson pressed for a continuing Australian presence in Antarctica, and in the 1950s Phillip Law used photographs from the United States' Operation Highjump to decide to build a year-round Australian station on the firm rock of Horseshoe Harbor. Law records that Mawson was not convinced, but Law prevailed and the two were later recorded examining a photograph of the new station. The Australian flag was raised at Mawson Station on February 13, 1954, and ten expeditioners over-wintered that first year, making Mawson the longest-operating station south of the Antarctic Circle.

In 1956 Antarctica's first aircraft hangar was built on the shore near Mawson. Today, the station uses a

Live Television
The first live television transmission was from Syowa Station in 1979. Syowa lies on Lützow-Holm Bay within an oasis where some of the coast and islands are ice-free, but pack ice often fills the bay.

Welcoming Committee
A line of emperor penguins strays from the main huddle to inspect an observer on the fast-ice breeding platform of Auster Rookery, near Mawson Station. By late summer, most chicks have safely fledged and are ready to head out to fend for themselves in the cold waters of the mighty Southern Ocean.

Historic Training Run
An expeditioner takes a team of huskies for a daily training run across the slick surface of Horseshoe Harbor at Mawson. Dogs were banned in 1994 when the treaty parties agreed to remove all non-indigenous species from the region.

blue-ice airstrip 3 miles (5 km) away, and light aircraft and helicopters operate from October to February. The 35 buildings house 20 people in winter and 60 in the summertime. Many of the original buildings are still standing, though now put to different uses—even the husky kennels remain, although unused. The main living area, consisting of bedrooms, a cinema, a mess, a bar, and recreation areas, is unceremoniously known as the Red Shed.

Syowa Station
The Japanese station Syowa, on the northern end of East Ongul Island, opened in January 1957 after the Japanese Antarctic Research Expedition of 1956. The

Norwegian cartographers who mapped the area from aerial photographs taken in 1936–37 thought that the two parts of this landform were one island, and called it *ongul*, Norwegian for "fishhook." However, when the Japanese expedition arrived, they found the "island" was split by a strait. Syowa has grown from the original three buildings to 47 structures today. The impressive main building is three stories high, topped by a domed skylight, and the station houses some 40 expeditioners in winter and up to 110 in summer. Each year it is supplied by an icebreaker that anchors nearby, and there is a sea-ice airstrip that is also quite close. Heavy sea ice is a constant problem and it has forced some breaks in operation. DM

Haakon VII Sea Region

This sector of East Antarctica covers about 6,270 miles (10,090 km) of the Haakon VII Sea coast between 20°W (the terminus of the Stancomb-Wills Glacier) and 44°38'E (Shinnan Glacier), and lies to the south of South Africa. It extends from Coats Land (west) to Enderby Land (east) and includes (from west to east) the Princess Martha, Princess Astrid, Princess Ragnhild, Prince Harold, and Prince Olav coasts.

Dronning Maud Land

The coastline of Dronning Maud Land—also known as Queen Maud Land—is skirted by a number of relatively narrow (30 to 125 mile/50 to 200 km wide) ice shelves, including Riiser-Larsen and Fimbul, and contains the fastest flowing glacier (1½ miles/2.4 km per year) in Antarctica: the Shirase Glacier. The East Antarctic Ice Sheet is as thin as 1½ miles (2.4 km) in this region, and forms a gently rising inland backdrop to a number of spectacular near-coastal mountain ranges, which are either ice-covered or pierce the ice sheet as rocky peaks. These range in height from 5,000 to 10,330 ft (1,500 to 3,150 m), and are challenging climbing destinations. A 1½ sq mile (3.9 sq km) area of ice-free cliffs called Svarthamaren, at 71°54'S, 5°10'E, 124 miles (200 km) inland from Princess Astrid Coast, is home to a colony of about 820,000 Antarctic petrels (including 250,000 breeding pairs)—the largest known seabird colony inland on the Antarctic continent, and a significant proportion of the world population of this species. The site also supports 500 to 1,000 pairs of snow petrels and about 80 pairs of south polar skuas.

Novolazarevskaya

Russia's "Novo" station is located at Schirmacher Oasis, 47 miles (75 km) inland and 3 miles (5 km) from India's Maitri Station. Schirmacher is a region of ice-free hills about 11 miles (18 km) long and dotted by meltwater ponds. The station, built on rock 335 ft (102 m) above sea level, opened in January 1961. The nine buildings house 30 over winter and up to 70 in summer. A resupply ship visits the coast, and supplies are either flown from the ship by helicopter or transported by tractor train, depending on the fast-ice conditions.

Maitri Station

The first Indian Antarctic Expedition took place in 1981, and India was admitted to the Antarctic Treaty in August 1983. The same year saw the construction of Dakshin Gangotri, India's first permanent Antarctic station, on one of the ice shelves on Princess Astrid Coast: it closed in 1989. Maitri opened on March 9, 1989. A modern station of innovative design, it is built on adjustable telescopic stilts. Winter staff have their own rooms in the main complex, and there is accommodation for 65 people when the summer huts are in use. Two doctors are in residence in winter and four in summer, and there is a permanent team of Indian Army engineers to service the equipment.

Like nearby Novolazarevskaya, Maitri sits on ice-free rock. India's National Center for Antarctic & Ocean Research, which opened in 1997, is in Goa—a world away climatically. But India has strong links with Antarctica: Maitri's geological interests include the study of Gondwana, when India and Antarctica belonged to the same landmass, and low-temperature engineering research in Antarctica is directly relevant to conditions in the Himalayan mountains.

SANAE IV

South Africa was an original signatory to the Antarctic Treaty, and has been conducting research in Antarctica since the International Geophysical Year of 1957–58. It has two stations on the sub-Antarctic Marion and Gough islands, and completed the current year-round station, SANAE IV, in the summer of 1997–98. The station is on the Vesleskarvet nunatak in Dronning Maud Land. The nunatak was first mapped by the Norwegians: Vesleskarvet means "the little barren mountain," but the South Africans call it "Vesles."

There have been year-round South African research stations on the Fimbul Coastal Ice Shelf of western Dronning Maud Land since 1962. SANAE IV is the first to be built on rock, not ice, and is located some 105 miles (170 km) from the coast, due south of SANAE III. Construction of the station began in the summer of 1993–94 using prefabricated panels made in South Africa, which were shipped to the Antarctic coast, and carried overland to Vesleskarvet by tractor trains. The station opened in January 1997. It is designed for 20 residents in winter and up to 80 people in summertime. It is resupplied twice each summer, and has a helipad and airstrip for flights to and from Germany's Neumayer Station and the coast.

Norvegia Sastruga
A sastruga, a ridge of snow formed by the wind, at Cape Norvegia, on the eastern edge of the Riiser-Larsen Ice Shelf, off Princess Martha Coast. It was not until the mid-twentieth century that the Dronning Maud Land region was explored and mapped extensively—largely using aircraft.

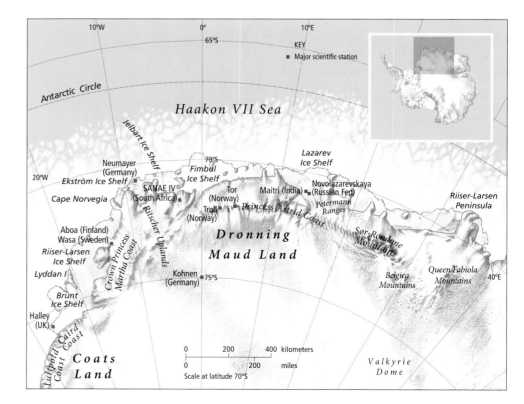

Accommodating Snow

Indicative of the extremity of the climate, Neumeyer Station has been dug into the Ekström Ice Shelf. The only above-ground structures are aerials, a wind generator, a balloon-launching facility, and an atmospheric research laboratory that can be jacked up as snow accumulates.

SANAE IV's smooth surfaces and rounded corners were designed after wind-tunnel testing. There are three main two-story buildings, 45 by 145 ft (14 by 44 m) in area and raised on stilts 13 ft (4 m) above the rock. They are linked by passageways at the lower level. The exterior is a dark color at the base to retain heat and discourage snow, while the roof is fluorescent orange for visibility—it matches the color of the station's vehicles.

Heavy Haulage

The task of constructing SANAE IV was considerable as all the materials had to be hauled from the coast by tractor trains and assembled above the ice and on top of a ridge. During summer, accumulated waste from the station is carried back to the coast and then shipped to South Africa.

Neumayer Station

Georg von Neumayer was a nineteenth-century German geophysicist who was a strong supporter of Antarctic exploration; German, Swedish, Belgian, and British expeditions have named Antarctic features after him. Germany established the first Georg von Neumayer Station in 1981 in the northeastern Weddell Sea on the Ekström Ice Shelf, a flat shelf 660 ft (200 m) thick.

By 1990 snow build-up and ice movement necessitated a move to a new station 6 miles (10 km) away. The present station opened in March 1992. It is 130 ft (40 m) above sea level, and is snow-covered: all that can be seen from a distance are the two towers at the station entrance, standing about 7 ft (2 m) above the snow. The station consists of two enormous parallel steel tubes about 300 ft (90 m) long, divided into living quarters, mess, hospital, workshops. Utilities are carried in a transverse tube, and there is a tunnel to the garage. In winter, a staff of nine passes nine months in total isolation, except for radio contact. In summer, the population may reach 50. The station is 5 miles (8 km) inland from its deepwater anchorage at the edge of the ice, and there is a nearby snow runway. Neumayer's principal research field at present is the role of the polar regions in world climate.   DM & RM

On the Edge

SANAE IV is sited on top of Vesleskarvet Cliff, which is part of the 70-mile (110-km) Ahlmann Ridge in Dronning Maud Land. The ridge runs south from the Fimbul Ice Shelf, where earlier South African stations were located.

The Polar Plateau

The term "polar plateau" refers to the high interior of the ice sheet. Around East Antarctica, the ice sheet surface typically rises from sea level to 6,560 ft (2,000 m) elevation about 660 to 1,000 ft (200 to 300 km) distance from the coast. This gentle rise, however, is large compared with the flatness of the plateau beyond, where the elevation only increases to 12,000 ft (3,660 m) over distances greater than 620 miles (1,000 km). Within a 9-mile (15-km) radius of Dome A, the summit of the East Antarctic Ice Sheet, the surface elevation varies by less than 10 ft (3 m).

Because of its great height, the Antarctic plateau is very cold and hence very dry. The average annual temperature at the highest point on the plateau is below –67°F (–55°C). Snowfall can be as little as the equivalent of 2 in (5 cm) of rain per year.

The first expeditions to cross the Polar Plateau were those of Amundsen, Scott, and Shackleton in their quests to reach the South Pole. In the late 1920s to 1940s, several aircraft surveys of the plateau were undertaken, mostly by the United States. During the International Geophysical Year (IGY) of 1957–58, a number of large-scale, over-snow surveys of the plateau were made, predominantly by the Soviet Union, New Zealand/Britain, and the United States. The same year, the United States and the Soviets established the permanent stations of Amundsen-Scott at the South Pole and Vostok. Other plateau stations, occupied for a few years only, were established during the IGY and in the decades after. These included the Soviet stations, Pionerskaya, Sovetskaya, and Komsomolskaya; Byrd and Plateau, both American; and two Japanese bases, Mizhuo, and the ice-core drilling site at Dome Fujii.

Today scientific parties continue to traverse the Polar Plateau conducting glaciological and geophysical investigations. Concordia, a year-round plateau station, was opened by France and Italy in 1997. Observations have been replaced with networks of automatic weather stations and automatic geophysical observatories, which relay data by satellite link. Much present knowledge of the Antarctic plateau comes from airborne geophysical surveys, using ice-penetrating radars and other instruments, and from sophisticated satellite sensors, which can measure the height of the ice surface to less than an inch, and map the movement of ice (which may be only a few feet per year).

Amundsen–Scott, South Pole

When Edmund Hillary and his party of New Zealanders arrived at the South Pole aboard their second-hand Massey Ferguson tractors on January 4, 1958, they were the first people to arrive overland at the pole since Robert Scott's team in 1912. However, Hillary was met by the American residents of the Amundsen–Scott Station, who had flown to the bottom of the world. There has been a continuously occupied United States station at the geographic South Pole since November 1956, and there are now over 30 flights there each year.

For decades, the most distinctive feature of the South Pole Station was The Dome, an aluminium geodesic dome, 165 ft (50 m) wide and 53 ft (16 m) high, enclosing three two-story buildings and a mass of equipment. Built in 1975, it was unheated and the floor was packed snow. The Dome has now outlived its usefulness. A new Amundsen–Scott Station—two U-shaped buildings on stilts, connected by enclosed walkways, and designed to house 250 people—was opened in January 2008. No flights to the South Pole are possible between mid-February and late October, so residents are as isolated as astronauts, but they can communicate by e-mail and telephone via satellite.

The station is 9,295 ft (2,835 m) above sea level, high enough to cause headaches in newcomers, and its recorded temperature range is –7.52°F to –117°F (–13.6°C to –82.8°C). The mean annual temperature is –56°F (–49°C), and the cold is exacerbated by an average wind speed of 18 ft per second (5.5 mps). Beneath the ground snow lies 9,345 ft (2,850 m) of polar ice sheet. The ceremonial pole that features in many photographs is a reflective globe encircled by flags of the 12 original Antarctic Treaty signatories.

Vostok Station

The lowest temperature ever recorded on the Earth was measured at Vostok: –128.56°F (–89.2°C) on July 21, 1983. The Russian station is situated high on the polar ice sheet, more than 11,500 ft (3,490 m) above sea level, and 870 miles (1,400 km) from the moderating effects of the ocean. The highest temperature recorded there was –7.6°F (–22°C). In June 1982, a nightmare scenario unfolded when a fire destroyed the station's

Woman Power
Frenchwoman Laurence de la Ferrière arrives at the South Pole on January 19, 1997, after a 54-day, solo skiing trek from Hercules Inlet on the Ronne Ice Shelf. In 2000–01, de la Ferrière was back in Antarctica, skiing from the Pole to Dumont d'Urville in Terre Adélie.

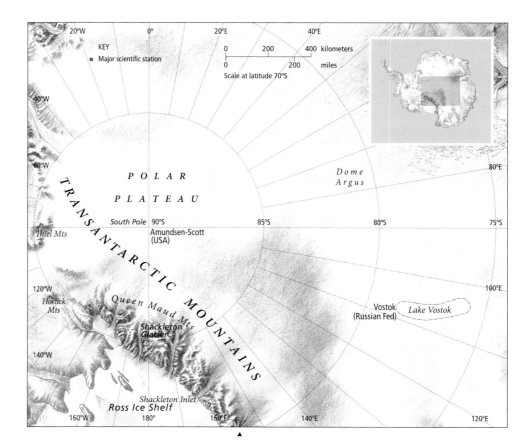

Marking the Pole
Because the ice on the Polar Plateau moves about 30 ft (10 m) toward the sea each year, the position of the South Pole moves slightly. Its ceremonial marker is repositioned annually.

Wintering Alone
New buildings replace the geodesic dome that served as Amundsen–Scott Base from 1975. During the long winter, temperatures are too low for aircraft to land. Telephone and e-mail links ensure remote contact with the outside world, but once the last flight departs, no new faces are seen until spring.

powerhouse, and 20 wintering expeditioners had to survive in a small building with only the warmth from an improvised oil heater until they could be evacuated by aircraft in November.

Vostok began operations on December 16, 1957, and ice-core drilling started in 1959. Except during the winter of 1994, the station has operated continuously ever since. The winter population is 13, rising to 25 in summer. The station is supplied by ski planes and by tractor trains from the coast, which take two months to travel from Mirny, 875 miles (1,410 km) away.

Below the station, more than 13,000 ft (4,000 m) beneath the ice, lies Lake Vostok, Antarctica's largest

"lake." Geothermal warming has produced a vast freshwater lake measuring 37 by 174 miles (60 by 280 km) and hundreds of feet deep. Closed off by ice for many thousands of years, this lake may harbor life forms long extinct elsewhere and provide valuable information about early climatic conditions. But if Lake Vostok is penetrated using contemporary technology, this precious resource will be compromised. Several years ago, drilling was halted just short of the lake— at ice that was on the surface about 400,000 years ago—although it has resumed recently.

Since 1964 Russia has operated a cooperative deep-drilling program with France, and since 1991 United States ski-equipped Hercules LC130s have flown winter teams from Mirny to the compacted ice airstrip near Vostok as part of a joint program for ice-core and lake investigations.

Concordia

In 1993, France and Italy signed an agreement to operate a wintering station at Dome C, called Concordia. Supported by the European Union and by ten European countries, its main goal is core drilling through deep ice. The first stage opened in December 1997. Consisting of three buildings, linked by enclosed walkways, that can house up to 45 expeditioners in summer, it operates all year round, with about 13 winterers.

Concordia is a remote inland base inside the polar vortex; in the southern spring, the ozone hole can be detected there. The station is 10,560 ft (3,220 m) above sea level. Its closest neighbor is the Russian station, Vostok, 350 miles (560 km) away. It is 590 miles (950 km) from the coast, 685 miles (1,100 km) from Dumont d'Urville, and 750 miles (1,200 km) from Terra Nova Bay. The station is serviced by tractor trains from Dumont d'Urville and by ski planes from Terra Nova Bay or Dumont d'Urville. DM & IA

Weddell Sea Region

Remote Region
A glacier flowing from the Ellsworth Mountains (right). These remote mountains were not mapped in detail until 1958–66, when the US Geological Survey and the US Navy conducted ground and aerial surveys.

The Filchner–Ronne Ice Shelves dominate the Weddell Sea. Together they cover an area of 211,000 sq miles (530,000 sq km) of floating glacial ice at the head of the Weddell Sea embayment, extending from 75°S to 83°S. They are bordered by Palmer Land to the west, Ellsworth Land to the south, and the Luitpold Coast (Coats Land) to the east.

Filchner–Ronne Ice Shelves
Taken together, the two ice shelves are the largest body of floating ice on the Earth, containing about 83,970 cubic miles (350,000 cubic km) of ice by volume.

Although about 10 percent smaller in area than the Ross Ice Shelf, their thickness is greater. This ranges from 492 to 1,640 ft (150 to 500 m) at the seaward margin (ice front) to almost 6,560 ft (2,000 m) near its inland margin, and the ice front flows northward at a rate of about 4,920 ft (1,500 m) per year. A number of research stations have operated since the 1950s, mainly on the Filchner Ice Shelf due to its relative ease of access compared with the Ronne Ice Shelf, which is typically locked in by sea ice.

Berkner Island, the second largest Antarctic island, forms a 3,200-ft (975-m) high ice-covered dome in the midst of the two ice shelves. The island is, in fact, termed an "ice rise," a mass of ice resting on rock and surrounded by ice shelf.

Record-keeping Rocks
The Dufek Massif in the Pensacola Mountains, near the southeastern edge of the Ronne Ice Sheet, contains a unique geological record of the Earth's history: a large body of igneous rock that may be associated with the rifting of Antarctica from Africa during the break-up of Gondwana.

Ellsworth Mountains
The region bordering the Ronne Ice Shelf's western margin is dominated by the Ellsworth Mountains, a 200 by 30 mile (320 by 48 km) range that rises spectacularly from the relatively featureless West Antarctic Ice Sheet and runs at right angles to the

Transantarctic Mountains. The Ellsworth Mountains play a major glaciological role in the region by damming the ice sheet flowing into the Ronne Ice Shelf, and in the process creating several outlet glaciers. The mountains contains Vinson Massif which, at 16,066 ft (4,897 m), is the highest point on the continent. The range also contains the next highest of Antarctica's mountains: Tyree (15,919 ft/4,852 m), Shinn (15,747 ft/4,801 m) and Gardner (15,370 ft/4,686 m).

Halley V
Britain was among the first signatories of the Antarctic Treaty, and it has a long history of Antarctic exploration and research. The first Halley Station, which opened in January 1956 for the International Geophysical Year, was erected on stilts over the Brunt Ice Shelf, but it had to be rebuilt in 1966, 1972, 1982, and 1989, as each successive station sank beneath the snow. In 1985, British scientists at Halley Station alerted the world to the springtime depletion of ozone in the Antarctic

Snow Drift
A modern traveler shares a little of the experiences of early explorers who found themselves caught in the pack ice. For them, however, there were no heated quarters to which they could retreat.

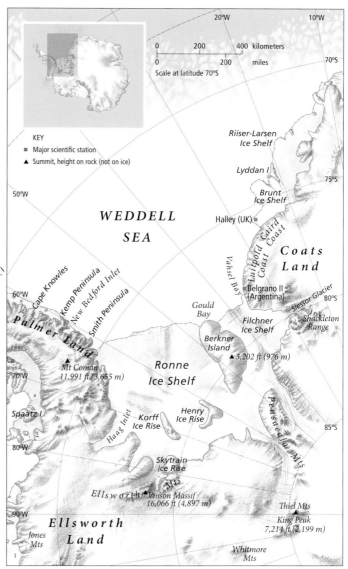

stratosphere, which led to an international agreement to ban chlorofluorocarbons (CFCs).

Halley V, about 8 miles (12 km) from the coast and 120 ft (37 m) above sea level, is the most isolated of Britain's Antarctic stations. Three of its buildings stand on telescopic stilts so that they can be raised above the rising snow, and a garage, which weighs more than 55 tons (50 tonnes), and an accommodation building are mounted on skis so they can be towed to a new location every year. The average winter population is 15, rising to a maximum of 65 in summer. Supplies are delivered twice a year, first by ship to the edge of the ice, then across the ice shelf by surface transport, and Twin Otter flights operate in summer from the nearby snow runway. During Halley's 105 days of total darkness each winter, the station staff enjoy magnificent auroral displays, and from May to February they are joined by a nearby emperor penguin colony, which has some 15,000 breeding pairs.

Belgrano II

Founded in January 1955, Argentina's Belgrano Base was the southernmost base in the world briefly until the station at the South Pole was built in 1956. The first Belgrano was on the Filchner Ice Shelf, but when it was buried in snow by 1978 it was

abandoned—a wise decision, as that bit of the ice shelf subsequently broke away and headed out to sea.

Belgrano II opened in February 1979. It is located on Bertrab nunatak, one of the few exposed areas of rock at the southwestern corner of the Weddell Sea near the Filchner Ice Shelf. Built on solid rock 165 ft (50 m) above sea level and 75 miles (120 km) inland, it has 12 buildings with a year-round population of 21. The region is noted for its violent storms, and the station experiences four months of darkness each winter; in summer the only wildlife are some gulls, petrels, and skuas. The station is serviced by an airstrip built on the surface of a glacier. DM & RM

Limited Shelf Life

Halley V stands on a steel platform supported by stout telescopic legs. The station is mechanically raised nearly 3 ft (1 m) a year to maintain a near-constant level above the surface snow of the Brunt Ice Shelf, which has already crushed and buried its four predecessors. Halley V is also being carried slowly out to sea by the inexorable motion of the floating ice shelf upon which it rests.

The Sub-Antarctic Islands

The vast Southern Ocean and its neighboring oceans are far from empty. Islands are scattered across these seas—but there is a lot of water in between. Most of these islands were discovered by chance and few are habitable. Many early "discoveries" were mistakes, and over centuries new islands have been added to—and others deleted from—nautical charts. For someone standing on the icy deck of a ship and peering through driving sleet or rain, a distant iceberg seen through the fog may look a lot like land.

Even defining what makes an island sub-Antarctic is something of a challenge. Islands that are truly sub-Antarctic lie south of the Polar Front. Cool temperate islands of the southern Pacific, Indian, and Atlantic oceans, including the Falklands, Macquarie Island, and Iles Kerguelen and Crozet, are covered here because all are important sites for Antarctic wildlife, and they are frequently visited and significant locations in the human history of Antarctica. The Auckland and Campbell islands, southeast of New Zealand, are comparatively warm and temperate; they are also called sub-Antarctic in some contexts, and they too are often visited on the way to the Ross Sea. The vegetation on sub-Antarctic islands ranges from mosses to small trees; life is gentler than on the Antarctic continent and the islands support a great variety of wildlife. DM

A Crowded Life
King penguin colonies, like this one at Salisbury Plain, South Georgia, are not as chaotic as they may seem. Breeding adults take over the center or rear of the colony, separate from the roosting and unoccupied birds at the edge. Chicks remain close together, where their parents left them, and during the winter adults and chicks tend to segregate.

Island Jewels
Though isolated and bleak, the sub-Antarctic islands are austerely beautiful, and are home to an impressive array of hardy wildlife that returns time and again to these rocky outcrops in a vast ocean.

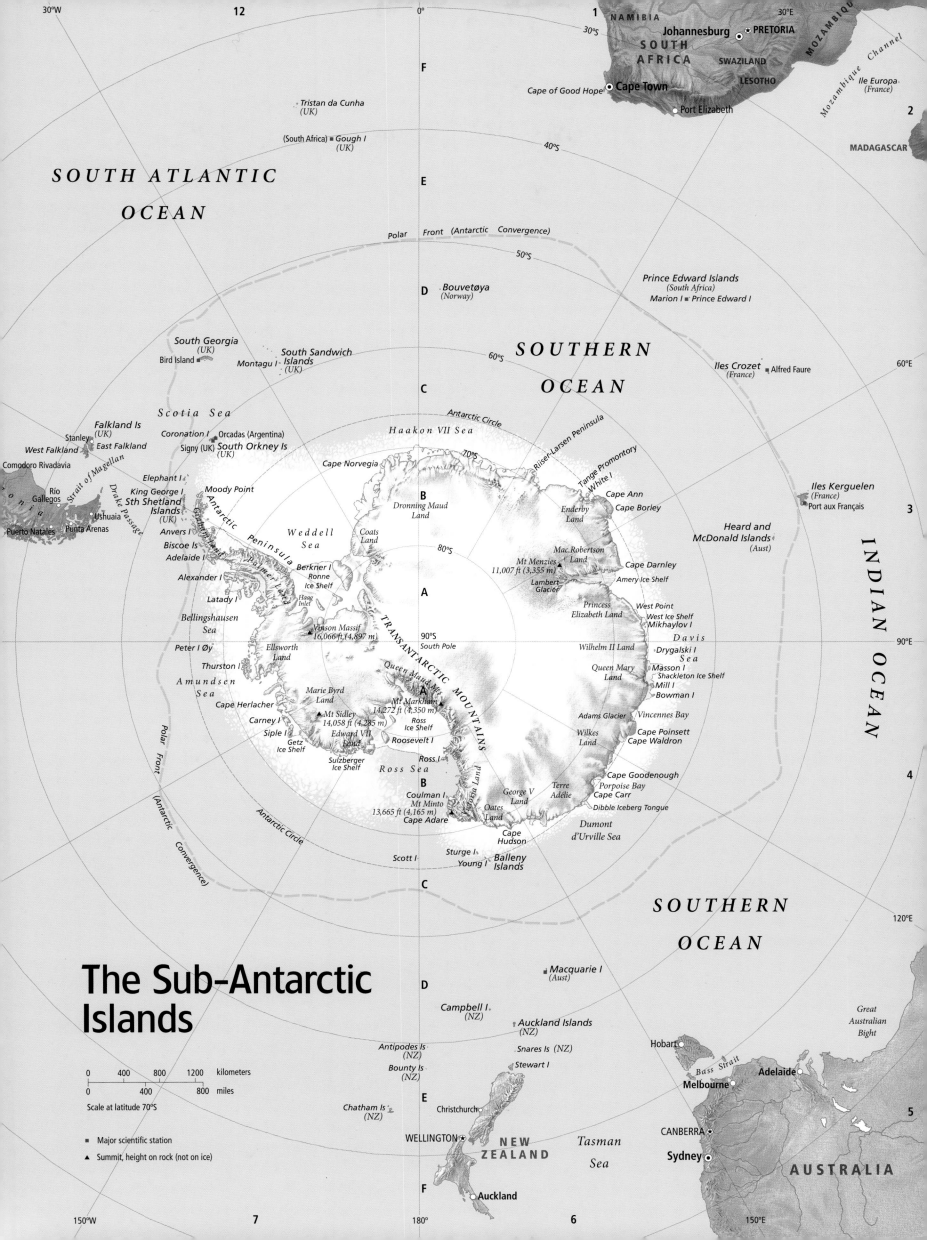

The Sub-Antarctic
Islands

30°W   12   0°   1

NAMIBIA   30°E

30°S

Johannesburg   ★ PRETORIA

SOUTH
AFRICA

SWAZILAND

MOZAMBIQUE

Mozambique Channel

LESOTHO

Cape of Good Hope ○ **Cape Town**

Ile Europa
(France)

○ Port Elizabeth

Tristan da Cunha
(UK)

2

40°S

MADAGASCAR

(South Africa) ■ *Gough I*
(UK)

SOUTH ATLANTIC

OCEAN

SOUTHERN

Polar Front (Antarctic Convergence)

50°S

D *Bouvetøya*
(Norway)

Prince Edward Islands
(South Africa)
Marion I ■ *Prince Edward I*

South Georgia
(UK)
Bird Island ▪ *Montagu I*

South Sandwich
Islands
(UK)

OCEAN

60°S

Iles Crozet
(France) ▪ Alfred Faure

60°E

3

C

Scotia Sea

Antarctic Circle

Haakon VII Sea

Iles Kerguelen
(France)
▪ Port aux Français

Falkland Is
(UK)
Stanley
West Falkland   *East Falkland*
Comodoro Rivadavia
Rio
Gallegos
Ushuaia
Puerto Natales   Punta Arenas

Coronation I   Orcadas (Argentina)
Signy (UK) **South Orkney Is**
(UK)

70°S

Cape Norvegia

B

Dronning Maud
Land

Riiser-Larsen Peninsula

Tange Promontory
White I

Cape Ann
Cape Borley

Enderby
Land

Heard and
McDonald Islands
(Aust)

INDIAN OCEAN

Strait of Magellan
Drake Passage

Elephant I
King George I
Sth Shetland
Islands
(UK)
Anvers I
Biscoe Is
Adelaide I

Moody Point

Weddell
Sea

Coats
Land

80°S

Mac.Robertson
Land
Mt Menzies ▲
11,007 ft (3,355 m)
Lambert
Glacier

Cape Darnley
Amery Ice Shelf

Alexander I

Berkner I
Ronne
Ice Shelf
Haag
Inlet

Princess
Elizabeth Land

West Point
West Ice Shelf
Mikhaylov I

Latady I

A

90°S
South Pole

Wilhelm II Land

Davis
Sea

Bellingshausen
Sea

Vinson Massif
16,066 ft (4,897 m)

Ellsworth
Land

TRANSANTARCTIC MOUNTAINS

Drygalski I

90°E

Peter I Øy

Queen Mary
Land

Masson I
Shackleton Ice Shelf
Mill I

Thurston I

Amundsen
Sea

Marie Byrd
Land

Queen Maud Mts

A

Mt Markham ▲
14,272 ft (4,350 m)

▪ Bowman I

Adams Glacier   *Vincennes Bay*

Cape Herlacher

▲ Mt Sidley
14,058 ft (4,285 m)

Ross
Ice Shelf

Wilkes
Land

Cape Poinsett
Cape Waldron

Carney I
Siple I
Getz
Ice Shelf

Edward VII
Land

Roosevelt I

Victoria Land

4

Sulzberger
Ice Shelf

Ross I

Ross Sea

Cape Goodenough
Porpoise Bay
Cape Carr
Dibble Iceberg Tongue

B

Coulman I
Mt Minto ▲
13,665 ft (4,165 m)
Cape Adare ▲

George V
Land
Oates
Land

Terre
Adélie

Dumont
d'Urville Sea

Cape
Hudson

Scott I ▪

Sturge I
Young I   *Balleny*
Islands

C

SOUTHERN

120°E

OCEAN

▪ *Macquarie I*
(Aust)

D

Campbell I
(NZ)

▪ *Auckland Islands*
(NZ)

Great
Australian
Bight

Antipodes Is
(NZ)

Snares Is (NZ)

Hobart ●

Bass Strait

Bounty Is
(NZ)

Stewart I

Adelaide

Melbourne

E

Chatham Is
(NZ)

○ Christchurch

CANBERRA ★

5

WELLINGTON ★

NEW
ZEALAND

Tasman
Sea

Sydney ●

F

★ Auckland

AUSTRALIA

150°W   7   180°   6   150°E

0	400	800	1200	kilometers

0	400	800	miles

Scale at latitude 70°S

▪ Major scientific station

▲ Summit, height on rock (not on ice)

South Georgia

South Georgia has been likened to a piece of the Alps dropped into the middle of the ocean—a fair description because the island is a summit of the Scotia Ridge, a line of mountains 2,700 miles (4,350 km) long, which links the ranges of the Antarctic Peninsula with the Andes through a submarine hairpin loop to the east. The South Sandwich and South Orkney islands are other exposed peaks along the ridge.

South Georgia has no near neighbors. Antarctica, South America, and Africa are 930 miles (1,500 km), 1,300 miles (2,100 km), and 2,980 miles (4,800 km) away respectively, and the closest substantial landmass is the Falkland Islands, which is 870 miles (1,400 km) from South Georgia.

After the 1982 Falklands War, South Georgia ceased to be a Falklands' dependency, and in 1985 became the British Dependent Territory of South Georgia and the South Sandwich Islands. Despite the title's equal billing, South Georgia's land area—1,350 sq miles (3,500 sq km)—is ten times larger than the 11 islands and several islets at the edge of the South Sandwich Trench which make up the South Sandwich Islands. These islands are a fog-enshrouded arc of very active volcanic, mostly ice-covered isles; they are rarely visited except by millions of breeding chinstrap, macaroni, gentoo, and Adélie penguins.

South Georgia is uplifted rather than volcanic; its main landmass, about 105 miles (170 km) long and 19 miles (30 km) wide in the middle, is surrounded by numerous offshore rocks and islets. Mount Paget, 9,626 ft (2,934 m) high, is the tallest of a spine of mountains running northwest to southeast. Although Antarctic pack ice does not extend this far north, the bays around the coast freeze in winter, and glaciers and ice fields cover two-thirds of the island. The south is higher and more rugged than the north, and in many places the mountains fall sheer to the sea. Because the mountains act as a barrier against the prevailing westerly winds, the northeastern side is much more sheltered than the southwestern side.

In recent years, South Georgia's only permanent population has been at Grytviken. Even before the Falklands War, there was a British Army presence, but the troops left in 2001, and were supplanted by a substantial scientific staff. Two buildings for fishery research and accommodation were erected at King Edward Point. In winter there are some 13 people here, most for more than two years, engaged in scientific research and support. In addition, there is always a government officer involved with fishing and licensing in winter, and with the tourist ships during the summer months. The island's museum and adjoining shop are staffed in the summer season.

No Wood for a Toothpick

The first person to see South Georgia is believed to have been London merchant, Antoine de la Roché, who was blown far off course while rounding Cape Horn in 1675.

It was next sighted in 1756 by the crew of the Spanish vessel, the *Léon*, but Captain James Cook was the first to step ashore here, disembarking from the *Resolution* on January 17, 1775. With "a discharge of small arms" that reportedly alarmed the penguins of Possession Bay, he claimed and named the land for King George III. Cook hoped this was the tip of the southern continent that he had traveled so far to find. He recorded his dashed hopes by calling the island's southernmost point Cape Disappointment (the first of three capes in Antarctica so named), noting: "the disappointment I now met with did not affect me much, for to judge by the bulk of the sample it would not be worth the discovery."

Even before this, South Georgia had not impressed Cook: "I did not think it worth my while to go and examine these places where it did not seem probable that any one would ever be benefited by the discovery ... The inner parts of the country were not less savage and horrible: the wild rocks raised their lofty summits till they were lost in the clouds and the valleys lay buried in everlasting snow. Not a tree or shrub was to be seen, no not even big enough to make a toothpick."

Cook did, however, report on the numerous seals, and in 1788 the first sealers arrived from Britain, followed in 1791 by others from the United States. Scant records indicate that Edmund Fanning, of Aspasia, New York, may have had the most lucrative harvest; in 1800 he took 57,000 seal skins from the

A Blonde in the Middle

Most Antarctic fur seals have medium to dark brown fur, but a small number are natural-born blondes. Pale, almost white fur seals are highly visible on dark rock and stand out in the crowd.

before Christmas, territorial adult males each jealously guard a stretch of beach. The males have departed by mid-January, but the remaining females rigorously enforce their own private space for their pups. Smaller numbers of leopard and Weddell seals are also found around South Georgia, and there are 100,000 elephant seals, although the huge dominant adult males leave the beaches early in summer.

It is thought that the population of fur seals has recovered very rapidly because the whales with which the seals competed for food have also been slaughtered to near extinction. Whales breed much more slowly than seals, so whale populations have been slower to recover. Humpbacks, minkes, and right whales are sometimes seen in South Georgian coastal waters.

War Zone

South Georgia was the flashpoint for the 1982 Falklands War. An Argentinian company had been given an option to collect scrap metal from abandoned whaling stations, but when an Argentinian naval ship arrived at Leith Harbor on March 18, 1982, and hoisted its flag, Britain registered an official protest. Thirteen days later, 22 British marines landed at King Edward Point.

On April 3 (a day after the Falkland Islands were invaded), the *Bahia Paraiso* sailed into Cumberland Bay and called on the island's administrator to surrender. In the ensuing three-hour battle, 200 Argentinian troops landed and captured the station and many of its staff, but only after two of their helicopters had been shot down and several soldiers killed. Subsequently, Royal Navy ships recaptured King Edward Point on April 25 and Leith the following day.

The British captives were held by the Argentinians for 12 days, and then released in Uruguay. The 185 Argentinian troops captured by the British were also released in Uruguay. On June 14, Argentinian forces on the Falkland Islands surrendered to Britain. DM

Shaded by Mountains

The southern end of South Georgia is cleft by Drygalski Fiord. Less than 1 mile wide and 7 miles long (1.6 by 11 km), this beautiful waterway is shaded to the east by the Salvesen Range, which rises to Mount Carse (7,650 ft/2,330 m), and to the west by Mount Sabatier.

more obviously carnivorous than other species of duck, has been observed eating seal carcasses. The South Georgia pipit, a drab, brown bird resembling a long-legged sparrow and the world's most southerly songbird, lives only on South Georgia, but rats introduced from ships have reduced its breeding range from all over the mainland to some of the offshore islands and isolated parts of the rugged south coast.

By the early twentieth century, seals had been hunted for their pelts to near extinction, but numbers have now risen again to some 3 million. More than 95 percent of the world's population of Antarctic fur seals is found on South Georgia and the little islands around it, and their numbers are continuing to increase by an average of 10 percent each year. In the weeks

South Orkney Islands

On December 6, 1821, two sealers, American Nathaniel Palmer and Briton George Powell, sailing together in the *James Monroe* and the *Dove* respectively, sighted these islands and charted them as Powell's Group. When another British sealer, James Weddell, arrived aboard the *Jane* in February 1822, he chose the current name because the islands occupy the same latitude in the south that Britain's Orkney Islands do in the north. Weddell's name won out in popular usage. However, Powell was not overlooked; one of the four main islands bears his name.

The entire group is approximately 375 miles (600 km) northeast of the Antarctic Peninsula and covers a land area of about 240 sq miles (620 sq km), almost 90 percent of it glaciated. Even in late summer, they can be surrounded by icebergs and sea ice, making landings difficult. Coronation Island, by far the largest, rises to the highest point: Mount Nivea, at 4,150 ft (1,265 m). Signy Island lies to the south; Powell Island

and Laurie Island are to the east. The landing beach at Shingle Cove in Iceberg Bay, Coronation Island, is backed by a towering glacier; once ashore, there is an easy walk up a rocky ridge to scattered Adélie rookeries.

In the nineteenth century, sealers annihilated virtually every seal in the vicinity, then whalers arrived in the twentieth century. An unsuccessful whaling station lasted from 1920 to 1926—the owner, Peter Sørlle, named Signy Island after his young daughter.

Territorial Claims

In 1903, the *Scotia* brought William Spiers Bruce and the Scottish National Expedition to the South Orkneys. From April 1 that year, a meteorological station operated on Laurie Island—Ormond House, above the beach in the lee of a hill, is now in ruins. Britain turned down Bruce's offer to continue to use the base as an ongoing meteorological station, but Argentina subsequently accepted his proposal. What is now known as Orcadas Station began operations on February 22, 1904, and is the longest continually operated Antarctic station; it can accommodate 14 people over winter and up to 45 in the summer months.

In March 1947 Britain set up Base H, a year-round meteorological station, at the site of Sørlle's whaling station at Factory Cove, Signy Island. Now known simply as Signy, the base is closed to unofficial visitors.

Britain's and Argentina's rival claims to the islands are held in abeyance by the Antarctic Treaty. DM

Penguin Haven
Coronation Island is a windswept rock in the middle of the Southern Ocean, but a few bare patches make it a haven for Adélie penguins which nest above the beach in Shingle Cove.

Serious Signage
Damp and desolate Laurie Island is a very long way from any of the cultural centers shown on the signpost in front of Orcadas Base. The station houses scientific personnel who continue the meteorological observations begun by William Bruce in 1903.

Glacial Beach
Glaciers form a grand backdrop to a narrow beach at Shingle Cove, part of Iceberg Bay on the south side of Coronation Island. On December 7, 1821, George Powell landed on Coronation Island and claimed the group of islands for Britain.

Falkland Islands

Albatross on Saunders
Several species of albatrosses are occasional visitors to the Falkland Islands. However, only the black-browed albatross calls the islands home. This medium-large albatross breeds in 12 locations on eight of the islands, nesting on remote cliffs and utilizing the updraughts for landing.

An Arch of Bone
Christchurch Cathedral in Stanley was built in 1872. The distinctive arch in front of the cathedral is fashioned from the lower jawbones of two sperm whales. It was brought from South Georgia to celebrate the centenary of the Falkland Islands in 1933.

The Falkland Islands lie north of the Polar Front and are not, therefore, sub-Antarctic islands, but they are a popular stop en route to South Georgia. In 1982 this archipelago hit the headlines when British troops fought to retain ownership against the invading Argentinian forces. Although both nations support their claims by invoking history, the Falkland Islanders are staunchly British. Union Jacks fly in many front yards, yet the Falklands still feature nightly on Argentinian weather reports—as Las Islas Malvinas—which record rain or snow for more than half the year.

Besides the large islands of West Falkland and East Falkland, there are at least 200 much smaller islands, adding up to a total landmass of 4,670 sq miles (12,170 sq km). The total population of the islands is about 2,700, some 2,500 people living in Stanley, the only town, in East Falkland. The war has left a legacy of mined beaches and fields behind barbed wire and warning signs near Stanley—finds of live ammunition are still made regularly.

Rich in Wildlife

West Point Island, on the western side of the group and owned by the Napier family for generations, is a convenient first landing site. Across the island, in a cleft high above the sea, rockhopper penguins and black-browed albatross nest in noisy profusion.

New Island, also on the west, is a nature reserve established by Tony Chater and Ian Strange and now run by a trust. The resident wildlife includes gentoo, magellanic, and rockhopper penguins, black-browed albatrosses, prions, cormorants, and fur seals. Carcass Island, at the northwest corner of the group, is named for nothing more sinister than one of the Royal Naval ships that first colonized the Falklands. Today's owners welcome visitors to view the island's wide range of bird life, including magellanic and gentoo penguins.

Saunders Island's Port Egmont was the first British settlement in the Falkland Islands; it was established in 1766. Some ruins from that era remain near the present owner's farmhouse. However, most visitors disembark in the north, where there are rockhopper and magellanic penguins on one shore of the spit and gentoo penguins on the other. Rockhoppers surf straight onto the rocks, even though there is a smooth beach close by. A colony of king penguins completes Saunders Island's array of flightless birds. On the grassy hill, where sheep graze among magellanic penguin burrows, are rockhopper rookeries and nesting pairs of black-browed albatrosses. DM

Squid Fishing
There are a number of squid fisheries around the Falkland Islands (shown), and southern New Zealand. Squid are usually taken at night by jigging: the squid are attracted to the boat by lights and snagged by barbs set in the water. The automatic jigging machines and lights are set alongside the boat.

South Shetland Islands

On the voyage to the Antarctic Peninsula from South America, the looming black volcanic crags and extensive glaciers of the more southerly South Shetland Islands provide a dramatic introduction to the Antarctic region. Many of the ships heading for the southern continent maneuver among the icebergs to make their first landfall on this part of the South Shetland archipelago; it may also be the last glimpse of Antarctica for vessels as they re-enter the Drake Passage on the homeward journey. Annexed by Captain William Smith in 1819, who named the group New South Britain, the string of islands was renamed the South Shetlands the following year.

Elephant Island

Captain George Powell probably named this island in 1821 because of the number of "sea elephants" he found there. Only 7 by 12 miles (12 by 20 km) in area, it lies at the northeastern end of the South Shetland archipelago. Even when conditions look good, ice in the bay and treacherous swells make landing tricky; it becomes almost impossible when seals and penguins are crowding the beach. A more convenient location for wildlife-watching, Cape Lookout in the south, is frequented by many fur and elephant seals, as well as a large population of penguins—chinstraps, and some gentoos and macaronis.

Ernest Shackleton brought his men to Elephant Island after they finally escaped the ice of the Weddell Sea in April 1916. Frank Wild, along with 21 others from the group, spent 137 days, in the deep winter, in this inhospitable spot until they were finally rescued on August 30. Shackleton's expeditioners made their first landfall at Cape Valentine in the northeastern corner of the island. Lack of high ground forced them to move 9 miles (15 km) west along the north coast to a small promontory, where a hurricane shredded the remaining tents the night that they arrived.

Their new residence was dubbed Cape Wild, a name that the Australian photographer Frank Hurley regarded as "at once an apt description and a tribute to a great-hearted comrade." Hurley described the campsite as "a precarious foothold on an exposed ledge of barren rock, in the world's wildest ocean … Our refuge was like the scrimped courtyard of a prison—a narrow strip of beach 200 paces long by 30 yards [27 m] wide … Behind us, the island peaks rose 3,000 feet [900 m] into the air, and down their riven valleys, across their creeping glaciers, the wind devils raced and shrieked, lashed us with hail, and smothered us with snowdrift … Inhospitable, desolate and hemmed in with glaciers, our refuge was as uninviting as it well could be."

The narrow 820-ft (250-m) gravel spit, rising to a tiny rocky headland of 20 ft (6 m), has since been downgraded to Point Wild.

Penguin Island

Rounding the northern end of King George Island, the first accessible landing sites are Turret Point on King George Island and Penguin Island, just ½ mile (0.8 km) away. Edward Bransfield named the latter in 1820 for the penguins he found there. Currently, about 8,000 breeding pairs of chinstraps and about half that number of Adélies hide away at the island's southern end. At the beginning of summer, the whole of this tiny island, about 1 mile (1.6 km) across, is usually still under snow. This snow cover and nesting southern giant petrels can make it

A Forbidding Shore
This view of Elephant Island from Dumont d'Urville's *Atlas* shows the western swell preventing the *Astrolabe* from closer approach in 1838. Like so many that come here only to be driven away by wind, waves, and fog, the French ships merely skirted the island.

A Rosy View
Approaching Elephant Island, Ernest Shackleton wrote: "Rose-pink in the growing light, the lofty peak of Clarence Island told us of the coming glory of the sun." A humpback whale enjoys the view close to where Shackleton "would have given all the tea in China for a lump of ice to melt into water."

Far from Home
Russians are used to winter, but this signpost at Bellingshausen Station (above) plaintively reveals that home is a long way away: Murmansk is 1,025 miles (1,650 km) in one direction, and Vostok is 2,718 miles (4,375 km) to the south; the balmy warmth of Havana is 7,140 miles (11,500 km) away.

Close Inspection
Only a small part of the Adélie penguin population (above left) is found as far north as the sub-Antarctic islands. Adélies are at home on snow like this: most of the species nest on the Antarctic Peninsula and elsewhere around the continent.

difficult to move onto the island. Later in the season, when delicate mosses, grasses, and lichens are exposed, there may also be fur seals to contend with on the beach. The island's summit is Deacon Peak, a 554-ft (169-m) volcanic cone.

King George Island

Largest of the South Shetland group and named in 1820 for the British monarch at the time, King George Island is roughly 43 by 16 miles (69 by 26 km), and 90 percent covered with ice. The glaciers of the island have retreated significantly during recent years.

Across the narrow passage from Penguin Island, Turret Point is distinguished by a cluster of tall rock stacks at either end of a gravel beach. The point marks the eastern end of King George Bay and was named by the *Discovery II* expedition in 1937. Katabatic winds rage off the nearby glaciers, and elephant and Weddell seals, as well as Adélie and chinstrap penguins, kelp gulls, Antarctic terns, and southern giant petrels may be observed here. Nesting Antarctic shags crowd one of the offshore stacks.

Admiralty Bay is the largest harbor in the South Shetland Islands, extending over 47 sq miles (122 sq km) and reaching depths of over 1,640 ft (500 m). This expansive anchorage was used by whalers in the early twentieth century—icebreakers are never needed to gain entry to it. Since 1996 the whole of Admiralty Bay has been an Antarctic Specially Managed Area.

Base G, the first permanent station on King George Island, was constructed by the British in 1947 at Keller Peninsula. Later the same summer, Argentina set up a hut 80 ft (25 m) away. Base G closed in 1961 and was removed in 1996; only its foundations remain. Brazil built Comandante Ferraz Station nearby in 1984, with living quarters for 13 in winter and 40 in summer. There are also Ecuadorian, Peruvian, and United States summer stations here; the American station, known as

Copacabana, lies within a designated Site of Special Scientific Interest, and is thus out of bounds. Polish Arctowski station, named after the *Belgica*'s geologist, Henryk Arctowski, has operated at Point Thomas since 1977. Upgraded in 1998, it can accommodate up to 12 expeditioners in winter and 40 in summer. Lichens and grasses make Point Thomas an attractive spot; wildlife here includes fur, elephant, Weddell, crabeater, and leopard seals, as well as Adélie, chinstrap, and gentoo penguins, and many flying birds.

James Weddell, sealing captain of the *Jane*, named Maxwell Bay in 1822 after Lieutenant Francis Maxwell. Lying at the southern end of King George Island, the bay is protected on its southern side by Nelson Island. The area is crowded with buildings, muddy roads, and an airport. So many bases are located here that it could well be called Antarctic Treaty Bay; it has even been the site of an international marathon for tourists.

Russia's Bellingshausen Station (1967) and Chile's Presidente Eduardo Frei Station (1969) are close neighbors. Air traffic from the Chilean Teniente Rodolfo Marsh Martin Station, behind the hill, passes low overhead, and one of the base's buildings sometimes operates as Antarctica's only hotel. China's Great Wall Station (1985) is in nearby Hydrographers Cove, and Uruguay's Artigas Station (1985) is to the northeast, on the far side of Suffield Point. The newest base is South Korea's King Sejong Station (1988) in Marian Cove. Argentina's Teniente Jubany Station was built at Potter Cove, near Maxwell Bay, in 1948.

Aitcho Islands

The thrusting, rocky spires of the tiny Aitcho Islands confirm the volcanic origins of the South Shetland Islands. The group lies at the northern end of English Strait, between Robert and Greenwich islands. Pronounced simply: aitch(h)-o, the group was named in 1935 by the *Discovery II* expedition in honor of the

A Fragile Shore
The relatively temperate climate of the South Shetland Islands supports a greater number of plants than anywhere else so near to the Antarctic mainland. But the plant life here is delicate— moss beds such as this one can bear the scars from a single footstep for many decades to come.

An Island Walk
Unlike other sub-Antarctic islands, it is easy to walk across Aitcho Island. However, it is no longer permitted to cross when snow cover has gone. The moss beds can easily be damaged by footprints. The presence of humans may also disturb nesting giant petrels.

The Westward Limit
For years, the remains of a BAS Otter fuselage slowly deteriorated beside the hangar on Deception Island; it has now been removed. The aircraft hangar marks the westward limit for visitors to Whalers Bay—the area beyond is protected to enable scientists to study plant re-establishment after eruptions and mudslides.

British Admiralty's Hydrographic Office, which is better known as the H.O.

Going ashore at Barrientos Island can often involve pushing through fringing kelp. The large rookeries of chinstrap penguins surround the landing site at the eastern end of the main island, while gentoos nest among the bones at Whalebone Beach. On the western side of the island, elephant seals wallow near another landing beach but, after the snow has melted, transit is not allowed because of the damage humans may do to the fragile environment and to the nesting wildlife.

Yankee Harbor, Greenwich Island
Yankee Harbor is a very sheltered bay on the southern side of Greenwich Island. The bay extends well back, protected by a crescent-shaped, narrow gravel spit about ½ mile (0.8 km) long. American sealers began visiting here in 1820, and a rusting tripod, for boiling blubber down to oil, marks the start of the spit. Here territorial skuas and feisty fur and other seals challenge visitors. Large populations of gentoo penguins and snowy sheathbills occupy the terraces above the beach.

Half Moon Island
Named for its crescent shape, this island appears on South American charts as Isla Media Luna. It is just over 1 mile (1.6 km) long, and lies on the north side of Livingston Island's eastern tip.

The landing site, near the eastern tip, is on a beach sheltered by crags. Here a small wooden boat wreck lies near a colony of nesting chinstrap penguins. The ridge behind the beach can be crossed when the penguin colony is not too densely occupied. Argentina's Teniente Camára Station (established in 1953 but not occupied regularly in recent years) lies to the west. The hill behind the station provides a sweeping view of the island, but during the nesting season it is claimed by aggressive skuas. On the beach in front of the station there are sometimes gentoo penguins, and fur seals arrive towards the end of summer.

Livingston Island
Livingston Island, only about 37 miles (60 km) long and between 2 and 20 miles (3 and 32 km) wide, has been known since 1820. Spain built the Juan Carlos I summer station on the south side of the island in 1988.

Hannah Point, on the eastern side of Walker Bay, is named after a sealing vessel that was wrecked there on Christmas Day, 1820. Landings are possible if wind and waves allow, but each shipload of passengers has the potential to do more damage to wildlife than at other, less crowded locations—access is denied until well after chicks have hatched. The top of the rise, where a thin, red band of pure jasper runs through the rocky ledge, commands a grand panorama of the north of the island.

The main landing point is near a rock, 50 ft (15 m) high, on the shoreline and surrounded by rookeries. A few nesting macaroni penguins sometimes reside among the multitudes of gentoo and chinstraps.

An alternative landing site at Walker Bay is well to the west, on a long, open beach, backed by a meltwater pond. The land behind rises to a towering ridge and

glaciers. Scientists who regularly camp here have set out bones, rocks, and minerals on two flat-topped boulders. Later in the summer, when the snow and ice have melted, the rocks reveal extensive growths of mosses and lichens. Antarctica's only flowering plants also eke out an existence in the rocky soil: the grass-like *Deschampsia antarctica* and the tiny *Colobanthus quitensis*. After midsummer, when the penguins start to leave the rookeries, it is possible to walk from one landing site to the other, though care must be taken not to alarm the giant petrels that nest nearby, or to disturb the still-crowded rookeries toward the headland. On a rise just above water level, at the western end of the rookeries, a dark brown streak usually marks the presence of a large wallow of molting elephant seals: a dramatic scene of perpetual squabbling and out-and-out blood-drawing fights.

Deception Island
Deception Island combines the novelty of sailing into an active volcano with the possibility of swimming in thermally heated water. With the possible exception of Maxwell Bay on King George Island, Deception Island has little in common with the rest of Antarctica, and is dominated by derelict buildings and obsolete whaling works. The island, which lies some 9 miles (15 km) south of Livingston Island, is nearly 8 miles (13 km) in diameter and rises to a height of 1,975 ft (602 m) at Mount Pond. Deception Island's distinctive horseshoe shape stands out on charts. Its name, which has been in use since 1821, probably came about because the harbor mouth is easy to miss.

Port Foster comprises the whole flooded caldera within the island. It is named after Henry Foster of the *Chanticleer*, who had conducted some scientific experiments there in 1829. Entering the caldera through Neptune's Bellows—a gap so named because of the violent wind that sometimes blows across the mouth of the entrance—is invariably exciting. Ships must negotiate a gap of less than 1,310 ft (400 m),

Penguin Playground, Human Challenge

Chinstraps (far left) frolic on the black-sand beach of Baily Head, on the north side of Deception Island's caldera. Heavy sea surges test the skills of Zodiac drivers trying to land on this beach, which is flanked by layered rocks called "tuffs," consisting of basaltic ash and other volcanic materials.

Warmed by Volcanic Action

The *Akademik Ioffe* (left) sails across the rim of the caldera into Port Foster. Several sea stacks, like that partly obscuring the ship, are found on the island. The volcanic warmth ensures Deception Island has less snow cover than its neighbors.

hugging the northern side (where cape petrels often gather) to avoid a submerged rock less than 7 ft (2 m) below the surface, almost in the middle of the Bellows. Jean-Baptiste Charcot named Ravn Rock in 1910 after the whale-catcher *Ravn*, which was based in Port Foster at the time he surveyed the island. The wreck of the *Southern Hunter*, a British whaler that ran aground in 1957, lies on the southern side of the entrance.

Once through the Bellows, ships generally turn hard to starboard to anchor in Whalers Bay. Above the black volcanic sand beach, which is often shrouded in fumes and vapor from natural vents in the sand, another missing chunk of caldera wall forms Neptune's Window. On a day with reasonable visibility, the Antarctic Peninsula, 56 miles (90 km) away, can be seen from this vantage point near the remains of a small dry dock by the water's edge. Some say that the young American sealing captain, Nathaniel Palmer, who was the first to enter Port Foster, may also have been the first person

to catch a glimpse of the Antarctic continent—from Neptune's Window in 1820.

From 1906, the Norwegian–Chilean factory ship, the *Gobernador Bories*, used the bay as a whaling base. Britain claimed the island in 1908 as part of the Falkland Island Dependency, and when the Norwegian Hektor whaling company wanted to set up a whaling station in 1911, Britain gave it a 21-year lease. This stipulated that Hektor must dispose of the 3,000 or so whale carcasses that littered the beach; the almost complete absence of whalebones around the bay testifies to the company honoring its contract.

When the whaling station was closed in 1931, after falling prices and new technology made it redundant, the equipment used to extract oil from the bones was abandoned. This unlovely collection of rusting metal looks like a nightmare version of Frankenstein's laboratory. During the whaling days, some 45 men died in these parts, and were buried in a cemetery behind the station. The cemetery was engulfed by a mudslide during volcanic eruptions of the 1960s; an empty coffin ejected by the slide poignantly marked the spot until 2000. Then the coffin disappeared, but has reappeared in subsequent summers.

Some of the Whalers Bay storage tanks contained fuel, and were destroyed by Britain during World War II

A Stunning Prospect

From the red building on the shore, which is part of Camara Station, to the rocky point on the right, Half Moon Island provides a grandstand view of Livingston Island (on the far right).

Baily Head

The broad, meltwater stream bed that runs through the Baily Head chinstrap rookery on Deception Island provides a useful highway for birds. The 550-ft (168-m) headland provides welcome shelter for the penguins that crowd the valley's floor and slopes.

to prevent German vessels using Deception Island as a refueling depot. From February 1944, the British occupied some of the Norwegian buildings as Station B, which functioned continuously until December 5, 1967, when a volcanic eruption forced a rapid evacuation.

The first building that was occupied by the British expeditioners of 1944 burned down in September 1946. The whaling station dormitory was then renamed Biscoe House, after the Royal Navy officer who explored Graham Land in 1832, and the building was used until

1969. The slab to the east of Biscoe House once held a prefabricated accommodation block, which was erected in 1966 and named after Sir Raymond Priestley, the geologist on Scott's last expedition, who later directed Britain's Antarctic operations.

The British occupation of Deception Island came to a violent natural end in the late 1960s. Volcanic rumblings over the years—including one in 1923 that had caused the water in Port Foster to boil and strip the paint from ships moored there—had indicated that

Deception Island was far from quiescent. In December 1967, two big eruptions wiped out Chile's Presidente Pedro Aguirre Cerda Station, and they caused the British and the staff of Argentina's Decepción Station to evacuate, picking up the Chileans as they went.

The British returned the next year, but only two and a half months later, on February 23, 1969, a further eruption accompanied by mudslides left the station in ruins. The British finally surrendered the wrecked base to the elements; it has deteriorated considerably and may be removed in the near future. The area around the abandoned Chilean station is now a Site of Special Scientific Interest, and Decepción Station operates irregularly in summer. The only other base on Deception Island—Spain's Gabriel de Castilla Station—is occupied in the summer months.

The last building to the west at Whalers Bay is the old British Antarctic Survey (BAS) aircraft hangar, which was erected in March 1962; the wings and tailplane of a BAS de Havilland Otter that was destroyed in the 1960s eruptions were stored there until 2004. Beyond this landmark, a Site of Special Scientific Interest protects the mosses and lichens that are slowly re-establishing themselves near the lake. Between the lake and hangar, the runway that was hacked out for Antarctica's first powered flight—on November 16, 1928, by Australian aviator Hubert Wilkins—is still discernible. It included two bends, a couple of gullies, and the certainty of a ducking in Port Foster if the take-off failed.

Deep inside Port Foster is Pendulum Cove, named for Foster's magnetic and pendulum experiments in 1829. The old Chilean station there is out of bounds, and Pendulum Cove's sole attraction is the probability of warm water near the shore, which is created by subterranean thermal activity when the tide is right. This pool is rarely deep enough for a human body to be fully immersed, and the surrounding water retains its extreme polar chill.

Antarctic creatures find the water of Whalers Bay uncomfortably warm, and they are reluctant to venture inside. In late summer, a few fur seals, or even the occasional Weddell or leopard seal, may haul out on the beach, and there are always one or two chinstrap or gentoo penguins about.

A full experience of Deception Island's wildlife, however, depends on the sea being calm enough for a landing at Rancho Point, known always by its unofficial name: Baily Head. Baily Head is worth any effort to get there. Fur seals sometimes crowd the long, black-sand beach, and multitudes of chinstrap penguins make it their domain. The small valley near the high, rocky headland extends inland and becomes a two-way penguin superhighway during the summer months. The birds move with great agility to and from a huge bowl-like amphitheater, which is completely covered with nesting chinstraps. If the weather is particularly good, it is possible to walk from here across the snow-covered ridge and down into Whalers Bay. DM

Rocks and Fossils
Looking eastward into Walker Bay on Livingston Island, the rocky buttress in the background lies beyond the far end of the island's walking trail. The large rocks below the glacier on the right hold a collection of fossils and rock samples.

Penguin Policemen
Chinstrap penguins are called "police penguins" in Russian—a reference to the strap-like pattern under the chin, like a Russian policeman's cap. The noisiest of the smaller penguins, they aggressively defend their territory.

Other Sub-Antarctic Islands

Although they are isolated, bleak, windswept, and often ice-covered, the remote sub-Antarctic islands support a variety of wildlife, including a number of unusual plant species.

Scott Island

Scott Island, named for Robert Scott, is a mere dot in the ocean just inside the Antarctic Circle at the entrance to the Ross Sea, about 1,310 ft long, less than 660 ft wide and about 165 ft high (400 by 200 by 50 m). It was discovered in 1902 by Lieutenant Colbeck aboard Scott's relief ship, the *Morning*. Scott Island had its moment in the sun—both literally and figuratively—in 1999–2000, when it was the first land on the Earth to witness the dawn of the new millennium.

Balleny Islands

Five islands and several islets form the Balleny group, straddling the Antarctic Circle south of New Zealand, and scattered across 100 miles (160 km). Sturge Island, the southernmost, is the largest, at 20 by 4 miles (32 by 7 km), and also the highest, with Brown Peak rising to 5,595 ft (1,705 m). Buckle Island lies midway along the group and measures 13 by 3 miles (21 by 5 km). Borradaile Island, to the north, is much smaller— 2 by 1 mile (3.2 by 1.6 km)—but is still larger than nearby Row Island, which is only about 1 mile (1.6 km) in diameter. At the northern end of the group, Young Island measures 19 by 4 miles (30 by 7 km). Tiny Sabrina Island is reserved for nesting birds—landings cannot be made there without special permission—but the shores of all the Balleny Islands are so sheer that few people have ever set foot on any of them. "One sight in bad weather of that sinister coast is enough to make a landsman dream for weeks of shipwrecks,

perils, and death"—so wrote Louis Bernacchi, after he sailed past the islands in 1900 on the *Southern Cross* expedition with Carsten Borchgrevink.

Campbell Island and the Auckland Islands

New Zealand has five sub-Antarctic island groups. The ones most often visited by ships sailing to the Ross Sea are Campbell Island and the Auckland Islands. These are basaltic volcanic islands, covered in peat, that stand on the submarine Campbell Plateau to the north of the Polar Front. They are cold, wet, and windy, but are ice-free (although shaped in the past by glaciers), and can support trees. Indeed, they have a wealth of vegetation that exists nowhere else, including many endemic vascular plants and the world's southernmost forests.

Campbell Island is the most southerly of the group. It lies about 435 miles (700 km) south of Bluff, the fishing port at the bottom of New Zealand's South Island. Campbell has an area of 44 sq miles (114 sq km), which is considerably less than main Auckland Island's 197 sq miles (510 sq km).

The Auckland Islands were first sighted on August 18, 1806, by Abraham Bristow, captain of the whaler *Ocean*, who named them after Lord Auckland: "my friend through my father." Later visitors included the Wilkes and Dumont d'Urville expeditions, only days apart, and Joseph Hooker, the celebrated naturalist, who visited in 1840 on the expedition led by James Clark Ross. Captain Frederick Hasselburg of the *Perseverance* (who first reported Macquarie Island) discovered Campbell Island on January 4, 1810, and named it after the owner of his sealing company. The seal populations of both island groups were decimated soon after their discovery.

Rising from the Sea
Haggits Pillar is Scott Island's only neighbor. The island was discovered on Christmas Day 1902 by the crew of the *Morning*, who thought that it "might have been a large discolored berg ... but it was soon observed to be a typical Antarctic Island."

Volcanic Islands
Almost completely covered in ice, the volcanic Balleny Islands are cloaked in cloud for much of the year. During the wintertime, pack ice surrounds these islands, which are seismically active still—a major earthquake was recorded here in recent years.

Albatross Havens
Auckland Island (foreground and left distance) and Adams Island (right) are the largest of New Zealand's sub-Antarctic islands. Adams is home to large breeding populations of great albatrosses, and several colonies of white-capped albatross breed on its steep slopes and cliffs.

Southern Forests

Clad in mosses and lichens, the southern rata forests of the Auckland Islands are unique. Rare species such as Hooker's sea lions and yellow-eyed penguins can be seen looking out from the forest, and overhead royal albatrosses soar toward their breeding areas on the nearby upland moors.

In 1849, Charles Enderby, of the famous British whaling family, set up an unsuccessful settlement at Port Ross, at the northern end of Auckland Island, but nothing of it remains today. Many early visitors were shipwrecked sailors, and wreck sites dot the coast. Both Campbell Island and the Auckland Islands were farmed, and native wildlife on many of the islands suffered as a result of introduced species. Rat eradication on Campbell Island has been spectacularly successful.

The Auckland Islands are home to the world's largest breeding populations of great albatrosses and shy albatrosses, and well over 90 percent of the rare Hooker's sea lion population also breeds there. Campbell Island has the largest breeding population of royal albatrosses; from the meteorological station at Perseverance Harbor, a boardwalk track leads up to the Col-Lyall Saddle, where the birds nest.

All the New Zealand islands are designated nature reserves; landing permits are required and visitors must be accompanied by a representative of the Department of Conservation. Visitors to Auckland Island can inspect the old settlement site at Erebus Cove, Port Ross. But most prefer nearby Enderby Island for the variety of rare wildlife there, and for the chance to walk through the mysterious rata forest with its bright red flowers—this is

the last opportunity to stand underneath a tree before Antarctica. Stella Hut, at Sandy Bay, is a small hut once used as a supply depot, and nearby are several modern huts that scientists use during the summer. The beach nearby is occupied by a colony of sometimes aggressive Hooker's sea lions, and vigilant observers may see the most elusive penguin of all, the extremely timid yellow-eyed penguin. Royal albatrosses nest on the slopes behind the rata forest.

Macquarie Island

Long, thin Macquarie Island—21 by up to 3½ miles (34 by 6 km)—lies about 930 miles (1,500 km) south of Tasmania, just north of the Polar Front and about halfway to the Antarctic continent. Its 50 sq miles (128 sq km) form the most remote state reserve of Australia's island state of Tasmania. The ocean around the island became an Australian marine park in 1999.

Like most other sub-Antarctic islands, Macquarie Island has no trees. It has a mean annual temperature of around 41°F (5°C), and some form of precipitation on 300 days a year. Its World Heritage listing in 1997 was largely geologically based, because it is the only island that consists solely of rocks from deep within the Earth's mantle and oceanic crust. It emerged from the

Sleeping Beauty

A white-capped albatross snoozes on its nest on Auckland Island. The lavish plant growth around it shows that some vegetation thrives in this challenging environment. The Aucklands group is renowned for its large flowering plants: here, the pink *Pleurophyllum speciosum* flowers just above the albatross.

sea some 600,000 years ago as the crest of a tall ridge, and continues to rise a fraction of an inch each year. Unlike the other sub-Antarctic islands, it is a plateau shaped by waves rather than by ice.

The first reports of Macquarie Island were given by Frederick Hasselburg, the captain of the sealer *Perseverance*, which arrived here on July 11, 1810. Hasselburg recorded the wrecked vessel "of ancient design" of an earlier visitor: the identity of this voyager has never been discovered. Hasselburg named the island after Lachlan Macquarie, then the governor of New South Wales.

Subsequently, every seal was taken for skins and oil, and boiling down penguins for oil continued well into the twentieth century. The destruction of the native wildlife continues from the rats, mice, cats, and rabbits that the sealers left behind.

The seals have returned—currently the island has a large population of elephant seals and three types of fur seals: New Zealand, Antarctic, and sub-Antarctic.

Penguin numbers have also recovered—there are now 850,000 breeding pairs of royal penguins, 5,000 pairs of gentoo penguins (the only ones found on the Pacific side of Antarctica), 100,000 pairs of king penguins, and large colonies of rockhopper penguins. Among other birds are several species of albatrosses and some 800 pairs of the Macquarie Island cormorant, an endemic subspecies of the king cormorant.

Bellingshausen called at Macquarie Island in 1820, Wilkes in 1840, Scott with the *Discovery* in 1901, and Shackleton in the *Nimrod* on his return from Antarctica in 1909. Douglas Mawson arrived in the *Aurora* in 1911 and established a radio relay station on Wireless Hill at the northern end of the island, which was used to send messages between the base at Commonwealth Bay and Australia. Australia's Macquarie Island Station was built on a spit below Wireless Hill on March 25, 1948, and has since operated continuously. Affectionately known as Macca, the station houses about 20 people during winter and 50 in summertime.

Pretty as a Picture
The pink flower heads of *Pleurophyllum speciosum* grow to the height of a small child, and its large corrugated leaves, which are designed to act as solar panels to soak up the sun's rays, are the size of serving platters.

One of a Kind
Campbell Island shags (above) belong to a large family of cormorants, many of which are confined to isolated islands of the sub-Antarctic. These shags are fine swimmers, hunting in large groups in sheltered bays or far out to sea, and covering great distances underwater in search of food.

A Diving Diva
Hooker's sea lion (right) is one of the rarest pinnipeds (mammals with feet that look like fins) in the world. It has a highly localized distribution—more than 90 percent of the population breeds on the Auckland Islands—and is vulnerable to any kind of disturbance. These animals are superb divers, and can plunge to depths of over 1,475 ft (450 m) in quest of their prey on or near the sea floor.

Heard and McDonald Islands
Heard Island lies 2,500 miles (4,000 km) southwest of Perth in Western Australia. The small, ice-free McDonald Islands are 25 miles (40 km) to the west of Heard, and the even smaller Shag Islands are 6 miles (10 km) to the north of it.

Like the much larger Iles Kerguelen, which lie 275 miles (440 km) to the north, these islands stand on the undersea Kerguelen Plateau. They were annexed by Britain in 1908, then transferred to Australia in 1947. They remain external territories of Australia.

Heard Island is roughly circular, and has an area of 142 sq miles (368 sq km), about 80 percent of which is ice-covered; much of the shoreline consists of ice cliffs. The island's major feature is the cone of Big Ben, Australia's only active volcano, topped by Mawson Peak, which, at 9,005 ft (2,745 m), is considerably higher than any other mountain in Australia.

Even in Antarctic terms, human contact with Heard Island is very recent. The first person to see the island is believed to have been the British sealer, Peter Kemp, aboard the *Magnet* on November 27, 1833. Two decades later, on November 25, 1853, the American sealer, John Heard of the *Oriental*, saw the island and subsequently had it named after him. Only a few weeks after that, in January 1854, William McDonald, the British captain of the *Samarang*, discovered the islands that now bear his name—the largest of them covers less than ⅓ sq mile (1 sq km). The first people to land on Heard Island were the crew of the American sealer, the *Corinthian*, captained by Erasmus Rogers, in 1855. Sealing reached a peak here in 1858, but by 1880 too few seals remained for viable hunting.

The *Challenger* collected scientific samples in 1874, and Drygalski visited in the *Gauss* in 1902. Mawson stopped for a week on the BANZARE voyages in 1929; the remains of his hut survive at Atlas Cove. Australia established a base on this site in 1947, but it closed in 1955 and visits since then have been sporadic. Since 1997, Heard Island has been a World Heritage-listed Wilderness Reserve. An Australian team cleaning up the old base in 2000–01 found that the glaciers have retreated and vegetation has dramatically increased.

These islands house the rare Heard shag and a subspecies of sheathbill, as well as more than 2 million macaroni penguins and a rapidly increasing population of king penguins—now more than 100,000 breeding pairs. These are also the only sub-Antarctic islands without any feral species. Because landings are difficult and the weather terrible, the Australian government's limit of 400 visitors each year has never been reached.

Iles Kerguelen
In 1771, Captain Yves-Joseph de Kerguelen-Trémarec sailed south with instructions to establish trade with the natives of the great southern continent. The ambitious Frenchman found neither. But on February 12, 1772,

A Thriving Community
Megaherbs Stilbocarpa polaris (the bright green leaves) and *Pleurophyllum hookeri* (gray-green leaves), along with the tussock grass, *Poa foliosa*, grow on Finch Creek on Macquarie Island.

he caught sight of a foggy, volcanic archipelago. He put some of his crew ashore briefly to claim the island for France, and sailed back to Mauritius with a fanciful tale of a land perfect for settlement which he had named Southern France. But he had miscalculated its position. In 1773 Kerguelen was sent back, with three ships and 700 men, to colonize his discovery. Again, he did not step ashore. Those who did soon decided that the land was worthless, and the fleet sailed on to Madagascar. When Kerguelen returned to France with this much more accurate report of the island's potential, he was court-martialed and imprisoned in Saumur, in the Loire Valley, for nearly four years.

When Cook came to Kerguelen in the *Resolution* and the *Discovery* on Christmas Day 1776, he claimed it for Britain, writing that: "I could have very properly called the island Desolation Island … but in order not to deprive M. de Kerguelen of the glory of having discovered it, I have called it Kerguelen Land."

The islands are now part of the French Southern and Antarctic Lands (Terres Australes et Antarctiques Françaises). They consist of the main island, measuring 75 by 68 miles (120 by 110 km), and some 300 small islets and reefs. Deep fiords thrust into the coast of the main island; the interior is heavily glaciated, and rises to Mount Ross, 6,070 ft (1,850 m) above sea level. The islands lie just to the north of the Polar Front, and the

weather there is typically wet and windy, although the sea remains ice-free.

Between 1791 and 1817, hunters wiped out the fur seal population, and then switched their attention to the elephant seals and whales. An attempt to farm sheep failed, but feral sheep now roam the island; at various times rabbits, cats (to control the rabbits), reindeer, cattle, rats, and even salmon and trout have been introduced. During World War II, German ships visited the island's fiords until Australia laid mines at some of the anchorages. France built Port-aux-Français, on the east coast, in the summer of 1949–50. Today it is a large, well-appointed base, with accommodation for 80 people in winter and 120 in summer and facilities that include a cinema, a sports center, and a chapel. It operates alongside a CNES (Center National d'Etudes Spatiales) satellite tracking station.

Splendid Isolation
A narrow isthmus links Macquarie Island to North Head, where Australia's National Antarctic Research Expedition base was built in 1948. Heavy seas sometimes breach the isthmus, leaving the small station even more isolated.

New Furs for Old
Elephant seals haul out onto Macquarie Island's Sandy Beach for their yearly molt. By huddling and rubbing against each other, the seals shed their old coats.

The most famous of the island's plant species is the Kerguelen cabbage (*Pringlea antiscorbutica*), which, as its scientific name suggests, contains the vitamin C needed in the human diet to ward off scurvy. The plant belongs to the same family as the common cabbage. It was discovered by James Cook and the first account of it was published by Joseph Hooker in the 1840s. The plant has adapted to wind pollination, rather than to insect pollination, because there is a lack of winged insects on the islands.

Iles Crozet

The Iles Crozet are the top of an underwater volcanic plateau; their highest point is Pic Marion-Dufresne, 3,575 ft (1,090 m) high, on Ile de l'Est. Like the Iles Kerguelen, these unglaciated islands belong to the Terres Australes et Antarctiques Françaises. The chain is made up of an eastern group that comprises Ile de la Possession, the largest of the whole group at 58 sq miles (150 sq km), and Ile de l'Est, and a western group 75 miles (120 km) away, which consists of the Ile aux Cochons, measuring 26 sq miles (67 sq km), and the much smaller Iles des Pingouins and Iles des Apôtres, together with other tiny rocks and shoals.

The French captain, Marion du Fresne, on board the *Mascarin*, discovered the Iles Crozet in 1772 and named them for his second-in-command, Jules Marie Crozet, who took over when du Fresne was killed by Maori in New Zealand in June 1772. For many years, the only human visitors to the islands were occasional sealers and shipwrecked mariners. Whaling stations and scientific bases were built on several of the islands, but now only Alfred-Faure Station, above Port Alfred on Ile de la Possession, is occupied year-round. The islands, declared a national park by France in 1938, have the largest colonies of king penguins in the world, with perhaps a third of the world's population of this species breeding there.

Bouvetøya

Bouvetøya is the tip of an inactive volcano that pierces the South Atlantic; at Olavtoppen, its highest peak, it is 2,559 ft (780 m) tall. About 3 by 5 miles (5 by 8 km), and 19 sq miles (49 sq km) in area, it is regarded as the world's most isolated landmass because, except for the diminutive Larsøya Island nearby, the closest land is the Antarctic continent, some 1,000 miles (1,600 km) to the south. The ice sheet that covers Bouvetøya ensures that nothing grows there. Wilhelm II Plateau, an ice-filled volcanic crater, dominates the center of the island, and the sheer sea cliffs of rock and ice make any landing very difficult.

This cloud-capped island was first sighted by Jean-Baptiste-Charles Bouvet de Lozier on New Year's Day 1739. Hoping that he had seen the tip of a southern continent, he named the northwestern point Cape Circoncision after the feast day it was discovered on. But after battling fog and ice for 12 fruitless days, he gave up hope of landing, or even circumnavigating the point to see if it really was a cape joined to a larger landmass. He made a mistake with the island's coordinates and placed it too far east, so that James Cook subsequently failed to find it in 1772 and 1775. American sealer Benjamin Morrell claimed to be the first to land there in 1822.

The island's position was accurately charted by Captain Lindsay, a whaler with the Enderby Company, in 1808, and George Norris, another Enderby captain, made a documented landing in 1825, claimed the island for Britain, and renamed it Liverpool Island.

Landings are so difficult that Bouvetøya has been largely left alone, and remains uninhabited. However, due to its interest in Antarctic whaling, Norway annexed the island in 1928. Britain relinquished its claim, and Norway declared the island a nature reserve in 1971. After several attempts to erect a weather station, the first automatic station was built there in 1977. DM

Endangered Habitat
The Iles Kerguelen, the largest areas of dry land in the southern Indian Ocean, attract more than 30 seabird species to their rocky shores. Some introduced species, particularly rabbits, have made life difficult for native fauna, especially burrowing petrels, whose nesting habitat has been greatly reduced.

A Great Liar
The enigmatic Benjamin Morrell was an American sealer who traveled widely in the sub-Antarctic islands and lived for some time on the Iles Kerguelen. He claimed to have been the first person to land on Bouvetøya, in 1822, but he later made other outrageous claims, and one contemporary called him "the biggest liar in the Southern Ocean."

Antarctic
Wildlife

Plants and Invertebrates

For land plants, life becomes harder the further south they grow. Plants can grow only where there is enough liquid water. They must also contend with freezing conditions, prolonged darkness, and extremes of light. As the latitudes rise, the plant life becomes less varied and abundant, with fewer species, fewer major plant groups, a lower biomass, a smaller proportion of ground covered by plants, and a shorter average plant stature. These changes resemble those along a moisture gradient from well-watered regions into arid deserts—a closer analogy than it might seem, because the Antarctic region is extremely dry. As well, low temperatures and extreme fluctuations in light levels at the high latitudes affect chemical processes such as photosynthesis, and plants are forced to take protective measures against these stresses.

Life on the Edge
In the dry, cold Antarctic environment, lichens (the circular dark patch on the nearest rock) benefit from the warmth absorbed and later released by dark rock surfaces. Snow melting at the rock surface provides much-needed moisture.

The Antarctic Zones

In terms of plant and animal life, the Antarctica region can be divided latitudinally into three distinct zones.

The sub-Antarctic islands are made up of six island groups near the Polar Front: Marion and Prince Edward islands, Iles Crozet, Iles Kerguelen, Macquarie Island, McDonald and Heard islands, and South Georgia. These islands support abundant but treeless vegetation, verdant in the summertime but brown when foliage dies in the winter. The islands to the south and southeast of New

Zealand (Snares, Auckland, Campbell, Antipodes, and Bounty islands) support similar vegetation, with the addition of upright shrubs and trees; they are in some contexts also called sub-Antarctic. All the major plant groups except cone-bearing trees (gymnosperms) occur in the sub-Antarctic zone. Microscopic invertebrates and larger invertebrates, such as snails, earthworms, flies, beetles, and spiders, are plentiful on the islands.

Maritime Antarctica consists of the west coast of the Antarctic Peninsula and its offshore islands, together with the islands south of the maximum extent of the sea ice: South Sandwich, South Orkney and South Shetland islands, and Bouvetøya. In the maritime zone, mosses and liverworts (bryophytes) are reasonably abundant and widespread; however, only two species of non-woody flowering plants (soft-stemmed angiosperms) grow, and then only sparsely. Microscopic invertebrates and larger invertebrates, such as springtails and free-living mites, live in soil and among rocks and vegetation.

Continental Antarctica is the "Great Circle" of East and West Antarctica—the Antarctic continent, with the exception of the west coast of the peninsula. Near the coast of continental Antarctica, mosses, lichens, algae, and liverworts can grow, but are generally sparse. On the high East Antarctic Ice Sheet at the South Pole itself, there are probably no plants at all. Microscopic invertebrates—protozoans, platyhelminths, tardigrades, nematodes, rotifers, and crustaceans—live only where moisture is available on land and in lakes, but there are no land-based vertebrates in the continental Antarctic zone: though seals, penguins, and petrels come ashore to breed, they must live and feed at sea.

Marine Plants of the Antarctic

For plants, the conditions in the ocean at high southern latitudes are much more favorable than on land.

In the intertidal zones and shallow waters off sub-Antarctic islands, some of the world's largest marine algae form dense kelp communities of high biomass. And even in the colder, shallow waters off continental Antarctica—below the depth of the sea ice and where the sea bed is not scoured by icebergs—the biomass, stature, and diversity of marine algae are greater than that of land plants.

Marine plants receive less light than land plants do because light intensity is much diminished by passing through water; this suggests that the low temperatures during the summer growing season and lack of liquid water are much more significant factors than low light intensity in the sparsity of Antarctic land vegetation.

Sub-Antarctic Islands

The sub-Antarctic islands vary greatly in altitude and area covered by permanent ice. They range from the unglaciated Macquarie Island, 1,420 ft (433 m) high and often fondly called "the great green sponge," to the heavily glaciated Heard Island, which reaches 9,006 ft (2,745 m) at its tallest point.

Growing Near Freezing Point

Many plants go on metabolizing and growing, albeit slowly, at temperatures well below 32°F (0°C). The moss *Bryum argenteum* photosynthesizes to at least 27.5°F (−2.5°C) and the lichen *Umbilicaria aprina* to at least 1.5°F (−17°C). Although water freezes at 32°F (0°C), dissolved organic and inorganic molecules, ions, and hydrated macromolecules lower the freezing point of cellular fluids in Antarctic plants.

Most Antarctic invertebrates avoid freezing by supercooling, surviving temperatures as low as −40°F (−40°C). The ability to supercool is achieved by having solutes, such as alcohols and sugars, in body cells. The mite *Maudheimia wilsoni* can withstand temperatures as low as −22°F (−30°C) without freezing. Alternatively, they may be freezing-tolerant, freezing at relatively high subzero temperatures, but because they have ice nucleators and cryoprotectants the tissues are not damaged. The nematode *Panagrolaimus davidi* is freezing-tolerant to −40°F (−40°C), and the rotifers *Philodina gregaria* and *Adenita grandis* to even lower temperatures.

In summer, dark-colored surfaces become surprisingly warm. The surface temperature of rocks facing the midday sun can rise to almost 86°F (30°C). The daily heat penetrates rocks to a depth of more than 3 ft (1 m), and is slowly released at night when the sun is low in the sky or just below the horizon. Plants growing near these rock surfaces benefit from the warmth, but such sites may also dry out very rapidly. In a large rocky oasis that has little permanent snow, such as the Vestfold Hills, the whole area has a slightly, but significantly, higher summer temperature than the neighboring ice-covered coastal edge of the East Antarctic Ice Sheet.

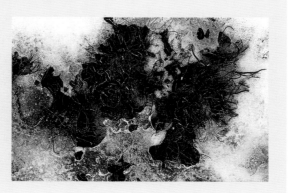

Large and Lush

Near sea level on Auckland, Campbell, and Macquarie islands, *Stilbocarpa polaris* grows to about 3 ft (1 m) tall. By midsummer, showy yellow flowers are abundant; at the end of the season, the seeds fall to germinate the following summer.

Antarctic Carpet

Lush growths of moss occur at sites where moisture and temperature conditions are favorable and there is some protection from wind-blown particles. Individual moss stems grow close together, forming a carpet that becomes wrinkled, folded, and broken by repeated cycles of freezing and thawing.

At sea level, the islands' climate reflects that of the surrounding Southern Ocean. Average air temperatures vary by only a few degrees from summer to winter, and they receive frequent precipitation as rain or snow/ice, so that plants have no extended periods without water.

From sea level up to an altitude of approximately 660 ft (200 m), the ground on the sub-Antarctic islands is completely covered by herbaceous flowering plants, bryophytes, and some ferns, except in places where there are rocky bluffs, screes, moraines, mobile sand sheets, fresh lava flows, landslides, or disturbance by penguins and seals. The tallest vegetation, which grows to a height of 7 ft (2 m), comprises tall tussock grass and large-leaved megaherbs (dicotyledons). In less favorable conditions, there is short herb vegetation about 1 ft (30 cm) high, including small dicotyledons, grasses and sedges (monocotyledons), mosses, and ground-hugging cushion plants.

At the higher altitudes, the lower temperature and increased exposure to wind reduce both the percentage of ground covered by plants and plant stature. Below 330 ft (100 m) on Macquarie Island, for example, the large herb *Stilbocarpa polaris* may have leaf stalks that

Feldmark Communities

Where bare ground exceeds 50 percent, polar (and alpine) vegetation is called feldmark (above). Vegetated areas may be scattered, or they may form stripes or steps-and-stairs terracing that marches spectacularly across hillsides roughly parallel to the contours.

Intertidal Giants

Luxuriant beds of bull kelp (*Durvillaea antarctica*, right) are attached to rocky sub-Antarctic shores in the intertidal zone. Many invertebrates live among the holdfasts that attach these plants to the shore. On treeless islands, these marine algae are the largest plants.

are 3 ft (1 m) long, each supporting a leaf blade up to the area of three dinner plates. However, at 1,000 ft (300 m), *Stilbocarpa* plants are rare, and they grow slowly to about 8 in (20 cm) in diameter only where they are sheltered from abrasive wind-driven sand, gravel, and ice pellets.

On islands that are glaciated, ice sets a limit to the distribution of plants, but in recent decades the ice has been retreating and much new land has become available for colonization by plants.

With the increasing altitude and the abrasion from winds, the proportion of flowering plants drops and the proportion of bryophytes rises. *Azorella selago* or *A. macquariensis* and other cushion-form flowering plants become most common—their tight mat of small, tough leaves is resistant to the wind and wind-driven abrasives, and their extremely long contractile roots, which radiate shallowly in the soil, keep them anchored securely. Under the most extreme conditions, a surface of apparently bare gravel may be held together by a fine but dense, threadlike network of liverworts, with the plants' minute, reddish leaves sheltering in the crevices between the small stones.

Maritime Antarctica

The Antarctic Peninsula represents a geographical and botanical transition zone between the far south of South America (Tierra del Fuego to Cape Horn) and the great circular sweep of East and West Antarctica. Although a fearsome barrier, the Drake Passage is the narrowest ocean gap between Antarctica and other landmasses plants might cross. The peninsula is a mountainous, glaciated spine, essentially devoid of plants except near sea level, and most of its offshore islands are of similar physiography. The western fringe of the peninsula near sea level is a relatively favorable environment for plants; the eastern fringe is colder.

The only two native flowering plant species growing in Antarctica—the small *Deschampsia antarctica* (hair grass) and the little cushion plant species, *Colobanthus quitensis* (pearlwort)—are to be found on the western

Cushion Plants

Several species of *Colobanthus*, known as pearlwort, grow as small, compact clumps on the sub-Antarctic islands. *Colobanthus quitensis* is one of only two species of flowering plants growing in maritime Antarctica.

Antarctic Plants and Invertebrates

Current estimates of numbers of species

PLANTS	Sub-Antarctic Islands	Maritime Antarctica	Continental Antarctica
Flowering plants	56	2	0
Ferns and relatives	16	0	0
Mosses and relatives	~400	~100	~30
Lichens	~300	~150	~125
INVERTEBRATES			
Millipedes (Myriapoda)	3	0	0
Mites (Acari)	52	24	21
Flies (Diptera)	44	2	0
Springtails (Collembola)	92+	?	24
Crustaceans (Crustacea)	41	10	14
Snails (Mollusca)	3	0	0
Annelid worms (Oligochaeta)	4	0	0
Nematode worms (Nematoda)	22	40	10
Tardigrades (Tardigrada)	40+	17	6
Rotifers (Rotifera)	102	46	41

Color for Protection
Lichen-covered rocks on Torgersen Island in the Antarctic Peninsula. The striking colors of Antarctic lichens—yellow, orange, and black—and of some Antarctic mosses—bronze and black—come from sun-screening pigments produced by the plants to protect their photosynthetic apparatus from damage by light of high intensity and by ultraviolet light.

side of the peninsula. By contrast, in Tierra del Fuego, just north of the Drake Passage, there are 417 species of vascular plants, of which 386 are flowering plants.

Antarctica's two species of flowering plants are found as far south as 68°S; at a similar latitude in the Northern Hemisphere, Iceland supports 329 species, and a further 10 degrees north, some 110 species of flowering plants grow in Spitsbergen. The suitability of soil and temperature, and adequate moisture, define the distribution of Antarctica's hair grass and pearlwort. During recent climate changes, their populations have

increased—a trend that is likely to continue with future warmer summer temperatures.

On the eastern side of the Antarctic Peninsula, and on Deception Island and Ross Island, volcanic activity encourages the growth of some species of bryophytes, which are otherwise known only from further north. Most sites are unsuitable for temperate species, but it appears that there is a constant rain of plant fragments over Antarctica from warmer regions to the north, and some of these plants manage to become established on steam-warmed ground where the temperature is right and moisture is available from condensed steam.

Continental Antarctica
In East and West Antarctica, plants are confined to snow-free and ice-free land, including mountains that project through the ice sheets, and to the land around the coastal fringes of the continent and of the Filchner, Ronne, Ross, and Amery ice shelves. In Victoria Land, volcanically heated sites can support bryophytes.

Apart from these areas, the only known plant forms in the continental Antarctica zone are algae, which grow on snow and ice, feeding on mineral nutrients from the snow and its trapped dust, and coloring the snow gray, green, orange, pink, or red.

Biologically, the most productive areas of Antarctica are the few small, ice-free patches of land around the coastal fringe. These areas, which are called oases, are breeding areas for marine animals, which carry ashore plant nutrients, particularly phosphorus. Humans also use the oases—for building bases and for tourism—and in the process bring in pollutants and substantially disrupt the local ecology.

Certain environmental conditions favor plant growth. Water is available at sites where there is adequate snow,

a northerly aspect that assists melt, and a micro-topography that channels and concentrates or retains meltwater. Heat is retained by rock surfaces when the sun is high in the sky and it is slowly released as the sun circles low each summer day. Also necessary are supplementary nutrients, such as nitrogen, provided by nitrogen-fixing lichens and cyanobacteria which are usually components of Antarctic vegetation, and the phosphorus brought from the ocean by marine birds.

Vigorous moss-dominated vegetation forming a green carpet is usually found in scattered small patches or in strips only, often following drainage lines. Extensive areas of abundant moss and/or lichen—exciting finds for botanists—are much less common. Two such areas in continental Antarctica are in the Windmill Islands oasis around Casey Station, and at Granite Harbor in southern Victoria Land.

By contrast, some areas have a remarkable scarcity of plants. In the Vestfold Hills, an oasis of 155 sq miles (400 sq km), a coastal strip about 6 miles (10 km) wide has very little vegetation, but there is abundant mossy vegetation farther inland, toward the edge of the ice sheet. Plant growth in the coastal strip is restricted by salinity, which was inherited over 5,000 years ago when ocean covered parts of this zone.

In Antarctica, as in hot deserts, the ultimate refuge of plant life is below translucent rocks or within the tiny pores in rocks. Here green algae and lichens grow in thin layers, receiving some light, warmth from the rock, and more moisture than they would in the open air.

The invertebrates that live among the Antarctic vegetation feed on dead plant material, soil bacteria, algae, and fungi. The carbon dioxide exhaled by the invertebrates and bacteria living among the stems of the Antarctic moss microforests increases the rate of photosynthesis in the moss leaves, thus enhancing the production of sugars during the brief growth period over the Antarctic summer.

Lakes are common in ice-free areas, the best known being those in the Dry Valleys of southern Victoria Land and in the Vestfold Hills of East Antarctica. Each of the lakes contains a population of microorganisms: protozoa, green algae, fungi, and cyanobacteria, as well as other

Tinted Snow

Unlike most Antarctic plants, snow algae are able to grow in snow itself, gleaning moisture from the snow as it melts. Pigments in the snow algae cells trap light for photosynthesis. En masse, the pigments color the snow—here, a pink hue.

Obtaining Water in the Driest Continent

Although Antarctica's volume of ice is huge, low precipitation and considerable evaporation into the dry atmosphere make it the driest continent. Plants grow where liquid water is available in summer, but the very low relative humidity dries them out quickly when there is no water. Most vascular plants (flowering plants, conifers, ferns) keep their cells moist by having leaves covered with a waterproof cuticle, and by controlling variable openings in leaf surfaces, called stomates. For vascular plants to produce the necessary sugars by photosynthesis, carbon dioxide gas, a raw material, must enter leaves via open stomates. But open stomates mean that water vapor can escape from moist cells inside to the dry air outside, leaving the plant vulnerable to death by desiccation in the absence of a water supply.

Mosses and lichens have no cuticles or stomates and desiccate quickly without water. But many of the mosses and lichens in Antarctica survive repeated cycles of wetting and drying by

suspending their metabolic activity when desiccated and resuming it when rehydrated. They also take advantage of any available water and light—for example, when a dusting of snow over their bodies melts in the midday warmth. Plants with this ability can exist outside the polar regions as well; they are known as resurrection or poikilohydrous plants.

The only plant communities known on the ice-sheet surface are transient patches of snow algae, found at low altitudes where some summer melt occurs. But the whole of the ice sheet contains airborne microscopic spores of fungi and other single-celled organisms, such as bacteria, and no doubt tiny reproductive structures and plant fragments—all of them probably trapped in snow in the same way dust particles occur throughout the ice sheet, and proving valuable markers of global arid periods in the past. These organic particles deep in ancient Antarctic ice may be another useful archive of data about the Earth's past environments. These great ice sheets also cover and seal lakes of liquid water, the largest being Lake Vostok. Fossils or living organisms occurring in their waters and sediment await discovery.

The Future of Antarctic Life Forms

The high-latitude environments are undergoing change, but the consequences for polar organisms are difficult to predict. Global climate change is bringing warmer

Communal Living

Around Adélie penguin nests, the green alga *Prasiola crispa* benefits from the nutrients in guano. It is one of the few plants that survives the continuous light of summer and months of winter darkness in continental Antarctica.

prokaryotes (simple organisms that survive in conditions which are too extreme for other forms of life, including volcanic soils with high concentrations of heavy metals, volcanically heated soils and water, and the extreme cold of Antarctic rock outcrops). The mix of organisms in the lake depends upon the chemical and physical properties of its water: some lakes are far more saline than seawater, others are as pure as distilled water; some are well mixed, others are stratified, with stable layers of different temperatures, light levels, oxygen content, and other chemical properties. Some non-saline, well-mixed lakes contain underwater forests of moss more than 3 ft (1 m) tall growing from the lake bed below the reach of the floating winter ice. The moss, *Bryum pseudotriquetrum*, is normally terrestrial, where it grows to a height of only a few inches. Sheets of photosynthetic cyanobacteria commonly form on lake beds, thin layer upon thin layer, forming mats that are 4 in (10 cm) thick or more; if lithified (turned to rock), they would be called stromatolites. Dried fragments of these sheets, like pieces of torn paper, accumulate as lines of flotsam on the shores of Antarctic lakes.

Coping with Light Fluctuations

Antarctic plants have to cope with low light levels, or even no light at all. They do this by storing enough carbohydrates to meet their very low respiratory needs during the lightless winter, thereby keeping their photosynthetic machinery intact during prolonged darkness. Photosynthesis resumes when the sun rises above the horizon and the plants rehydrate, even when buried by snow. Low light merely prolongs the no-growth period of winter, or causes very slow growth.

Extremely high summer light, with its ultraviolet component, is a more substantial problem, apparently exacerbated by the combination of high light levels and low temperatures. Light absorption by the photosynthetic pigments (chlorophylls) occurs equally well at high and low temperatures, and generates energized, highly reactive states in these pigment molecules. This energy is used in the biochemical reactions that convert carbon dioxide gas and water into sugars. However, these biochemical reactions slow down as temperature falls, and the unused energy damages the pigment complexes that absorb it, so that high light at low temperatures may inhibit or harm plants' photosynthetic machinery. High ultraviolet levels in the incoming light, especially when there is severe ozone depletion in spring and early summer, compound the problem. Antarctic plants, and many elsewhere, respond by secreting large amounts of colored sun-screening chemicals that absorb much of the damaging light.

temperatures, changed precipitation, and changed ice cover. The depletion of the ozone layer over both the Antarctic and Arctic has resulted in increased levels of ultraviolet light.

On sub-Antarctic islands, introduced plants and animals are other threats: introduced plants compete with native ones, often growing faster in disturbed habitats; wild rabbits damage vegetation and promote soil erosion; cats prey on ground-nesting birds; and invertebrates that have been accidentally introduced in building materials influence native vegetation by consuming litter and live plants and competing with the indigenous species.

Human activity has also altered Antarctic habitats. Toxic waste and fuel spills from scientific bases cause serious habitat destruction for Antarctic animals and plants. And, finally, tourism may become a problem: tourism is now ship-based, but serious environmental damage to plant and invertebrate habitats is inevitable if tourism ever becomes largely land-based. PS

Unusual Adaptations
Beneath the frozen surface of an Antarctic lake (right of photograph) is liquid water in which populations of microorganisms live. Some of them photosynthesize, some ingest other organisms (heterotrophs), and some photosynthesize during summer when light is available but are heterotrophic during the dark winter.

Epic Landscape
Glaciers cover 80 percent of Heard Island (above). Its topography ranges from ice-capped volcanic peaks to lush sub-Antarctic vegetation growing on rocky cliffs, sand and gravel beaches, peninsulas, and lagoons.

Maritime Diversity
Unlike most of the Antarctic Peninsula zone, Petermann Island's flora (left) is relatively diverse. The flowering plant *Deschampsia antarctica* grows here, as do green and snow algae, and several moss and lichen species.

Life in the Southern Ocean

Albatross in Motion

These huge birds (above) travel vast distances, effortlessly gliding on long, narrow wings using the updrafts of the Roaring Forties, a belt of strong westerly winds that occurs between about 40°S and 50°S. Albatrosses are generally large and live for many decades. With their relatives, the petrels, they form one of two main groups of birds in Antarctica; the other is the penguins.

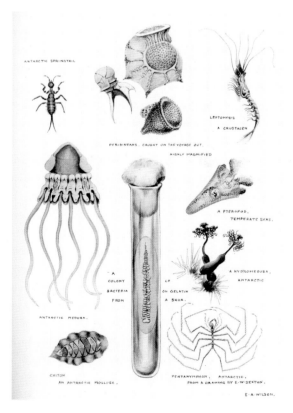

Journey of Discovery

On Robert Scott's inaugural journey to Antarctica in 1901–04, the explorers were delighted to discover a rich and unique fauna. These illustrations, from Scott's *The Voyage of the Discovery*, published in 1905, are of some representatives of the invertebrate groups discovered on the expedition.

The first explorers to go to Antarctica, more than two centuries ago, commented on the abundant wildlife—whales, seals, penguins, and winged seabirds. The rich, ochrous color on the underside of the sea ice also intrigued them. They recognized that it came from high concentrations of microscopic, single-celled plants called diatoms, which, they concluded, represented the base of a rich food web.

From that period up until about 20 years ago, the Southern Ocean was thought to be one of the richest marine ecosystems on the Earth, with a simple food web in which diatoms fed shrimp-sized crustaceans called krill, and fish, whales, seals, and birds consumed the krill. We now know that this is an oversimplification. Although the Southern Ocean contains some extremely productive areas, it is no more productive, overall, than more nutrient-poor parts of the world's oceans, and its biology is just as complex as that of warmer waters. In fact, the ocean is several interconnected ecosystems rather than one large one.

Biological Habitat

The extent of the Southern Ocean is not clearly defined: it takes in the southern parts of the Atlantic, Pacific, and Indian oceans. Depending on exactly how its northern boundary is defined, it accounts for 10 to 20 percent of the global ocean. It is dominated by the huge eastward flow of the Antarctic Circumpolar Current—the Earth's largest current, which provides the main connection between the three adjacent oceans. Near the Antarctic continent, the coastal current flows in the opposite direction, toward the west. The shape of Antarctica, with its large indentations—the Ross and Weddell seas—the northward-jutting Antarctic Peninsula, and the varying contours of its sea floor, change the breadth and speed of the circumpolar currents and give rise to localized currents. Oceanic boundaries between these currents and Antarctica's coastal features determine sea-ice conditions and the level of biological activity. As a consequence, the marine life around Antarctica has a very patchy distribution.

The Southern Ocean and its Antarctic Circumpolar Current developed after the break-up of Gondwana, some 40 million years ago. Fossil evidence indicates that the Circumpolar Current was established about 27 million years ago, and the Polar Front (also known as the Antarctic Convergence) had formed by roughly 22 million years ago. Sinking warmer northern water

Humpback Diving

Humpbacks (right) are readily identified by their low, broad blow, huge flippers, and also the way they arch their backs and show their flukes when diving. During summer, they feed in Antarctic waters, and in early winter they migrate to their breeding grounds in tropical waters. Populations were drastically depleted by whaling—recovery is slow.

In Search of Food

Adélie penguins breed on and around the Antarctic continent. They will travel 125 to 200 miles (200 to 300 km) from their nesting sites in search of food—krill and other small crustaceans, and also fish. Their breeding success rate is poor when the sea ice is extensive.

and colder southern water meet in the circumpolar zone. It is not uncommon for the temperature of the surface water to drop by about 3.6°F (2°C) across the front. The Polar Front is one of the major oceanic boundaries and is a significant biological barrier.

The Southern Ocean south of the Polar Front can be divided into three zones that differ significantly in their biology, because of environmental differences.

The permanently open ocean zone (POOZ) is nutrient-rich but it has relatively low levels of primary production. The main water flow is the Circumpolar Current. The phytoplankton (single-celled plants) here are generally tiny—nanoplankton, micro-fractions in size. They are grazed by several groups of animals, but generally not by Antarctic krill (*Euphausia superba*).

The most productive zone of the Southern Ocean is the seasonal ice zone (SIZ), south of the POOZ. It is ice-covered in winter but is essentially open water in the summer months. Phytoplankton blooms are common in the shallow, less saline water produced by the retreating ice edge in spring and early summer. The dominant phytoplankton are blooms of *Phaeocystis* and large diatoms, on which various groups of planktonic animals graze. Especially in the more southern parts, krill are abundant. Baleen whales, crabeater seals, penguins, and other birds feed on massive krill stocks.

The most southerly ocean region, the coastal and continental shelf zone (CCSZ), is also known as the

permanent pack ice zone—although the region is not covered by pack ice at the time of its maximum retreat. The ice that remains is often fast ice. The phytoplankton blooms may be intense, but they are generally short-lived. Antarctic krill are uncommon, and are replaced by their smaller relative, *Euphausia crystallorophias*. Because of the absence of krill, birds and mammals are less abundant than they are in the SIZ.

The temperature at which seawater freezes depends principally on its salinity. Around Antarctica, its freezing point is close to 28.4°F (–2°C); it is rarely warmer than

Algal Blooms

Shieldlike scales cover this tiny single-celled alga. It occurs throughout the world's oceans where the water temperature is above 35.6°F (2°C). It can be so abundant that the sea appears milky white, or so extensive that blooms are visible from space.

Algae Under Sea Ice

For long periods, the surface of the ocean around Antarctica is frozen. In some areas, algae grow underneath the ice in profusion. These plants are an important winter food source for herbivores. When the summer melt occurs, the algae are released for the ocean's deeper dwellers.

Influencing the Climate

As well as being the basis of the food web on which essentially all the other marine organisms depend, marine microbial components play a role in influencing climate. The Southern Ocean is a major oceanic site for the absorption of atmospheric carbon dioxide. And it is principally the activities of microorganisms that determine whether carbon dioxide is rapidly re-released into the atmosphere, or retained in the ocean for hundreds or even thousands of years. Additionally, some marine microorganisms produce sulfur-containing compounds, which, when released into the atmosphere, form aerosol particles that promote the formation of clouds. Clouds play a significant role in climate by reflecting the sun's energy back into space. The Southern Ocean is the source of some 17 percent of the total global production of these sulfur compounds.

39.2°F (4°C). As a result, Antarctic marine organisms experience constantly low temperatures.

For most of the year, the surface of the sea around Antarctica is frozen to a depth of up to about 7 ft (2 m), with an average thickness of less than 3 ft (1 m). The freeze begins in March and peaks in September, when the sea ice covers almost 8 million sq miles (20 million sq km)—nearly two-and-a-half times the total area of Australia. The sea ice is a major habitat for microscopic organisms, including diatoms. They proliferate on its underside, as well as at the boundary between the sea ice and the snow which settles on its surface and which depresses the ice into the water. Features in the ice, such as cracks, are refuges for the small animals that graze the microorganisms. In addition, some species of seals and penguins breed on the sea ice.

Our understanding of sea-ice ecology is limited. The amount of light in the water under the ice is effectively too low for photosynthesis by phytoplankton. However, there are environments favoring algal growth on, within, and on the underside of the sea ice. Some extremely

high concentrations of algae can develop in these places. On sunny days, oxygen is produced by sea-ice algae faster than it can diffuse away, and it is trapped as bubbles under the ice, surrounding the organisms that produced it. Oxygen at such concentrations is toxic and is likely to inhibit photosynthesis. The algae and its associated microbial community of protozoa (single-celled animals) and bacteria on the underside of the sea ice are effectively the only food available to grazers while the sea is covered with ice (nine or ten months per year). The under-ice algae on fast ice supports fish as well as crustacean grazers—copepods, krill, and amphipods. The algae living on the underside of pack ice (which drifts with the currents) is grazed by krill. The distribution of sea-ice algae and its associated microorganisms is extremely patchy. Generally, there is more biological activity on the underside of sea ice where the snow has blown away and more light can penetrate. In addition, the age and the texture of the sea ice influence the composition of the species and the abundance of these microbial communities.

As the sea freezes, cold, highly saline water is excluded from the forming ice, which has a lower salinity than the seawater from which it is produced. The cold, saline water sinks to the sea floor and flows to the Northern Hemisphere, as Antarctic Bottom Water.

During the spring and summer, melting sea ice in the SIZ produces a layer of less saline surface water. Phytoplankton bloom in this stable, shallow, nutrient-rich layer, which receives 24 hours of sunshine per day and produces localized areas capable of sustaining biological diversity and richness.

There is concern that global warming may reduce the amount of sea ice around Antarctica, and therefore diminish the production of Antarctic Bottom Water. If vertical mixing of the oceans lessens, it could lead to stagnation of the bottom water, depleting it of oxygen and making it uninhabitable by most living things. Any decline in sea ice would also mean less habitat for the organisms that use it, as well as reducing the relatively less saline water that promotes the ice-edge blooms.

Protists

At the level of the single cell, the distinction between plant and animal becomes clouded. These days they are grouped together and called protists.

The phytoplankton are microscopic, floating, single-celled plants. The 200 or more species identified from Antarctic waters are startlingly diverse in size, shape, lifestyle, and their food value to grazers. The smallest species measure about 1 micrometer (0.001 mm), and they vary in shape from nearly spherical to hairlike, up to about about ⅛ in (3 mm). Phytoplankton are divided into three categories: picoplankton (0.2 to 2 micrometers), nanoplankton (2 to 20 micrometers), and microplankton (20 to 200 micrometers). Generally, the nanoplanktonic and picoplanktonic organisms are more abundant and diverse than microplankton.

Cells of one group of phytoplankton, the diatoms, are encased in intricately sculptured, glassy walls, and others are covered with finely patterned scales. Some of them can swim; others drift enclosed in mucus.

Like all other plants, phytoplankton require light for the process of photosynthesis, in which they convert carbon dioxide and water into sugars and oxygen. Very little light is available in winter Antarctic waters, which are also covered with ice and snow. In summer, when the sun shines for 24 hours and the sea ice is decaying, phytoplankton grows rapidly. Summertime blooms of microplankton are an important food source for krill.

The concentrations of the major nutrients needed for phytoplankton growth—nitrogen and phosphorus, and, in the case of diatoms, silica—are high in the Southern Ocean, yet concentrations of phytoplankton are much lower than in most other parts of the world's oceans with similar nutrient levels. The Southern Ocean is one of the oceanic areas described as High Nutrient–Low Chlorophyll (HNLC): it is very windy, and the wind mixes its surface. Phytoplankton can be wind-mixed to depths greater than 330 ft (100 m), where there is not enough light for photosynthesis. Though concentrations of major nutrients are high, those of micronutrient iron (essential for phytoplankton growth) are very low over

much of the ocean; the lower the light is, the more iron phytoplankton needs. Thus, iron and light appear to act together to limit phytoplankton growth.

There are also single-celled animals (protozoa) that feed on the phytoplankton, bacteria, and detritus in the water. They, too, are highly diverse in shape, size, and lifestyle. Some of them filter their food from the water, others glide over surfaces grazing as they go, and others are voracious hunters. Some plants are able to hunt for their food. At the other extreme are protozoa, which retain the photosynthetic systems from the phytoplankton they consume, and they use their photosynthetic products as food. Grazing by plants and photosynthesis by animals is called mixotrophy. Grazing by the protists removes a

Pastures of the Sea
Phytoplankton are floating, microscopic, single-celled plants that make up the base of oceanic food webs. This photomicrograph is at 500 times the magnification of living phytoplankton. It shows a diversity of shapes and sizes of phytoplankton, here mostly diatoms. Some diatoms are needle-shaped, others form long, ribbonlike chains.

Swaying Anemone
A *Phyctenactis* anemone, surrounded by kelp, sways in the currents off the Antarctic Peninsula. Though anemones look like plants—in the way they cling to the substrate—they are actually animals. They are filter-feeders, which trap detritus from the water as the current passes over them.

Diatom Disks
Diatoms are the most abundant phytoplankton in the Antarctic seas. These are cells of the diatom *Porosira pseudodenticulata* seen through a scanning electron micrograph. When these cells divide, they form short chains that are easily broken up. The glassy silica walls accumulate as thick deposits on the bottom of the ocean.

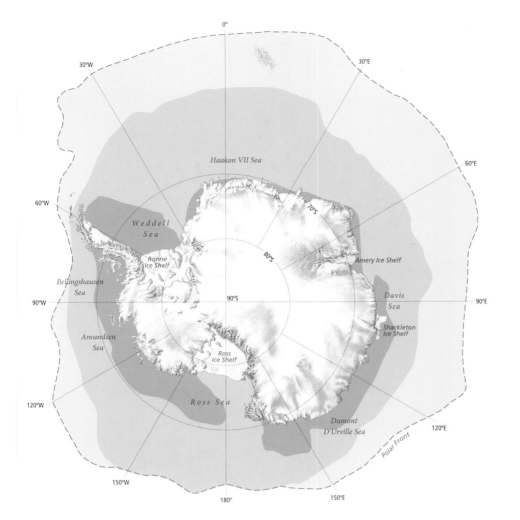

Haakon VII Sea

Weddell Sea

Ronne Ice Shelf

Bellingshausen Sea

Amundsen Sea

Ross Ice Shelf

Ross Sea

Amery Ice Shelf

Davis Sea

Shackleton Ice Shelf

Dumont D'Urville Sea

Polar Front

Biological Zones

The Polar Front is a major oceanic boundary in the Southern Ocean between the warmer northern waters and the cold Antarctic waters. South of the Polar Front is an ice-free zone surrounding the seasonal ice zone, which is ice-covered most of the year. Close to the coast are regions that are permanently ice-covered.

Permanently open ocean zone (POOZ)
Seasonal ice zone (SIZ)
Coastal and continental shelf zone (CCSZ)

The Smallest

The smallest, but most abundant, biological components of the marine environment are bacteria and viruses. A fluorescent stain makes them visible under the microscope—the bacteria are the larger bright dots. There are 10,000 times more viruses than bacteria in the Southern Ocean.

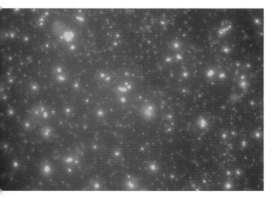

substantial amount of the smaller phytoplankton and bacteria. Bacteria feed on dissolved organic material which is produced by all organisms, particularly the phytoplankton. As the protozoa are eaten by krill and other grazers, they and the bacteria represent another pathway in the food web between phytoplankton and the grazers: this is called the microbial loop.

Bacteria and Viruses

Bacteria make up a major component of the food web of the Southern Ocean. They occur in concentrations of about 500,000 per 1 pint (600 ml), and they break down detritus, remobilizing the nutrients. They play several other roles which can have ecosystem-wide ramifications. Some bacteria species live alongside living phytoplankton, where they can influence nutrient concentrations. Bacteria dissolve mucus on the sticky surfaces of diatoms; this alters the diatoms' ability to form clumps. In addition, bacteria dissolve silica, the main constituent of the cell walls of diatoms, which, in turn, promotes the growth of diatoms. Bacteria secrete digestive enzymes to dissolve organic material, which they may then absorb as food. However, these enzymes also break up material, which would have otherwise sunk to the deep ocean. Thus bacteria are crucial to marine biological processes, playing a pivotal role in the grazing food chain, the microbial loop, particle sinking, and carbon fixation and storage.

The smallest biological entities found in the ocean are viruses. Viruses consist of genetic material (DNA or

RNA) enveloped in a protein coat. They are not living, in the sense that they cannot metabolize, or reproduce, by themselves. They infect a host, hijack its reproductive system, and produce masses of new viruses that can infect yet more host cells. The abundance of viruses in the Southern Ocean is similar to that elsewhere in the world, about 5 billion per pint (600 ml). These viruses probably infect all types of marine organisms, influence the species composition of microbial communities and nutrient cycling, and are involved in transferring genetic material between the host organisms. Virus infection accounts for about 50 percent of the mortality of marine bacteria, and may inhibit phytoplankton production by up to 80 percent. Viral destruction of bacteria may be one of the major sources of dissolved organic matter in the sea, which is, in turn, consumed by bacteria.

It is now known that much—often more than half—of the total flux of matter and energy passes through the microbial loop. This appears to be an organic matter continuum, from small, simple molecules of amino acids (which are the building blocks of proteins) and sugars through colloids to transparent polymer particles. It has been suggested these organic constituents of seawater interact to form a very dilute gel with a structure that has microscale hotspots (sites of heightened microbial activity). Microorganisms, like larger plants and animals, do not live by themselves but in communities with other organisms, each having its own role.

Krill and Other Grazers

Krill—food for fish, birds, and mammals, and the target species of a commercial fishery—are one of the central elements in the Southern Ocean ecosystem. There is probably a greater mass of Antarctic krill (*Euphausia superba*) than of any other single species on the Earth. Krill has a patchy offshore, circumpolar distribution. Its abundance and distribution vary from year to year, with profound consequences for its predators.

Krill are a group of around 80 different species of crustaceans, the euphausiids, which resemble shrimps, and live in the open ocean. In the Antarctic context, the term "krill" refers to Antarctic krill, which is the largest and the most abundant of the five euphausiid species living in the Southern Ocean. The dominant species in Antarctic coastal waters is *Euphausia crystallorophias*, a much smaller species than *E. superba*.

Remarkably little is known about krill. Estimating their abundance is fraught with difficulties because of their very patchy distribution and ability to avoid capture in research-size nets. The latest estimates, based on surveys using sophisticated echo-sounders and a re-examination of historical information on their range, is between 65 and 330 million tons (60 and 300 million tonnes). Some studies indicate there may have been declines in some areas over the past two decades, but whether these are part of a cycle is not known.

In order for a sustainable krill fishery to be properly managed, it is essential to know how many krill there are, how old they grow, and the extent to which they are food for other animals. Only in the past decade has it been established that they live for five to ten years and are sexually mature at three years of age.

Spawning is believed to take place near the surface, from late December to March. The females can produce several broods of up to 10,000 eggs, depending on food availability. The eggs sink to depths of about 3,300 ft (1,000 m), where they hatch; the embryos go through a series of non-feeding larval stages before rising to arrive near the surface as larvae needing to feed. This descent and ascent process is believed to take between three to five months. In their second summer, they develop into juveniles and they can form swarms. At the beginning of their second winter, juvenile krill are about 1 in (25 mm) long, and look like miniature adults.

Krill form swarms, which can contain around 30,000 individuals per 35 cub ft (1 cub m) and are easy targets for whales, seals, and penguins—and fishing trawlers. Often krill swarms consist of all males, of all females, or of all juveniles. A single swarm of krill encountered near Prydz Bay in East Antarctica, calculated to weigh about 63,000 tons (57,000 tonnes), was composed entirely of large and sexually immature males. Such swarms color the sea reddish brown. Occasionally, large "superswarms" form; one seen in 1981 off Elephant Island covered about 175 sq miles (450 sq km) and was estimated to contain more than 2.2 million tons (2 million tonnes) of krill. This swarm contained both sexes; however, there appeared to be some segregation by sex and age.

Adult Antarctic krill are about 2½ in (6 cm) long. All crustaceans, including krill, have a shell (exoskeleton).

They grow by molting the old shell, expanding the new one as it is forming, and then growing into it. This is done repeatedly during the life of the animal until it reaches maturity—so the larger the crustacean, the older it is. With Antarctic krill, however, it is not possible to tell their age from their size. For much of the year their principal food—phytoplankton—is scarce because of the wintertime absence of sunlight for photosynthesis. In laboratory studies, krill have coped with more than 200 days of starvation. Rather than rely on accumulated fat reserves, they use up muscle protein. They continue to molt about once a month, but instead of growing, as they do when food is abundant, they shrink. Also, at the end of the summer months, mature krill lose their sexual

Krill Swarm

Antarctic krill characteristically form dense swarms with tens of thousands of animals per cubic yard. When these swarms are close to the sea's surface, the water appears a brick-red color. Such swarms are highly attractive to predators. Swarms vary in size, and superswarms can cover many hundreds of square miles.

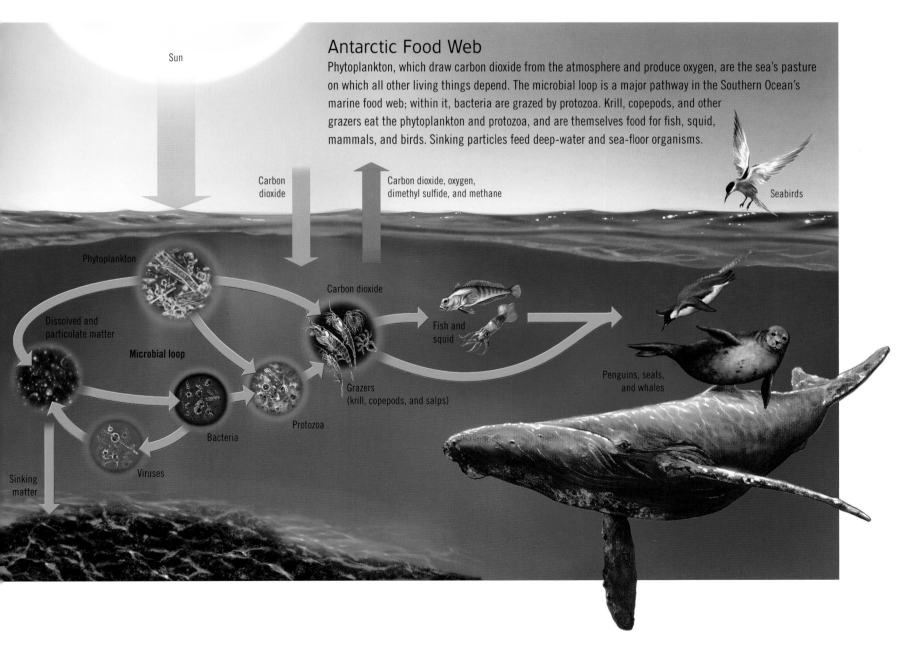

Antarctic Food Web

Phytoplankton, which draw carbon dioxide from the atmosphere and produce oxygen, are the sea's pasture on which all other living things depend. The microbial loop is a major pathway in the Southern Ocean's marine food web; within it, bacteria are grazed by protozoa. Krill, copepods, and other grazers eat the phytoplankton and protozoa, and are themselves food for fish, squid, mammals, and birds. Sinking particles feed deep-water and sea-floor organisms.

Sun

Carbon dioxide

Carbon dioxide, oxygen, dimethyl sulfide, and methane

Seabirds

Phytoplankton

Carbon dioxide

Dissolved and particulate matter

Microbial loop

Fish and squid

Penguins, seals, and whales

Grazers (krill, copepods, and salps)

Protozoa

Bacteria

Viruses

Sinking matter

characteristics, which they regain in spring for spawning in summer. So a five-year-old mature krill, when it has lost its sexual characteristics and has shrunk, cannot be distinguished from a sexually immature two year old.

Because of the ecological importance, biomass, and resource potential of krill in the Southern Ocean, other zooplankton tend to have received much less attention. These include other euphausiids, copepods, amphipods, pteropods (floating snails), chaetognaths (arrow worms), salps, and larval fish.

Of these, copepods and salps can play key roles and reach high abundances. It has been estimated that non-krill zooplankton account for 75 percent of the total zooplankton biomass in the Scotia Sea, north of the Antarctic Peninsula, and copepods have been found to constitute 60 to 87 percent of the summer zooplankton biomass in Admiralty Bay on King George Island. Some recent studies demonstrate that krill's contribution to the zooplankton biomass has been overestimated and that there is dramatic variability in the distribution and abundance of zooplankton geographically, as well as seasonally and interannually.

Copepods, like krill, are crustaceans. More than 70 species are recognized in Antarctic waters. They are generally microscopic, and only a few species are more than $\frac{1}{32}$ in (1 mm) long. Some of these graze on protists, others are carnivores, and yet others are omnivorous. Estimates indicate that copepods may consume more than eight times as many protists as krill do. Copepods are mainly grazed on by fish, including the Antarctic herring or silverfish (*Pleuragramma antarcticum*), which, in turn, are eaten by seals, penguins, and other seabirds.

Salps are planktonic tunicates (sea squirts). They feed by pumping water through a fine mucus net filter, and they are able to take in a wide size range of particles. They are delicate, gelatinous organisms and have a life

history alternating between solitary organisms and colonies that reproduce asexually. The production of colonies coincides with the springtime phytoplankton bloom. As a solitary salp can produce hundreds of colonies, high concentrations of these organisms can develop rapidly in spring and summer in response to food availability. Their high localized abundance and high filtration rates enable them to consume between 10 and 100 percent of the protists in these areas. Thus salps reduce concentrations of food for other planktonic grazers. These observations, plus discovering areas that are dominated by krill are usually devoid of salps, and that areas in which salps are abundant are low in krill, have led to a reappraisal of the ecological role of salps in the Southern Ocean.

Krill reproduction and the survival of larvae also appear to be affected by blooms of salps. Investigations point to an interaction between sea-ice extent and the dominance of either krill or salps. Patterns of oceanic circulation and the sea-ice conditions have a profound influence on the biology. It appears that in years when the southern boundary of the Antarctic Circumpolar Current tends to be offshore, a greater quantity of sea ice develops. In these conditions, krill populations are dominant and salps are farther offshore. However, when the current's southern boundary is close to the Antarctic coast and the sea ice is less extensive, salps are more abundant. At least in the Antarctic Peninsula region, there is evidence, spanning the past 50 years, of fewer and fewer winters having extensive sea-ice cover, which correlates with the increase in air temperatures over this time. As decreased krill abundance is related to a decrease in sea-ice extent, there is concern about the food availability for krill predators. Although salps are not a major constituent in the diets of the vertebrate predators, they do play another important role in the Southern Ocean. Because their feces sink quickly, they contribute to the transport of carbon from the surface waters to the deep ocean. This vertical flux of carbon in the world's oceans is one of the major pathways in the global carbon cycle—atmospheric carbon dioxide,

Life's Cycle
An Adélie penguin (left) feeds its chick. Adélies prefer to feed on krill, although they will eat fish, amphipods, and cephalopods as well. Krill graze on microorganisms and are themselves food for whales, seals, birds, and fish.

Dinnertime for Seals
A Weddell seal devours its prey, an Antarctic cod. These seals live farther south than other seals. They stay close to their breeding sites, feeding on a wide range of prey, varying from large to small fish, squid, and tiny crustaceans, such as amphipods and euphausiids.

Antarctic Fish

By far the most successful Antarctic fish are those in the suborder Notothenioidei, which includes Antarctic cod, dragonfish, and icefish. Of these, the cod and icefish are the major targets for commercial fisheries. This suborder has radiated to fill ecological niches from close inshore to a depth of about 6,560 ft (2,000 m) on the continental slope. This one group, therefore, occupies habitats shared by a much greater taxonomic diversity of fish in other parts of the world. They are principally bottom-dwelling because they lack a swim bladder—the gas-filled organ commonly used by open-water fish in temperate and tropical waters in order to achieve neutral buoyancy.

Antarctic cod

Icefish

Dragonfish

taken up by phytoplankton, is transported to the deep ocean, where it is only slowly recycled back to carbon dioxide by the activity of deep-sea organisms.

Antarctic Fish

More than 20,000 species of fish occur worldwide, but only about 120 are found south of the Polar Front—a major oceanic barrier to the movement of fish species, especially those that live in coastal and shallow habitats. Also, the Antarctic continental shelf is deeper than the shelves around other continents, and has no shallow connection to these continental shelves. This isolation has played a major role in the evolution of Antarctic fish and in the composition of their communities. Fish that live in Antarctic coastal bottom waters are by far the most diverse and abundant, accounting for more than 60 percent of species and 90 percent of the abundance of the Antarctic fish fauna.

In marked contrast to other parts of the global ocean, the Southern Ocean contains extremely few true schooling pelagic (open-sea) fish—the few which do occur there have evolved from bottom-dwelling groups to a partial or full-time life away from the bottom. The sea-ice zone is the habitat of adults of many of these, but they spend their first years of life in deep water. The Antarctic herring or silverfish (*Pleuragramma antarcticum*) is pelagic and often, especially when juvenile, associated with sea ice. It feeds on copepods, euphausiids, and juvenile fish, and is, in turn, eaten by marine mammals and birds. Lacking a swim bladder, it achieves neutral buoyancy with fat deposits. *P. antarcticum* reaches a length of approximately 10 in (25 cm) and it is an abundant and important organism in the marine food web. Myctophids (lanternfish), ¾ to 12 in (2 to 30 cm) long, are the other important pelagic group of fish in Antarctic waters. They are called lanternfish because of the groups of luminous organs (photophores) on their head and body. Opportunistic feeders, they consume

copepods, euphausiids, and other grazing species, fish eggs, and juvenile fish. During the day, lanternfish can be found at depths of about 660 ft (200 m) but migrate near the surface at night. They have a swim bladder.

Unlike most Antarctic invertebrates, body fluids of fish are less saline than seawater—their freezing point is about 30.2°F (–1°C). They have all evolved antifreeze compounds called glycoproteins, which depress the freezing point of the water in their body fluids. These antifreezes work by binding to the surface of forming ice crystals, thus preventing their growth.

Most Antarctic fish are less than 12 in (30 cm) long, but a few species grow to more than 5 ft (1.5 m) and weigh more than 110 lb (50 kg). They generally grow slowly, take three to eight years to become sexually mature, have long lifespans, and low metabolic rates. They produce only a few large, yolky eggs, so their reproduction rate is generally low. These characteristics put the Antarctic fish at serious risk of overfishing. Marbled rockcod (*Notothenia rossii*), mackerel icefish (*Champsocephalus gunnari*), and the Patagonian toothfish (*Dissostichus eleginoides*) have all suffered the predictable pattern of exploitation— initially high catches that rapidly decline to an uneconomic point, whereupon fishing is transferred to a new target species, and the pattern is repeated once again.

Antarctic Squid

Squid are almost certainly one of the most important components of the ecosystems of the Southern Ocean. Because of their ability to avoid capture in nets, very little is known of their abundance, longevity, and their lifestyles. Most species that are netted are not the squid taken by toothed whales, and are much smaller than

Benthic Antarctic Cod
Most Antarctic fish live close to the ocean bottom. They are slow-moving (unless threatened), grow slowly, have long lives, become sexually mature late in life, and have low reproductive rates.

Long and Hungry

The large nemertean or proboscis worm (*Parborlasia corrugatus*), pictured on the seabed below the diver, can grow to several feet in length. It is a carnivore that feeds by enveloping its prey in its proboscis. Although it has a soft body and would appear to be an easy target for other predators, it protects itself by exuding acidic mucus.

A Spiky Find

D. G. Lillie, who did biological work on Scott's 1910–12 Antarctic expedition, noted: "almost every trawl brought up quantities of siliceous sponges covered with glassy spicules." The skeletons of some sponges are composed of silica fibers called spicules, which provide support for the sponge and deter predators. In recent years, it has been discovered that the glassy spicules also function as optical fibers, transmitting light to algae that adhere to the spicules deep within the sponge.

those found in dietary studies of whales. There is a strong relationship between a squid's body and its beak size; some of the best information on squid has come from beaks recovered from the stomachs of seals and toothed whales. Up to 18,000 squid beaks have been recovered from inside a single sperm whale.

Some 70 species of squid occur south of the Polar Front; about half of these are endemic. Many species apparently produce egg masses in water deeper than 3,300 ft (1,000 m). Young squid grow rapidly, reach sexual maturity in a year, spawn, and then die. Some species, however, may live for several years.

Squid form a major part of the diet of seals, toothed whales (except orcas), and birds, particularly albatross. It is difficult to tell how much squid is consumed by predators, but crude estimates put it at more than 44 million tons (40 million tonnes) per year, and suggest that the total squid biomass of the Southern Ocean is about 110 million tons (100 million tonnes). Squid eat voraciously, some species consuming up to 30 percent of their body weight a day. They were once thought to be major consumers of krill, but recent investigations indicate that myctophids and deepwater fish are their major prey. As these fish feed mainly on copepods, it is emerging that squid may represent an important link in the food web between the copepods and toothed whales, seals, and seabirds.

Although squid are fished around New Zealand and the Falkland Islands, at present there is no commercial squid fishery in Antarctic waters. Squid populations can fluctuate dramatically because of their short life cycles, and the fact that they spawn only once, then die soon after. They are especially susceptible to overfishing. Any commercial fishery would have to be carefully managed to ensure sustainability and to prevent adverse impacts on predators of squid.

Benthic Communities

The coastal and continental shelf areas of Antarctica contain rich benthic (bottom-dwelling) communities, which are generally dominated by animals feeding on particles of matter raining down from surface waters.

The Antarctic continental shelf is generally deeper and narrower than those of other continents. The huge amount of debris deposited on the seabed around

Antarctica is only inorganic (rock) material transported by glacial activity, rather than the biologically derived organic and inorganic deposits surrounding the other continents, which are carried by rivers and wind action. Ice exerts a major influence on the Antarctic benthic communities in several ways. These include shading the light available for benthic algae, scouring by sea ice and icebergs, and the formation of ice on the shallow seabed. The water temperature on the continental shelf is generally low and stable. Light for photosynthesis and the fallout of material from the surface waters is a sharp seasonal pulse. Unlike the coastal waters in the other oceans, scouring by sea ice ensures that the intertidal community around most of Antarctica is sparse.

One of the best studied benthic environments at high latitude is McMurdo Sound. Generally, the area, which is regularly scoured by ice down to a depth of about 50 ft (15 m), is almost devoid of algae, except for those capable of fast growth in the short summer. Animals are mobile foragers, such as starfish, various worms, sea urchins, crustaceans, and fish. However, below this zone is a rich and diverse flora and fauna. In shallow coastal environments, benthic microalgae make a substantial seasonal contribution to primary production. Some 700 species of seaweeds have been reported from Antarctic waters, of which about 35 percent are endemic. Seaweed and bottom-dwelling animal communities vary with latitude and water depth. Seaweeds are sparse below about 130 ft (40 m) and are virtually absent below 330 ft (100 m), where there is scarcely enough light for them to photosynthesize. Many of the seaweeds found south of 32°F (0°C) do

not occur north of this boundary, and vice versa. The ice-scoured zone below that is colonized by a diverse array of filter-feeders, which include anemones, soft corals, molluscs, ascidians (sea squirts), and tube worms. Here, too, are mobile scavengers, including sea urchins, fish, starfish, and pycnogonids (sea spiders). Below about 100 ft (30 m), sponges are the dominant animals; they cover up to 55 percent of the bottom in McMurdo Sound. The sponges vary widely in size and shape and grow very slowly. The largest are shaped like volcanos, and are up to 7 ft (2 m) tall and 5 ft (1.5 m) across. This sponge community provides homes for many mobile species. Anemones, bryozoans, molluscs, tube worms, and ascidians also occupy the zone.

Areas around Antarctica where the sea floor is sandy or muddy tend to contain burrowing animals. Densities of these animals are among the highest found anywhere in the world.

As is the case in temperate and tropical waters, predator–prey interactions are thought principally to determine the composition of the Antarctic benthic communities. However, in Antarctic waters, icebergs scour the continental shelf, which has a depth of about 1,640 ft (500 m). Statistically, this scouring would be expected to happen at least once every 230 years, although depth and proximity to active iceberg calving areas would mean much more frequent scouring of some parts of the continental shelf. Areas scoured by icebergs look like ploughed fields with parallel grooves, some bordered by raised embankments. The passage of the iceberg strips all organisms from the bottom. There is a sequence of recolonization of scours with

mobile animals, including molluscs, sea urchins, and fish. They are followed by animals that are attached to the bottom, including tube worms and sea squirts. Habitat destruction by scouring and by subsequent recolonization may explain the great differences in community composition between the habitats on the Antarctic continental shelf.

The Antarctic sea floor is covered with microscopic algae, protozoa, and other animals and bacteria as well. The protozoa and bacteria live in benthic sediments. The benthic microalgae are generally subdivided into two groups: those that live within the top few layers of sediment, and those called epiphytes which live attached to other living things, such as seaweeds, or to the surface of rocky seafloors. These algae can be extremely abundant, especially in some habitats, such as beds of sponges with long, needlelike spicules to which the algae are attached and which discourage grazing animals. The benthic microalgae are thought to make significant contributions to primary production in shallow coastal areas and in providing nutrients to the benthic animals, as well as contributing to the organisms living in the overlying water column.

The benthic bacterial concentrations are generally many times higher than those in the overlying water column, but similar to those found in sediments in other parts of the world. Bacterial activity in sediments is usually confined to the top 2 in (5 cm). In Antarctic marine sediments, the rates of bacterial processes in breaking down organic material and then recycling it back to carbon dioxide are thought to be similar to that in the deep oceans.   HM

Prickly Relatives
The common Antarctic sea-urchin eats algae that it scrapes off rocks as it crawls. The large, many-armed starfish is a carnivorous predator; the bulge in its center suggests it is has just eaten.

Spider of the Sea
Sea spiders from Antarctica (above) are not spiders at all but crustaceans. They span as much as 8 in (20 cm) across. Like other Antarctic bottom-dwelling invertebrates, they grow slowly and become larger than their warm-water relatives. It is thought their growth rate is controlled more by food availability than the low sea temperature.

Fierce Predators
The seabed communities of Antarctica are surprisingly diverse and colorful. Many starfish species live in coastal areas, and some, such as this *Odontaster validus* (left), are found in large feeding aggregations. Although they do not move quickly, they are voracious predators and prey on other starfish much larger than themselves.

Seals

ORDER Pinnipedia

Seals are a significant and highly visible part of the wildlife of Antarctica. They are classified into three families: Phocidae, the true seals, or hair seals; Otariidae, the eared seals; and Odobenidae, the walrus (which is an exclusively Arctic animal). Of the 19 true seal species, five are found in Antarctica. The crabeater, Weddell, Ross, and leopard seals each have their own genus, while the fifth, the southern elephant seal, shares its genus with the northern elephant seal (*Mirounga angustirostris*).

The eared seals include sea lions and fur seals. There are five sea lion genera, and each has a single species. Named for the long mane of fur that covers the beefy necks of the males, sea lions have a broad distribution, generally preferring temperate coasts but occasionally occupying equatorial niches or sub-polar zones. The nine fur seal species, distinguishable from sea lions by a thick waterproof pelt under long guard hairs, form two groups: the northern fur seal forms its own genus, and the eight southern fur seals have a second genus, *Arctocephalus*. Eared seals are less polar in their habits than many true seals, but most southern fur seals live in, or occasionally reach, the sub-Antarctic or Antarctic regions.

Seal Physiognomy

Seals are well adapted for ocean life. The true seals have a thick layer of blubber for insulation in water, whereas the fur seals have to rely on their dense fur

for insulation. The hind limbs of all seals are long, thin, and flattened, rather like paddles. Their faces are short, with reduced (in eared seals) or non-existent (in true seals) external ears. The bodies of seals are smooth and sinuous, and they swim by undulating their bodies and using their flippers. Most have a thick, heavy neck, but the vertebrae are only loosely interlocked, making them extremely flexible for a mammal. Agile and very strong, seals can withstand the impact of waves and currents and maneuver around ice and rocky shores.

Seals have slitted nostrils that close under water—another adaptation to marine life. When the nostrils are in the relaxed position they are closed, and the water pressure increases the force keeping them closed. This means that when resting in water a seal does not need to concentrate on keeping its nostrils closed. Instead,

Fat but Warm
Elephant seals use blubber as their primary insulation. This may look awkward on land, but it does not interfere with swimming. The seals often cover 45 to 50 miles (70 to 80 km) per day when traveling to feeding locations.

Seal Family Tree

Little is certain about the origins of seals. Many scientists place them in their own taxonomic order, Pinnipedia, while other authorities consider them a suborder of Carnivora. Some scientists believe that the closest living relatives to all seals are the weasels, badgers, skunks, and otters, although both modern and fossil cranial material suggest that all seals are descended from a bearlike ancestor.

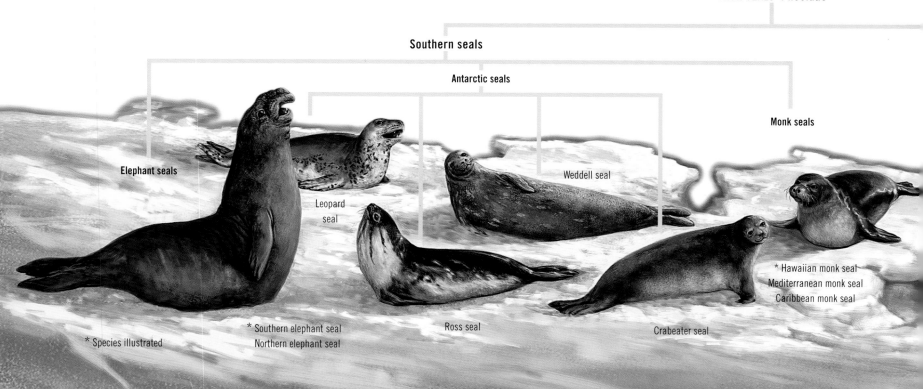

TRUE SEALS Phocidae

Southern seals

Antarctic seals

Monk seals

Elephant seals

Leopard seal

Weddell seal

* Hawaiian monk seal
Mediterranean monk seal
Caribbean monk seal

* Species illustrated

* Southern elephant seal
Northern elephant seal

Ross seal

Crabeater seal

in order to breathe, the seal must actively contract a pair of muscles. While they are at sea, seals hold their breath for extended periods—elephant seals, for example, have been recorded holding their breath for two hours. When a seal does open its mouth under the water, its tongue and soft palate close off the back of the throat, which means it can take a mouthful of food without swallowing seawater. In addition, the larynx presses against the epiglottis to stop any food and water from entering the lungs.

Eared seals have powerful shoulder muscles to support their strong front flippers, which they use as paddles in the water and as feet on land. They can also bring their hind flippers forward and use them like legs to run across land. True seals, however, have lost this facility, and when swimming they use their hind limbs to undulate their hindquarters, where their main muscles are. This gives them a highly efficient swimming motion, but their front flippers do not have the strength or reach to be effective on land, and their rear flippers cannot bend forward in the way that the flippers of eared seals can. True seals do not run, but they undulate across land, humping their body up and then throwing it forward with surprising speed.

Underwater Sight

Seals' eyes have several adaptations for seeing under water, including a flattened cornea and pupils that can open very wide to admit as much light as possible. In land-dwelling animals, the curved surface of the eye refracts light to project an image onto the retina in the back of the eyeball and the lens is used for focusing only, but in water the lens directs the image to the back of the eyeball. All seals have large, well-developed eyes,

Out for the Count

Antarctica poses unique problems in estimating population sizes. If you count those on land, what about the seals in the water? How do you count the seals that are hauled out on sea ice far away? The solution is to survey long, narrow strips of sea from a ship or an airplane and count the seals on ice. Seals prefer to haul out at certain times of day. The time is noted and the proportion of seals presumed to be on the ice is factored into the estimate. The ice type and cover are also taken into account, as the seal densities can be related to the area of pack ice available for them to haul out onto. Final estimates are likely to be low, as the proportion of seals actually on the ice will never be 100 percent; the best available estimates place the crabeater seal population at about 14 million, but if you change the assumptions, the same counts suggest that the number may easily be as high as 40 million.

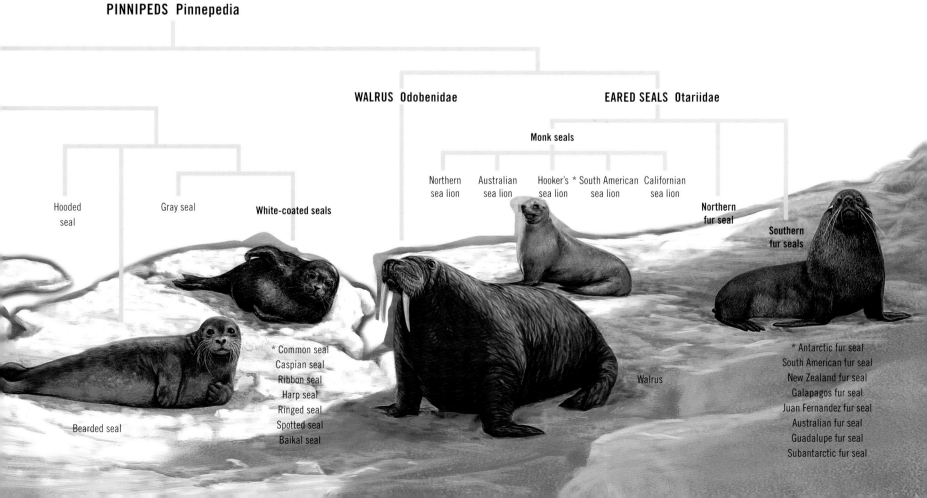

PINNIPEDS Pinnepedia

WALRUS Odobenidae

EARED SEALS Otariidae

Monk seals

Northern sea lion

Australian sea lion

Hooker's sea lion

* South American sea lion

Californian sea lion

Northern fur seal

Southern fur seals

Hooded seal

Gray seal

White-coated seals

* Common seal
Caspian seal
Ribbon seal
Harp seal
Ringed seal
Spotted seal
Baikal seal

Bearded seal

Walrus

* Antarctic fur seal
South American fur seal
New Zealand fur seal
Galapagos fur seal
Juan Fernandez fur seal
Australian fur seal
Guadalupe fur seal
Subantarctic fur seal

Ice Fishing

A Weddell seal (above) with a large Antarctic toothfish which it caught in 1,000-ft (300-m) deep water under this hole in the fast ice near Cape Armitage, Ross Island. Fish provide the main food for Weddell seals, which also eat squid, bottom-living prawns, and octopus.

Still Tied to the Land

As is true of all true and eared seals, this fur seal has a flexible, streamlined body with reduced appendages, which makes it well adapted to life in the water. But seals, unlike whales, are still connected to land. Although a few seal species appear to mate in the water, most mate on land, and all come ashore or onto ice to give birth.

Bachelor Rights

A young bachelor southern elephant seal rests on top of a weaner (above). Sometimes seals of different ages and sexes come ashore to rest and molt at the same time. A peaceful scene like this one may erupt into a short-lived aggressive outburst at any time. When caught in the fray, a smaller individual often becomes the victim of a larger seal's temper tantrum.

which gather light more effectively than small ones do. Elephant seals and Ross seals, both deep divers, have very large eyes with large lenses and they can see well in deep water. Seals have retinas like those of terrestrial carnivores but lack cones, the sensory cells necessary to detect color. Their eyes consist entirely of rods— sensory cells that function well in low light. The eyes of elephant seals have a modified photo-pigment like that of deep-sea fish, perhaps to discern the bioluminescence of their prey, deep-sea squid.

Eating for Energy

Seals are active hunters, and as a group they prey on everything, from plankton to higher vertebrates. But each species occupies its own ecological niche and exploits a particular type of prey. Crabeater seals, for example, eat krill primarily, whereas Ross seals inhabiting the same regions take mostly squid and fish.

Some species—for example, some of the southern fur seals, many southern sea lions, and the elephant seals—ingest rocks and sand. Rocks the size of a small orange have been recovered from seals' stomachs, and one southern elephant seal had 80 lb (35 kg) of rocks in his gut. These rocks, which are known as gastroliths, are found in such quantities that it is probable they are ingested intentionally, rather than by accident, and may provide stabilization while diving or ballast to minimize the energy required for deep dives, or they may be used for breaking up food or for grinding up parasitic worms. Seals that are breeding or molting are much more likely to have gastroliths, possibly because they need to fill up their stomachs while they cannot feed at sea.

Large Mouth, Large Prey

All seals are carnivores, and the size of the mouth of each species is related to the size of its preferred prey. Adult leopard seals like this one usually feed on a range of prey that includes fish, penguins, and seals.

Mating and Gestating

A female becomes pregnant immediately after mating, but the embryo stops growing after a few days and does not implant in the uterus for several months. It then develops in the normal way, and at the normal speed. Delayed implantation means the pup is born almost a year after mating, although the period of fetal growth is generally eight to nine months. Thus, birth and the next mating occur around the same time, so the pup is born when it has the best chance of survival and the female has the shortest possible breeding and pupping period.

Females occasionally adopt orphaned pups if their own die. Some pups find a foster mother after a normal period of nursing by their blood mother, and so they are suckled for much longer than normal and thus gain many advantages from their resultant extra size and energy stores. Most foster mothers adopt a single pup, usually only when they lose their own pup at a very young age. In a few seals, however, their mothering instinct is so strong that they will nurse so many orphans they cannot produce enough milk to feed them properly. Seal milk is a very high-energy drink, with a fat content of 30 to 60 percent and at least 5 to 15 percent protein. LW

Eared Seals

There are two main groups of eared seals, the fur seals and the sea lions. They are physically very similar, although sea lions have rounder snouts and shorter front flippers, and are slightly larger. The feature most often used to distinguish the two groups, however, is the fur seal's thick, almost waterproof pelt, which has given the animal its English name.

This pelt, which is commercially valuable, consists of two types of hairs. Each long, stout guard hair has about 50 shorter, softer underfur fibers growing from the same hair shaft. This gives the seal about 300,000 hairs per sq in (46,000 per sq cm) of skin. (A human head contains about 100,000 hairs per sq in/15,000 per sq cm.) Sebaceous glands in the skin of the seal produce enough oil to transform the fine hairs into a waterproof barrier that traps air, so that only the tops of the guard hairs get wet. In addition, the strong guard hairs serve to support the fine underfur. In this manner, the fur seal's pelt is custom-made to provide insulation when the animal is diving. Sea lions lack this underfur, and historically their pelts have little financial value.

Fur seals and sea lions probably divided into two unique groups only about 2 million years ago—which is a relatively short period in evolutionary terms—and the two groups retain many similarities. For example, their social and reproductive behaviors are very similar, and natural hybrids have been observed where their populations overlap in the wild.

Flushing the System

Like all mammals, seals need water—yet many do not seem to drink it. Some seals get enough water from eating fishes that contain very little salt. The many species that live on invertebrates consume salty food, but seal kidneys are so effective that they can excrete salt in extremely concentrated urine and even extract freshwater from the seawater. Male eared seals are especially likely to drink seawater during their long breeding fast. Although they can derive some water through the oxidation of stored fat, they must replace water lost through panting, urination, and sweating. Presumably, the net gain in freshwater from drinking seawater, while very small, is enough to be worthwhile.

While Mum's Away

Hooker's sea lion pups are very active animals, and when their mothers go to sea to forage, they take their pups to vegetated areas behind the beaches. In large colonies, the pups will gather together in groups and play in the zone behind the beach, scrambling around in stream beds and along rocky outcrops.

Sensitive Seal

The long and sensitive whiskers of this bull New Zealand fur seal (left) help it to detect movements made by prey. Whiskers also seem to assist seals to find their way in dark or murky waters.

On average, eared seals live for around 18 years. Males probably live shorter lives than females because they must vigorously compete with the other males for breeding rights. It is a stressful process for the males, and many do not survive past early reproductive age.

Male fur seals and sea lions are much larger and heavier than the females. The males have much bigger and particularly solid-looking chests that are covered with longer fur to increase their apparent size. They use these big chests for displaying to rival males in order to intimidate them. This robust chest offers protection on the rare occasions that displays degenerate into active fighting between competing males.

Living Spaces

The Otariidae occur from tropical, through temperate, to polar environments, but most commonly in temperate or sub-polar conditions. Tropical populations of fur seals and sea lions tend to be quite small, whereas sub-polar populations are usually very large. These environments tend to be more productive, with nutrient-rich waters that encourage large populations of prey species, which eared seals can exploit on a seasonal basis. On average, tropical environments are less productive and so have smaller populations of potential prey.

All seals replace their hair, or fur, in a process known as molting. This is particularly important for the fur seals, which rely on the waterproofing in their soft underfur for temperature regulation. Fur seals and sea lions molt very

Tailor-made

All seals have flippers adapted more for swimming than walking—the foot bones are elongated and flattened, and the tissue surrounding the toes is broadly webbed. Here, the seal uses its small claws to scratch itself in the same manner as a dog, making both the length and the flatness of the hind foot clearly visible.

Awaiting Lunch
A New Zealand fur seal pup (right) peers out from its hideaway above the high-tide mark. After the mother has suckled her pup for about ten days, she mates, usually with the bull controlling the territory. She then leaves on her first feeding trip, returning a few days later to suckle her pup again.

Waterproof Jacket
Waterproof and nearly windproof, the coat of a fur seal is made up of two layers—an outer layer of strong guard hairs and a velvety underfur about 1 in (25 mm) thick.

gradually so they can continue to swim throughout this process without compromising the insulative capacity of their fur. Northern fur seals have the slowest known molt in the Otariidae; it can take up to three years for their fur to be completely replaced. More generally, the process takes about one month.

Claiming Territories

Eared seals come together to give birth and to mate in breeding colonies on favored beaches. The males will generally arrive before the females and spend the first few days fighting to establish territories, and they then maintain rank by vocalizing, posturing, and threatening. These breeding territories are often defined by natural boundaries, such as rocks or crevices, which allow easy identification of interlopers and provide clear-cut zones for posturing. In many species, the males return to the same area for as long as they can continue to win their place—usually only two or three years.

All eared seal bulls must go without food during the breeding period, because to leave would mean a new series of dangerous, exhausting battles to re-establish

dominance. Leaving the colony also means the male runs the risk of missing the female's receptive period. This is one reason why male otariids are so much larger than females—their size gives them energy reserves that allow the larger males to remain on the beach for a long period, and to father more offspring than smaller males, which must abandon their posts to feed.

The territory of a breeding male can contain more than 20 females, so the rewards for winning a territory are considerable. Superfluous males usually hover on the edges, trying to engineer sly opportunities to mate, although in some species bachelors form non-breeding herds with the youngsters from other beaches. The females arrive some days after the males and give birth a few days after that. They then stay with their pup for a week or two, nursing their young while the males jostle for position and control of females. Females then come into estrus and mate.

In many eared seal species, the female will bellow loudly when a male tries to mount her. If that male is not the local territory-holder, her call will notify him that another is trying to mate with one of his females. Calling helps to make sure that her pup will be fathered by the strongest male, increasing the pups' likelihood of being dominant adults and, in turn, parenting many pups.

A mated female stays with her pup for a couple of days and then starts to alternate between going to sea for several days to feed and returning to nurse her pup. This pattern of absence and attendance means that the period of pup dependence can last for many months. In some sea lions, it takes females more than a year to wean their pup. In polar species, lactation is restricted to summer, and can take as few as four months.

Female Power

Breeding fur seal males typically defend a particular patch of beach—gambling that it will fill with females. Sea lions are less obviously territorial, but still work hard to isolate and maintain exclusive access to a group of breeding females.

The number of pups a male fathers depends on the quality of his territory and length of tenure. The quality of a territory is reflected in the number of females that it attracts which, in turn, depends on the location of the

Playtime

Eared seals can be playful, surfing,
tossing and dragging rocks, sticks,
or other animals, chasing each other,
wrestling, rolling over, waving their
flippers, or biting their own tails. The
purpose seems to be the formation
of social bonds, and possibly also the
development of motor skills.

territory and the distribution of females across it. The
distribution of females is partially influenced by climate
and access to water. The length of time a male retains
his tenure depends on his size and age, as these factors
affect his competitive and fasting abilities. Experience
counts: young males start with poor breeding territories
and acquire better ones over time, until they, in turn,
submit to a younger and fitter contender.

The females can influence their breeding success
by choosing their territory, sometimes moving through
territories of several males before settling. The males of
many species try to herd the females, usually without
much luck because females will return to the territory of
their choice. Of all the otariids, South American sea lion
bulls are possibly the most successful herders.

The mother and her pup bond by nuzzling and
vocalizing. When she goes to sea to feed for the first time
after giving birth, the pup stays close to where it suckled;
the returning mother locates her pup by sound and smell.
When she calls, all nearby pups respond, but she accepts
only her own offspring. If the pup does not call, the mother
bites it or tosses it away, a behavior possibly developed to
stop her from feeding greedy or orphaned pups. LW

Diving Data

All female fur seals go to sea to feed several days at a time when raising a pup. While at sea,
the females need to be adroit and effective swimmers to pursue krill, fish, and squid. For
most female fur seals, the mean duration of a dive is two to four minutes, with a maximum of
five to seven-and-a-half minutes. The mean depth of most dives is 100 to 150 ft (30 to 45 m),
with a maximum depth of 330 to 660 ft (100 to 200 m), and a mean number of 15 to 20 dives
per hour. Most experimental data on fur seals tend to be on females because they are easier
to catch than the males, and much easier to follow when they haul out to the breeding beaches.
The females also gather in very large numbers, making it possible to collect a large amount of
data on many individuals relatively easily.

Fur Seals

ORDER Pinnipedia
FAMILY Otariidae

Big, Beautiful Eyes
This fur seal's large eyes are designed for seeking out prey in dim waters. A seal's eyes are a slightly different shape to a land mammal's, to see underwater.

Fast and Accurate
Otariids are high-speed predators able to modify the depth of dives to match their prey. They usually make steep dives through schools of krill or fish.

Of the eight southern fur seals of the genus *Arctocephalus*, five permanently occupy or regularly visit sub-polar or polar regions.

South American Fur Seal
Arctocephalus australis
OTHER NAMES Southern fur seal, Falkland fur seal
LENGTH Female 4½ ft/1.4 m; male 6 ft/1.9 m
WEIGHT Female 110 lb/50 kg; male 350 lb/160 kg
STATUS Up to 450,000, sensitive to El Niño events.
 IUCN: Lower Risk—least concern. CITES: Appendix II

South American fur seals range up the coasts of South America from Lima to Rio de Janeiro, throughout Tierra del Fuego, and to the Falkland Islands. They breed on islands and remote coasts, congregating in small groups with the females well spaced. Pups are born in spring and early summer, the peak time being November and early December. Mating occurs six to eight days after the female gives birth. The females nurse for at least eight months, and sometimes up to two years. Most of the population is non-migratory, with some limited dispersal in winter, but most of the females remain near the colonies all year.

This species is unusual in that noisy gangs of young males—usually about ten, but sometimes as few as two or as many as 40—regularly raid the breeding territories, trying to herd females away or mate with them under the protection of the gang. These raids may occur every two hours during the peak of female receptiveness.

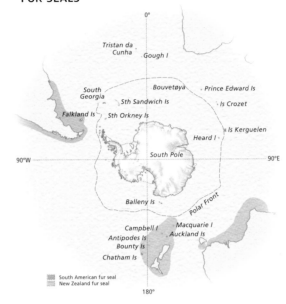

SOUTH AMERICAN
AND NEW ZEALAND
FUR SEALS

South American fur seal
New Zealand fur seal

New Zealand Fur Seal
Arctocephalus forsteri
OTHER NAME Western Australian fur seal
LENGTH Female to 5 ft/1.5 m; male to 8 ft/2.4 m
WEIGHT Female to 110 lb/50 kg; male to 410 lb/185 kg
STATUS 85,000–135,000; increasing at most sites.
 IUCN: Not listed. CITES: Appendix II

Populations are found in New Zealand and the nearby sub-Antarctic islands, in South Australia and Western Australia, and recently at Macquarie Island. These seals breed on islands and remote coasts in small to very large groups, with the females being well spaced. The males have massive necks and thick manes, which are wider than the rest of their body, and which are used in display threats when establishing territories: they undulate the neck and utter guttural barks to intimidate. The species is non-migratory, with some dispersal after breeding and a return just before breeding. Most of the females stay near the colonies all year. They eat squid, octopus, fish, and even seabirds. Being largely non-migratory, males spend much of the year together. Male hierarchies are based on size—the biggest males are the most dominant outside the breeding period. During breeding, however, dominance is territorially based. The females pup in late spring and summer, with births peaking in the second half of December. Mating occurs eight to nine days after birth, and peaks in late December to early January. Pups are weaned at about 12 months.

Antarctic Fur Seal
Arctocephalus gazella
OTHER NAME Kerguelen fur seal
LENGTH Female 4 ft/1.2 m; male 6 ft/1.8 m
WEIGHT Female 88 lb/40 kg; male 420 lb/190 kg
STATUS More than 4 million; increasing. IUCN: Lower
 Risk—least concern. CITES: Appendix II

This colonial fur seal species breeds on islands south of the Polar Front, including the South Shetlands, the

South Orkney and the South Sandwich islands, South Georgia, Bouvetøya, and Heard and McDonald islands, and, to the north of the Polar Front, on Prince Edward and Marion islands, Iles Kerguelen, and on Macquarie Island. Antarctic fur seal colonies vary considerably in terms of size and spacing.

Both males and females move away from breeding colonies in winter. When on land, they sometimes travel inland to lie on top of tussock grass behind the breeding beaches. They can move at 12 mph (20 kph) on beach flats, and they can easily outmaneuver humans through the tussock grass.

This species makes comparatively long trips during the breeding season, in which they eat cephalopods, krill, and fish. While at sea feeding, females make more than 400 dives of between 15 and 330 ft (5 and 100 m) in five to six days. They can travel 36 miles (60 km) in over eight hours before making their first dive, and then regularly dive for four minutes at a time, to an average depth of about 100 ft (30 m).

Antarctic fur seals give birth in late spring to early summer, peaking in early December. Mating occurs six to seven days after the birth of the pup, and peaks in mid-December. Females nurse for four months or less, and the pups appear to be the ones to initiate weaning. The fur seal pups are extremely active and they begin practicing swimming in January, and by March have become skilled in the water.

There is some overlap in the ranges between the Antarctic fur seal and the subantarctic fur seal species, and hybrids have occurred in the wild.

South African Fur Seal

Arctocephalus pusillus

OTHER NAMES Cape fur seal, Australian fur seal, Tasmanian fur seal, Victorian fur seal

LENGTH Female 6 ft/1.8 m; male to 8 ft/2.4 m

WEIGHT Female to 265 lb/120 kg; male to 800 lb/360 kg

STATUS To 2 million in southern Africa, 50,000 in Australia. IUCN: Lower Risk—least concern. CITES: Appendix II

The South African fur seal ranges over southwestern Africa, from Angola to Algoa Bay, and to southeastern Australia and Tasmania. It breeds on islands in medium to large colonies, with females being relatively densely packed. As is true of other fur seals, it is polygynous, with the female group sizes ranging from five to more than 50. The species is non-migratory, but shows some seasonal movement and some long-distance dispersal in winter. Individuals have been found as much as 930 miles (1,500 km) from the nearest colony. The feeding females generally stay at sea for about five days before returning to the colony. Unusually for an *Arctocephalus* species, mature bulls will tolerate the presence of young males more than one year old.

Pupping takes place from the late spring to early summer, and it peaks around late November and early December. The peak mating period then follows five to six days after birthing. South African fur seals generally nurse their offspring for eight to ten months, sometimes longer if no new pup is born.

Subantarctic Fur Seal

Arctocephalus tropicalis

OTHER NAME: Amsterdam Island fur seal

LENGTH: Female to 5 ft/1.5 m; male to 6 ft/1.8 m

WEIGHT: Female to 110 lb/50 kg; male to 330 lb/150 kg

STATUS: Slightly more than 310,000. IUCN: Lower Risk—least concern. CITES: Appendix II

This species breeds on islands and island groups north of the Polar Front, including Tristan da Cunha, Gough and Marion islands, Iles Crozet, Iles Amsterdam, and Macquarie Island. The colonies vary in size and tend to be well spaced. Their diet ranges from crested penguins and squid at Iles Amsterdam to krill, cephalopods, and fish at Marion Island. The males move out to sea after breeding; the females and pups remain in the colonies. Females give birth in late spring to summer, peaking in the middlle of December, and mating occurs eight to 12 days after the birth. The pups are normally nursed for nine to 11 months. Non-territorial males form bachelor colonies with immature seals, and spend 95 percent of their time completely inactive. The breeding males actively herd females and defend their territories. LW

ANTARCTIC, SOUTH AFRICAN AND SUBANTARCTIC FUR SEALS

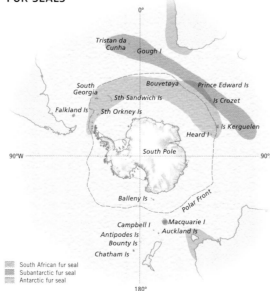

Distinctive Features

The eared seals are generally smaller than the true or hair seals, and they are all very lithe and supple. The small external ear that characterizes the group can be seen on this Antarctic fur seal, as can the thick coat that made this species a target for sealers.

Agile on Land

Fur seals are extremely flexible. When moving fast, they usually bound forward with both their hind limbs together, a motion well suited to moving around in their rocky habitats.

Sea Lions

ORDER Pinnipedia
FAMILY Otariidae

Hooker's Sea Lion
Phocarctos hookeri
OTHER NAME New Zealand sea lion
LENGTH Female to 7 ft/2 m; male to 8 ft/2.4 m
WEIGHT Female 510 lb/230 kg; male 880 lb/400 kg
STATUS Less than 14,000. IUCN: Vulnerable.
 CITES: Not listed

Hooker's sea lions breed on the sub-Antarctic islands of New Zealand, including Auckland, Campbell, and Snares, with some births occurring on Stewart Island and on the New Zealand mainland. They gather in large groups on remote coasts, with the females remaining in very dense groups. The males are highly polygynous, each having a harem of at least 15 females to oversee, and sexual dimorphism is pronounced. There is no migration, but some dispersal; the non-breeding males occur on both Macquarie Island and the South Island of New Zealand.

When they are foraging, the females dive for two to three days without a break, making dives as deep as 1,500 ft (457 m) and staying down for as long as 12 minutes. Pupping occurs in December or early January and it peaks in late December. Mating takes place in early January, normally six to seven days after birth. The pups quickly become strong and active, swimming to nearby islands at two months. They may suckle for up to a year, but they begin feeding independently long before they are fully weaned.

South American Sea Lion
Otaria flavescens
OTHER NAME Southern sea lion
LENGTH Female to 7 ft/2 m; male to 8 ft/2.4 m
WEIGHT Female to 310 lb/140 kg; male to 770 lb/350 kg
STATUS Approximately 265,000, sensitive to El Niño
 events. IUCN: Lower Risk—least concern. CITES:
 Not listed

The South American sea lion breeds on islands and remote coasts of South America, from Peru south through Tierra del Fuego to Uruguay, and occasionally farther north to the coast of Brazil, and is also found in the Falkland Islands. It forms medium to large colonies, with the females densely packed. South American sea lions consume fish, squid, crustaceans, and, occasionally, young fur seals. They are non-migratory, but disperse up to 1,000 miles (1,600 km) after breeding, with males covering greater distances, and they are known to enter freshwater rivers. During the breeding season, males sequester up to 18 females, and will attempt to take them by force. Successful males may fast and get very little sleep for up to two months. Pupping occurs in mid-December to early February, peaking in mid-January. Mating is at its peak six days after birth, and mothers nurse their young for six months, or up to two years if no new pup is born. LW

Strong Differences
Hooker's sea lions show a strong sexual dimorphism from birth. Males are born bigger than females, and older males become progressively darker. Less than 20 percent of the males will breed successfully during the mating season.

Friends or Enemies?
Hooker's sea lions (above) are very sociable and spend much of their time ashore in groups as tightly packed as sardines in a tin. Unlike the New Zealand fur seal, these sea lions favor sandy beaches over rocky shorelines. They often move a long way inland and sometimes climb high on hillsides.

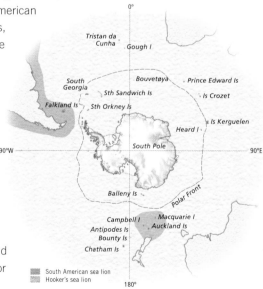

**HOOKER'S AND
SOUTH AMERICAN
SEA LIONS**

Room for One More

A Hooker's sea lion pup (right) leaves its mother's side when it is as young as two days old, and within ten days congregates with other pups. About 6 percent of the females nurse more than one pup at a time—her own and a foster pup.

Danger in Numbers

In 1998, due to a bacterial infection that spread through the breeding colonies (below), 53 percent of the Hooker's sea lion pups died, as well as more than 20 percent of other groups, such as juveniles and adult males. Widespread death from a single factor—hunter, predator, infection, or even weather—is a risk for species that live in close-knit colonies.

True Seals

The true seals, sometimes called hair seals to distinguish them from eared seals, belong to the family Phocidae. There are five Antarctic species of true seal: the crabeater seal, the leopard seal, the Weddell seal, the Ross seal, and the southern elephant seal. All Antarctic true seals are grouped together with the Northern Hemisphere monk seals and the northern elephant seal and called Monacine seals. This grouping is based primarily on dental and skeletal similarities.

Very streamlined, true seals are well designed for life in the oceans, and some are truly ice-dwelling. To cope with the extreme conditions in which they survive, they have well-developed layers of blubber rather than the thick fur that insulates the fur seals—even the thickest waterproofed fur would be inadequate to the challenges of truly polar conditions.

True seals are more at home in the water than they are on the land. They swim using their strongly muscled hind limbs, applying their power in a lateral stroke with their toes expanded. Their entire hind end undulates laterally, and they straighten out their body in a recovery stroke, with their toes bunched together so that they minimize the drag from the water. This is an extremely powerful swimming stroke.

True seals are not as competent on land as they are in the water because they have lost the ability to bring their rear flippers forward and use them as legs, in the way of eared seals. Instead, true seals lumber along on the ground, using their whole body for movement and often their front flippers as levers. Those that live on ice have longer, stronger claws than land-based seals, and these long claws are extremely effective for clambering up on ice and moving around on ice ledges.

Streamlined Design

Their wonderfully streamlined bodies make true seals extremely efficient in water. They have fewer external parts, which would be difficult to heat in such extreme conditions, than do fur seals. For example, instead of external ears (pinnae), they have small ear openings, called meatuses, which lie flush with the sides of their head. The meatus can be less than ½ in (12 mm) wide in northern seals, but in Antarctic seals it is less than ⅛ in (3 mm) wide and is concealed and protected by hair.

Their genital organs are concealed within the contours of their body. Their testes are situated inside the skin muscles, but outside the main body muscles. This differs slightly from genital streamlining in whales,

Dental Problems

Weddell seals only live for around 22 years. A contributing factor may be their practice of maintaining holes and ramps in the ice by sawing and grinding, thus causing damage to their teeth. Dental problems, such as broken, worn, and abscessed teeth, can eventually lead to an early death.

Mother and Child

A female Weddell seal remains very close to her pup for the first two weeks after birth, then she occasionally goes to sea and leaves the pup alone. The pup may be given its first swimming lessons when less than ten days old, and will be fed for a total of six or seven weeks before having to go to sea alone.

where the testes are completely internalized within the body muscles. When the sexual organs are not in use, all that can be seen is a small hole in the belly of the animal. Antarctic true seals have only a single pair of nipples, whereas fur seals and sea lions have two pairs.

Mothering Behavior

Unlike the Otariidae, true seals have a very short period of pup dependence—between four days and several months, depending on the species. The main task of the female during this time is the efficient production and transfer of milk from mother to pup. Because their milk is highly nourishing, this accelerates the nurturing process: true seals produce milk that ranges from 5 to 14 percent protein and 40 to 60 percent fat, with fat levels usually increasing toward the end of the nursing period. It is not surprising that suckling pups can gain as much as 25 percent of their birth weight each day. Most mothers fast while nursing, alternately resting and then feeding the pup several times a day. This is an unusual pattern for mammals—because of the amount of energy required to feed young mammals, most mammalian mothers need to at least double their food intake when nursing. Due to this fasting, female seals lose a great deal of weight while nursing. They wean their young abruptly, simply by going to sea, leaving the pup to fend for itself.

A mother's hunting skills are of great importance to her pups, their survival being dependent on the foraging success of the mother over the whole preceding year. The chances of a pup surviving after weaning depend on a combination of weaning mass (amount of weight it has managed to gain during the nursing period) and the availability of prey while it is learning to catch its own food. Each year, a new cohort (the complete group of youngsters born to a population in that year) takes to the water at approximately the same time, when they must survive—or not—without any maternal guidance.

In many Antarctic seal populations, the survival rates of various cohorts appear to have some relationship to El Niño/Southern Ocean Oscillation (ENSO) events—for example, cohorts of Weddell seals and crabeater seals have shown fluctuations over four to six years that have matched the occurrences of ENSO events. The numbers of leopard seals on sub-Antarctic islands also demonstrate a correlatation with the unusual pack-ice movement that is associated with ENSO events.

Rich Milk

Elephant seal mothers care for their young for about 20 to 25 days. During this time, the pup grows from about 90 lb (40 kg) to about 265 lb (120 kg).

Warm and Fluffy
Waterproof and nearly windproof, a Weddell pup's coat is made up of lanugo, very fine fur in a thick layer for insulation. As it grows, it relies more on its blubber to keep warm.

Analyzing Seal Diets

Working out what a true seal eats is not easy. Marine species, unlike land animals, are extremely difficult to observe when they are feeding.

One way of finding out a seal's diet is to examine the contents of its stomach, intestines, or feces, and identify the hard remains left after digestion. Bones, crustacean exoskeletons, squid beaks, and shells—indeed, any hard items—can be employed to determine exactly what an animal has swallowed.

To count fish species, scientists most often count sagittal otoliths, which are small bones from the head of fish. From these they can also identify the species, as well as estimate the size, of the fish. This technique can provide estimates of the quantity of fish consumed and of the approximate weight of each type of prey. However, the method has its shortcomings. It can only provide information about what an individual animal has eaten recently, and it is likely that the animal's diet will change when it goes out to sea. In addition, different hard items break down at different speeds, so that the contents of an animal's stomach at any given time might be misleading. It is also highly likely that this type of information will hold for a limited region or season, and it might even be heavily influenced by individual animals and their personal preferences, since individual seals are known to prefer different food types when a range of choices is available.

Diving Adaptations

Rather than the usual mammalian number of 13 pairs, seals have 15 pairs of ribs. This allows extra space for their lungs, which are slightly larger than those of other mammals. A human diver breathes deeply before diving and takes a large breath before actually going under the water. This is a disadvantage, because air in the lungs makes the diver more buoyant so that he or she needs more energy to descend.

More seriously, pressurized air in the lungs, or a scuba tank, can lead to a dangerous decompression problem, known as the bends, when surfacing after a dive: nitrogen dissolved into the blood under pressure can be rapidly released from solution into the diver's bloodstream, which can be fatal. When a seal dives, however, the pressure of other organs collapses its diaphragm against its lungs, forcing any remaining air out of its lungs, protecting the seal from risk of the bends.

Seals hyperventilate before diving, but they store the oxygen in their blood and muscles and then expel the air. Seals have a greater quality and quantity of blood than do similar-sized terrestrial mammals. For example, they have proportionally about 60 percent more blood than a human. In addition, seal blood contains a higher proportion of hemoglobin, which carries the oxygen. The combined advantages of these two features mean that seals can carry about three-and-a-half times more

Storing Oxygen
Weddells have very large spleens, where they store red blood cells for release into circulation during deep dives.

Baby Fur

Most true seal pups have no insulating blubber, but they do have very fine, thick baby fur, called lanugo, that is very long in comparison with that of the adult seals. At birth, the baby seal has virtually all the hair follicles that it will have as an adult, so these are much closer together than they will be when the animal has grown to full size. This combination of features means that the pup has a more effective insulating coat of fur than its parents, and it very seldom gets cold. If a pup does begin to get cold, it can produce some additional heat by increasing its metabolic rate.

oxygen per unit of body weight than humans can. About 25 percent of a true seal's weight is blubber, which is effectively inert tissue that does not require a large amount of energy to maintain. If the blubber component is discounted, seals are effectively able to store nearly five-and-a-half times as much oxygen as humans can on a per-weight basis. Seals also store a much greater quantity of oxygen in the high amounts of myoglobin in their muscle tissue. The muscles of a Weddell seal, for example, carry approximately ten times as much oxygen as human muscles do, and they are so red that they look almost black.

When a seal dives for long periods, its heart rate drops from 60 to 70 beats per minute to as low as 15 beats per minute. The brain and the heart receive fully oxygenated blood for as long as possible, while blood flow to the viscera, skeletal muscle, skin, and flippers falls by 90 percent.

In extreme cases, tissues may run out of oxygen and acquire energy by converting glucose to lactic acid—a process that is known as glycolysis. When a Weddell seal dives, lactic acid does not start to build up for 25 minutes, which indicates that oxygen is available in these tissues for at least that long. If a Weddell seal spends 45 minutes under the water, approximately 20 minutes of that time is spent drawing on energy derived from glycolysis, and it must rest on the surface for at least 105 minutes to recover. Recuperation involves an increased heart rate to return oxygen to the body, and the removal of lactic acid from body tissues.

Maintaining Body Temperature in Antarctica

Under normal conditions, the environment of Antarctica is much cooler than the body temperature of a seal, which is around 100°F (38°C), and therefore it must use a range of mechanisms to maintain its vital organs, particularly its heart and its brain, at this temperature. Seals have developed some interesting adaptations to minimize heat loss and conserve energy, including the ability to constrict their blood vessels and cut off the supply of warm blood to skin lying against the ice. This skin becomes very cold, close to freezing point, but the seal's layer of blubber effectively insulates the critical internal organs, which remain heated by warm blood, without the constant energy drain of trying to maintain 100°F (38°C) in the exposed skin.

Paradoxically, Antarctica receives a great deal of solar energy in high summer, and seals coming ashore on warm days would overheat very rapidly if it were not for a special mechanism. In mammals, blood normally flows outward from the heart through the arteries to arterioles and the capillary bed, and is returned through venules to the veins and back to the lungs, where its oxygen is replenished. But seals have an alternative blood flow system: they can dilate special blood vessels that are very close to the surface of the skin and that bypass the capillary bed. These bypasses, known as arterio-venous anastomoses, allow the animals to direct warm blood to their skin surface very rapidly, enabling them to disperse excess heat into the environment and thus quickly return the cooled blood to their body core, where it can draw off more heat and return to the skin surface to repeat the cooling process. A re-circulation system such as this is essential in animals that are as well insulated as seals.

Wonderful Whiskers

Whiskers, known scientifically as vibrissae, are extremely important to seals. They are prominent in most species, visibly clustered on the upper lip and above the eyes, with a single pair on top of the nose. There are many nerve endings at the base of the whiskers, and they are extremely sensitive to motion.

In marine mammals, vibrissae can pick up minute changes in the flow of water—even a minor disturbance in the water caused by a fish's tail while it is swimming may be detected through a seal's whiskers. When a seal is chasing prey, the vibrissae are pushed forward, which possibly makes them more effective at detecting fine differences in motion. This may also enable the seal to predict the exact location of a fish by changes in motion rather than through sight, which is often of limited use, especially at greater depths.

The experimental trimming of the vibrissae of seals has shown that seals without their whiskers find it much harder to catch fish in the dark than do seals that have their whiskers intact. LW

A Hazardous Life
The scavenged remains of a seal, possibly injured by an orca or leopard seal while learning to swim. Upon coming ashore to recover, too tired to defend itself, the seal may have been pecked to death by giant petrels and other scavenging birds.

Cooling Off
Elephant seals are well insulated against the cold water, but conversely may overheat in the hot Antarctic sun. They have a special internal mechanism to combat this, but will also employ other measures to keep cool, such as covering themselves in damp sand, which cools by evaporation and, after it is dry, serves as a sunshade.

True Seal Species

ORDER Pinnipedia
FAMILY Phocidae

Nothing to Hide

The fur of crabeater seals is very stiff, and, since it is not used as a primary source of insulation, it lacks the underfur layer that fur seals have. A crabeater's coat is mostly medium brown when new, and fades over the year to a light yellow or even white-blonde by summer. Older crabeaters often have very scarred hides—healed wounds from attacks by leopard seals and possibly orcas.

Crabeater Seal

Lobodon carcinophagus
LENGTH 7½ ft/2.35 m
WEIGHT 500 lb/225 kg
STATUS Estimates range from 10 to 75 million, but probably between 15 and 40 million. IUCN: Lower Risk—least concern. CITES: Not listed

The crabeater seal is a truly Antarctic seal species. They populate the outer fringes of the pack ice, and range among the icebergs and smaller floes. Recent evidence suggests that they may also like the deep ice close to the continent. Crabeater seals are exploratory: sometimes lost individuals have been discovered wandering north to New

Zealand and Australia, and as far as the River Plate in Argentina. Within their main habitat, they may be found as solitary individuals, in pairs, triads, or small groups, and occasionally in very large groups. Though crabeater seals do not truly migrate, they exhibit some seasonal movement with the annual expansion and retreat of the polar ice.

Crabeaters have an enormous area to call home. In summer, the Antarctic pack ice covers an area of about 1½ million sq miles (4 million sq km). During winter, this increases to more than 8½ million sq miles (22 million sq km). Crabeaters have been observed at population densities ranging from three to 18 animals per sq mile (one to seven per sq km) but under the right conditions, on good firm ice, up to 1,000 will gather in a small area.

Despite their common name, crabeater seals do not live on crabs: they are specialist feeders living mainly on krill. They also eat a small proportion of fish and squid, as well as other invertebrates. Their teeth are highly adapted for their diet, each tooth has a similar shape: extremely convoluted with a few twisted gaps. The complex cusps are designed to strain krill from the

**CRABEATER
SEAL**

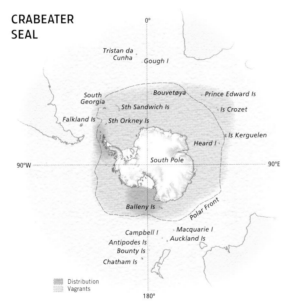

Distribution
Vagrants

Well-fed Crabeater

A crabeater seal stretched out on an
ice floe in sheltered Antarctic Peninsula
waters. The rusty red staining around
the seal's mouth is from the pinkish
krill that makes up the bulk of the
animal's diet.

water; it is thought that a seal catches many krill in one
mouthful, closes its teeth, pushes its tongue up toward
the roof of its mouth, as baleen whales do. Seawater is
forced out through the teeth, then the krill are swallowed.

Up to about 78 percent of all crabeater seals bear
the scars of attacks by leopard seals or orcas. Leopard
seals target the smaller and younger crabeaters. Pups
under six months probably do not survive most leopard
seal attacks, and while older, stronger crabeaters may
survive, the leopard often inflicts permanent damage.

Like all the true seals, crabeaters cannot bring their
hind limbs forward to assist with locomotion on land.
Nevertheless, they can move at 15½ mph (25 kph), and
are probably the fastest moving seals on ice. They use
a sinuous, galloping movement, in combination with
alternate backward thrusts of their front flippers and
side-to-side movements of their hindquarters.

Crabeater seals do not breed in colonies as otariids
do; instead, they breed on the pack ice in isolated pairs.
Their breeding system is one of serial monogamy, in
which the males search out and find a female, remain

with her until she comes into estrus, mate with her, and
then move on in search of another female. Generally,
breeding crabeater females are dispersed so that there
is at least ⅔ mile (1 km) between mating pairs. Females
give birth in spring (late September through to early
November), peaking in October. They fast for the four
weeks they are nursing, during which they lose about
50 percent of their body weight; the pups gain almost
220 lb (100 kg) while being nursed. The nursing females
are very aggressive toward males. They nurse the pups for
around 28 days, then come into estrus and mate in late
October and November, when they stop lactating.

During the lead-up to mating, the males fight for the
opportunity to attach themselves to a mother and her
pup. The successful male protects the female and her
pup from predators, as well as from other males. It is
not known if the male forces the pup to wean by driving
it away when the female comes into heat, or if she weans
it before coming into heat. At any rate, at the point the
pup is weaned, the male becomes aggressive, separating
the mother from her pup, then mating.

Leopard Seal

Hydrurga leptonyx

OTHER NAME Sea leopard

LENGTH Female 10 ft/3 m; male 9 ft/2.8 m

WEIGHT Female 83 lb/370 kg; male 73 lb/325 kg

STATUS 220,000 to 440,000. IUCN: Lower Risk—
least concern. CITES: Not listed

The leopard seal is the second largest of the five true
seals in Antarctica, after the southern elephant seal. It
is a true hunter, possessed of powerful jaws and a long
sinuous neck, which it uses to great effect, retreating
backward before it strikes its prey like a snake. This
powerful animal can leap onto ice floes 7 ft (2 m) above
water level—an exit speed from water of around 20 ft
(6 m) a second. Leopard seals would rather be on ice
than on land, but they will haul out of the ocean onto
land if ice is not available.

The leopard seal's main habitat is the Antarctic pack
ice, and its icebergs and smaller ice floes. The animal
is circumpolar, ranging through the southern oceans. It
has often been observed on sub-Antarctic islands, and
in Australia, New Zealand, Chile, Tristan da Cunha, and
southern Africa. Individuals have drifted as far north as
Raratonga in the Cook Islands.

But despite this broad range, leopard seals are not
truly migratory, even though they do disperse northward
from time to time; probably this is for reasons that are
associated with the availability of food and the location
of the pack ice. Generally speaking, the mature leopard
seals prefer to remain in the Antarctic pack ice, while
youngsters can be seen throughout the sub-Antarctic
region from June to October each year. Adults ranging
from three to nine years of age tend to congregate at the
Antarctic coast in summertime.

Leopard seals are known for their hunting prowess,
and this is one of the few seals that sometimes prey on
warm-blooded animals. They take crabeater seal pups
and also penguins. They usually capture adult penguins
in the water. They rarely hunt on land, where they are
relatively cumbersome. There is an extraordinary record
of four leopard seals consuming 15,000 Adélie penguins
in 15 weeks; one of the animals was found with 16 adult
penguins in its stomach.

Despite their reputation, most leopard seals do not
regularly hunt warm-blooded animals. Their usual diet
consists of 50 percent krill, 20 percent penguins, 15
percent other seals (mostly crabeater pups), 9 percent
fish, and 6 percent squid. Most juvenile leopard seals
do not possess very sophisticated hunting skills and they
probably rely on krill for food. However, they are not as
expert at catching the krill as the highly successful and
specialist crabeater seal.

The leopard seal's teeth provide a clue to its diet. In
shape they are similar to the crabeater's, with extremely
convoluted lobed cusps that are used as krill strainers.
However, unlike the crabeater's, the leopard seal's teeth
have very sharp edges and points, which are used for
grasping seals, penguins, and fish.

Leopard seals are solitary animals. The only groups
that they ever form are mother–pup pairs and temporary
mating pairs. It is not known whether leopard seals are

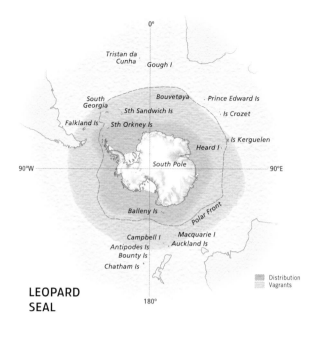

**LEOPARD
SEAL**

serially monogamous, like crabeater seals. They breed
in the pack ice, with breeding females widely separated
from each other. The females, which are slightly larger
than the males, pup in the spring and the early summer
(from September to December), with the births peaking
in November. The mothers nurse their pups for about a
month, and then mate when they stop lactating, which
is generally in late December.

Fast and Ferocious

This leopard seal displays teeth well
adapted to seizing and tearing flesh,
but also for straining krill—which is
an important part of its diet. Leopard
seals are fast, fierce hunters and they
take a wide variety of prey, including
penguins, fish, squid, and even other
seals—usually crabeater pups.

Rich Pickings
During the latter part of the summer, many leopard seals hunt by waiting offshore for young penguins and seals to enter the water. Some leopard seals return to the same location each winter, while others just drift around.

Staying Alone
Leopard seals are closely related to crabeater and Weddell seals, but have a sleek reptilian grace their relatives lack. They do not form large breeding colonies, and their seasonal movements are not well known—although some individuals disperse northward during winter, and young animals seem more likely to move about than adults.

Weddell Seal

Leptonychotes weddelli

LENGTH Female 11 ft/3.3 m; male 10 ft/3 m

WEIGHT 880–1,1100 lb/400–500 kg

STATUS Approximately 800,000. IUCN:Lower Risk—
least concern. CITES: Not listed

The Weddell seal is the most southerly breeding of all
mammals. They tend to inhabit areas of coastal fast ice
(ice that is connected to land) rather than the moving
pack ice. Although Weddell seals favor the coastal fast-
ice areas of the Antarctic continent and of the nearby
islands, they are occasionally found in South America,
Australia, and New Zealand.

Particularly during winter, these seals breathe and
enter the water through holes in the ice. They need to
make or maintain these holes themselves, cutting the
ice with incisor teeth, and enlarging and maintaining
the openings with their canine teeth. They will not begin
to cut a hole in ice that is more than 4 in (10 cm) thick,
but once they have made a hole they will maintain it
until the ice reaches a thickness of 7 ft (2 m).

Outside the breeding season, Weddell seals are not
particularly gregarious, but nor do they actively avoid
each other, and in good conditions they will comfortably
occupy ice at densities of 38 to 90 animals per sq mile
(15 to 35 per sq km), while maintaining a good distance
between each other.

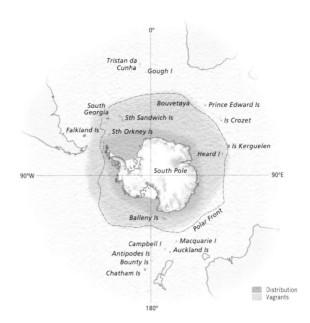

WEDDELL SEAL

However, during the breeding season, the females
gather in small, loosely knit groups of as many as 100
individuals to give birth in the very stable fast ice around
the southern continent. They share the breathing holes
that occur along cracks in the ice, while breeding males
wait in the water beneath these entry points.

The males defend an aquatic territory against other
males. Indeed, by defending his entry hole, a single
male can monopolize a large number of females. It is

Home Bodies

Ross and leopard seal populations
tend to move with the flow of the ice—
presumably to stay in their favorite
zone within the pack ice—but Weddell
seals seem to prefer to stay put on the
inshore fast ice.

possible that male Weddell seals are a little smaller than the females so they have an advantage when they are maneuvering to defend their underwater territory.

During the pupping season, their breathing holes become scenes of fierce displays of male aggression. The breeding males guard their holes against each other and they drive the younger animals away from the area. The sub-adults remain in the background until the breeding season ends; at this point, tension eases somewhat, and adult males permit the young to come back into the breeding area. The groups gathered at the breathing holes do not really disperse after breeding, but they do eventually spread out over a larger area as more cracks appear in the ice with the warmer weather, creating new breathing holes.

Their peak pupping period is from September to November; however, pupping can take place at any time from late August onward. The lactation period lasts for approximately 50 days, much longer than for most other phocids. This means that the lactating Weddell females must feed during the nursing period, unlike the other Antarctic true seal species. The males that have won territory wait for females to wean their pups, and then mate with them when they enter the water.

Weddell seals are believed to consume a diet of fish, cephalopods, krill, and other invertebrates. They locate and catch this prey using a range of diving behaviors.

Some dives can be quite long (between 20 and 73 minutes), and comparatively shallow. These dives are generally aerobic (using oxygen stored in the blood and muscle), but a small proportion are longer than their aerobic limit of 25 minutes. Other dives are deeper and may last for 5 to 25 minutes, and be 1,300 ft (400 m) deep. Weddell seals often feed at a depth of 650 to 1,300 ft (200 to 400 m) to catch Antarctic cod—prey that is well worth the effort, as these fish may reach as much as 66 lb (30 kg) in weight. When diving, the seals swim at about 115 ft (35 m) per minute. Thus, a normal dive covers the surface equivalent of about 2 to 4 miles (3 to 6 km), although some Weddell seals also have been known to travel up to 7 miles (12 km) from their access hole in the ice.

The diving patterns change over the year, probably according to the availability and distribution of the seals' prey. In the summer, Weddell seals hunt fish and squid throughout the day, diving to a mid-depth of about 500 to 1,000 ft (150 to 300 m). In spring and autumn, when their main target is krill, in the course of a day's diving the depths fluctuate considerably, from between 330 and 1,600 ft (100 and 500 m). Weddell seals have good underwater eyesight and make much deeper and riskier dives in daylight. It is likely that in the winter dark they dive for only short periods within familiar areas.

A Mother's Fast
A Weddell seal pup weighs 55 to 66 lb (25 to 30 kg) at birth and when weaned, at six or seven weeks, is about 220 lb (100 kg). The full-grown females are slightly larger than the males, and nurse their pups longer than other seals.

Seal Song

Most seals vocalize from time to time, either above or below the water. Underwater, their utterances vary considerably, some species being almost completely silent, others extremely vocal. The noises they make include barks, whinnies, clicks, trills, moans, hums, chirps, belches, growls, squeals, and roars, as well as many other sounds not so easily described. Some sounds are similar to those that whales use for echolocation. However, suggestions that seals echolocate are still extremely speculative. Rather, many of the reported instances of underwater vocalization by seals appear to be related to establishing and maintaining breeding territories.

Weddell seals are particularly noisy underwater and produce a constant stream of varied sounds, such as whistles, buzzes, tweets, trills, chirps, and growls. Male Weddell seals produce more complex vocalizations than do females and these sounds can be detected as much as 20 miles (30 km) away. The vocalizations can even penetrate the fast ice and so might well play a part in territorial displays.

Seals also chirp, whine, bark, hum, growl, squeal, and roar when they are above the water. Not surprisingly, communal species are much more vocal on land than solitary species, and seals that live in colonies spend much more time vocalizing than those that live widely spaced apart. When individual animals from noisy communal species are isolated from other individuals, they fall almost completely silent. However, if they are returned to a group situation, they become extremely noisy again, which suggests that their vocalizations are a form of direct communication.

Ross Seal
Ommatophoca rossii

OTHER NAMES Singing seal, big-eyed seal

LENGTH To 8 ft/2.4 m

WEIGHT 500 lb/220 kg

STATUS Thought to be approximately 220,000.
IUCN: Lower Risk—least concern. CITES: Not listed

Just Wandering About

Ross seals don't appear to migrate, but they live so deep in the ice that it is hard to be certain. Populations do, however, move over large areas, and surveys often show very patchy distributions within the ice.

Ross seals live deep in the Antarctic pack ice, and they are patchily distributed right around the southern polar regions. They challenge the Weddell seal for the title of southernmost mammal. The Ross seal is a true pack-ice species and only rarely hauls out on land. Scientific

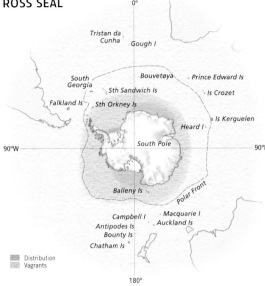

ROSS SEAL

Distribution
Vagrants

knowledge of the species is limited: its preferred habitat in dense pack ice is difficult for humans to easily reach in ships or from land.

At least part of the Ross seal population appears to undertake some seasonal movement, presumably in response to movements of the pack ice and fluctuations in food availability. It is believed that individuals may maintain separate underwater territories, and that they may communicate with each other from within those territories, across great distances. Ross seals can dive to considerable depths in pursuit of their favored prey, which includes cephalopods, fish, krill, and bottom-dwelling invertebrates.

The most distinctive physical feature of the Ross seal—a feature that is unique among seals—is the dark, longitudinal streaks on its throat and on the sides of its head. The Ross seal is not as heavy as most of the other true seals in Antarctica, but it does have a thick neck and throat that show this striped coloration very clearly. It has very long back flippers—the longest of any seal in proportion to its body size. Its body hair and vibrissae (whiskers) are very short. It has a small mouth with very highly developed muscles of the tongue and pharynx, together with a short hard palate at the top of the mouth and a very long soft palate behind that.

When they are approached, Ross seals open their mouths very wide and fill their soft palate with air. They then use their inflated soft palate and their pharynx as resonating chambers, pushing the back of their tongue against the soft palate to produce a series of remarkable trills and thumps, without needing to expel any air at all.

Breeding takes place in the pack ice, where the females are solitary or accompanied only by their pup, or sometimes are part of a very small group of females which remain widely separated. Not much is known about the Ross seal's mating behavior, but it is known that the females are slightly larger than the males, that the pups are born in the late spring and early summer (November and December), and that they learn to swim almost immediately. The seals' peak mating period is in December, about one month after the females have pupped, and it is believed that mating marks the point at which the new pup is weaned.

Southern Elephant Seal

Mirounga leonina

OTHER NAME Southern sea elephant
LENGTH Female 7–10 ft/2–3 m; male to 17 ft/5 m
WEIGHT Female 880 lb/400 kg; male to 8,160 lb/3,700 kg
STATUS 750,000 in 1985; some populations are
declining. IUCN: Lower Risk—least concern.
CITES: Appendix II

The southern elephant seal is the largest seal species.
Both sexes are solid-looking animals with thick layers of
blubber, but the bulls are very bulky, with exceptionally
thick necks and chests, and are over nine times larger
than the females. Despite their bulk, both sexes are
surprisingly quick and nimble, even when they are out
of the water. Around their necks, mature bulls usually
show a great deal of heavy scarring from fighting, and
cows may have some small scars on the head and neck
incurred during mating.

The male elephant seal has an inflatable nose. This
developed nose is a sign of adulthood, and extends fully
only when the bulls are completely mature, at about
eight years of age. At this time, the tip of the nose can
hang down below the animal's mouth, and its nostrils
point straight down. The nose has a massively enlarged
nasal cavity, which a breeding male can fill with air. The
whole nose can then be made erect by increasing the
blood pressure to the area. At this point, a large raised
cushion along the ridge of the nose is formed. The

**SOUTHERN
ELEPHANT SEAL**

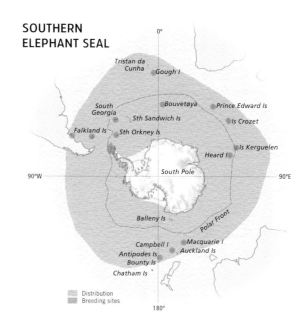

enlarged cavity may act as a resonating chamber for the
male's roaring, which he uses to intimidate other bulls.

Southern elephant seals spend most of their lives at
sea in Antarctic and sub-Antarctic oceans, where they
consume a diet of cephalopods and fish.

They breed on islands situated on both sides of the
Polar Front, on the Argentinian coast in Patagonia, and
on the Antarctic mainland—though rarely. They occur
in three discrete groups: a southwestern Atlantic Ocean
population, centering on South Georgia and including
the Falkland Islands and Signy Island; a population in

Dominance Fighting

Fighting is the last resort of the males
after a great deal of posturing and
roaring. Bulls rise on hindquarters and
slam their chests into opponents,
reaching for vulnerable spots, such as
the throat or head, to inflict a bite. They
are able to do much harm to each other,
but their thick necks, which are covered
in a layer of fat, are well protected.

Misled by a Nose

The male has the big inflatable nose
that gives elephant seals their name.
It grows as the male matures, and
reaches full size at about eight years,
but even then it is only fully enlarged
during the breeding season. When an
elephant seal snorts and grumbles, his
sounds are directed into the resonating
chamber of his mouth and pharynx. This
makes him sound big and tough, and is
intended to scare off other bulls without
any actual fighting.

A Well-deserved Rest

Elephant seals are great divers, and of their time in the water, 93 percent is spent underneath it. The average dive duration for elephant seals is 22 minutes, the longest recorded dive being 120 minutes. The average depth of their dives is 1,300 ft (400 m), but they regularly go down to ⅔ mile (1 km).

the southern Indian Ocean, based predominantly on Heard Island and Iles Kerguelen; and a population that breeds on Macquarie Island, south of Australia and of New Zealand. Non-breeding animals are regularly spotted along the coasts of Antarctica. One traveling southern elephant seal was tagged in South Georgia and some time later was spotted off South Africa—a massive swim of about 3,000 miles (4,800 km).

The adult males are the first to come ashore at the beginning of the breeding season, in late winter to early spring (from late August to September). The bulls stay on the beaches for more than two months, fighting to establish and maintain dominant hierarchies. The adult females begin to arrive a few weeks later, in September or October, and they remain there for four to five weeks. Within less than a week of coming ashore, the females give birth to the pups that were conceived the previous year. When ashore, the females form large aggregations, known as harems, which can contain more than 100 individuals, although the average is generally about 40. Neither sex feeds while ashore, and both of them lose vast amounts of weight.

There is generally plenty of space when the earliest females come ashore, and they begin in small, peaceful harems, managed by the dominant bulls. But once they have given birth, and more females come ashore, the seals become much more aggressive. As the season progresses, more cows arrive and more pups are born. The beaches become thronged with females and pups fighting to claim space among the disorderly hundreds of animals. Meanwhile, the bulls are fighting constantly for dominance as they attempt to keep control over as many cows as possible.

Newborns are suckled by their mothers for only 20 to 25 days, but during that period they gain between 70 and 230 lb (30 and 105 kg), often tripling their birth weight. The cows produce as much as 1¼ gal (4.7 L) of milk each day and may lose up to a third of their body weight during their short nursing period. Right at the end of their nursing period, 17 to 22 days after giving birth, the cows come into estrus. Sometimes the bulls may weigh up to 25 times more than the pups. In their amorous enthusiasm, these large males scarcely seem to notice the pups, and can accidentally trample on the

youngsters—some of whom may die. A few days after mating, the females permanently abandon their pups and go back to sea to feed and to regain the weight lost during the breeding season. At this time, the cows are pregnant, but the implantation and development of the embryo is delayed for four months, which ensures the pups will not be born for 11 months, when the mothers return to the breeding beaches the following season.

After their mothers have returned to sea, the pups congregate around the edges of the beaches. Together they learn to swim and to hunt, first in nearby shallow waters and gradually farther out to sea. They leave the beaches when they are two months old, usually about a month after the departure of their mothers.

Bull southern elephant seals retain dominance and control of their females by fighting other males to keep rank. When a challenger appears, he will announce his arrival by making a huge roar in an attempt to intimidate his rivals with noise and bluster. A fight will ensue if the answering roars from the resident bull, known as the beachmaster, do not frighten off the intruder. The bull will rear up and strike at his opponent's throat and neck

with his teeth. Such conflicts, an awesome sight, often leave the opponents bloodied. However, the seals' thick necks, which are well protected in a layer of blubber, usually prevent any serious injuries.

Early in the breeding season, big experienced bulls dominate the beaches—a powerful bull being able to control a breeding territory with as many as 60 females. As the harem grows with the arrival of more females, a second male will snatch the opportunity to mate with new females as they come into estrus, and on densely populated beaches, where more than 130 females may haul out, a third bull may be able to gain access to the newly arrived females.

A strong, dominant bull can keep his position at the top for the whole season, but this requires a great many battles, which can be exhausting. Before the breeding period draws to a close, the weaker males, particularly those that are older or younger, or have been injured, are likely to lose to a vigorous bull in his prime. But as soon as the new bull takes up his dominant position, he, in turn, is certain to be challenged by other males, each awaiting his chance to make the top spot.   LW

Vocal Reminder

A female southern elephant seal (above) loudly claims her breeding spot on the crowded and noisy beach. She may be one of a harem of 100.

Not Built for Walking

Young pups (top) are comparatively light and flexible, but as they grow, they quickly become wrapped in fat. Adult elephant seals carry a lot of blubber, and have limited ability to move on rough ground. When not in the water, they prefer to remain near the sea, on gravel, sand, or mud, and in flat tussock areas near beaches.

Whale Family Tree

All cetaceans evolved from a common ancestor. Surprisingly, their nearest living relatives are terrestrial grazing animals such as cattle. The fossil record shows cetaceans present in recognizable form 50 million years ago. The most significant evolutionary event in the order Cetacea, the divergence between baleen and toothed whales, took place about 38 million years ago.

ODONTOCETI
Toothed whales

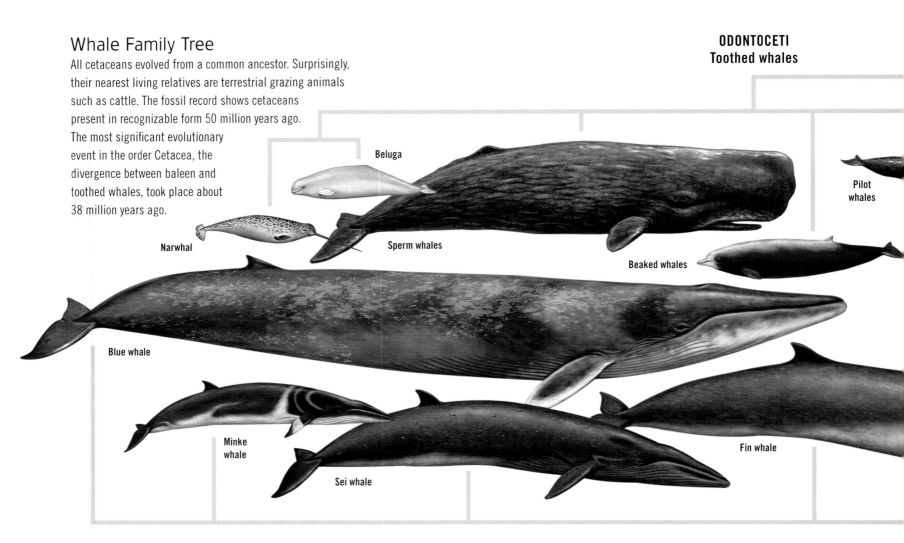

Beluga

Narwhal

Sperm whales

Beaked whales

Pilot whales

Blue whale

Minke whale

Sei whale

Fin whale

Whales

ORDER Cetacea

Whales, dolphins, and porpoises belong to the order Cetacea, from a Greek word meaning "large sea creature." And many are indeed large—few, if any, dinosaurs ever reached the size of a blue whale, which, at 100 ft (30 m) long, is one of the largest animals ever known to exist. Not all cetaceans are large, however; some dolphins reach 4 ft (1.2 m) only, but are still classified as whales. There are two suborders of whales: Mysticeti, the baleen whales, and Odontoceti, the toothed whales. Of almost 90 known whale species, 11 are baleen feeders and 76 toothed species, which are generally smaller than the baleen whales. Dolphins and porpoises are toothed whales.

All the whales are carnivorous, but different species favor different types of food. One of nature's paradoxes is that huge baleen feeders mainly live on zooplankton, filtering great quantities of these minute sea creatures from the water through comblike plates in their mouths called baleen. Toothed whales use echolocation to hunt larger animals, such as fish and squid, and in Antarctic waters orcas hunt warm-blooded prey, including seals, penguins, and even other cetaceans.

To Dark and Icy Depths
With flukes raised, a sperm whale begins a deep vertical dive to feed on squid. Sperm whales reach phenomenal depths—some 10,000 ft (3,000 m) or more—and can remain submerged for more than two hours. But most dives are much shallower and shorter. Whales use loud clicks both to communicate with each other and to echolocate their elusive, fast-moving prey.

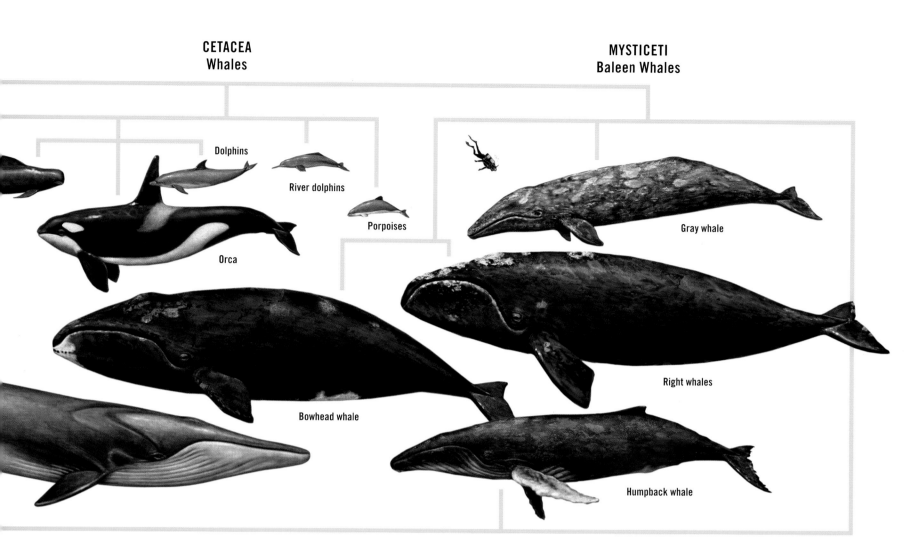

CETACEA
Whales

Dolphins

River dolphins

Porpoises

Orca

Bowhead whale

MYSTICETI
Baleen Whales

Gray whale

Right whales

Humpback whale

Whales inhabit all the world's seas, from shallow tropical waters to the deepest and coldest oceans. Their lives revolve around two major requirements—feeding and breeding. For most baleen whales, the feeding season is determined by the annual cycle of productivity in Antarctic waters. During the summer season, sea ice is at its minimum extent, and zooplankton, including krill, are most abundant and easily found. During winter, when ice covers the sea, krill are more difficult to find, and the majority of baleen whales head north to warmer tropical or temperate waters to breed. The reason they do this is still not certain, but it is thought likely that warmer water allows calves to channel their energy into growth in early life, when their insulating blubber is thin. However, it is now known that some minkes and orcas, at least, remain in sea ice during winter, and may breed there. Because warm seas tend to be food-poor, most baleen whales have evolved a cycle of feast and famine, storing enough energy in blubber during summer to fuel them during their travels and breeding.

Living in Water

Whales are mammals, and like all mammals they must breathe air. But because they have adapted over tens of millions of years to a totally aquatic existence, they differ from land mammals in many ways. Two of the important differences are how they breathe and the related issue of how they give birth in water.

Most mammals breathe through their mouth as well as through their nostrils. In an adaptation to aquatic life, however, whales have an airway that is separate from

their mouth, and their nostrils (the blowholes) are on the top of the head to allow efficient breathing when surfacing. This also means baleen whales can breathe while simultaneously filling their mouths with water and food, which may be particularly important for suckling calves. Unlike humans, whales are voluntary breathers: they hold their breath and choose when to breathe. The breath is often exhaled explosively—the blow—followed by a very rapid inhalation before diving again. Newborn calves are often assisted to the surface for their first breath—in some species, by "aunts," who accompany the mothers during birthing.

In other adaptations to the aquatic lifestyle, whales have virtually no hair, no sweat glands, no external male sex organs, no external ears, and no hind limbs. They do not have gills or scales as fish do, but they do have some fishlike characteristics—a streamlined, torpedo-like form, for example. Unlike fish (which have vertical fins and tails that move from side to side), whales have horizontal tail fins, which are called flukes; these move up and down. Cetaceans have evolved to move through the water with minimal turbulence compared with most of the other mammals.

Temperature Control

Being warm-blooded, cetaceans need to maintain a constant body-core temperature. Water quickly draws heat from warm bodies, and some whale species that migrate to polar waters inhabit seas with temperatures as low as 28.4°F (−2°C). To cope with this, they carry a warm coat of blubber—a mixture of fat, oil, and connective

Seasonal Mover
A southern right whale swimming near the Auckland Islands, a winter breeding area where mating and calving take place. In the summer, southern rights migrate southward to the rich feeding grounds in the Southern Ocean.

Skimming the Surface

Southern right whales spend most of their time near the ocean surface where krill and other small crustaceans swarm. When feeding, the whale opens its enormous mouth and swims slowly, continuously filtering prey through its very long baleen as it moves along. Among the fattest of whales, the thick blubber coat that warms southern rights in cold waters also equips them for their buoyant life at the surface.

tissue—which lies underneath their skin. In some whale species, blubber may make up as much as 30 to 40 percent of their total body weight, and is the energy store that fuels their long winter migrations, including the production of calves and milk. Humpback whalers in New Zealand and elsewhere caught whales on their northward migration to breeding areas while they were heavy with blubber, but ignored them at the end of the winter on their southward migration when the whales were thin and hungry. Many species that migrate to warm tropical waters to give birth may do so simply for the benefit of their infants: the calves have time to grow a layer of blubber before encountering colder waters for the first time. On the other hand, some whale species (including orcas, bowheads, and narwhals) live all their lives in icy waters and bear their calves there, and these calves are quite capable of surviving the conditions.

Whales also have a heat-exchange mechanism in the circulatory system that helps to regulate their body temperature. In mammals generally, warm blood flows to their extremities and cooler blood, stripped of its oxygen, then returns to the lungs to be replenished. The whales' blood system is different; they have interwoven inward and outward blood vessels. When they are cold,

they can remove the heat from the blood traveling to the extremities and return it back to the core to keep their vital organs warm. When a whale becomes too hot, it can direct warm blood to its body surface, especially the fins. The colder water absorbs the heat, and the cooled blood lowers the temperature of the internal organs.

An advantage of their large size is that larger bodies conserve heat more efficiently because of comparatively low ratios of skin surface to body mass. Animals with a larger body mass have proportionally less surface area to lose heat through than small animals. One possible cause of their evoluting to such sizes is to enable the whales to spend more time in cold, food-rich waters.

Cetacean Anatomy

Whales moved into the oceans millions of years ago and they have evolved adaptations that are not seen in land animals. For example, they have no gall bladder or appendix. They do not have carotid arteries; instead, small blood vessels (retia mirabilia) carry blood to the brain and may help maintain consistent pressure in the blood flow to the brain, especially during deep dives.

Some whales have become so large because, living in water, they are not as subject to constraints of gravity

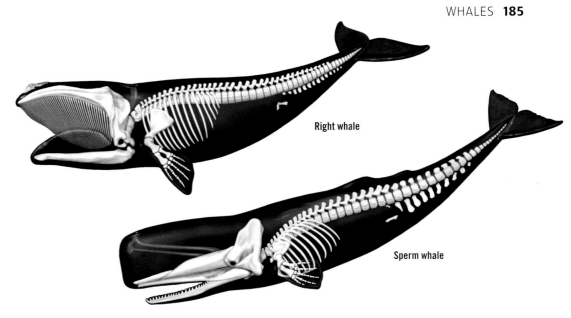

Right whale

Sperm whale

Baleen and Toothed Whales
The main difference between baleen and toothed whales relates to feeding.
Baleen whales (top) have flexible plates for filtering small schooling prey.
Toothed whales (below) have sharp teeth for grasping larger individual prey.
Baleen whales have retained two nostrils; toothed whales have only one.

as terrestrial animals. When large whales strand, they may be crushed under their own weight. Whale bones are spongy and filled with oil, which may increase their buoyancy, although most whales will sink when dead. Their skeletons are flexible in the vertical plane, but they generally lack sideway flexibility. The neck vertebrae fuse during maturation to support the head as it travels forward through the water. The head itself has become elongated during the course of evolution, to support the feeding structures, such as baleen or an array of teeth. The front limbs have become flattened paddles, which are used for steering and control; however, their bone structure still shows their common mammalian ancestry in the form of five "fingers." Toward the whale's tail, the vertebral column becomes more tapered and flexible, permitting the vertical movement used for locomotion. The flukes themselves are not supported by bone, but by muscle and firm, elastic connective tissue. Whales do not have any trace of leg bones, and the vestiges of the pelvic girdle (the hips) have dwindled to two small bones that are embedded in the body wall rather than attached to the skeletal structure. In land mammals these bones support the hind limbs, but in whales their only purpose is to support the muscles of the male's

external reproductive organs, which are usually hidden in a slit that runs along the underside of the body.

Taking to the Water
Prehistoric cetaceans are among the earliest recorded marine mammals: fossil cetaceans have been found on all continents. Aquatic dinosaurs died out 65 million years ago, leaving oceanic niches. Mammals took to the water, and cetaceans were already present 50 million years ago. There are fossils of animals with webbed hind feet—semi-aquatic precursors of the whales—and there is some evidence that whales descended from a wolflike, hoofed land mammal with large teeth, which is also thought to be an ancestor of cattle and other even-toed hoofed mammals.

Cetaceans gradually became more efficient in water; by 38 million years ago, toothed and baleen whales had evolved their main characteristics, and 25 million years ago, different species had evolved within the two groups. About 15 million years ago, climatic cooling produced the nutrient-rich polar seas of today: larger whales seem to have evolved as a result of these conditions. Modern groups of whales had largely emerged by approximately 7 million years ago. LW

Survival Skills
Insulating blubber enables the minke whale to live throughout the year in ice-packed Antarctic waters. One of the smallest baleen whales, minkes spend their lives locating krill swarms beneath the ice, and at the same time playing a cat-and-mouse game of survival with their own predators—orcas. Minkes can hold their breath for only 20 minutes or so, and they may use echoes of their own calls to detect vital breathing holes.

Baleen Whales

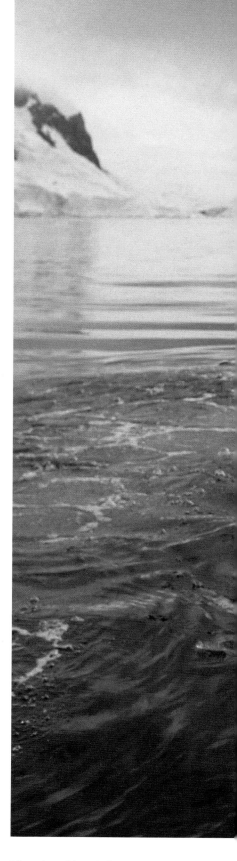

Breathing Easy
A rorqual surfaces, showing the twin blowholes characteristic of baleen whales. Nostrils on the top of the head enable these whales to breathe while mostly submerged. Many fast-moving species also have a raised splashguard (seen here), so they can breathe while swimming, even in rough seas.

Only around 15 percent of whale species are baleen whales—but what they lack in numbers they make up for in their size. Even the smallest baleen, the pygmy right whale, reaches 20 ft (6 m), and the mighty blue whale grows to about 100 ft (30 m).

Unlike the toothed whales, which have only a single blowhole, baleen whales have two blowholes, which are protected by a raised splashguard, known as coaming, that prevents water from entering the blowholes while the whales are breathing.

Baleen whales are named for the comblike baleen plates, which hang from their upper jaws and which are used to strain their prey from seawater.

Baleen whales are divided into two groups: rorquals (groove-throated whales) and right whales. Right whales have enormously long baleen, and continuously strain prey as they swim, whereas most rorquals rapidly engulf huge mouthfuls of food, extending their throat capacity by ballooning out their throat pleats, creating a huge sac containing water and food. The muscular pleats are then contracted, squeezing water through the filtering baleen while their prey remain inside.

In order to catch enough food, these whales need enormous jaws that can provide enough room for the filtering baleen. This requirement has resulted in major modifications to the shape of the baleen whale skull, which is very different to that of the toothed whale. The head of some species, such as the right whale, may be as much as a third of its body length.

Each species of baleen whale has a different color, thickness, and number of baleen, depending on their feeding habits. Coarse baleen filter out the larger prey, whereas finer baleen filter smaller prey. Right whales, which generally hunt very small prey, have long, very fine baleen as well as a very bowed upper jaw in order to accommodate the length of baleen.

Baleen whales feed on zooplankton, but they also devour small fish, particularly schooling species, such as herring, mackerel, capelin, pilchards, and sardines. These species, or bottom fish such as cod and sand lance—and even the occasional squid—make up the bulk of baleen whale food in regions, especially in the Northern Hemisphere, where small schooling fish are abundant. However, in the cold southern waters of the Antarctic, zooplankton, particularly krill, is the baleen whales' principal food source.

The distribution of krill—and of baleen whales—is determined by the dynamic oceanographic processes influencing phytoplankton production. These minute sea plants—and krill—live under and among sea ice during the winter and, as the ice melts back in spring and summer, migratory baleen whales follow its retreat to feed. It was long considered that the southern oceans were largely empty during winter, but recent studies have shown that phytoplankton, krill, and minke whales (which are small enough to live in the open cracks in the ice) are present year-round.

Whale Talk

All cetaceans produce sound for communication, and at least some for echolocation. Toothed whales, which are known to echolocate, use higher frequencies than do the baleen whales, which have not yet been proven to echolocate. The higher frequencies travel only short distances, whereas the very low frequencies may travel hundreds or even thousands of miles. The sounds some dolphins produce are too high-pitched for the human ear, whereas the deep rumbles of fin and blue whales are below our hearing range. In between, most whale species produce sounds that can be heard by the human ear. Many of us are familiar with the haunting songs of humpback whales, or with the whistles and the clicks of bowriding dolphins.

Size Matters

Female baleen whales are larger than males; a mature female blue whale, for example, can be 7 ft (2 m) longer than a male. This sexual dimorphism is thought to be linked to the migratory habits of the whales: females, which travel at fast speeds during long winter migrations, must have the energy reserves not only to travel from feeding grounds to breeding grounds while they are fasting, but also to produce and suckle a calf while they are doing so. The greater size of females enables them to carry more energy supplies, in the form of blubber. LW

Massive Mouthfuls
Humpback whales surface-feed on krill. Keeping their mouths wide open, the whales lunge through prey. Humpback whales often form pairs in the feeding grounds and they may cooperate to herd and capture their prey.

Food Strainer

The comblike horny material growing from their upper jaws is the most distinctive feature of baleen whales.

Hearing Under Water

Sound travels about four-and-a-half times faster in water than it does in air, and much farther; vision, on the other hand, is poor, especially in deep or murky water. So it seems likely that hearing plays an important role for species that live mainly under water, conveying information rapidly and from great distances.

As whales' internal ear bones are well developed, hearing is probably important to them. However, whales may not hear as humans do—that is, through their ear canals. In toothed whales, a waxy plug blocks a narrow ear canal that does not appear to be connected to the eardrum, suggesting that their hearing mechanism is not like that of humans. Scientists suspect that sounds may be transmitted to the whales' internal ears through soft tissues, possibly the fatty deposits in their chins, or through their skull bones. It may be that under the water the tissues or bones receive and transmit sound more effectively than would the ear canal.

The internal ears of a whale are protected by foamy surroundings that reflect sound, and this may form a sophisticated stereophonic sound system that allows whales to use echoes to pinpoint a mate, an enemy, or their prey.

Right Whales

SUBORDER Mysticeti

FAMILIES Balaenidae, Neobalaenidae

Right whales were so named by whalers because they were the "right" whale species to hunt: they inhabited coastal waters, were slow swimmers, they floated when dead, and they provided valuable baleen and high-quality oil. All these factors contributed to the mass slaughter of right whales in the nineteenth and early twentieth centuries.

The five right whale species—the north Pacific right, the north Atlantic right, the bowhead, the southern right and the pygmy right—inhabit the Earth's cooler seas. The bowhead is restricted to the waters of the Arctic, and the two northern rights species to their respective ocean basins. The southern right and the pygmy right whales share the cool temperate waters of the southern oceans, although the southern rights penetrate farther south into polar waters.

A right whale has a deeply curved jawline to provide room for its long, slender baleen plates. This feature— and its large head, which makes up to a third of its body length—means that it is easy to identify. It uses its fine baleen to filter from the water copepods, krill, and other small crustaceans, usually preferring to skim-feed near the surface with its mouth open. Unlike the rorquals, right whales have no throat pleats, and therefore cannot expand their throats to maximize a mouthful of food.

They are quite stocky, have no dorsal fin, and are usually almost completely black-brown, except for white patches, called callosities, on the head and white belly patches. Only the southern right whales and the two northern right whale species grow callosities, so their presence is a positive identification of a right whale.

Callosities are raised, horny layers of skin that may be heavily infested with barnacles, parasitic worms, and whale lice (cyamid amphipods). They are present from birth, and males have more than females; possibly they function as weapons against other males. Whalers gave names to the various patches—the largest, on the tip of the upper jaw, they called the bonnet. Others were variously described as the beard, the eyebrow, and the lip patch, and any further patches were called islands. But each right whale is imprinted with a unique pattern of callosities and can be recognized by it, rather like humans can be by fingerprints.

Mother and Calf

A southern right whale and her white calf (on her left). The color will darken to a milk-chocolate brown during its first year of life, but never to the brown-black of its mother.

Breaching

A southern right whale breaching. It is believed the main reason for breaching is communication with other whales— the thump can be heard over ⅔ mile (1 km) away. Breaching may also be to remove parasites or dead skin, to observe objects above the surface, or simply to express exuberance. Whales may breach when threatened by a predator, another whale, or a boat that approaches too closely.

Inseparable

The close bond between southern right whale mothers and calves made them easy targets for nineteenth-century whalers. The newborn calf will not leave its mother's side, but after a few months it will begin to venture farther afield, in preparation for weaning at 12 months. Whale watchers can once again observe southern right mothers and their calves at breeding grounds, as the species slowly recovers from near-extinction in the 1800s.

Southern Right Whale

Eubalaena australis
FAMILY Balaenidae
LENGTH Female 46 ft (14 m); male 45 ft (13.7 m)
WEIGHT Female 25 tons/23 tonnes, male 24 tons/ 22 tonnes
STATUS Approximately 7,000, increasing. IUCN: Lower Risk—conservation dependent. CITES: Appendix I

The southern right whale is often consigned to the same species as *Eubalaena glacialis,* the north Atlantic whale, but the two species never mix, and recent research has made it clear the groups probably separated 2 million years ago. Southern right whales range from Antarctic to temperate waters. The species reaches the northern end of its distribution in winter, which they spend off the coasts of the southern parts of Africa, South America, New Zealand, Australia, and the sub-Antarctic islands. During their breeding season in winter, they come very close to land, preferring shallow protected bays. Then they go south to feed in the plankton-rich sub-Antarctic waters in summer. Migration routes of southern rights are shorter than most other baleen whales because they do not migrate to tropical waters. But as their numbers are increasing, their migratory range also seems to be expanding into both warmer and colder seas—there have been increased sightings along the Antarctic sea-ice edge, and they may now be re-occupying habitat they have avoided since the whaling era.

Right whales are comparatively slow swimmers and seldom exceed 5 mph (8 kph). They are often regarded as placid animals, but are capable of energetic displays, such as lobtailing (slapping their flukes on the water), breaching (leaping from the water), and other social behaviors. Calves splash about boisterously, breaching or lobtailing repeatedly, and pester their mothers for a feed. Southern right whales may also use subtle changes in posture to signal information: they might warn an approaching whale, for example, simply by shifting their body so their flukes—their main defensive weapon—face the intruder. A further weapon in their armory is explosive blowing to signal alarm or annoyance when another whale—or boat—comes too close.

Right whales have an extensive repertoire of sounds, which includes a loud bellow that is sounded when their blowholes are above water and can be heard several hundred feet away. Southern rights also produce low-frequency sounds, which have been recorded during travel, courtship, and play, but these are not prolonged and repetitive like humpback songs. However, they do make grouped sounds, called belches or moans, ranging from 50 to 500 Hz, for up to a minute. The purpose of all these different noises is not yet known, nor is it understood why they appear to be most vocal at night.

Pygmy Right Whale

Caperea marginata
FAMILY Neobalaenidae
LENGTH To 20 ft/6 m
WEIGHT Up to 3⅓–3⅗ tons/3–3.5 tonnes
STATUS Unknown but not rare. IUCN: Lower Risk—least concern. CITES: Appendix I

Pygmy right whales appear to be part way between a right whale and a rorqual. They have a comparatively slim build, more reminiscent of rorquals, and shorter baleen than other right whales. They also have a dorsal fin (unlike other right whale species), a comparatively small head for their body size, and four fingers instead of five. Another feature they have in common with rorquals is throat grooves, but whereas rorquals have many, they have only two; other right whales have none. They do have the characteristic bowed upper jaw of right whales, and very similar coloration, being dark brown or black on top, with a slightly paler underbelly.

Very little is known about this species, which is clearly not very closely related to other right whales. It is the smallest of the baleen whales, and is thought to be mostly pelagic, ranging through the cold and temperate waters of the southern oceans. It has very fine bristles in its baleen, which suggests it prefers very small prey. Pygmy rights were long thought to be solitary, but have been sighted in large, sociable groups. They are not thought to migrate seasonally, and perhaps remain in temperate waters year-round. Available strandings data suggest younger animals roam near the coasts of South America, South Africa, Australia, and New Zealand. LW

Courtship and Mating

No whale species mates for life: both males and females will have several partners. In the first stage of courtship, the female signals that she is in estrus and ready for a mate—possibly by releasing hormones in her urine, but this is speculation. The male humpback signals his availability by singing, but once the female has chosen him, he falls silent, perhaps to avoid attracting the attention of other males, who will attempt to replace him through aggressively competitive behavior such as fluke slapping. The mating pair are not as rough with each other, but they do engage in flipper slapping, which is believed to help synchronize them. When right whales mate, it is common for a second or even third male to be present.

Rorquals

SUBORDER Mysticeti
FAMILY Balaenopteridae

enormous: an adult minke, at 33 ft (10 m), is only fractionally larger than a newborn blue whale, which may grow to 100 ft (30 m). Most of the rorquals migrate seasonally between warm waters, where breeding and calving occur, and the cool temperate to polar waters, where they feed. The predictability of the timing and destinations of rorqual migrations made it easy for the nineteenth- and twentieth-century whalers to locate aggregations of the larger species, which were heavily hunted as a result.

Rorqual Life Cycles

Whalers believed that the various species migrated at slightly different times of the year, with the blue whales leaving Antarctic waters first, followed by the fins and humpbacks, and the sei whales being the last to depart. While they are migrating, rorqual whales may segregate according to both age and sex. In humpbacks, lactating females leave first, followed by immature animals, adult males, non-breeding adult females, and finally pregnant females. Blue whales and minkes tend to feed near the edge of the sea ice, but fin and sei whales generally feed farther north. Humpbacks may be found between the ice and ice-free offshore areas.

The yearly cycle of the rorqual whale centers around migration. Conception and birth generally fall within the three to four months the whales spend in warm waters. Rorquals usually mate when in warm waters during the winter, and then migrate to their polar or cool temperate feeding grounds, where they spend three to six months feeding. With the summer over, they migrate back to the tropics where, 11 to 12 months after mating, the females give birth to a single young. After about three months, the newborns accompany their mothers on the migration south, nursing as they travel. Whale calves learn to feed themselves in the rich southern feeding grounds, where there is plenty of prey on which to hone their hunting skills. Some species, such as blue whales, abandon their young at this point, after seven months of life, whereas others, including humpbacks, do not do so until they have returned to their wintering grounds and the young are nearly one year old. In later life, calves often winter in the area in which they were born.

Toothed whales use clicking sounds, generated by forcing air between their blowhole and their larynx, to "sound out" their environment, but the baleen whales lack the physical features required for this form of echolocation. Some rorquals utter repeated sounds, at a frequency that may be effective for detecting large, slow-moving objects, such as schools of fish and krill swarms. It is also possible that minkes have developed their own unique form of echolocation.

A threat that rorquals and other whales may face in the future is the increased exploitation of Antarctic krill as a food source or as fertilizer.

Drawn by Curiosity

Inquisitive humpbacks investigate the photographer's boat in the placid waters of the Antarctic Peninsula. As the whale on the left is spy-hopping (raising its head above the surface to get a better view), its companion prepares to dive under the boat. Of all the whales, humpbacks are the most curious; if they have no other pressing business—such as feeding, courtship, or travel—they will spend considerable time around and under vessels.

Rorquals are the most abundant and diverse group of baleen whales. The name "rorqual" is derived from the Norse word for a tubed or furrowed whale, which refers to the grooves, or pleats, that run the length of the throat and that allow it to expand when the whale is feeding. With their mouths open wide, rorquals engulf huge mouthfuls of water containing prey, and then pump out the water while retaining the prey inside their filtering baleen, which is much shorter than that of right whales. The rapid expansion of throat pleats to accommodate their food may give rorquals the appearance of giant tadpoles when seen from the air. Otherwise, most of the rorquals are slender and fast-moving, except the stocky, slower humpback, which is unique in many respects, including its exceptionally long flippers and its head tubercles.

There are six species in the family Balaenopteridae. The minke, blue, fin, sei, and Bryde's whales form the genus *Balaenoptera*; the humpback has its own genus, *Megaptera*. The size difference among *Balaenoptera* is

Temper Tantrums

The bond between a whale mother and her calf is a strong one, and parent and offspring regularly nuzzle and touch each other. Mother whales will place themselves between their young and any perceived threat, such as a boat or an orca, to protect them. But life is not easy for the mothers of young whales. Should they refuse their calf a feed, they may be subjected to behavior that any parent would recognize: head butting, fluke slapping, and very large temper tantrums. Whale mothers occasionally have to resort to using their pectoral fins in order to restrain their unruly young pups.

Sleek and Fast

A sei whale (right), recognizable by its tall, hooked dorsal fin, surfaces to breathe. Midway in size between blues and minkes and one of the most diverse feeders of the baleen group, sei whales sometimes exploit copepod swarms in cool temperate or sub-Antarctic waters. At other times, they move south into icy waters to feed on krill. Very little is known about their social organization, migrations, or breeding grounds.

Breathing Space

Minke whales in Antarctic waters are superbly adapted to living in heavy pack ice. They feed on krill, constantly moving between breathing holes that open and close under the influence of winds and currents. This whale has raised its snout through brash ice, which is no real obstacle. Sometimes the breathing holes are so small that the minkes must rise through the ice vertically to draw breath.

Minke Whale

Balaenoptera bonaerensis

OTHER NAMES Lesser rorqual, piked whale, pikehead
LENGTH Female 33 ft/10 m; male 26 ft/8 m
WEIGHT Up to 14¼ tons/13 tonnes
STATUS Northern Hemisphere: 60,000–80,000; Southern Hemisphere: 500,000–1.4 million. IUCN: Lower Risk—conservation dependent. CITES: Appendix I

Minkes are found in all of the world's oceans, and they are equally at home in the Arctic, the Antarctic, and the tropics. Many inhabit the open ocean, but they are just as capable of living in the heaviest of Antarctic sea ice.

Growing to only about 33 ft (10 m), the minke is the smallest of all the rorquals. It is very sleek in shape and its head is sharply pointed. Its body is dark gray on top, graduating through various shades to a pale under-side, and it often has a whitish patch on its flipper. At birth, it is only 9 ft (2.7 m) long, which is about the same size as some of the smaller toothed whales, including the pygmy sperm and the dwarf sperm whales. Unlike the larger rorquals, who breed only every two or three years, there is some evidence to suggest that minkes probably have an annual breeding cycle.

Minkes mainly eat krill, but will also eat fish and the occasional mollusc. In Antarctic waters, they consume a larger proportion of krill than do other minke populations, due to the plentiful krill supplies in the Antarctic ecosystem. Sightings of minke whales in winter sea ice suggest they are perfectly capable of feeding in the wintertime, when krill is thought to be relatively difficult to catch. They may herd prey into tight aggregations before engulfing them in great mouthfuls. At times, hundreds of minkes may gather to feed on very large aggregations of krill. Minkes are sometimes drawn to ships, racing in from a distance to swim ahead or alongside of the ship, before quickly losing interest and drifting off.

Because they are relatively small, minkes were the last of the Antarctic rorquals to be hunted. Now they are pursued by the Japanese whaling industry for "scientific whaling," and it is possible that commercial whaling may soon resume in the Antarctic. Norway has begun limited whaling in its waters.

Sei Whale

Balaenoptera borealis

OTHER NAMES Pollack whale, short-headed sperm, Japan finner, sardine whale
LENGTH Female 56 ft/17 m; male 50 ft/15 m
WEIGHT 22–27½ tons/20–25 tonnes
STATUS Population thought to be increasing since mass whaling ended—now estimated at 25,000 to 70,000 in Southern Ocean. IUCN: Endangered. CITES: Appendix I

Sei whales are a worldwide species that occupy cold, temperate, and tropical waters—although some tropical sightings have been confused with the tropical species, Bryde's whale (*B. edem*). The two species are of similar

size and appearance, both being dark gray on top and paler below. Some Southern Hemisphere sei whales are slightly larger than their northern counterparts.

Although sei whales will devour almost anything they can find, they seem to prefer smaller species such as copepods and amphipods. They will consume larger crustaceans such as krill, as well as fish and squid, but they are basically adapted to hunt smaller species. One of the favorite techniques of the sei whales is skimming, in which they swim along the surface, twisting through the water while they skim zooplankton into their open mouths. When they are feeding in this way, they do not take individual mouthfuls of water, but instead push water through the baleen constantly by the pressure of swimming. This technique is also used by right whales.

Seis are often found away from the coastal edges of the seas, in the open ocean, where they gather along oceanic fronts in pods of three or more. Although most rorqual species have been observed to form groups of ten or less normally, it may be that the small group size is a consequence of whaling, and that rorqual groups may have been larger before mass whaling. However, on their feeding grounds, group size may be a result of prey availability and patchiness: for example, humpback whales are not usually seen in groups of more than two, despite a considerable population increase in recent decades, and this appears to be an optimal number of animals to feed on small, scattered prey swarms. The pod sizes of rorquals may become larger as populations increase with conservation measures.

Fin Whale

Balaenoptera physalus

OTHER NAMES Common rorqual, finner, finback, razorback
LENGTH Female 85 ft/26 m; male 69 ft/21 m
WEIGHT Up to 88 tons/80 tonnes
STATUS Northern Hemisphere population estimated at 40,000; Southern Hemisphere: 5,000–30,000. IUCN: Endangered. CITES: Appendix I

Fin whales are dark gray above, and have swirling patterns known as chevrons on their backs, and white underneath. They are the largest whale after the blue whale, but are slimmer, and not particularly heavy given their size. It is the only whale with a distinctive asymmetrical body coloration, with its right lower jaw being white, and the left side dark. The baleen plates also follow this pattern. This coloration may be an advantage when feeding as it possibly helps to confuse their prey. Fin whales were called "razorbacks" by whalers because of their ridged tailstocks.

Fin whales range from polar to tropical waters. There are clearly discrete populations, which are highly migratory, and many individuals have been identified returning to mate in their mother's breeding grounds. Fin whales are normally seen in pods of six to ten, but they are also often found in pairs. The size of the group may be related to the amount of food in the area, or to courtship and mating.

The fin whale likes fairly deep water, and is usually found outside the continental shelves. However, there have been recent sightings in waters less than 660 ft (200 m) deep. Faster than most rorquals, the fin whale can swim at a fast 23 mph (37 kph) for short periods; however, it can also maintain a steady pace of 14 mph (22 kph) for weeks. This very active whale also dives for longer than some other species, and will submerge for up to 30 minutes—particularly when feeding. Dives may reach 1,640 ft (500 m), which is exceedingly deep for baleen whales. The fin whale feeds on a wider range of prey species than the blue whale. In the Northern Hemisphere, it consumes krill and small fish, such as capelin and herring.

Why So Blue?

Seen from space, the Earth is blue. Seen under water from above, a blue whale is also blue. This is puzzling, as the true skin color of a blue whale, when viewed above the water surface, may vary from silver to almost black. Yet once they slip beneath the water there is an instant transformation, as if a light has been switched on.

Blue Whale

Balaenoptera musculus

OTHER NAME Sulfurbottom

LENGTH Female 102 ft/31 m; male 100 ft/30 m

WEIGHT 110–132 tons/100–120 tonnes, possibly up to 165 tons/150 tonnes

STATUS Unknown population, but probably less than 10,000. IUCN: Endangered. CITES: Appendix I

The blue whale may be the largest animal ever to have lived; *Brachiosaurus*, the biggest known dinosaur, only reached a length of 75 ft (23 m), small compared with the blue's 100 ft (30 m), although a recently discovered relative, *Sauroposeidon*, may have been larger. The largest accurately measured blue whales were both females. The longest one had a body length of 110¼ ft (33.6 m); the heaviest weighed 210 tons/190 tonnes and was 90½ ft (27.6 m) long. The larger animals live longer than the smaller ones, and the blue whale is no exception; some individuals may have lived 110 years.

Blue whales are a mottled slate-gray, but the reason for their name becomes apparent when they submerge and their skin appears a pale, luminous turquoise-blue. In Antarctic waters their skin even takes on a yellowish brown hue from the diatoms that grow on them. Their main features are their size, the position of their dorsal fin, their generally paler color, and spotted appearance.

Although it is so large, the blue whale is sleek and slender. Like most rorquals, it is streamlined and is a comparatively fast swimmer, having a torpedo-shaped body with no unnecessary appendages that could drag. In both males and females, a slit along the underbelly contains the sexual organs. In their internal structure, the testes of a mature blue whale male are identical to those of other mammals, but may be as long as 30 in (75 cm) and as heavy as 82 lb (37 kg), and the penis can be as much as 10 ft (3 m) long and 12 in (30 cm) in diameter. This, the largest of whales, produces very large babies—about 23 ft (7 m) long.

The blue whale is a cosmopolitan species, living in all the oceans, and ranging from polar to equatorial waters. When in Antarctic waters, it will follow the ice edge in its search for krill swarms. It goes deeper into the Antarctic ice than all other baleen whales, except minkes, and like most it migrates from the poles to the tropics at summer's end. Blue whales seem to gather in low numbers, which may be the effect of slow recovery from whaling.

Several breeding stocks of blue whales may mingle in the polar regions while they are feeding, but they part company as they return to their various breeding grounds. Some blue whales may live in warmer waters and not migrate to the poles at all. While some blue whales are feeding in temperate waters, such as those off southeast Australia, others are feeding along the ice edge much farther south.

Blue whales are regarded as fast swimmers, and they have been timed at up to 30 mph (48 kph) when alarmed. They are relatively shallow divers, but feeding dives to 1,180 ft (360 m) have been recorded, and they are capable of submerging for about 20 minutes. When they surface, their blow can reach a spectacular 33 ft (10 m). When traveling, they may blow eight to 15 times over a five-minute period before diving again.

Blue whales, the largest of all animals, prefer one of the smallest for their food, specializing almost entirely on krill. To satisfy the needs of its huge body, an adult must catch a total weight of around 4 tons/3.6 tonnes of krill each day. Their energy requirements are so huge that they probably must search out tropical breeding areas where food is abundant.

Like many rorquals, the blue whale employs the gulping technique to feed. A single mouthful of water and food may weigh as much as 50 tons/45 tonnes. Gulp feeding may happen at any depth, depending on the prey, and the whale might modify the technique to concentrate the amount of food, forcing its prey back against the surface of the water, an iceberg, or land.

Humpback Breaching
Humpbacks are easily identified by their low blow, huge flippers, and the way they arch their backs when diving.

Humpback Whale

Megaptera novaeangliae
LENGTH 50 ft/15 m
WEIGHT Generally 27½–33 tons/25–30 tonnes; up to 40 tons/36 tonnes
STATUS Possibly 20,000–30,000. IUCN: Vulnerable. CITES: Appendix I

Humpback whales are much heavier in build than other rorquals and are slow swimmers and, as a result, they were heavily hunted by whalers. Although only 50 ft (15 m) long, they can weigh up to 40 tons (36 tonnes). A distinctive feature is their long pectoral fins—much longer than those of any other rorqual— which reach about one-third of their body length. These fins make them very maneuverable and, despite their stocky appearance, they are graceful acrobats underwater.

Their coloration can vary from all black to all white, but they tend to be dark above and white below. The undersides of their tail flukes, and often of their flanks, are pigmented with distinctive patterns, enabling the photo-identification of individuals. They hunch their back when they dive and there are a series of lumps (tubercles) around their head. They are one of the few whales to grow any hair; young humpbacks may have a single hair growing in the center of each tubercle.

Humpbacks are about 14 ft (4.3 m) long when born in the tropics in late autumn or early winter. They must feed and grow fast to be ready for their first migration. Like most baleens, the mother and calf is the basic unit of humpback society, which tends to be fluid. Groups form and split according to the abundance of food or dynamics of the breeding season when groups of males compete boisterously and violently for the females. In feeding areas, they are found in ones or twos, reflecting the scattered, patchy nature of the krill they feed on.

Migration

Many baleen whales undertake long migrations from cold feeding waters to warmer breeding grounds. Humpbacks travel from near the poles to tropical waters. The longest migration route of any mammal is that of some Antarctic humpbacks, which feed on the western side of the Antarctic Peninsula and migrate northward across the Equator to breeding grounds off Colombia. Antarctic humpbacks breed in the tropical waters off all the southern continents, taking about three months to commute from Antarctica at a speed of about 1 to 3 mph (2 to 5 kph). Their leisurely migration is punctuated with rests, courting, and occasional feeding. How the whales find their way over such vast distances is not known for certain, but several possible navigational aids have been suggested. They may use the sun and stars as guides, or natural features such as landmasses, and wind and swell patterns. Like birds, they may have magnetic navigation systems that detect anomalies in the Earth's magnetic field. Or perhaps they "hear" the shape of the sea floor by listening to sounds made by other marine life or echoes of their own sounds.

Humpbacks, known as the "singing whales," sing complex, repetitive songs, each of which may last for ten to 15 minutes. The individual sounds, or "units," of song have been described as yaps, snorts, chirps, eees, ooos, groans, and whooos. The units are divided into "phrases," which are organized into "themes"—and the themes are organized into songs. Songs may be repeated for hours, and have been heard at distances of at least 20 miles (31 km).

Only mature males are known to sing, and song is most often heard in their breeding grounds. Each breeding population has its own song; over time, mostly in winter, new phrases are added and elaborated and old ones omitted. At any one time, all the whales in a group sing much the same song, which evolves constantly, for reasons unknown.

Humpback songs are thought most likely to be displays to attract females. They may serve to synchronize ovulation in females and mark out "territories" of competing males. The singers often stop singing when they are joined by another whale, probably a responsive female. LW

Photo-ID
A diving humpback whale shows its distinctively pigmented tail flukes. These unique pigment patterns make it possible to study the life histories and migrations of individual animals.

Toothed Whales

The best known of all whales must be Moby Dick, the great white sperm whale immortalized in the novel by American writer Herman Melville. Moby Dick was a giant male sperm whale, and his is the tall, blunt head profile that often features in children's books and drawings of whales.

The sperm whale is the largest of all the toothed whales and can reach 66 ft (20 m) in length. However, its vast impressive size and shape are unique among toothed whales. The dolphins, porpoises, and beaked whales that make up the bulk of toothed whale species are generally much smaller, some of them barely more than 3 ft (1 m) long, and most under 17 ft (5 m). They also have a very diverse range of habitats, from steamy jungle rivers to ice-covered seas, and many of them live their lives rarely seen by humans.

Toothed whales hunt larger prey than do the baleen whales, feeding on fish, molluscs, birds, and mammals; indeed, almost any animal they can catch—up to and including baleen and other toothed whales. Each of the toothed species has its preferred hunting patterns and prey: some home in on cephalopods, such as octopus, squid, and cuttlefish; others specialize in grubbing up bottom-dwelling worms and crustaceans. Dolphins use their conical teeth to grab fish and swallow them whole, whereas porpoises have shearing teeth, which they use to rip prey apart before swallowing it.

Like baleen whales, toothed whales have a smooth, torpedolike body but, due to their generally smaller size, they are far more flexible and agile than their more ponderous baleen cousins, and are far less likely to be encrusted with barnacles and whale lice. The males generally grow larger than the females, the opposite to baleen whales. This sexual dimorphism may extend to include differences in other features—for example, the size and shape of the dorsal fin. As well as using their teeth to grasp prey, the toothed whales may use them for fighting, and sexually mature animals of both sexes are often savagely striped with scars, which have been inflicted by the teeth of other whales.

Head Pieces

Most toothed whales have a melon, a large, lens-shaped pocket of fatty oil, at the front of their head. The melon is more highly developed in the species that feed in very deep water where light does not penetrate, or in muddy river water where visibility is poor. It is thought that the melon acts as a lens to focus the echolocation sounds that are generated inside the skull into beams of sound of varying widths and intensities. Returning echoes are received through oil-filled channels in their lower jaw. Echolocation is a sense still dimly understood—humans have nothing comparable. What is known is the sense is used for hunting prey, for navigation, and possibly for investigating the emotional state of companions.

The jaws of many toothed whale species project in a beak shape in front of the melon, a feature that is very highly developed in beaked whales and most dolphins. Unique among vertebrates, toothed whales have only one external blowhole. They originally had two nostrils, but these have merged within the whale's head, and a single nasal passage reaches the skin.

All toothed whales have teeth at some stage in their life cycles, but these teeth are often vestigial or highly modified, particularly in the beaked whales, which may have just one pair of teeth in the lower jaw. This raises the question of how these animals with such minimal dentition are able to secure their prey. The answer is that they use a sudden, powerful suction once they are near enough to their prey for it to be effective. At the other end of the scale, the species with functional teeth may have more than 260 of them. These are usually simple conical pegs, but some toothed whale species have evolved highly specific tooth shapes that are of little practical use in normal feeding. Male narwhals are the most dramatic example of this development. They have one normal tooth and one extremely long, spirally twisted tusk, and the adult males have been seen using these to "joust" with each other. Many beaked whales have similar, but less spectacular, dental equipment.

Battle-scarred

The flukes of this sperm whale show tooth rakes, probably caused by an orca or one of its relatives, such as a false killer whale or a pilot whale. Whales of many species show similar scarring. It is likely that orcas nip individuals to "test" them. A spirited response sends the orca into retreat. Feeble resistance, however, may be the cue for a more determined attack.

Supportive Species

Most toothed whales are gregarious, spending their whole lives in tightly knit social groups. These groups are critically important to the whales, who seem unable to function in isolation. It is believed that many mass whale strandings occur because a group refuses to abandon a sick or dying individual.

In the heyday of whaling, a whole pod would surround an individual wounded by whalers, who would then harpoon all of them one by one with very little effort. Sperm whales, in particular, were known for this remarkable nurturing behavior, surrounding injured or weak individuals with their flukes, their most effective weapon, facing outward in a protective pattern called the marguerite formation, so-named because it looks like the petals of a flower.

Echolocation

Toothed whales share with some bats the remarkable faculty of echolocation—the emission of sounds that bounce off surrounding objects and provide the animal with a "picture" of its environment; it has been dubbed "biological sonar." The whales use a click sound, which is loud enough to generate an echo and short enough to be repeated rapidly in order to detect any change in the environment. The clicks may be widely spaced when seeking prey and become much more rapid as the item of interest is approached. The species that require a clear image of their immediate surroundings use high frequencies, which are effective at short range, whereas species that need to scan the more distant environment employ lower frequencies.

Bottlenose dolphins can echolocate to a distance of 2,500 ft (750 m)—possibly farther. Even in murky conditions, the dolphins use echolocation to catch prey, avoid predators, and map the sea floor. Captive dolphins have been able to distinguish different types of metal in balls, and can pick out hollow metal balls from solid ones. Echolocation operates in a very narrow band, so the animal must swing its head to scan a broad area.

Toothed whales seem to echolocate by forcing air through the passages between blowhole and larynx. The skull may operate as a parabolic reflector, and the melon, which can change shape, may focus the sound beam to a chosen distance. The echo is registered by oil-filled sinuses in the whale's lower jaw or, possibly in some species, by the teeth.

It appears whales echolocate only when something interesting—such as food—is around. Sperm whales

may echolocate constantly during feeding dives, though they are occasionally quiet. They usually echolocate at a steady rate, but sometimes the click speed accelerates rapidly—probably because the whale has located prey and is seeking more detailed information about its size, shape, speed, and direction.

It is possible that echolocation is such an effective faculty that whales can use it to "X-ray" other whales to examine their stomach contents, or in order to detect their mood. It may also be possible that some toothed whale species could use an extremely loud echolocation click to stun their prey into immobility.   LW

Take a Deep Breath
Mother and calf orcas are spyhopping in an icebreaker channel in McMurdo Sound, coming up in open pockets of water to breathe.

Dancing on Water
A dusky dolphin displays its distinctive striped flank as it porpoises to breathe. Porpoising is the most efficient way to breathe while moving at speed.

Dolphins and Porpoises

SUBORDER Odontoceti

FAMILIES Delphinidae, Phocoenidae

Oceanic dolphins—family Delphinidae—make up one of the world's most successful groups of whales. Over 30 species are distributed across almost every oceanic environment, as well as many of the world's rivers. Some dolphins have quite limited geographic ranges, whereas others have broad or even worldwide distribution. Dolphins occupy both coastal and open ocean habitats, and both shallow and deep waters. In deep water, they may pursue surface-feeding fish or hunt at great depths. Dolphins range in size from small river dolphins, at 5 ft (1.5 m) long, to the orca, or killer whale, which is almost 33 ft (10 m). Generally, dolphins are the fastest and most agile of all the cetaceans, and most species have a long, beaklike snout and a sleek, streamlined body.

Fundamentally, dolphins are social animals. They often hunt cooperatively to herd schools of prey, and they join forces to attack their prey and avoid predators. Under the right conditions, with a rich source of prey, pelagic dolphins may gather in the hundreds, even the thousands. These large dolphin gatherings string out over several miles, swimming in smaller subgroups as they move between feeding grounds, but reassembling into a tight group when the opportunity to feed arises.

Porpoises—family Phocoenidae—are generally smaller and stockier than dolphins, with a more rounded body and no real beak. They have a long, shallow, triangular or rounded dorsal fin, and spade-shaped teeth. All the porpoises hunt cephalopods and small fish, using echolocation and sight. Not as gregarious as dolphins, they occur in much smaller groups. One of the north Pacific species, Dall's porpoise, is believed to be the fastest of all the cetaceans, though they are preyed on by orcas—and by humans.

Dolphin Characteristics

Dolphins exhibit a wide range of color patterns, which to some extent reflect their way of life. Animals that live in turbid riverine or coastal waters and form small social groups are often relatively drably patterned; whereas the animals that hunt in the open sea in larger groups are often more strikingly patterned. This patterning may help to deceive their prey, avoid predators, or aid in the recognition of other group members.

Beached but Saved
A pod of long-finned pilot whales swim off the coast of New Zealand, where they had been stranded on a beach.

Living Together
The dusky dolphin has large, heavily falcated (curved) dorsal fins, and often travels in tight groups.

Female at the Center

This orca (right) can be identified as an adult female by the size and shape of her dorsal fin. An individual whale can be recognized by the unique shape of its saddle—the patch of white pigment behind the dorsal fin. Adult females are at the center of orca society. The smallest stable groups consist of an adult female and her calves, some of them with calves of their own.

Determined Hunter

An orca in its element—the Antarctic sea ice, where it hunts penguins, fish, seals, and whales. Orcas have been seen rushing at ice floes where seals are resting, creating a bow wave that rocks the floe and tips the hapless prey into the water. They are formidable predators and capable of complex, coordinated attacks on animals ranging in size from sardines to blue whales.

Most dolphins have many functional teeth in both their upper and lower jaws. These teeth are designed for grasping prey and preventing it from escaping before being eaten. Dolphins are capable of swallowing surprisingly large prey. Most species prefer to hunt either fish or squid, and fish-eating dolphins often have narrower heads than squid-eating dolphins, Some species are more generalist feeders, taking whatever food presents itself. The bottlenose dolphin is a good example of this.

Some dolphin species, such as striped or hourglass dolphins, are oceanic nomads, having no fixed territory, and constantly roaming in large groups over vast distances in search of seasonally available prey, and foraging along oceanic fronts or upwellings. At the other extreme, river dolphins and coastal bottlenose dolphins, for example, form much smaller groups and are quite restricted in their range: they may inhabit a single river or bay, or short stretch of coast.

Dolphins make a variety of sounds that range from low-pitched chirps to ultrasonic clicks, which they use for echolocation, and possibly to stun their prey. Sounds at higher frequencies carry only short distances beneath the water, but most dolphin species have little need to communicate over long distances as their social cohesion keeps group members fairly close to each other.

A Social Species

Each dolphin species has its own pod structure. Many of them operate a flexible, open-group policy where individuals come and go freely and often move from group to group. Other species live in very stable groups of related animals, with very little change over periods of years. Previously, it was thought that several species formed harems. Now we know that many toothed whale societies are matrilineal; that is, the older females form a stable nucleus, around which the younger adults and immatures gather. Adult males in many species form their own groups, usually approaching female groups only for mating. In other species, the males remain with their mother's group for life, and mating occurs between such groups when they meet. Promiscuous mating is the general rule for dolphins, and the male–female and the male–calf bonds are weak; the bonds between the mother and the calf, and between female and female are usually much stronger.

Dolphins
FAMILY Delphinidae

Orca
Orcinus orca

OTHER NAMES Killer whale, blackfish, grampus
LENGTH Female 23 ft/7 m; male 33 ft/10 m
WEIGHT Female 6 tons/5.4 tonnes; male 11 tons/ 10 tonnes
DISTRIBUTION AND HABITAT Worldwide; found in all ocean habitats, from the tropics to the polar ice, but most common near rich food sources in high polar latitudes.
STATUS Unknown population, but considered locally abundant. IUCN: Lower Risk—conservation dependent. CITES: Appendix II
APPEARANCE These are the largest of the dolphins, and they are huge by dolphin standards; their size, and their black-and-white, pandalike pattern make them impossible to miss. Mature males are almost twice as large as adult females, and they have an exceptionally tall dorsal fin, and very large fins and flukes.

Working Together

Dusky dolphins, like most dolphins, are gregarious animals, sometimes forming temporary groups of as many as several thousand, although they are usually found in much smaller, more permanent groups of 15 or more. When they locate schools of fish, individuals are thought to summon others with exuberant leaps, because cooperative herding increases the chances of a good feed for all. Then, well-fed, they lounge around in small groups, relaxing and socializing.

The orca, or killer whale, is one of the most successful dolphin species, with populations occurring in all the world's seas and oceans. Orcas have 44 conical teeth with which they seize their prey. They are often referred to as killer whales, nomenclature somewhat undeserved because every other cetacean species also kills to eat— and they have never been known to attack people in the wild, even when provoked. Unfortunately, the reverse is not true: orca whales are often killed because of their propensity for stealing from human fisheries. To such an efficient hunter as the orca, taking fish from longline hooks is as easy as picking grapes.

Adult orcas need to consume an average of 150 lb (68 kg) of seafood each day. Orcas usually hunt in packs, and will show a remarkably disciplined division of labor, such as when they are attacking large whales. Some populations have discovered how to hunt seals resting on shore by surfing in and out on waves—this is probably the closest any modern whale gets to going on land (unless they are stranded). No matter the target, orcas are efficient, cooperative hunters. Depending on their location, they will feed on whatever is available, including fish, squid, sharks, stingrays, shellfish, birds, dugongs, seals, dolphins, porpoises, and baleen whales up to, and including, sei and blue whales. No predator is known to prey on orcas, however.

The basis of orca society is a matrilineal group that consists of a mature female and her offspring, and often their offspring. Such groups of resident animals form a series of progressively larger groups, up to about 60 individuals. Transients form much smaller groups of up to four, consisting only of a female and her immediate offspring. So far, there have been no studies of social organization in Antarctic orcas, but there is evidence of two types, which have been documented from British Columbia and Alaska: residents, which inhabit a home range and mostly feed on fish; and transients, which range much more widely, and favor marine mammals as prey. There may also be an offshore population.

Females are physiologically capable of bearing a calf every three years, but they normally produce one only every eight years. The young measure 7 ft (2 m) at birth, and calves remain in their birth pod with their mother for many years. The females generally live for 80 to 90 years and bear about five calves; males have a shorter lifespan of about 60 years.

Commerson's Dolphin

Cephalorhynchus commersonii

OTHER NAME Southern dolphin

LENGTH 4½ ft/1.4 m

WEIGHT 110 lb/50 kg

DISTRIBUTION AND HABITAT The southern oceans off both the southern coasts of South America and into the Atlantic Ocean.

STATUS Unknown population. IUCN: Data Deficient. CITES: Appendix II

APPEARANCE These dolphins are dramatically black and white, with a black head, a black dorsal fin, and a black tail. A white patch runs from just behind the head down to the belly. They have no melon, and a very short beak.

Caring Role
Long-finned pilot whales are one of the few mammals to undergo menopause. The role of post-menopausal whales may be that of teacher and carer.

Long-finned Pilot Whale
Globicephala melas
LENGTH Female 16½ ft/5 m; male 20 ft/6 m
WEIGHT Female 2 tons/1.8 tonnes; male 4 tons/ 3.5 tonnes
DISTRIBUTION AND HABITAT North Atlantic Ocean and the southern oceans; possibly all cold–temperate waters.
STATUS Probably common. IUCN: Lower Risk— conservation dependent. CITES: Appendix II
APPEARANCE They have a square, bulbous head, and a strongly sickle-shaped dorsal fin. The back and flanks are black, and their chin and belly are a grayish white that darkens slightly as they mature.

Southern Right Whale Dolphin
Lissodelphis peronii
LENGTH 6 ft/1.8 m (possibly larger)
WEIGHT 132 lb/60 kg
DISTRIBUTION AND HABITAT An offshore species, known to inhabit Southern Hemisphere cold–temperate waters and Antarctic waters south of Argentina.
STATUS Unknown population, but probably common. IUCN: Data Deficient. CITES: Appendix II
APPEARANCE They are black on top with a large, wide, white belly patch running up their flanks, including their flippers, head, and tail. They have no dorsal fin.

Peale's Dolphin
Lagenorhynchus australis
OTHER NAME Whitesided dolphin
LENGTH 7 ft/2 m
WEIGHT 253 lb/115 kg
DISTRIBUTION AND HABITAT Cold to temperate waters off south coasts of South America and the Falklands.
STATUS Population unknown. IUCN: Data Deficient. CITES: Appendix II
APPEARANCE Round snouts and short beaks. They are black or dark gray on the dorsal surface, with a light gray stripe on the flanks and a fine white line above the gray patch that enlarges to a wider pale stripe.

Hourglass Dolphin
Lagenorhynchus cruciger
OTHER NAME Whitesided dolphin
LENGTH 5¼ ft/1.6 m
WEIGHT 220 lb/100 kg
DISTRIBUTION AND HABITAT Southern Hemisphere, Antarctic and cold–temperate waters.
STATUS Unknown population, presumed abundant. IUCN: Lower Risk—least concern. CITES: Appendix II
APPEARANCE Hourglass dolphins are black or dark gray and white in color, and derive their name from the hourglass-shaped patch on their black flanks. They have rounded snouts and short beaks. Often grouped with *L. obscurus*.

Dusky Dolphin
Lagenorhynchus obscurus
OTHER NAME Whitesided dolphin
LENGTH 5¼ ft/1.6 m
WEIGHT 220 lb/100 kg
DISTRIBUTION AND HABITAT Southern Hemisphere, Antarctic and cold–temperate waters.
STATUS Unknown population. IUCN: Data Deficient. CITES: Appendix II
APPEARANCE Dusky dolphins are similar to *L. cruciger*, with which they are often grouped. They have a rounded snout and short beak; complex coloration— pale below, dark gray above, with a distinctive pale forked blaze running forward on the flanks, and a pale trailing edge to the dorsal fin. The distinctive pattern on its flanks ensures instant recognition by its own kind. This, and regular contact with its own group, are vital protection measures.

Porpoises
FAMILY Phocoenidae

Spectacled Porpoise
Australophocaena dioptrica
LENGTH 7 ft/2 m
WEIGHT 187 lb/85 kg
DISTRIBUTION AND HABITAT A Southern Hemisphere species, possibly circumpolar, that occupies cold– temperate waters off South America and the Falkland Islands, South Georgia, Heard Island, Macquarie Island, and the Auckland Islands.
STATUS Unknown population. IUCN: Data Deficient. CITES: Appendix II
APPEARANCE Spectacled porpoises have black backs and white bellies, with a distinctive black outline around their lips and eyes giving the appearance of spectacles—hence their common name.

Burmeister's Porpoise
Phocoena spinipinnis
LENGTH 6 ft/1.8 m
WEIGHT 155 lb/70 kg
DISTRIBUTION AND HABITAT Ranges from Brazil southward into sub-Antarctic waters; prefers coastal habitats.
STATUS Uncertain, but possibly as low as 500 animals. IUCN: Data Deficient. CITES: Appendix II
APPEARANCE Dark black-gray on top; lighter below. LW

Bold and Curious
The dramatically marked hourglass dolphin is sometimes mistaken for a small orca. Ranging farther south than other dolphins, almost to the edge of the sea ice, it is one of the least studied cetaceans. Hourglass dolphins often spend hours accompanying ships.

Seldom Seen
Southern right whale dolphins rarely come close to land and are never seen for long at sea. They roam the Southern Ocean in search of squid and small fish, porpoising in a welter of white water in a motion similar to bouncing. They have no dorsal fin and can be confused with penguins at a distance.

Sperm Whales

SUBORDER Odontoceti

FAMILIES Physeteridae, Kogiidae

There are two families of sperm whales. The sperm whale—the largest of all toothed whale species—belongs to the family Physeteridae; the dwarf and pygmy sperm whales, which are anatomically similar but much smaller, belong to the family Kogiidae. These two are as small as dolphins but several characteristics make it clear that they are more closely related to the sperm whale than to any of the smaller toothed whales.

The most significant shared feature of these three species is the spermaceti organ—a highly specialized melon—which gives the sperm whales their distinctive blunt profile. The three also share highly unusual nasal passages. In all three sperm whales, the blowhole is located far to the left side of the head; however, in the two smaller species, it is at the back of the head, close to the normal position, whereas in the great sperm whale it is at the front of the melon. All three of the sperm whale species have functional teeth in their small lower jaw, and all are very deep divers, feeding on squid that live at great ocean depths.

Sperm Whale

FAMILY Physeteridae

Physeter macrocephalus

OTHER NAMES Catchalot, spermaceti whale

LENGTH Female 43 ft/13 m; male to 66 ft/20 m

WEIGHT Female up to 22 tons/20 tonnes; male up to 55 tons/50 tonnes

DISTRIBUTION AND HABITAT Worldwide; observed in all the world's seas and oceans from the Antarctic, through cold and temperate zones, to tropical waters.

STATUS Unknown population, probably abundant. IUCN: Vulnerable. CITES: Appendix I

APPEARANCE Sperm whales are very dark gray-brown with a wrinkled appearance over much of their body, and have characteristic white scarring from deep-water squid.

One of the largest of the extant whales, the sperm whale has a dorsal hump, rather than a dorsal fin. Its most recognizable feature is its enormous rectangular head, which may take up to one-third of its total body length and which is most fully developed in the mature males.

The blow of the sperm whale has a forward angle of 45 degrees, a feature that readily identifies it at sea. It has no teeth in the upper jaw and a relatively short, narrow lower jaw. The teeth in the lower jaw are the largest functional teeth known in the animal world, and may be up to 10 in (25 cm) long. Sperm whales are one of the most sexually dimorphic whale species: a very well-grown male may weigh three times as much as a mature female and can grow to twice the length. As well as being much smaller than the males, females have a less developed spermaceti organ—the huge melon that gives the male such a distinctive square shape. In

a male sperm whale, the large melon can be up to one-quarter of the animal's total body length.

The Largest Brain on the Earth

Behind the spermaceti organ, the sperm whale's huge skull contains a brain that can weigh more than 20 lb (9 kg). In comparison, the weight of a human brain is only 3 lb (1.3 kg). Relative to its body size, the sperm whale's brain has probably been as large as this for more than 30 million years, whereas humans reached their present brain capacity only 100,000 years ago. The whale's brain is very convoluted, with a highly developed cerebral cortex—which has been cited as evidence of high intelligence. However, intelligence is extremely difficult to compare across cultures, and it is probably impossible across species. Many scientists believe that the size of a sperm whale's brain may be necessary for the animal to deal with the complex echolocation information that it must process, but it is not possible to rule out an intelligence of a different order than our own.

The shape of a sperm whale's skull is asymmetrical, unusually for a mammal. The spermaceti organ in its forehead is filled with a huge quantity of waxy oil, and nasal passages run between air sacs at the front and back of this organ. Spermaceti oil was the most highly prized product of the nineteenth-century commercial whaling industry (except for ambergris, an aromatic waxy deposit also obtained from sperm whales). The oil was used as a fine machine oil, and is still unrivaled as a lubricant for missile inertial guidance systems.

There are several theories about the function of the spermaceti organ. According to the buoyancy-control theory, sperm whales can remain motionless at a chosen depth, or swim vertically up or down, by regulating the temperature, and thus the density, of the waxy contents of the spermaceti organ, heating it with body-warmed air and blood or cooling it with seawater drawn into the nasal passages. Another theory holds that sperm whales use their spermaceti organ for long-range echolocation. It is thought that when air moves through the sperm whale's right nasal passage, it generates sound at the entrance to the forward air sac, and is then recycled back along the left nasal passage, and in the process probably amplified en route.

For echolocation or communication, sperm whales use low-frequency clicks. The clicks might be arranged into organized patterns, known as "codas," which are thought to mediate social interactions between animals that come together after a separation. They also make "creaks"—very fast click trains that rise in frequency—which may be used to track their prey at close range. Whereas the sounds of sperm whales appear relatively simple when compared with the songs of humpback whales, it is highly likely their giant brains are capable

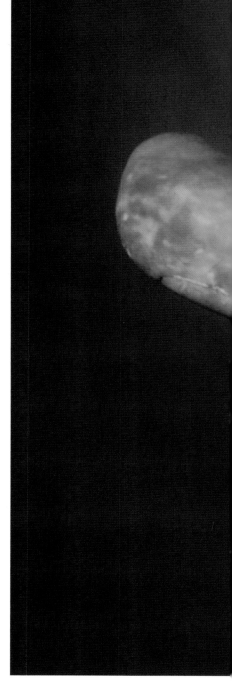

Deep-sea Diver

A sperm whale shows its characteristic blunt head, short underslung lower jaw, dorsal hump, and wrinkled skin. Sperm whales breathe, travel, socialize, and rest at the surface, but the serious business of feeding is carried out far below, in total darkness and under unimaginable pressure. Even sperm whales with broken or malformed lower jaws can feed successfully, suggesting that prey (mainly squid and octopus) are swallowed by suction.

of processing information of a richness and complexity that we cannot begin to imagine.

Hunting

Sperm whales normally swim at about 6 mph (10 kph), but can reach speeds of up to 20 mph (30 kph). In the summer, mature males migrate slowly to the cold polar waters, where they live as solitary hunters and dive in search of squid to ocean depths unmatched by any other mammal. They may ambush or pursue their prey through the deep water before surfacing. Sperm whales feed mainly on squid and octopus, but they will also opportunistically grab any deep-water species, including sharks and fish. They have been tracked to depths of 8,200 ft (2,500 m), and there is evidence that they dive even deeper after prey. They usually patrol waters that are too deep for light to penetrate: totally blind sperm whales have been captured in perfect health, and with food in their stomachs. Clearly, these whales must have evolved a highly effective technique for detecting and catching prey without the aid of vision, and echolocation is the obvious answer.

It is the stuff of legend that sperm whales devour giant squid. Most of their squid victims are relatively small—3 ft (1 m) or less. But body scars suggest that these whales do regularly engage in battle with larger specimens, and

A Species at Risk

Before species conservation became a matter of concern, adult male sperm whales were hunted almost to extinction for the commercially valuable spermaceti oil from their melon and ambergris from their gut, and the huge males of the species are no longer found. Whalers took the biggest mature males and left females and immature males. When these young males reached breeding age (about 27), they had not acquired the social skills necessary to reproduce, so the birth rate of the species—already quite low—fell even further. The young males seem to have been learning, though, because the number of sperm whale births is increasing.

a squid 34-ft (10.5-m) long has been retrieved from the stomach of one sperm whale.

Family Life

Female and young sperm whales stay in nursery groups of some 20 to 40 animals, most of which are related through the mature females, and they stay in temperate and tropical waters. The sperm whales have very low reproductive rates, calving in the summer or autumn and producing one offspring every four to six years. The young, which are 13 ft (4 m) at birth, are weaned into juvenile pods, but the mothers may suckle their last offspring for up to 15 years. As the whales mature, the females rejoin their nursery groups, and the males form bachelor herds, which contain fewer numbers as the whales grow, until the really large males tend to be very widely spaced, though they may still communicate with a group. During breeding, each nursery group is joined by up to five mature males—about one male for every ten females. Recent research suggests that the males stay with a nursery group for only a few hours and then depart in search of another group where the females are ready to mate. Males compete for females, and only 25 percent are successful.

Sperm whales feed mainly on squid, including giant squid. They hunt for their prey around the edges of the continental shelves, oceanic islands, and underwater seamounts, and they are often to be found in waters that are at least 3,280 ft (1,000 m) deep. Sometimes the sperm whales follow their prey into shallow coastal waters, where they are prone to stranding.

Pygmy and Dwarf Sperm Whales

FAMILY Kogiidae

Adult pygmy and dwarf sperm whales feed on deep-water cephalopods, fish, and crustaceans, whereas the younger members of these two species hunt similar prey from shallower waters. The smaller and younger *Kogia* species are unable to hunt successfully in deep waters, but by the time they are mature and have reached their full size, they have learned to plumb deep waters for their prey. Both species have a largely offshore distribution and keep to small groups, ranging from two (usually a mother and calf) to a maximum of ten. They are generally quiet and inactive while on the surface of the water.

When threatened by predators, pygmy and dwarf sperm whales exhibit a unique protective mechanism that involves releasing a cloud of reddish brownish feces, and then thrashing their flukes in order to spread the cloud as much as possible. They then make their escape into the murk. Curiously, this tactic is also used by squid on which the whales prey.

Pygmy Sperm Whale

Kogia breviceps

OTHER NAMES Short-headed whale, small whale, lesser sperm whale

LENGTH 9 ft/2.7 m

WEIGHT 880 lb/400 kg

DISTRIBUTION AND HABITAT Worldwide; from cold and temperate waters to the tropics.

STATUS Unknown population, probably widely distributed. IUCN: Lower Risk—least concern. CITES: Appendix II

APPEARANCE The head of a pygmy sperm whale is blunt, rather like that of a shark. They share some features with sperm whales, but have a dorsal fin.

Dwarf Sperm Whale

Kogia simus

OTHER NAMES Owen's pygmy sperm whale, rat porpoise

LENGTH 9 ft/2.7 m

WEIGHT To 600 lb/272 kg

DISTRIBUTION AND HABITAT Worldwide, from colder and temperate to tropical waters.

STATUS Unknown population, probably widely distributed. IUCN: Lower Risk—least concern. CITES: Appendix II

APPEARANCE This species is difficult to distinguish from the pygmy sperm whale, and was once consigned to the same species. It is now thought to be smaller, with a rounder nose and a larger dorsal fin.　LW

Avoiding the Bends

Before a feeding dive, sperm whales lie at the surface blowing repeatedly—roughly one blow for every minute they will be submerged. Dives can last well over an hour. Though the whale plunges to extraordinary depths, it does not get the bends because it does not breathe oxygen under pressure. Its lungs empty soon after diving, and it relies on oxygen stored in its blood and muscle.

Beaked Whales

SUBORDER Odontoceti

FAMILY Zipiidae

The scientific name for beaked whales comes from the Greek *xiphos*, meaning sword. The species are small to medium in size. They have a single pair of throat grooves, unlike most toothed whales which do not have any. A very successful group, there are around 20 species, most of which are less than 20 ft (6 m) long.

Beaked whales are generally deep-water animals, and they are very reclusive. Populations could be quite large, especially for the southern bottlenose whale and Arnoux's beaked whale, which live in Antarctica in the summer; however, they are seldom seen, and scientists know very little about their behavior. On occasions, they have been spotted in groups of 50 or more, which suggests that they are gregarious. Beaked whales feed on squid. Males have one or two tusklike lower pairs of teeth, but females have no real teeth. Males seem to use these teeth for fighting among themselves.

Arnoux's Beaked Whale

Berardius arnouxii

OTHER NAME New Zealand beaked whale

LENGTH 33 ft/10 m

WEIGHT 11 tons/10 tonnes

DISTRIBUTION AND HABITAT Circumpolar throughout the southern oceans.

STATUS Unknown population, probably common in Antarctica. IUCN: Lower Risk—conservation dependent. CITES: Appendix I

APPEARANCE These whales are blue-gray or brownish, with slightly darker flippers, flukes, and back. The older males tend to be paler and can be heavily scarred. This is one of the few beaked whale species with functional teeth.

Southern Bottlenose Whale

Hyperoodon planifrons

OTHER NAMES Flower's/flat-headed/Antarctic bottlenose whale, Pacific beaked whale

LENGTH 25 ft/7.5 m

WEIGHT 8¾ tons/8 tonnes

DISTRIBUTION AND HABITAT Throughout the Southern Hemisphere, particularly in Antarctic waters, but occasionally reaching into the tropics.

STATUS Unknown population, thought to be common in Antarctica. IUCN: Lower Risk—conservation dependent. CITES: Appendix I

APPEARANCE Southern bottlenose whales are an unusual metallic gray on top and paler below, and they may be heavily scarred. Females are larger than males.

Cuvier's Beaked Whale

Ziphius cavirostris

OTHER NAME Goose-beaked whale

LENGTH Female 25 ft/7.5 m; male 23 ft/7 m

WEIGHT 3⅓ tons/3 tonnes

DISTRIBUTION AND HABITAT Worldwide, in cold to tropical waters; believed to live in small, deep-water pods, feeding on squid and deep-sea fish.

STATUS Population unknown. IUCN: Data Deficient. CITES: Appendix II

APPEARANCE Cuvier's beaked whales can be very light yellow-brown or gray-blue. Their heads are slightly concave and the beak is less developed than in most beaked whales.

Mesoplodon Species

COMMON NAMES Gray's beaked whale, scamperdown whale (*M. greyi*); Hector's beaked whale (*M. hectori*); strap-toothed whale (*M. layardii*)

LENGTH *M. greyi*—20 ft/6 m; *M. hectori*—13 ft/4 m; *M. layardii*—16 ft/5 m

WEIGHT *M. greyi*—5½ tons/5 tonnes; *M. hectori*—2⅕ tons/2 tonnes; *M. layardii*—4 tons/3.5 tonnes

DISTRIBUTION AND HABITAT Cold–temperate Southern Hemisphere waters, in singles or groups up to three; deep divers, submerging for up to two hours.

STATUS Population unknown. IUCN: All Data Deficient. CITES Appendix II

APPEARANCE Seldom seen, these are hard to distinguish. The color patterns of some species are unknown; they are generally dark gray above and paler below, often with extensive scarring. Squid eaters, they have few functional teeth, and may capture prey by intense suction. LW

Mystery Whale

Little is known about the southern bottlenose whale. Although it is believed to be reasonably common in the Southern Ocean, it is rarely seen, and it is difficult to identify unless observed at close range. These enigmatic animals are thought to migrate to warm waters during the winter months. They dive deep to feed on squid, but their surface behavior is inconspicuous.

Seabirds

Breeding Colors
In the breeding season, the imperial shag (above) has a bright blue eye-ring and yellow caruncles at the base of its bill. It builds a nest from seaweed and pieces of grass, which it cements together with guano.

Of the nearly 10,000 bird species on the Earth, only about 3 percent are considered seabirds, meaning those that obtain virtually all their sustenance from the sea. Seventy percent of the Earth's surface is covered in water, and this is the world of the seabirds. Many seabirds, and a few land birds, make their homes in Antarctica or on the many sub-Antarctic islands scattered around the Southern Ocean. To stand on the deck of a ship in the Drake Passage, the wild waterway that separates the southernmost tip of South America from the Antarctic Peninsula, is to marvel at the birds that wheel above and dive into waves. Humans cannot imagine a life away from terra firma—so how do these birds survive on an apparently featureless ocean?

At Home at Sea

Unlike marine mammals, no seabirds have evolved the ability to reproduce without coming ashore somewhere to find a mate, build a nest, and lay eggs. Most seabird species breed only once each year (some only every other year), spending four to five months in the process.

Tied to the land for less than half of each year, seabirds are truly at home when at sea. Most seabird species are relatively long-lived and take a long time to reach sexual maturity, so that once a young bird fledges and strikes out on its own, it may be years before it alights on land again. There is no freshwater at sea, and the food, while abundant, is often found in widely separated patches or hidden deep beneath the surface. Finding ways to meet these challenges has produced the diversity of seabirds known in the world today.

Several groups of birds have adapted to life at sea and there is no single story of seabird evolution.

Of the flying birds around the world, 113 species are Procellariiformes, or tube-nosed birds, expert fliers that roam huge expanses of the oceans to find food; they are an old group with great diversity that evolved in the Southern Hemisphere. Descended from ancestral Procellariiformes, another 50 species of seabirds are the Pelecaniformes: the pelicans, gannets and boobies, tropicbirds, cormorants and shags, and frigatebirds, which tend to live in temperate or tropical habitats and

stay closer to shore than tube-nosed seabirds. Another 107 species come from a larger group of waterbirds, the Charadriiformes—these are the skuas, gulls, terns and noddies, skimmers, and auks. Primarily birds of the Northern Hemisphere, they have spread throughout the world's oceans. Several other scattered species can also claim the title seabird: a few duck and geese species of the order Anseriformes, and grebes (Podicepediformes) and loons (Gaviiformes).

Seabirds wander more than land birds, and trans-equatorial crossings have occasionally led to the rise of closely related species pairs, with one species in the north and the other in the south—for example, Arctic/Antarctic terns, lesser black-backed/kelp gulls, great/Antarctic skuas, and northern/southern fulmars. Some of these pairs are literally sister species, whereas others are the result of convergent evolution. GM

Attracting a Mate
Two young wandering albatrosses (below) engage in their complex courtship dances, which include outstretched wings, bill pointing, and preening.

Eliminating Salt

Seabirds face a problem not usually encountered by land birds: when they spend months at sea, there is no freshwater for them to drink. It was known for many years that bird kidneys were less efficient than human kidneys, and no one understood how birds dealt with excess salt. Anatomists knew that all seabirds have two small glands lying in a small groove above their eyes, but it was not until the late 1950s that the animal physiologist Knut Schmidt-Nielsen found the explanation: the function of these glands is to get rid of excess salt. The salt glands are ten times more efficient at removing salt than bird kidneys. Salt is picked up by the circulating blood, and as it passes through the salt glands it is excreted as a highly saline solution that drains into the nasal cavities. In most birds, this solution drips from the nostrils to the end of the bill, and the characteristic head-shaking of most seabirds is to get rid of these drips of salt. Many kinds of land birds possess rudimentary salt glands, but they are tiny compared with those of seabirds.

Distinguishing Traits
Together, a northern giant petrel (on top) and a bloodstained southern giant petrel (on the bottom) scavenge the carcass of an elephant seal. The northern giant petrel is identified by the characteristic red tip on its bill.

Albatrosses

ORDER Procellariiformes
FAMILY Diomedeidae

Economical Flight
A royal albatross hangs in the wind near its breeding colony. By taking advantage of the constant winds, albatrosses use no more energy on long foraging flights than they do sitting on their nests.

Mating for Life
Royal albatrosses normally take several seasons of courtship dancing to choose a mate: a very important decision. They will stay with the same mate throughout their long lives unless one of them fails to return to the breeding site.

Albatrosses have an almost mythical significance for sailors: the early venturers into the Southern Ocean believed that to shoot an albatross would bring bad luck, and biologist Robert Cushman Murphy said: "I now belong to the higher cult of mortals, for I have seen the albatross." Despite their remote habitats, commercial interests imperiled albatrosses for many years. In the late nineteenth and early twentieth centuries, many species were brought almost to the verge of extinction by hunters for the feather trade. Today, the main threats are overfishing in some regions, resulting in severe reductions in the birds' food supply, and the ongoing threat of illegal longline fishing operations. Every year thousands of albatrosses are hooked and drowned when they try to seize baits from the hooks of longlines. New regulations for fishing equipment and procedures have dramatically reduced the number of birds killed, but illegal operations do not usually follow the approved guidelines and they continue to decimate the albatross and petrel populations.

The earliest fossil of the family Diomedeidae dates from about 50 million years ago, in the late Eocene, and it has been identified as an albatross or petrel. The first fossil that is clearly an albatross comes from about 32 million years ago, when the seas were changing and there was an important diversification in the bony fish and squid. Most modern albatrosses are squid-eaters, and it is believed that the increasing diversity and abundance of squid sparked the development of more albatross species. Albatrosses evolved in the Southern Hemisphere and then spread to all the world's oceans, but they have always been more diverse in the Southern Hemisphere. Only three species live in the northern Pacific Ocean today, and none live in the north Atlantic, although fossils from a former resident have been found there. Fossil history indicates that probably there were never any species larger than the wandering albatross of today, but several extinct species were smaller than the smallest extant albatrosses.

The distinctive feature of albatrosses is their size: the wandering albatross, which has a wingspan of up to 11½ ft (3.5 m), is the largest flying bird in the world. Albatrosses traditionally numbered 14 species that belonged to two genera: 12 *Diomedea* species, and the two smaller and darker *Phoebetria* species. However, recent genetic research suggests that 24 different taxa (many formerly designated subspecies) should be given species status in their own right.

There are many questions remaining regarding the proposed species designations; however, the new genera designations seem appropriate. Here we follow the traditional taxonomy and retain *Diomedea* for the great albatrosses (the royal and wandering albatrosses) and *Phoebetria* for the sooty albatrosses, but use the genus *Thalassarche* for the (smaller) mollymawks.

The Next Generation
A white-capped albatross preens its small chick. Once the egg hatches, the parents tend the chick for three weeks.

Built for Gliding

Most albatrosses have typical seabird plumage—dark back and upper wings, and light body and underwings. The great albatrosses, which are born conspicuously darker than the adults and slowly grow more and more white feathers, may take up to seven years to develop full adult plumage, whereas the two *Phoebetria* species remain dark as adults. Albatrosses have long wings relative to their body size, and their upper-wing bones and vertebrae are hollow yet strong, reinforced with struts that reduce their body weight and make gliding easier. Even with these adaptations, the long wings are awkward, and they are not very maneuverable. They soar majestically over the waves, seldom flapping their wings. Albatrosses are clumsy and slow on land and, without sufficient wind speed, they must land on the ocean. They are the most oceanic of the seabirds and cover vast distances. Even during the breeding season, when they must return regularly to their nest to feed chicks, they may travel 5,000 miles (8,000 km) in a week in search of food. Modern satellite-based studies show that, on their long journeys, albatrosses choose curving paths around the ocean to avoid both the calm regions and the areas of the highest wind speed.

Most albatross breeding colonies are on exposed sloping hillsides of isolated islands at high southern latitudes, where there are persistent strong winds to help the birds take off. The three largest species only breed every other year, but many of the smaller species return to their breeding sites to mate each year. Most albatrosses are sexually mature at four to six years of age, but do not normally attempt to breed until they are seven to nine years old, and the largest species may wait until they are ten or more. Courtship is protracted, taking perhaps two to four years of visiting the breeding colony to form a bond with a bird of similar age. Once paired, albatrosses usually stay together for life; if one fails to return, its mate will try to find a new partner, but this may take two or more seasons.

Albatrosses have a marvelous repertoire of courtship displays. They stand motionless with wings outspread, or point skyward with their bills, or do an exaggerated head-swaying walk. They clack their bills together and, when they eventually form a pair bond, they usually preen each other vigorously. Pairs of the sooty albatross and the light-mantled sooty albatross also perform one of the most beautiful duet flights; they swoop above the cliffs where they will nest in perfect synchrony, turning and banking in an aerial *pas de deux* that consolidates their bond and prepares them for breeding. Southern albatrosses build a substantial nest of mud, grass, and moss, returning year after year and adding to it to create a tall pedestal or a massive mound. Northern species build a more rudimentary nest: a simple scrape in the sand surrounded by twigs. The waved albatross of the Galapagos Islands makes no nest, but lays a single egg onto the ground and rolls it round during incubation.

The male of each pair arrives early in the breeding season and immediately begins to defend his territory. While he is waiting for his mate to arrive, he rebuilds the nest or starts a new one. The female arrives a few days later, and they enact their courtship displays and settle

Following Ships

Seabirds are opportunistic and optimistic, and they have learned that a ship usually means food. Most ships discharge kitchen scraps and sewage into the sea, so a ship's wake often rewards a following bird. Fishing boats especially attract large flocks of seabirds, because they use bait and throw their unwanted catch back overboard, as well as the offal of the catch that they have kept and cleaned on board.

But even non-fishing ships attract seabirds. Many smaller seabird species live on minuscule prey, such as krill and amphipods, and the waves created by ships' propellers throw these small crustaceans and fish to the surface, attracting birds that rely on these prey.

Ships also create wind vortexes that are kind to birds. Albatrosses and petrels soar on the almost constant winds of the Southern Ocean, riding the eddies and updrafts generated by the swell of the open ocean. There may be vast distances and no obstructions, but wind swirling around and over a moving ship creates powerful lift that can assist seabirds.

The birds take advantage of the windward flow, where the wind rises over the ship, and on the leeward side they travel a little farther away to avoid the dead air near the ship.

into their nest. After a bout of copulation, both the birds return to sea to build up their energy reserves. They return to the nest just before egg-laying, and the female lays a single egg. All Procellariiformes lay only one egg in each breeding season.

The parent birds share the incubation and feeding duties. The male takes the first incubation stint while the female goes to sea to feed. Her trip may last a few days or a few weeks, during which she may travel up to 9,300 miles (15,000 km), and when she returns the male goes foraging. Observations suggest that the males travel south and forage around the Polar Front, whereas females tend to fly north, where they are more likely to encounter the hazards of longline fishing.

The parents take turns bringing food back to the nest in their stomachs and regurgitating it to the chick. The chick pecks and bites at the adult's bill and, as the adult begins to regurgitate, the chick maneuvers its bill crosswise inside the adult's bill so the food is transferred directly into its mouth. The parents feed the chick until it is considerably heavier than they are—up to 166 percent of typical adult weight. The extra weight, mostly fat, provides the energy to grow feathers, and to help the chicks survive their first days alone at sea.

Albatrosses either alight on the sea or plunge into the ocean surface in quest of squid, fish, krill, and other crustaceans; however, they cannot dive very far below the surface. Many of them feed in the early morning and evening when squid rise to the surface.

Courtship Dance

Starting at age three, young wandering albatrosses return to the islands where they were reared to find a mate. But it may be four to six seasons before they successfully find a partner and breed.

Begging for Food

After the end of the brooding stage, parents feed their chick every three or four days. When the parent returns and approaches with food, the hungry chick eagerly bites at the bill of the parent to beg for food. When the parent opens its bill, the chick reaches inside and across the mouth of the parent to receive a meal of regurgitated fish and squid.

Wandering Albatross

Diomedea exulans
LENGTH 42–53 in/107–135 cm
WINGSPAN 98–138 in/250–350 cm
STATUS About 21,000 breeding pairs; 100,000 birds.
IUCN: Vulnerable. CITES: Not listed

This is the largest of the albatrosses. As a juvenile, its feathers are brown, but it becomes progressively whiter as it matures. First the body whitens and then the upper wings, starting with white spots on the mid-wing close to the body, and slowly whitening along the entire wing except the tips. The tails retain some black.

The wandering albatrosses' breeding cycle occupies just over a year, and the birds spend the year following breeding at sea. This means that, at best, they can raise one chick every two years. Wandering albatrosses breed in loose aggregations on gentle hills with good exposure to the wind. The eggs are incubated for 78 to 79 days before the fluffy white chick hatches. Chicks stay in the nest and are fed by both parents for an average of 246 days before they are ready to venture out on their own. Toward the end of the nestling period, older fledglings are left alone for long periods. They will leave the nest to walk around, and on windy days they exercise their wings with short, and at first very shaky, hovering flights. Eventually, they launch for good and head out to sea for their first three years.

Wandering albatrosses are found all around the Antarctic in the open ocean, away from all land except for their own nesting areas. They range over the entire Southern Ocean, as far north as 27ºS and as far south as 67ºS, a short distance from the ice edge. The most abundant form nests throughout the sub-Antarctic.

Although traditionally considered a single species containing five subspecies, according to recent genetic studies the wandering albatross is actually five distinct species, each containing unique characteristics in its plumage. The four recognized new species nest on Gough Island, Ile Amsterdam, and the Auckland and Antipodes islands, and each of the island populations has its own home range.

Wandering albatrosses feed primarily on squid and fish, which they seize once they land on the surface of the water. They eat more carrion than other albatrosses, and, although they are habitual ship-followers, they tend to soar at a farther distance from ships than the black-browed albatrosses and other petrels.

WANDERING ALBATROSS

(map) Distribution / Breeding sites — Tristan da Cunha, Gough I, South Georgia, Falkland Is, Bouvetøya, Sth Sandwich Is, Sth Orkney Is, Prince Edward Is, Is Crozet, Is Kerguelen, Heard I, I Amsterdam, South Pole, Balleny Is, Polar Front, Campbell I, Antipodes Is, Bounty Is, Chatham Is, Macquarie I, Auckland Is

Studies on South Georgia have found that female wandering albatrosses tend to fly northward to forage, whereas the males head south. The north-flying females are more likely to encounter longline fishing boats, and so are more likely than males to get caught and drown. The chicks of drowned adults do not survive because a single parent cannot provide enough food for a growing bird. The loss of more females than males to longline fishing causes a greater impact on breeding success than the simple loss of females does. Widowed males searching for a new mate often pester already paired females, and the confusion and harassment sometimes causes breeding females to fail.

Royal Albatross

Diomedea epomophora
LENGTH 42–48 in/107–122 cm
WINGSPAN 120–138 in/305–350 cm
STATUS About 10,000–20,000 breeding pairs.
IUCN: Vulnerable. CITES: Not listed

Closely related to the wandering albatross, the royal albatross is very similar in appearance. Immature royal albatrosses, however, are mostly white, and only young juveniles have black on their tails. Royal albatrosses also have distinctive black lines along the cutting edges of their upper bill, although these can be seen at close range only. The adults of the southern subspecies retain the dark tops of the wings.

Royal albatrosses are most common in the Pacific, but can also be seen throughout the Southern Ocean. They are divided into two subspecies (or species in the new taxonomy)—the northern royal (*D. e. sanfordi*), which breeds on the South Island of New Zealand and on the Chatham Islands, and the southern royal (*D. e. epomophora*), which breeds on Campbell Island and the Auckland Islands. In the non-breeding season, the two subspecies mingle at sea.

The royal albatross's breeding behavior is exactly like that of the wandering albatross, though the precise timing may differ, according to the place where they breed and the conditions they find there.

Like wandering albatrosses, royal albatrosses feed extensively on squid, which they capture by surface-seizing. But unlike wandering albatrosses, they are not habitual ship-followers, although they readily join other petrels to feed on offal discarded from fishing boats.

Devoted Parents
Sitting on its egg high up on Auckland Island, this white-capped albatross and its mate will tend the egg constantly. Males usually incubate for a few days longer than the females, but each sits for four- to 13-day shifts, patiently waiting and fasting in all weather.

Resting at Sea
A yellow-nosed albatross swims calmly on the sea. The usual image of albatrosses is soaring on the wind, but they regularly alight on the water to feed or rest. They have no trouble taking off as long as there is a wind blowing.

ROYAL ALBATROSS

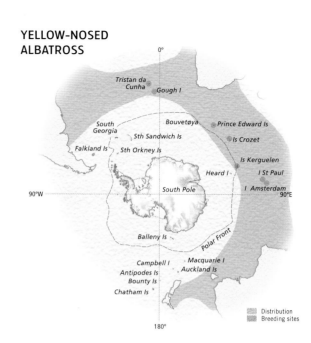

YELLOW-NOSED ALBATROSS

Yellow-nosed Albatross
Thalassarche chlororhynchos
OTHER NAME Yellow-nosed mollymawk
LENGTH 28–32 in/71–81 cm
WINGSPAN 79–101 in/200–256 cm
STATUS 80,000–100,000 breeding pairs. IUCN: Lower Risk—near threatened. CITES: Not listed

The yellow-nosed albatross, with its dark wings and its white body, looks like the other mollymawks, but its gray head is lighter than that of the gray-headed albatross, and it has a whitish cap.

The bird's underwings have more white than those of the black-browed and gray-headed albatrosses but not as much as those of the white-capped albatross. It is slimmer in build than other mollymawk species and its long, thin bill is black, with pronounced lines of yellow above and below.

In the revised taxonomy, the two subspecies are defined as species in their own right. *T. c. bassi*, the eastern subspecies, breeds in the west Indian Ocean, around Prince Edward Island and Iles Amsterdam and Crozet; the western subspecies (*T. c. chlororhynchos*) breeds around Gough Island and the Tristan da Cunha group of islands in the south Atlantic. These albatrosses feed mainly on squid, and they also take crustaceans and offal. Around fishing boats, they typically grab a morsel of offal or bait and then fly away from the other scavengers to consume it.

Yellow-nosed albatrosses start their breeding cycle in August and September, which is earlier than other Southern Hemisphere species. They nest singly or in loose groups on fairly flat ground, in nests of mud and grass. The parents incubate the single egg for 71 to 78 days, and then feed the chick for about 130 days, when it fledges. After the breeding season, the birds disperse around the south Atlantic and Indian oceans between latitudes 15ºS and 50ºS.

Yellow-nosed albatrosses occur along the coasts of South Africa and Argentina, and in the region south of Australia and New Zealand. In recent years, they have been observed along the coast of North America as far to the north as Canada.

Bonding

A white-capped albatross preens its mate. The most faithful of birds, albatrosses take a long time to mature sexually and they chose their lifetime partners with care. When they finally form a pair bond, they will preen each other vigorously.

Mutual Preening

Allopreening, where one bird preens another, is important social behavior among albatrosses, particularly for strengthening pair bonds. While adults that have chicks are at sea finding food for their growing chicks, failed breeders and non-breeders continue to seek mates or reconfirm their pair bond.

Black-browed Albatross

Thalassarche melanophris
OTHER NAME Black-browed mollymawk
LENGTH 33–37 in/83–93 cm
WINGSPAN 94 in/240 cm
STATUS 550,000–600,000 breeding pairs; more than 2 million birds. IUCN: Lower Risk—near threatened. CITES: Not listed

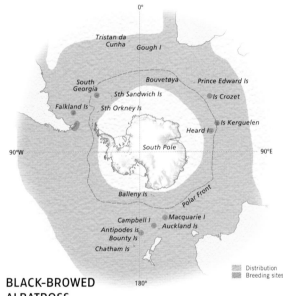

BLACK-BROWED ALBATROSS

Probably the most common of the albatrosses, the black-browed albatross is one of the most recognizable. Its body is white with a black saddle between its wings, its upper wings are black, and its underwings are white with a broad black margin. It has a yellowish bill, and a white head with clearly marked black eyebrows. From a distance, and at rest, it can be mistaken for the kelp gull, but its soaring flight, which is in elegant contrast to the flapping motion of the kelp gull, identifies it as an albatross. It feeds on crustaceans—mainly krill—as well as squid and fish, but it also devours carrion, joining other petrels at fishing boats to scavenge the offal.

Black-browed albatrosses fly over the whole of the Southern Ocean, and have been spotted close to shore as well as far from land. Their range extends along the cold-water currents of the west coasts of South America and Africa, and in the high southern latitudes they are regularly seen along the Antarctic Peninsula coast in the Gerlache Strait, where there may be considerable ice. They breed annually, the females laying a single egg in

a substantial nest of mud and grass. Breeding colonies can consist of as many as 100,000 pairs of birds, and are often shared with shags and penguins.

Little is known about their migratory patterns after breeding, but there is a general movement northward in the autumn. Stragglers are seen in the north Atlantic regularly: there have been over 40 sightings in Britain.

White-capped Albatross

Thalassarche cauta
OTHER NAMES Shy albatross, white-capped mollymawk
LENGTH 35–39 in/90–99 cm
WINGSPAN 87–101 in/220–256 cm
STATUS 800,000–1,000,000 birds. IUCN: Lower Risk—near threatened. CITES: Not listed

The white-capped albatross is the heaviest and most thickset of the mollymawks. The undersides of its wings are almost entirely white, with a thin border of black, and it has characteristic black "thumbprints" on the underwing near the body. Its head is a very pale gray, and its dark eyes and eyebrows make it look stern. Its bill is light gray with a yellowish tip.

There are three subspecies of *Thalassarche cauta*, distinguished by the coloration of the head and bill. The Tasmanian shy albatross (*T. c. cauta*) breeds around Tasmania and the Auckland Islands; the Chatham albatross (*T. c. eremita*) on the Chatham Islands; and Salvin's albatross (*T. c. salvini*) on Snares and Bounty islands in the southern Pacific and Iles Crozet.

White-capped albatrosses normally breed annually, starting in September, and their breeding cycle is the same as those of other mollymawks. They eat mostly squid, as well as fish and crustaceans. Despite one of their common names, shy albatross, they often feed with other seabirds, and they have been seen following whales. They compete successfully with birds that crowd around fishing boats.

After the breeding season, white-capped albatrosses are found throughout the sub-Antarctic latitudes in the Pacific and Indian oceans, with the exception of the Chatham Island subspecies, which is rarely seen far from its island home. Unlike other albatrosses, white-capped albatrosses are frequently seen over continental shelves, or even close to shore.

WHITE-CAPPED ALBATROSS

Slow and Steady

Gray-headed albatrosses are biennial breeders. Their chicks grow slower and are fed longer than other mollymawks, but they do have consistently higher fledging success.

Large Babies

Light-mantled sooty chicks are guarded and fed by their parents for the first 19 to 21 days. They grow up to 140 percent heavier than their parents but, by the time they make their first flight, have dwindled to 91 percent of adult weight.

Gray-headed Albatross

Thalassarche chrysostoma

OTHER NAME Gray-headed mollymawk

LENGTH 32–33 in/81–84 cm

WINGSPAN 71–87 in/180–220 cm

STATUS Approximately 500,000 birds. IUCN: Vulnerable. CITES: Not listed

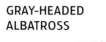

GRAY-HEADED ALBATROSS

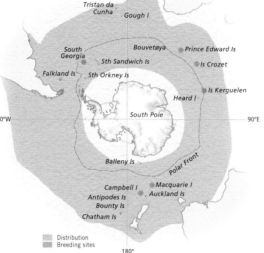

Except for its gray head that gives it a hooded appearance, this species is very much like the black-browed albatross. Its bill is black with yellow lines along the top and bottom, although this is difficult to see in flight. Its flight patterns, food, and lifestyle are similar to those of the black-browed albatross. The two species often share colonies, and the structure and timing of their reproductive cycles are also similar, except the gray-headed albatross feed its chicks for three or four weeks longer so, if it succeeds in raising a chick, it is only able to breed every second year.

Gray-headed albatrosses prefer colder water than most other albatrosses, and they stay away from land except at their breeding sites. They eat mostly squid, and follow fishing vessels less than other albatrosses. Like most albatrosses, they are at risk from longline fishing, competition from local squid fisheries, and introduced animals at some of their breeding sites.

Sooty Albatross

Phoebetria fusca

OTHER NAME Dark-mantled sooty albatross

LENGTH 33–35 in/84–89 cm

WINGSPAN 80 in/203 cm

STATUS 80,000–100,000 birds. IUCN: Vulnerable. CITES: Not listed

Light-mantled Sooty Albatross

Phoebetria palpebrata

LENGTH 30½–31 in/78–79 cm

WINGSPAN 72–86 in/183–218 cm

STATUS About 150,000 birds. IUCN: Lower Risk–near threatened. CITES: Not listed

The two albatrosses constituting the genus *Phoebetria* can be distinguished from other albatrosses by their narrow, pointed wings and long, wedge-shaped tails.

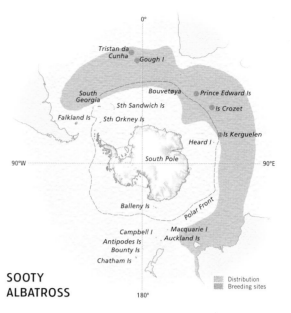

SOOTY ALBATROSS

Returning to shore only for breeding, they nest in loose colonies among vegetation on cliffs or steep slopes. If they fail in breeding one year, they can try again in the following year but, if they succeed in raising their chick, they can only breed every other year. The breeding areas of the two species overlap in some places: where this happens, they lay their eggs at different times. They feed mainly on squid, but also take fish, crustaceans, and carrion, usually as penguin carcasses.

Sooty albatrosses are uniformly dark, with a partial white eye ring and a yellow line along the cutting edge of their lower bill. They are found through the southern Atlantic and Indian oceans from Argentina to Tasmania, and they nest on Gough and Prince Edward islands, on Iles Crozet, and on the islands of the Tristan da Cunha. Sooty albatrosses on Nightingale Island in the Tristan da Cunhas are under threat from humans; each year as many as 3,000 chicks are taken from a colony there.

Identical to the sooty albatross except for an ash-gray collar and mantle and a blue line on the lower mandible, light-mantled sooty albatrosses prefer colder water. They nest around New Zealand, on Iles Crozet and Kerguelen, the Prince Edward islands, and South Georgia, breeding farther inland than the sooty species, and during the non-breeding season are much more widespread and occur through the Southern Ocean. GM

LIGHT-MANTLED SOOTY ALBATROSS

Fulmars

ORDER Procellariiformes
FAMILY Procellariidae

Petrel Heads
Cape petrels nest on narrow ledges on rocky cliff faces, often in large numbers. A continuous, burbling chatter comes from these cliffs as the gregarious birds vocalize, purr, and cackle to each other.

Armed Against Attack
Southern fulmars nest in the open on cliff ledges and steep slopes. They can protect themselves and their chicks from attacks, often by skuas, by spitting or spraying sticky stomach oil onto attackers. They even spit on people who approach too close.

The Smallest Petrel
The snow petrel is the smallest of the fulmarine petrels and the ecological equivalent of the ivory gull of the Arctic. They nest farther inland than any other petrel, and to combat the cold their downy feathers cover the base of the bill and extend well down the legs.

This family consists of a diverse group of petrels—small to medium-sized seabirds, dark with long wings that enable them to use the wind to fly great distances. The nostrils on the top of their bill join into a single tube with a visible septum separating them. All petrels lay only one egg at each breeding attempt, but incubation times are variable. They often leave their egg unattended while they forage. Because of these breaks in incubation, petrel embryos withstand cold better than those of most other birds. The period between hatching and fledging also varies due to changes in local feeding conditions and the parents' ability to raise their chick.

Family Origins

The first group of petrel-like birds diverged from the divers, or loons (order Gaviiformes), about 50 million years ago, not long before the Diomedeidae (albatrosses) diverged from the petrels. The first Procellariidae fossil evidence dates from 40 to 50 million years ago and is believed to be from a shearwater, which likely originated in the Southern Hemisphere and later spread northward. Petrels are found in all the world's oceans, but there are many more species in the Southern Hemisphere than in the north.

This group presents many taxonomic puzzles that are exacerbated by lack of information about their ecology and breeding behavior. However, it can be split into four main groups—the fulmarine petrels, prions, gadfly petrels, and shearwaters—based on differences such as bill shapes, feeding habits, and flight patterns.

Fulmarine Petrels

Ranging from the Arctic to the Antarctic, the fulmarine petrels are the most diverse of the fulmars. All have relatively long, broad wings, but they vary considerably in size and in plumage. There are two pairs of sibling species—the northern and southern giant petrels, and the northern and southern fulmars—and three further species—the snow petrel, the Antarctic petrel, and the

cape petrel. The giant petrels fly with slow, powerful wingbeats, whereas the smaller species have faster wingbeats; all are good gliders. Fulmarine petrels feed on squid and crustaceans, mainly surface krill. Giant petrels, fulmars, and cape petrels are ship-followers.

Prions

Prions are small petrels of the genus *Pachyptila*, plus the blue petrel, the only member of genus *Halobaena*. (The blue petrel has been placed in each of the four procellariid groups, but its color and other structural features link it with the prions.) Prions are similar in appearance and habits, sharing a buoyant, erratic flight. All are a light blue-gray, with white underparts, dark primary feathers, an M-shape on their upper wings, and except for the blue petrel (which has a white tip) have black tips on their wedge-shaped tails.

Gadfly Petrels

Traditionally, this group comprises 23 to 34 *Pterodroma* species and two *Bulweria* species, but the relationships within the group are poorly understood. Several of the *Pterodroma* species are known only from specimens taken long ago and recently rediscovered. Gadfly petrels have long, wedge-shaped tails with relatively short wings. They flap and glide, rising in high, swooping arcs in flight, and they tend to bend their wrists while flying. They rarely land on water, but they pluck crustaceans, squid, and fish from the surface; they do not dive for food. Most nest in small colonies or scattered pairs and, because of their isolated breeding areas, breeding sites of several species are unknown. Many of the colonies are threatened by predation and introduced species.

Shearwaters

Shearwaters are a very old, complex group consisting of four species in the genus *Procellaria*, some 15 to 20 in *Puffinus*, and two in *Calonectris*—the latter two species

rarely venturing into the Antarctic region. The range of many shearwaters is only partially in Antarctic or sub-Antarctic waters; several of them migrate to Northern Hemisphere waters. Shearwaters are medium-sized to large petrels with long bills, long, narrow wings, and short, rounded tails. The smaller birds alternate bursts of rapid wing beats with stiff-winged banking glides near the ocean surface, dipping into troughs and rising over the crests of the waves. The flight of the larger species is more relaxed and deliberate. Their plumage is dark above, and brown or white below. They feed on squid, fish, and crustaceans on the surface or from shallow dives. Only the four *Procellaria* species commonly venture into Antarctic waters, and none of the species has a wholly Antarctic distribution.

Southern Giant Petrel

Macronectes giganteus
OTHER NAME Southern giant fulmar
LENGTH 34–39 in/86–99 cm
WINGSPAN 73–81 in/185–205 cm
STATUS Approximately 36,000 breeding pairs. IUCN: Vulnerable. CITES: Not listed

SOUTHERN GIANT PETREL

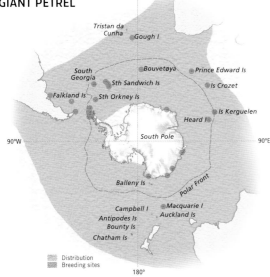

The southern giant petrel, with its close relative the northern giant petrel, is the largest of the Procellariidae. These two petrel species are sometimes mistaken for small albatrosses, but they do not glide as much, their bodies are heavier and are more hunched, and they have thicker bills because of the size of the tubular nostrils on the bills' topside.

Southern giant petrels have heavy, yellow-green bills with pale green tips. Most adults are brownish gray, with dirty white heads, necks, and upper breasts; juveniles are sooty black with a few white flecks on the head. A small percentage of southern giant petrels are white morphs; their plumage is all white with scattered black flecks. White morph juveniles have the same plumage as adults, so are difficult to distinguish in the field. Their circumpolar range extends south to the edge of the Antarctic ice and north into the subtropics. They breed on exposed hilltops and plains on sub-Antarctic islands throughout the region in small, loose colonies, building

nests of small stones or grass. Their breeding season begins in October, with both parents taking turns to incubate a single egg for 55 to 66 days. After around four months, the chicks fledge and leave the breeding site to roam the Southern Ocean for six or seven years until they are mature and ready to breed.

Giant petrels are probably Antarctica's most important scavengers, feeding on carrion, especially on the carcasses of seals, penguins, and petrels. They can be aggressive, and will drown or batter to death other birds, such as cape petrels or immature albatrosses. They also surface-feed on squid, krill, and fish. They are avid ship-followers in quest of refuse or offal, and many are caught on the hooks of longline fishing boats. Partly for this reason, populations appear to be dwindling, but this apparent decline may also be due to their extreme sensitivity to disturbance at their nest sites by humans.

Northern Giant Petrel

Macronectes halli
OTHER NAME Northern giant fulmar
LENGTH 32–37 in/81–94 cm
WINGSPAN 71–79 in/180–200 cm
STATUS 7,000–12,000 breeding pairs. IUCN: Lower Risk—not threatened. CITES: Not listed

The coloration of the northern species resembles that of the dark-phase southern giant petrels, although they are never as light on the head and neck as their southern cousins and there is no white form. They are slightly smaller than the southern species, but the physical feature that distinguishes them from the southern giant petrel is their bill, which is horn-colored with a pinkish or dark red tip. Like the southern species, they have a circumpolar distribution, but they generally stay north of the Polar Front, breeding mostly on South Georgia and Iles Crozet and Kerguelen, and in the New Zealand region. They either nest alone or in small colonies, and their breeding season starts in August. Interestingly, the sexes eat different diets, the males preferring to feed on the carrion of seals, penguins, and seabirds, and the females mostly going to sea to catch live prey, especially krill and squid.

NORTHERN GIANT PETREL

Antarctic Vultures
Because of their size and strength, giant petrels are the only birds that can gain access to the larger carcasses of seals. While feeding, the dominant bird threatens all others with outspread wings and erect, fanned tail. Once he takes his fill, the next dominant giant petrel will get its turn feeding.

Ocean Scavengers
A northern giant petrel feeds on a dead penguin in shallow water. The giant petrels, especially the males, are the top scavengers throughout the Southern Ocean. They dominate all other petrels and skuas at a carcass.

Prominent Nostrils
The southern fulmar is a good example of the tube-nosed seabirds with nostrils that develop into prominent tubes on the top of the bill. Studies show that the tubes facilitate a well-developed sense of smell. The clear drop of liquid hanging from the bill is the highly saline excretion from the salt glands.

Inland Nesters
Antarctic petrels usually feed in open water around the pack ice, but they nest on cliff ledges, sometimes more than 155 miles (250 km) from the coast.

Southern Fulmar
Fulmarus glacialoides
OTHER NAMES Silver-gray fulmar, Antarctic fulmar
LENGTH 18–20 in/46–50 cm
WINGSPAN 45–47 in/114–120 cm
STATUS Abundant, but no accurate estimates.
IUCN: Not listed. CITES: Not listed

The southern fulmar is the largest of the petrels after the two giant petrels, easily recognized by its pale gray upper parts and white underparts. Its wings have a dark trailing edge, a white flash at the base of the primaries, and there are variable amounts of black at the wingtips. Southern fulmars are circumpolar, and normally frequent cold-water areas on the edges of the pack ice. Outside the breeding season, they range widely around Antarctic waters. They feed mainly on krill, other crustaceans, and fish (Antarctic silversides), with varying amounts of squid, carrion, and offal. They are primarily surface-feeders, but sometimes make shallow dives.

They nest in colonies on sheltered ledges on cliffs and steep slopes. Breeding sites are restricted because of the birds' need for cliffs, but there are huge colonies on the South Sandwich and the South Orkney islands. Breeding begins in November, and the parents take turns incubating the single egg for about 46 days. The chick fledges after 48 to 56 days.

SOUTHERN FULMAR

Antarctic Petrel
Thalassoica antarctica
LENGTH 16–18 in/40–46 cm
WINGSPAN 40–41 in/101–104 cm
STATUS Many millions of birds. IUCN: Not listed.
CITES: Not listed

This is one of the most numerous of Antarctic birds. Like the cape petrel, for which it is often mistaken from below, its underbody and wings are white, the face and throat black. But its upper side is quite distinctive: the head, body, and the front half of its wings are dark brown to dark gray; the back half of its wings are white, as is its tail except for a black tip. The female is slightly smaller than the male.

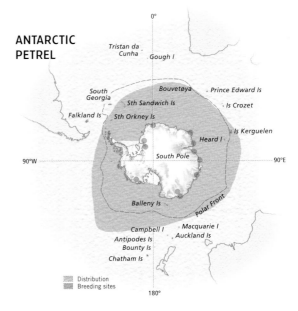

ANTARCTIC PETREL

Antarctic petrels are difficult to observe: they stay mostly within the boundaries of the pack ice, although some do fly as far north as the Polar Front in winter. They are truly Antarctic birds, feeding in open water near the pack ice and favoring areas with icebergs. They are sometimes found in association with other seabirds and whales, and may roost in large numbers on icebergs. They nest on cliff ledges and begin their breeding in November. Approximately 35 nesting colonies have been identified, some as far as 155 miles (250 km) inland, but the known colonies cannot account for their estimated numbers. They feed mainly on krill, and to a lesser extent on fish and squid, which they surface-seize, although they also plunge-dive and dip for food.

Cape Petrel
Daption capense
OTHER NAMES Pintado petrel, cape pigeon
LENGTH 15–16 in/38–40 cm
WINGSPAN 32–36 in/81–91 cm
STATUS Several million birds. IUCN: Not listed.
CITES: Not listed

One of the most familiar of the fulmarine petrels, the cape petrel has an unmistakable upper-wing pattern of white splotches on black, and the larger white patches on its back and tail are often flecked with black. Its tail is white with a broad, black terminal band. Underneath, it is white with a black head and throat.

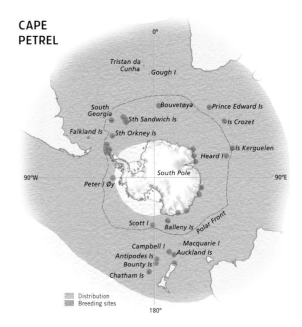

CAPE PETREL

wing, mainly on krill with some fish, squid, and carrion. They also land on the surface and feed by surface-seizing or shallow diving. They are circumpolar, nesting extensively on the southern continent and on some of the islands in high latitudes. Some remain around the continent all year, but the main influx of breeders arrives from mid-September to early November.

Snow petrels, along with south polar skuas, breed farther south than any other birds in the world, and in some of the harshest conditions on the Earth. They nest in small to large colonies in crevices on cliffs or among boulders on scree slopes, mostly near the coast, but sometimes as much as 215 miles (345 km) inland and at elevations as high as 7,870 ft (2,400 m). In the bitter weather, they must dig snow out of the crevices where they nest. Incubation averages 45 days, and the young are brooded for eight days after hatching; the parents then go to sea to bring food back to their chick. Chicks leave the colony at 42 to 50 days old.

Almost Adult
Already sporting the checkered wings of the adult, this cape petrel chick still has some nestling-stage down. Soon it will be off on its own for six years before returning to its natal area to breed.

Cape petrels have a circumpolar distribution that ranges from the subtropics to the edge of the Antarctic continent. In winter they stay far from shore, and from April to August they leave the Antarctic region and move northward into the sub-Antarctic.

During the breeding season, they stay closer to their Antarctic home, and forage mostly in inshore waters. In November, they return to their nesting cliffs to breed, building simple nests on small ledges. Adults take turns to incubate the single egg for 41 to 50 days. The young fledge in 47 to 57 days, and those that do survive have a life expectancy of 15 to 20 years.

Snow Petrel
Pagodroma nivea
LENGTH 12–16 in/30–40 cm
WINGSPAN 30–37 in/75–95 cm
STATUS Possibly several million birds. IUCN: Not listed.
  CITES: Not listed

Fairies of the South
Apsley Cherry-Garrard, biologist on Scott's 1910–12 expediton, called snow petrels "the fairies of the south," a poetic name for one of the most beautiful of Antarctic birds.

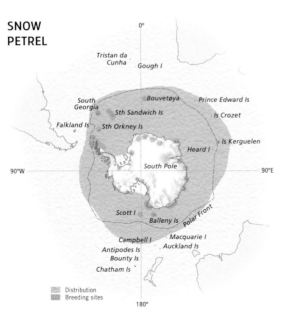

SNOW PETREL

Snow petrels have pure white plumage, dark eyes, and a dark bill. They forage in areas with pack ice; with their white camouflage, they appear to simply materialize out of nowhere, flitting by with a dancing flight and then disappearing into the distance. They dip-feed on the

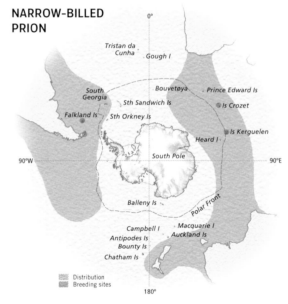

NARROW-BILLED PRION

Narrow-billed Prion
Pachyptila belcheri
OTHER NAMES Slender-billed prion, thin-billed prion
LENGTH 9¾–10 in/25–26 cm
WINGSPAN 22 in/56 cm
STATUS More than 1 million breeding pairs.
  IUCN: Not listed. CITES: Not listed

The narrow-billed prion is one of the more distinctive of the prions. It has the blue-gray body typical of all prions, but the black on the end of its tail is just a small central spot instead of a broad terminal band. Its slender bill, black eye line, and clearly marked white stripe above and behind the eye give it a striped facial pattern that may be distinguished in flying birds at sea.

These birds are gregarious. They feed mainly at night on small crustaceans—mostly amphipods, but they also take krill and small fish, using their thin bill to pick prey from the water individually. Although feeding at sea for most of the year, during the breeding season they feed inshore close to their breeding colonies. They breed in burrows on the Falkland Islands, on some islands off southern Chile, on Iles Crozet and Kerguelen, and possibly on Macquarie Island and South Georgia.

Antarctic Prion

Pachyptila desolata

LENGTH 10–10½ in/25–27 cm

WINGSPAN 23–26 in/58–66 cm

STATUS Many millions of birds; abundant and widespread. IUCN: Not listed. CITES: Not listed

The Antarctic prion is a fairly large bird, with the widest distribution of all the prions. It can be identified by the broad black tip on its tail and its relatively extensive, dark blue-gray collar. It feeds mainly on krill and other small crustaceans, although it will also take some fish.

The bird's bill is adapted for filter feeding, and it has developed a specialized behavior, called hydroplaning, to take advantage of this feature: it scurries along the surface of the water into the wind, with its body resting on the water but partially supported by the wind flowing under its outstretched wings. Holding its mouth open at the surface, or with its head entirely submerged, it filters its prey from the water as it swims, sometimes swinging its head from side to side.

ANTARCTIC PRION

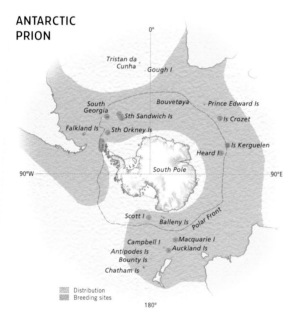

They range from latitude 40°S to the pack ice, and are circumpolar, apart from a gap in the southern Pacific Ocean. They breed on islands, from New Zealand west to South America, and nest in burrows in very dense colonies. Breeding begins in November, or later if there is too much snow. After their chick fledges, at 45 to 55 days, all birds leave, dispersing north over a broad area.

Broad-billed Prion

Pachyptila vittata

LENGTH 10–12 in/25–30 cm

WINGSPAN 22½–26 in/57–66 cm

STATUS Several hundred thousand birds. IUCN: Not listed. CITES: Not listed

The broad-billed prion has a fairly restricted range, nesting on Gough Island and the islands of the Tristan da Cunha group in the Atlantic, on Marion Island, and on the sub-Antarctic islands off the coast of New Zealand. It is highly gregarious, but rarely follows ships. Broad-billed prions are

BROAD-BILLED PRION

difficult to identify in flight, but up close their large size and massive bills distinguish them from other prions.

They live on small planktonic crustaceans and feed by hydroplaning with their specialized bill, like Antarctic prions. Weather at their breeding sites is comparatively mild, and breeding begins early, in July or August. They nest in burrows, occasionally sharing with another pair of birds. The parents incubate the egg for 50 days, and then spend another 50 days with the chick.

Fairy Prion

Pachyptila turtur

LENGTH 9–11 in/23–28 cm

WINGSPAN 22–23½ in/56–60 cm

STATUS Several million birds. IUCN: Not listed. CITES: Not listed

Little is known of the distribution of fairy prions, but they are probably circumpolar. They have the standard prion plumage, with few distinguishing features. Highly colonial, they breed on most sub-Antarctic islands in burrows on cliffs and rock falls, or in grassland with limited vegetation. Breeding begins in September, the egg is incubated for about 55 days, and the chick is fed for 43 to 56 days. And then all birds leave the colony, probably for the subtropical waters off the Australian and South African coasts.

FAIRY PRION

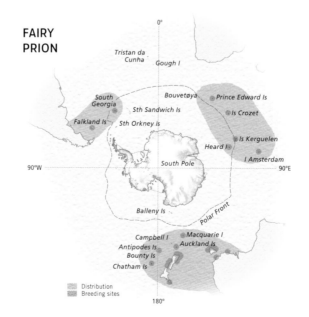

Protective Measures

A broad-billed prion emerging from its burrow. Skuas are their major predator on land because they can dig prions right out of their nests. The concrete mouth of this burrow may provide some protection.

Blue Petrel

Halobaena caerulea

LENGTH 10–12½ in/26–32 cm

WINGSPAN 24½ in/62 cm

STATUS Several million birds. IUCN: Not listed.
CITES: Not listed

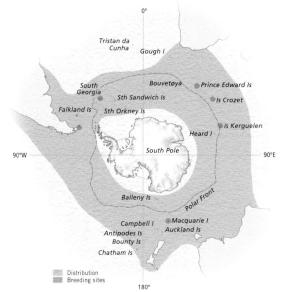

BLUE PETREL

Distribution
Breeding sites

Daylight Courtship

Unlike other prions, which shun the daylight hours, fulmar prions regularly perform courtship displays during the day while perched in the open. Predatory skuas are uncommon in many of their breeding areas, but prions will mob a skua if it dares to come near a nest.

Fulmar Prion

Pachyptila crassirostris

LENGTH 9½–11 in/24–28 cm

WINGSPAN 23½ in/60 cm

STATUS 100,000 breeding pairs. IUCN: Not listed
CITES: Not listed

One of the least known of the prions, the fulmar prion is so similar in appearance to the fairy prion that it is sometimes classified as a subspecies.

The fulmar prion has a limited range and breeds on Heard, Auckland, Snares, Chatham, and Bounty islands in October, where its breeding colonies occupy rock crevices and cracks in coastal cliffs and boulder slopes, sometimes alongside albatrosses. It is largely sedentary, and some adults roost in their nests all winter. They are affected by the presence of cats, rats, and pigs on some of the islands on which they breed.

FULMAR PRION

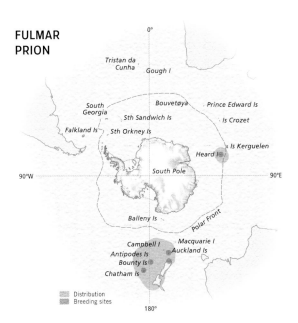

Distribution
Breeding sites

Small Differences

With its species-identifying white band visible on the end of its tail, a blue petrel (right) takes off from the surface. These petrels are usually classified with the prions, although they are slightly larger and have a narrower bill.

In appearance, the blue petrel resembles the prions, except for its white tail tip. Its flight is a little slower and less erratic because it is slightly larger. It has a typical circumpolar distribution and flies over all the oceans, from the temperate latitudes along the coasts of Africa, Australia, and South America to near the Antarctic continent. A marine pelagic bird, it does not venture into the pack ice in the far south.

Blue petrels feed mostly on crustaceans in some parts of their range, but take more fish in other parts. They take food mainly by surface-seizing and dipping, but will also dive and plunge. They breed in colonies on sub-Antarctic islands that have soil and vegetation, nesting in long burrows which they excavate in soft soil under tussocks. They begin breeding in September and raise their chicks for 43 to 60 days. Many of the adults are sedentary, staying close to their breeding sites in the winter and even visiting the sites intermittently. The young birds disperse more widely, traveling throughout the range for several breeding seasons before they are sexually mature and ready to breed.

Great-winged Petrel

Pterodroma macroptera

OTHER NAME Gray-faced petrel

LENGTH 16 in/41 cm

WINGSPAN 38 in/97 cm

STATUS About 400,000 breeding pairs. IUCN: Not listed. CITES: Not listed

The great-winged petrel is uniformly dark with a stubby bill and black feet; however, some have a patch of light gray feathers around the base of their bills, hence their other common name, the gray-faced petrel. They are almost circumpolar in the sub-Antarctic, being absent only from the western Atlantic zone. They feed mainly at night on squid, which they may track down by their bioluminescence. They start breeding in April in scattered pairs or small colonies on Iles Kerguelen, New Zealand and its islands, and the south coast of Australia. At the end of the breeding period, little penguins or one of the shearwater species often move into their burrows after the great-winged petrels have left.

White-headed Petrel

Pterodroma lessoni

LENGTH 16–18 in/40–46 cm

WINGSPAN 43 in/109 cm

STATUS About 100,000 breeding pairs. IUCN: Not listed. CITES: Not listed

The white head and underbody contrast with its dark underwings. The white of the throat and head becomes gray on the mantle, darkening toward the tail, and its broad, dark eye line gives it a masked appearance at sea. They range throughout the sub-Antarctic region from about latitude 30°S to the pack ice in the south. They breed on most sub-Antarctic islands except South Georgia, but are less common in the southern Atlantic. They start breeding in October, dispersing around the Southern Ocean in May.

Kerguelen Petrel

Pterodroma brevirostris

LENGTH 13–14 in/33–36 cm

WINGSPAN 31½–32¼ in/80–82 cm

STATUS Several hundred thousand birds. IUCN: Not listed. CITES: Not listed

Kerguelen petrels are uniformly dark on the upper parts and slightly lighter below, with plumage so shiny it seems to have white or silvery patches. The birds' large head and steep forehead give them a hooded appearance.

Kerguelen petrels are abundant throughout their range, and are circumpolar right from the sub-Antarctic region down to the pack ice. The birds begin to breed in August, and the chicks fledge by the end of December.

Soft-plumaged Petrel

Pterodroma mollis

LENGTH 12½–14½ in/32–37 cm

WINGSPAN 32½–37½ in/83–95 cm

STATUS Tens of thousands of birds. IUCN: Not listed. CITES: Not listed

The soft-plumaged petrel is mostly dark above and white on the breast and belly. The crown, nape, and upper back are usually slate brown, as is the partial or complete collar around the upper breast. The lores, chin, and throat are white. The underwings look dark from a distance, but at close range white highlights can be seen, and the upper wings are marked with a dark "M," like that of the prions. They are found around the sub-Antarctic region (except in the Pacific Ocean), on the islands around New Zealand, and westward through the Indian and Atlantic oceans to the South American coast. They breed on most sub-Antarctic islands within their range, except on South Georgia and the Falklands. They commence their breeding season in September, and the chicks fledge in March.

A Precarious Existence

The most important problems for great-winged petrels are introduced cats and rats that prey upon them, and rabbits that evict them from their burrows.

Away from the Colony

With an unmistakable white body, dark underwings, and a black eye-patch, the white-headed petrel is easily recognized at sea. These petrels usually nest in large colonies but, like most of the gadfly petrels, they are often seen alone when flying at sea away from the nesting area.

Separate Species

The soft-plumaged petrel is one of the gadfly petrels, a group of different species that are sometimes classified together. Indeed, gadfly species are very similar, but the fact that they are distinct is demonstrated by four species nesting on Gough Island without interbreeding.

Mottled Petrel

Pterodroma inexpectata
LENGTH 13–14 in/33–35 cm
WINGSPAN 29–32¼ in/74–82 cm
STATUS 10,000–50,000 breeding pairs. IUCN: Lower Risk—near threatened. CITES: Not listed

The mottled petrel can be distinguished from the other *Pterodroma* species by its white underbody and tail, with contrasting dark belly. Its tail, throat, and chin are usually white, and the rest of the lower body dark. The underwings are mostly white, with a dark line along the lower leading edge of the wing to the white primaries. The bird's upper body is gray, and its face is white with various amounts of gray mottling.

The mottled petrel ranges southward from around the Drake Passage to Prydz Bay in East Antarctica, and northward through the central part of the Pacific Ocean into the Bering Sea. They breed in October, forming dense colonies, and use burrows or rock crevices on the islands around New Zealand. During the breeding season, the adults probably feed in the south near the Antarctic pack ice. After breeding, they quickly fly north to the Bering Sea to feed. They are vulnerable because they suffer from predation by introduced mammals and wekas (New Zealand wading birds with a thievish and aggressive disposition).

White-chinned Petrel

Procellaria aequinoctialis
OTHER NAMES Shoemaker, cape hen
LENGTH 20–23 in/51–58 cm
WINGSPAN 53–58 in/134–147 cm
STATUS Probably several million birds. IUCN: Vulnerable. CITES: Not listed

The white-chinned petrel is a dark, heavily built seabird with white chin markings, which are either small and inconspicuous, or extend up the cheeks in a curving line to partially encircle the eye, which gives the petrel a spectacled appearance. Its bill is thin in relation to the heavily built head and neck, and it is normally ivory in coloration with some black on the top ridge.

They are circumpolar and, even though they breed on most of the sub-Antarctic islands, the majority—several million birds—breed on South Georgia. They

begin breeding about October in burrows on grassy slopes or boggy ground, and disperse widely around the Southern Ocean after breeding. The birds gather around ships in their thousands, and they are sustaining severe losses getting hooked and drowning while seeking bait.

Gray Petrel

Procellaria cinerea
OTHER NAMES Brown petrel, pediunker
LENGTH 19–19¾ in/48–50 cm
WINGSPAN 45–51 in/115–130 cm
STATUS Poorly known; probably hundreds of thousands. IUCN: Lower Risk—near threatened. CITES: Not listed

The gray petrel has a whitish underbody contrasting with ash gray underwings. It has a dark cap and a light mantle that darkens toward the tail and the primary feathers on the wings. Gray petrels return to their breeding colonies in February/March, which is unusual for a southern petrel. They incubate the egg for 52 to 61 days, then feed the chick for 110 to 120 days. Usually solitary at sea, they will feed in larger flocks in productive spots. They are common around New Zealand and its islands, and also breed on Gough Island and the Tristan da Cunha group. Although not seriously affected by commercial fishing, they have suffered high losses from predation. Wekas introduced on Macquarie Island probably extirpated them there, and in other places rats take many chicks.

Consummate Divers

White-chinned petrels are heavily built birds, often with such inconspicuous white markings that they look to be all dark. They are particularly vulnerable to depredation around fishing boats because they are better divers than most other petrels and will dive after longline baits, often pursuing the baits far below the surface.

Sooty Shearwater
Puffinus griseus

OTHER NAME Muttonbird

LENGTH 16–20 in/40–51 cm

WINGSPAN 37–43 in/94–109 cm

STATUS Several million birds. IUCN: Not listed. CITES: Not listed

Sooty shearwaters are dark brown with slender bodies, long bills, and long, backswept wings. The birds' only markings are variable amounts of light brown, gray, or white on the underwing. They have the largest range of all shearwaters, and are found throughout the world's oceans, except in the Arctic basin and the tropical parts of the Indian Ocean.

They generally prefer cold offshore water, where they feed on shoaling fish and squid. They readily dive from the wing and are good underwater swimmers. Sooty shearwaters dive to an average of 128 ft (39 m) depth when feeding, and have been recorded down to 220 ft (67 m). They are aggressive feeders and often displace competing species by flying through in air-feeding flocks, breaking up rival flocks. In addiiton, they regularly cover the targeted prey swarm in dense rafts, thus reducing access for other species, particularly for plunge-feeding birds.

Sooty shearwaters wander in and out of unclaimed burrows early in the nesting cycle, possibly inspecting for quality, but once selected and occupied, they will defend the entrance to their burrow from other birds. Sooty shearwaters start breeding in October, and nest in large and dense colonies, primarily on New Zealand's sub-Antarctic islands: Snares Island alone has 2,750,000 pairs. The incubation period is 53 to 56 days, and chick rearing takes another 86 to 109 days.

After breeding, they travel great distances around the globe, migrating into the north Pacific and Atlantic oceans. In the northern Pacific Ocean, thousands of the birds are killed by entanglement in gill nets. While they

SOOTY SHEARWATER

are still abundant, several factors are affecting their population sizes. Fledgling success and adult survival can be quickly compromised by land-based predators: a 50 percent decline in some populations over 20 years has been linked to increased land-based predation. A 90 percent decline over an eight-year period in another area is strongly related to increased water temperatures.

Historically, the chicks have been a source of food, fats, and oil for the New Zealand Maori. Currently, the sooty shearwater is the only seabird that is permitted to be commercially harvested in New Zealand; 250,000 young are taken each year.

Great Shearwater
Puffinus gravis

LENGTH 17–20 in/43–51 cm

WINGSPAN 39¼–46½ in/100–118 cm

STATUS Total population too large to count. IUCN: Not listed. CITES: Not listed

One of the easiest of the shearwaters to recognize at sea, the great shearwater has pale underparts, with a

Migratory Habits
The sooty shearwater only breeds in the Southern Hemisphere, but after breeding it crosses the Equator and spreads across the world's oceans.

white neck and throat, and a dark cap. White bands across the base of the tail and the back of the neck emphasize its dark cap. The dark feathers on the back and top of the wings have light edges, which give it a scaled appearance.

Great shearwaters are colonial and nest in burrows or in rock crevices. Their breeding season begins in October and the pairs incubate the egg for 53 to 57 days. The chick fledges at about 105 days, and the birds disperse throughout the Atlantic Ocean, flying as far north as Greenland, Iceland, and Scandinavia. Great shearwaters feed mainly on fish, squid, and fish offal, following fishing trawlers in large numbers and plunge-diving from high above the surface. They also follow their prey underwater.

Great shearwaters breed only in the Falkland Islands (a few hundred pairs), Gough Island, and the islands of the Tristan da Cunha group. Each year a few thousand adults and around 50,000 chicks are harvested from Nightingale Island by Tristan locals; however, there is concern this current exploitation is too severe, despite the large overall population.

Little Shearwater

Puffinus assimilis
LENGTH 10–12 in/25–30 cm
WINGSPAN 23–26 in/58–67 cm
STATUS Little data; widespread and locally abundant.
 IUCN: Not listed. CITES: Not listed

The little shearwater is very small, dark above and white below. The underside of its wings and body is white, and the white of its throat extends to the base of the bill and around the back of the eye. They fly low, with a flutter and glide pattern, and "shear" less than other species. Nearly circumpolar in subtropical and sub-Antarctic waters, they also occur in the Azores and Cape Verde islands and surrounding ocean. Adults appear to be sedentary; those that range widely are mostly juveniles under breeding age. Because they breed in such widely separated places, the season is highly variable but is generally in local summer months. They nest in burrows or rock crevices among boulders, incubate the egg for 52 to 58 days, and care for the chick for 70 to 75 days. Threats are human egg harvesting and predation by introduced species. GM

Mass Feeding
A flock of sooty shearwaters mass to feed on small shoaling fish, squid, and crustaceans. Wherever the food is concentrated, seabirds congregate in impressive numbers to take advantage of the abundant—but usually short-lived—bounty.

Storm-petrels

ORDER Procellariiformes

FAMILY Hydrobatidae

Storm-petrels probably diverged quite early from the other petrels. There are two subfamilies, the Hydrobatinae in the Southern Hemisphere, and the Oceanitinae in the Northern Hemisphere, but it is highly probable that all the storm-petrels originated in the Southern Hemisphere and colonized the north.

About 7 to 8 in (18 to 20 cm) long, storm-petrels have shorter and broader wings than other petrels. They dart and weave at wave level in agile and restless flight. In a strong wind they may glide slowly, with wings raised in a V-formation and legs dangling. They often dip their feet into the water as they fly, so that they seem to be scurrying along the surface. They have highly developed salt glands, long tubular nostrils, and a keen sense of smell for detecting food. Most are black to dark brown on their upper parts, with some gray or white on their face, rump, tail, flanks, or belly. Their flight feathers are mainly dark because of a high concentration of melanin pigment, which toughens their feathers to withstand the ravages of the salt and wind.

Although small, storm-petrels are usually found far out to sea, except during the breeding season, and they are relatively long-lived, with life spans ranging from six to 20 years. They inhabit all the world's seas except those in the highest northern latitudes. They make their nests in remote predator-free areas in cliff crevices, boulder fields, or scree slopes, and come to their colonies only at night, except in high southern latitudes with long hours of summer daylight. Colonies range from a few dozen nests to tens of thousands scattered over large slopes.

Unless they repeatedly fail to reproduce, storm-petrels are monogamous. Both sexes incubate the egg in stints of two to eight days, although they may leave the egg unattended for long periods. The chick fledges after seven to 11 weeks. Those chicks that survive the first couple of years join flocks of non-breeders, visiting the breeding colony for several years before becoming mature enough to breed themselves.

Wilson's Storm-petrel

Oceanites oceanicus

LENGTH 6–7½ in/15–19 cm

WINGSPAN 15–16½ in/38–42 cm

STATUS Many millions of birds. IUCN: Not listed. CITES: Not listed

Wilson's storm-petrel is dark except for a simple white band across its rump, and sometimes lighter bars above and below the wings. In flight, its legs project beyond its tail. It prefers cold waters above continental shelves, feeding mostly on crustaceans such as krill, on floating carcasses, and on fish offal and rubbish from ships. It is a trans-equatorial migrant, and has one of the longest recorded migrations. It breeds only on sub-Antarctic islands and around the Antarctic continent, but in the

WILSON'S STORM-PETREL

non-breeding season it flies to the north Atlantic, to the Gulf of Alaska in the north Pacific, and throughout the Indian Ocean—amazing journeys for such small birds.

Other Storm-petrels

Three other storm-petrels breed in the Antarctic and sub-Antarctic waters.

The black-bellied storm-petrel (*Fregetta tropica*) is widespread in the Southern Ocean. It has the storm-petrel's typical dark back and white rump, but the white extends to most of its belly and breast, and its belly is bisected by a thick black line. It breeds on islands and isolated offshore stacks in order to avoid predation by introduced cats, rats, and mice. In the non-breeding season, it travels throughout the Southern Hemisphere, and may even reach the northern parts of the Indian Ocean.

The gray-backed storm-petrel (*Garrodia nereis*) has a dark back, a white belly, and a gray band on its rump and tail. It occurs throughout the Southern Ocean, but is more common around New Zealand and the nearby sub-Antarctic islands in the Indian Ocean and western Pacific. It stays closer inshore over the continental shelf than the other storm-petrels, probably only visiting deeper waters when it disperses.

The white-faced storm-petrel (*Pelagodroma marina*) takes its name from the prominent white pattern on its face. Its back is gray brown, and its underside mainly white. Six subspecies are scattered on remote island groups throughout the Atlantic, southern Pacific, and Indian oceans. Although they are very common—with about a million in New Zealand and its outlying islands alone—they are very vulnerable to the destruction of their breeding areas by livestock, and to exploitation by fishing. In 1970, over 200,000 of the birds were found dead on Chatham Island, their legs entangled in the filaments of large flatworms (*Distomium filiferum*). GM

Widely Traveled

Perhaps the most abundant seabird in the world, Wilson's storm-petrel breeds in Antarctica but migrates throughout the world's oceans after breeding. The birds cannot rest on the wing; instead, they actively flap and glide very close to the water. Despite their relatively small size, they perform one of the longest migrations of any bird species.

Black Stripe

In seeming contradiction to their name, black-bellied storm-petrels appear to have a white belly when seen on the wing. However, although it is hard to see in flight, they have a thick black line down the center of their belly.

Diving Petrels

ORDER Procellariiformes
FAMILY Pelecanoididae

There is sparse fossil evidence about the origins of diving petrels, but they most probably evolved in isolation somewhere around the southern parts of South America or Australia, and then spread around the Southern Ocean. Whereas most tube-nosed seabirds are supremely adapted to living an aerial life on the open ocean, diving petrels have evolved in a different direction—they have adapted to "flying" under water, and could be considered flying penguins.

To adapt to swimming, diving petrels developed heavy, stocky bodies with short, rounded wings. Like those of most aquatic birds, their feathers are dark above and light below, with very little patterning. Their short, black bills have a small, elastic pouch like that of a pelican for food storage—hence their family name, Pelecanoididae. To counter buoyancy in the water, their bones are denser than those of their flying cousins.

These adaptations act against the birds when they fly, so they must work very hard when in the air. Most of the petrels glide on wind currents, flapping their wings intermittently, but diving petrels are a constant whir of wings, and their bones are so dense that they would drop like stones if they stopped flapping. Their wings beat so quickly that they appear as a blur alongside a small, bullet-shaped body. They normally only fly short distances, and keep very close to the surface. They do not skip and dance over the waves like a shearwater, but fly directly into waves and pop out on the other side.

Diving petrels are strictly marine, but because they need so much energy to fly, they are most often found inshore rather than far out to sea. Most of them tend to stay close to their breeding areas, although the South Georgia and Peruvian diving petrels are sometimes seen farther offshore. At sea, they are usually seen singly or in small groups, roosting on the water when not actively feeding. Typically, they dive rather than fly to escape approaching ships.

Diving petrels begin visiting their breeding areas well before breeding starts. They nest in burrows they dig on sloping ground, sometimes in colonies of thousands of pairs. Normally, they only come ashore under cover of darkness in order to avoid the gulls and skuas that prey on them. They will spend the night inside their burrows, calling during the night. There are always a great many birds flying overhead during the breeding season. Once paired, the females spend some time feeding intensively at sea, and then lay a single egg. Both parents take it in turn to incubate the egg for seven or eight weeks in daily shifts—the shortest incubation shift of any of the tube-nosed seabirds—because their feeding areas are so close to their breeding grounds. The parents feed the chick for about eight weeks before it goes to sea on its own. Chicks that survive their very difficult first year join the flocks of immature birds wheeling over the breeding colony in the following year.

The South Georgia diving petrel and the common diving petrel—two of the most common sub-Antarctic diving petrel species—are both small, dark birds with white bellies and with no distinctive plumage characteristics, and it is virtually impossible to distinguish them at sea. Occasionally, however, diving petrels become stranded on a ship during the night, providing the opportunity to examine one in the hand for the only reliable means of identification, the bill shape.

South Georgia Diving Petrel

Pelecanoides georgicus
LENGTH 7–8 in/18–21 cm
WINGSPAN 12–13 in/30–33 cm
STATUS More than 4 million breeding pairs. IUCN: Not listed. CITES: Not listed

The South Georgia diving petrel has a wide-based bill tapering to a point in a continuous curve. It breeds on most sub-Antarctic islands. On Iles Kerguelen, where its range overlaps with that of the common diving petrel, the South Georgia species forages farther from the island. It feeds mainly on the planktonic crustaceans, especially krill and amphipods, as well as some small fish and squid. There are over 2 million pairs on South Georgia, and their breeding areas tend to be in gravelly substrate.

Common Diving Petrel

Pelecanoides urinatrix
LENGTH 8–10 in/20–25 cm
WINGSPAN 13–15 in/33–38 cm
STATUS More than 4 million breeding pairs. IUCN: Not listed. CITES: Not listed

The bill of the common diving petrel, the feature that distinguishes it from the South Georgia diving petrel, looks long and narrow because the sides are parallel along the length of the bill until near the tip.

The species is found throughout the sub-Antarctic region. The Kerguelen diving petrel (*P. u. exsul*), one of six subspecies of the common diving petrel, ranges from South Georgia east to the Antipodes Islands. It is most abundant around Iles Kerguelen, where there may be as many as 1 million breeding pairs. GM

Busy Birds
Common diving petrels seemingly stop only when they get to their nest burrows at night. At sea, they are a constant whir of wings as they dive through the waves and pursue their prey. They will rest on the water, but dive quickly out of sight when a boat approaches.

Cormorants

ORDER Pelecaniformes
FAMILY Phalacrocoracidae

Juvenile Plumage
It takes a juvenile Antarctic shag three to four years after it fledges before it becomes sexually mature and develops the bright blue eye ring and the yellow caruncles of the breeding adult.

A Busy Colony
Imperial shags (right) nest in crowded colonies that are sometimes very large. Once the chicks begin to hatch, these large colonies are a scene of bustling activity, with arriving and departing parents trying their best to keep their hungry chicks fed.

Cormorants are medium to large aquatic birds with webbing across all their four toes. Although they are excellent divers and swimmers, they are rather awkward on land.

They have heavy, elongated bodies and long necks, heads, and bills. Cormorants have a pouch of loose skin on the throat, called a gular pouch, like pelicans. They use it for holding food and for thermoregulation. On warm days, cormorants are often to be seen fluttering their throats to increase the evaporation on the interior of their gular pouch.

Because they are heavy, cormorants swim quite low in the water, sometimes with just their head and neck raised. Although they normally stay submerged for just a few seconds while foraging, they can dive for more than 90 seconds and reach depths of 80 to 165 ft (25 to 50 m). Like all seabirds, cormorants have oil glands to waterproof their feathers, but their feathers also have a special structure that allows their plumage to become waterlogged to adjust their buoyancy for diving. They quickly shake off most of the water when they return to land, but it is common to see cormorants standing on low rocks or on the shoreline with wings outstretched as they dry off their feathers.

Cormorants do not rest on water, so they require dry perches. Because they need to dry their feathers, they are almost always within sight of land, and most do not migrate. Many species are also restricted to relatively shallow water because they forage for small bottom-dwelling fish and invertebrates. Some of the species follow coastlines seasonally in search of food, but all the island forms are entirely sedentary.

Most cormorants are very dark brown to black, with a metallic green or blue sheen to the feathers. All the Antarctic and sub-Antarctic shags have some white on their necks and bellies. At the approach of the breeding season, mature adults develop white, hairlike plumes and brightly colored facial patches: these are important during courtship, but are lost after breeding. Juveniles are uniformly drab brown with brown eyes, developing the typical adult plumage and eye color as they mature.

Cormorants breed in busy and noisy colonies, often sharing their breeding sites with other species, such as gannets, boobies, or penguins. Their colonies are on cliffs, low ledges, rocky slopes, or even on sand. Each year, after attracting a female, the male cormorant goes in search of seaweed, mud, or grasses. He flies back with his bill filled with nesting material, and the female plasters it to the nest with mud or excrement. Nests are reused year after year, and many become quite large. Females lay two to four eggs. Both parents take turns on the nest, and chicks hatch in laying order after 23 to 25 days. Then the real work begins: with two to four chicks on each nest, all raising their heads and calling for food, the din is unrelenting.

Imperial Shag
Phalacrocorax atriceps
OTHER NAME Blue-eyed shag
LENGTH 27–30 in/68–76 cm
WINGSPAN 49 in/124 cm
STATUS About 10,000 breeding pairs. IUCN: Not listed. CITES: Not listed

The imperial shag is widespread in the southern part of South America, the Falkland Islands, Prince Edward Islands, Iles Crozet, Heard, and Macquarie islands. It has a large white patch on its back, which is visible only when the bird is in flight. During the breeding season, the imperial shag develops a brilliant cobalt blue eye ring and bright yellow-orange caruncles at the base of its bill. Some populations also develop fine filoplumes that accentuate their upper cheeks or eyebrows. They lay two to five (usually three) eggs in substantial nests made of grasses, seaweed, and excrement.

Antarctic Shag
Phalacrocorax bransfieldensis
LENGTH 29½–30 in/75–77 cm
WINGSPAN 49 in/124 cm
STATUS 12,000 breeding pairs. IUCN: Not listed. CITES: Not listed

Antarctic shags are a little larger than imperial shags, although they are similar in appearance. They are found on the Antarctic Peninsula year-round, except during short foraging flights to open water. Living in such a cold climate, they do not hold their wings out to dry, as the other cormorants do, and they possess an extra-thick coat of down under their contour feathers. Antarctic shags normally congregate in small groups, often on the edges of penguin colonies.

Kerguelen Shag
Phalacrocorax verrucosus
LENGTH 25½ in/65 cm
WINGSPAN 43 in/110 cm
STATUS 6,000–7,000 birds. IUCN: Not listed. CITES: Not listed

The smallest of the blue-eyed shags is the Kerguelen shag. It has black upperparts, blue-black head and neck, and white underparts. The bird is restricted to Iles Kerguelen, and is the only shag of the islands of the southern Indian Ocean that has remained a separate species; all others have been classified with the imperial shag. The Kerguelen shag feeds in shallow water within about 4 miles (6 km) of the coast of Iles Kerguelen, close to its breeding ground. It normally breeds in small colonies of up to 330 nests, but some colonies have expanded to more than 400 nests. GM

Waterfowl

ORDER Anseriformes

FAMILY Anatidae

Most members of the family of ducks, geese, and swans are found in temperate and tropical environments; only four species, all belonging to the genus *Anas*, are regarded as resident in the sub-Antarctic and Antarctic environments, but a few other species visit during the southern summer.

Waterfowl are direct, rapid flyers, the smaller ducks having more rapid wing beats than the larger geese and swans. On land, waterfowl move slowly and awkwardly, but with surprising speed if they are disturbed. All the species have webbing connecting their three front toes, and are excellent swimmers. They feed on plant and animal material, partly by foraging on land but mostly by dabbling—upending in shallow water and reaching for the bottom. All of the waterfowl species, except for the magpie goose, shed all their flight feathers at once, soon after breeding, and this process renders them flightless for a few weeks.

Waterfowl are territorial during the breeding season, but quite gregarious at other times. They usually nest on the ground near water and hidden in tall grass or shrubs. When making a nest, the female lines a small depression on the ground with down from her breast to create a soft, warm spot for her eggs. Swans and geese will often attack any intruders, but the smaller ducks usually stay on their nests when approached, relying on their camouflage for protection. If they are approached too close, they will burst from their nests in a startling profusion of wings.

A Larger Pintail
The yellow-billed pintail (above) is larger by a third than the South Georgia subspecies, and is partially migratory.

Carnivorous Duck

This yellow-billed pintail is restricted to South Georgia. These birds are common around freshwater lakes and protected bays, and generally feed on vegetation, algae, and aquatic plants. On South Georgia, however, where there are few aquatic insects, yellow-billed pintails are carnivorous and, along with other scavengers, feed on carcasses of seals.

Gray Duck

Anas superciliosa

OTHER NAMES Black duck, brown duck

LENGTH 23 in/58 cm

WINGSPAN 35 in/89 cm

STATUS About 10,000 breeding pairs. IUCN: Not listed. CITES: Not listed

The gray duck is common in Australia, and it is also found north on other Australasian islands and into Malaysia and Indonesia, as well as in the sub-Antarctic islands of Macquarie, Chatham, Snares, Auckland, and Campbell. Rather than being gray, it is a drab brown color, with a scaled appearance to its feathers, and has a striking facial marking of black and buff stripes. Unlike many of the *Anas* ducks, females are very much like the males in coloration, although a little smaller. Macquarie Island gray ducks are timid and they are active mainly at twilight. They pair over an extended period between the autumn and spring, and the timing of their breeding varies across their range.

Yellow-billed Pintail

Anas georgica

OTHER NAME South Georgia pintail

LENGTH 17–21½ in/43–55 cm

WINGSPAN 28 in/71 cm

STATUS Subspecies *A. g. georgica*: less than 2,000 birds. IUCN: Not listed. CITES: Not listed

Generally brown in color with a scaled appearance, the yellow-billed pintail has a chestnut crown, a long brownish tail, and a bright yellow bill that is marked with black. It normally feeds on aquatic insects, emergent vegetation, algae, and small invertebrates in shallow water by dabbling, and it dives frequently, especially for its favorite prey, which includes fairy shrimp. Of the two subspecies, *A. g. georgica* is resident and sedentary on South Georgia, where it lives around protected bays, and freshwater lakes and ponds surrounded by tussock grasses. Flocks form in the winter, although pairs may

stay together. Their breeding season begins in the early springtime, and the unmated males will court paired females in an attempt to lure them from their current mates. The females lay up to five eggs, incubating them until they hatch, and the male helps to raise the chicks.

Kerguelen Pintail

Anas acuta eatoni

LENGTH 14–18 in/35–45 cm

WINGSPAN 25½–27½ in/65–70 cm

STATUS 5,000–10,000 breeding pairs. IUCN: Vulnerable. CITES: Not listed

Resident in a number of the sub-Antarctic islands, this is a subspecies of the northern pintail (*A. acuta*), one of the most common ducks throughout its large range. Like many island species, the Kerguelen pintail is a little smaller than its more widespread northern relative, but similar in appearance. It is not officially threatened, but in southern regions feral cats are taking their toll.

Speckled Teal

Anas flavirostris

OTHER NAME Yellow-billed teal

LENGTH 14–18 in/35–45 cm

WINGSPAN 30 in/76 cm

STATUS 40–50 birds on South Georgia. IUCN: Not listed. CITES: Not listed

In 1971, a small breeding population—a mere 40 to 50 birds—of speckled teal was discovered in Cumberland Bay on South Georgia. Whether the birds were naturally or intentionally introduced to the island is unknown. A very adaptable species, the speckled teal is abundant throughout its range—southern South America and the Falkland Islands—and occurs in freshwater, brackish, and marine habitats. Still restricted to Cumberland Bay on South Georgia, the bird is grayer and less reddish than South Georgia's very similar yellow-billed pintail and has more pronounced spotting on its breast, which helps to distinguish the two species. GM

A Sheltered Life

Small populations of several species of waterfowl have made their home on sub-Antarctic islands. The Auckland Island teal (left) is mostly found in coastal waters and sheltered bays, where it feeds on aquatic invertebrates. It usually feeds in shallow water by probing or dabbling near the shore.

Skuas

ORDER Charadriiformes

FAMILY Stercorariidae

This family is divided into two genera: the larger *Catharacta* genus and the smaller, more agile *Stercorarius* genus, sometimes known as jaegers. Both of the groups probably originated in the Northern Hemisphere, splitting from the gulls about 10 million years ago, but the *Stercorarius* have remained wholly Arctic, whereas the *Catharacta* skuas are found in high latitudes in both hemispheres.

Skuas have heavier bodies than their gull relatives, and their bills are stronger and more hooked, reflecting their more predatory lifestyle. Their predominantly dark plumage has conspicuous white patches on the top and bottom of their wings, and they employ powerful wing beats in flight. With strong, hooked claws on their toes, as well as full webbing between them, the skuas are true seabirds of prey. Like the land birds of prey—the raptors and owls—they have hard, scaly plates called scutes on their legs, and females are larger than males by 11 to 17 percent, depending on the species.

Skuas may live for 38 years, and are very faithful to both their breeding territories and their mates. They are migratory during the winter, arriving at breeding areas in the early spring, just as the penguins are returning to their colonies. They are mature by the age of three, but cannot normally claim a territory until they are six or seven. Most pairs produce two eggs. If both the eggs survive, the first egg hatches two or three days before the second one. The older chick will pick fights with the younger, sometimes killing it, but more often preventing it from getting enough food. The younger chick will die of starvation or will stray into other territories in search of food—usually to be killed and eaten by adult skuas. Parents work hard to raise their chicks, but the dangers of living around other predatory skuas and the trampling feet of penguins take their toll, and their reproductive success is usually very low.

Two *Catharacta* skuas breed in the Antarctic or sub-Antarctic regions. Antarctic skuas nest on most of the sub-Antarctic islands, on cold temperate-zone islands south of New Zealand, and on the northern part of the Antarctic Peninsula. On the peninsula they overlap with south polar skuas, which breed exclusively around the coast of Antarctica. Where the two species overlap, they regularly interbreed, even though the Antarctic skuas normally will not allow south polar skuas to nest around penguin colonies, which forces them to forage almost exclusively at sea for fish and krill. Farther south, away from the overlap with Antarctic skuas, most south polar skuas nest in loose colonies near to penguin or petrel colonies, on which they prey. Outside the breeding season, they mainly fish by plunge-diving at sea.

Herbert Ponting, photographer on Robert Scott's British Antarctic Expedition of 1910–12, called skuas "the Buccaneers of the South." They prey on storm-petrels and other petrels, as well as on the eggs and

chicks of penguins and other seabirds. In addition, they protect their breeding territories savagely, even diving at and striking human intruders.

Skuas are also kleptoparasites: they attack seabirds bringing food back to their mates and chicks, and steal the food. Although this behavior is not unique to skuas, in this species it is developed to the extreme.

Guarding the Home
A pair of south polar skuas perform the "long call" display to protect their territory and their food. The display lets all other nearby skuas know that the pair are at home and are ready to fight for their territory.

Safe Haven

South polar skuas (above) normally lay two eggs. For the first couple of weeks, the chicks are protected and brooded regularly, but not constantly. But accidents caused by aggressive penguins and predation by other skuas often result in pairs failing before their eggs have hatched.

Antarctic Skua
Catharacta antarctica
OTHER NAME Brown skua
LENGTH 25 in/63 cm
WINGSPAN 47–63 in/120–160 cm
STATUS 13,000–14,000 breeding pairs. IUCN: Not listed. CITES: Not listed

Antarctic skuas are large, powerful birds with heavy bodies and a fierce demeanor. They are dark brown with yellowish streaks in their neck and nape feathers. Most populations are fairly small, but bigger populations occur on larger islands which harbor more prey species. The birds are generalist predators which adapt to local conditions, and are also scavengers: their numbers tend to increase around human settlements.

ANTARCTIC SKUA

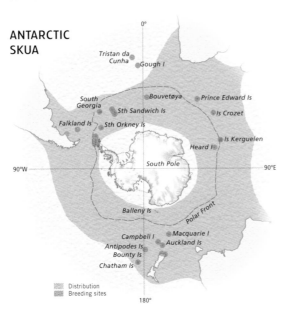

South Polar Skua
Catharacta maccormicki
OTHER NAME McCormick's skua
LENGTH 21 in/53 cm
WINGSPAN 51–55 in/130–140 cm
STATUS 5,000–8,000 breeding pairs. IUCN: Not listed. CITES: Not listed

SOUTH POLAR SKUA

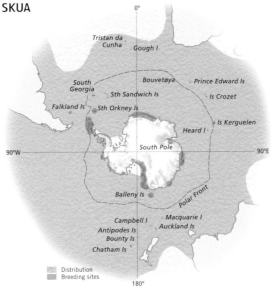

Smaller than the Antarctic skua and slightly more agile when in flight, the south polar skua is the grayest of the *Catharacta* skuas, occurring in pale, medium, and dark color forms. There is usually a contrast between their lighter nape and their darker back. They have a wide range, extending to the Antarctic continent. GM

Predatory Birds

This brown skua from Auckland Island shows the rich brown typical of the larger species. Skuas are related to gulls, but are much more predatory and have very strong bills with pronounced hooks on the tip. Many brown skuas, particularly on the sub-Antarctic islands around New Zealand, nest in trios or larger groups, with one female and multiple males.

Sheathbills

ORDER Charadriiformes
FAMILY Chionidae

Sheathbills—medium-sized, pigeon-like shorebirds—belong to the same order as skuas, gulls, and terns. They probably evolved around the tip of South America from a ploverlike ancestor. Sheathbills are the only shorebird family that is found entirely within the Antarctic and sub-Antarctic regions. Today, there are just two species in the family, both of the genus *Chionis*. They differ in appearance, especially around the face, but are very similar in habits.

Both species inhabit coastal areas and rely on the sea for some of their foraging, feeding on algae, limpets, and other small invertebrates in the intertidal zone. The snowy sheathbill inhabits the milder regions around the Antarctic Peninsula; some migrate northward in winter. The lesser sheathbill occurs on sub-Antarctic islands. These birds are the ultimate scavengers of Antarctica, eating virtually every form of organic matter they can find. They forage in penguin and cormorant colonies for unattended eggs or small chicks, and for dropped food, expelled stomach linings of penguins, and for the feces of other birds and seals. They also congregate around concentrations of seals in the early spring, feeding on placentas and the inevitable seal carcasses, picking at scabs and wounds on adult seals, and scavenging on the nasal mucus of elephant seals.

They are the smallest of the Antarctic scavenger/predators, which limits their predatory activities to the smallest of penguin chicks and eggs, but it also means that they can move about the colonies more freely than larger predators without arousing aggression. They are expert at stealing food from penguins, cormorants, and albatrosses by leaping against adults who are feeding their chicks and quickly scooping up spilled food. They are very agile, and easily dodge the penguins and seals that lunge for them as they forage. Their fearlessness also brings them into close contact with humans at settlements and scientific bases, where they scavenge absolutely anything they can find. Unfortunately, they sometimes eat dangerous items, such as grease from machinery and even small batteries.

Sheathbills make their nests in crevices or under protective overhangs near penguin colonies, and they time their own breeding to coincide with that of the penguins. They attempt to breed at three years old, but most fail. Once they succeed in breeding, they remain faithful to their territory and their mate, and are strongly territorial during the breeding season. They normally lay two or three eggs in a rough nest built of tussock, algae, moss, and old bones, situated in crevices or under an overhang to protect them from marauding skuas and from the weather. In January, once their chicks have hatched, the parents deliver food by the billful, rather than regurgitate it. At about 50 days, when the chicks become independent, the adults gather in small flocks to forage and to roost.

Snowy Sheathbill
Chionis alba
OTHER NAMES American sheathbill, greater sheathbill, pale-faced sheathbill, yellow-billed sheathbill
LENGTH 13–16 in/34–41 cm
WINGSPAN 29½–31½ in/75–80 cm
STATUS About 10,000 breeding pairs. IUCN: Not listed. CITES: Not listed

With their thick, entirely white plumage and their sturdy bodies, snowy sheathbills are well adapted to the cold Antarctic environment. Their yellow-green bill sheaths and pink facial caruncles make their faces look untidy. They live and breed around the Antarctic Peninsula and the islands of the Scotia Arc. Many snowy sheathbills migrate to Patagonia and Tierra del Fuego in winter, but on Signy Island and the Falklands some stay near their breeding areas throughout the winter.

Lesser Sheathbill
Chionis minor
OTHER NAME Black-faced sheathbill
LENGTH 15–16 in/38–41 cm
WINGSPAN 29–31 in/74–79 cm
STATUS 6,500–10,000 breeding pairs. IUCN: Not listed. CITES: Not listed

This bird has black sheath and caruncles, and pink eyerings. Its leg color varies from pink to black, depending on the subspecies. Four separate subspecies occur on sub-Antarctic islands in the Indian Ocean: one on Iles Kerguelen; one on Iles Crozet; one resident on Marion and Prince Edward islands; and one on Heard and McDonald islands. The sheathbills nest in association with penguin colonies, and king penguins provide a host colony all year round, so the sheathbills can scavenge around their colonies all through winter. Rockhoppers and macaronis abandon their colonies in winter to forage at sea, which means the sheathbills have to fend for themselves, foraging for terrestrial and marine invertebrates. Those without winter host colonies tend to have lower reproductive success rates. GM

Not Fussy Eaters
Snowy sheathbills, garbage collectors of Antarctica, eat anything small enough for them to swallow, including spilled food, eggs of any other species, excrement, carrion, and even the mucus from seals' noses.

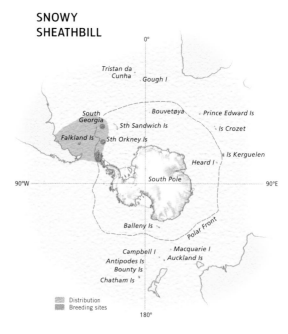

SNOWY SHEATHBILL

Gulls

ORDER Charadriiformes
FAMILY Laridae

It is probable that gulls diverged from a shorebird ancestor about 65 million years ago. Fossil records suggest that they originated in northern Europe and Asia, and then spread throughout the world. Nowadays they occur on all coasts and in most inland areas.

Gulls are fairly uniform in shape, with heavy bodies, long wings, and webbed feet. Most have a rounded tail and a stout bill that is either straight or slightly hooked. They are generally white-bodied, with a darker mantle that varies from black to pale silvery gray. The wingtips of most species are black, with white patches near the tips. Juveniles typically have brown plumage speckled with black or buff, and they molt within a few months into their first winter plumage, which may resemble adult winter plumage. Some species take four years or more to develop full adult plumage.

Most gulls breed in colonies and form strong pair bonds, usually heterosexual and monogamous, but in some colonies species may include a small proportion of female pairs—perhaps because of a lack of breeding males. Both females mate with already paired males and lay fertile eggs, and as a result their clutches are often twice as big as usual. Within their colonies, gulls are territorial and defend their own nests, but the entire colony will attack any approaching intruder.

Most live on sea coasts, but some species live well inland and nest near freshwater. Their salt glands grow or shrink as required: those of inland gulls are very small during the breeding season, but when the birds must migrate in winter to coastal areas, these glands enlarge to cope with food that contains more salt.

Gulls have a variety of habitats, and are omnivorous and opportunistic feeders. They eat virtually anything and frequent refuse tips around human habitations, and they can obtain food in a variety of other ways as well. On the wing, they catch food, such as insects or food dropped by others. They hover above water and drop down to pick items off the surface, surface-feed, or plunge-dive to seize food from below the surface. Most species patrol shores and tidal areas for invertebrates and to scavenge whatever they can find; more than half the food they eat is snatched up from the ground.

Kelp Gull

Larus dominicanus

OTHER NAMES Dominican gull, southern black-backed gull

LENGTH 21–25½ in/54–65 cm

WINGSPAN 50–56 in/128–142 cm

STATUS More than 1,085,000 breeding pairs; 10,000–20,000 in Antarctica and sub-Antarctic regions. IUCN: Not listed. CITES: Not listed

The only gull species commonly seen in Antarctica, kelp gulls are most abundant in New Zealand, although they

KELP GULL

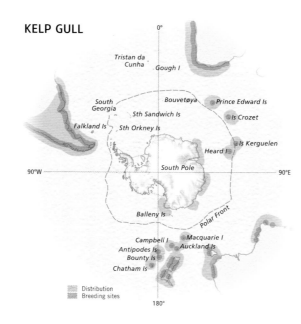

are also found on sea coasts throughout the Southern Hemisphere. In Antarctica, they occupy the coasts and islands of the Antarctic Peninsula throughout the year; in the winter months, they may move away from the coast in search of open water, but they do not migrate north into South America. These large, white-headed gulls have a black mantle and upper wings, and snow white tails. Their straight yellow bill has a red spot at the tip of its lower section.

Kelp gulls are generalist predators and scavengers. Their diet varies, depending on location and availability. They eat molluscs, fish, birds, arthropods, amphibians, reptiles, worms, and even small mammals. They also feed on carrion and scavenge on rubbish dumps, and on sewage. In the Antarctic region, kelp gulls specialize in the Antarctic limpet (*Nacella concinna*), which they seize from rocky shores at low tide or glean from the rocks as they swim parallel to the shore. They swallow the limpets whole, digest the flesh, and regurgitate the shells in piles that resemble stacks of miniature dinner plates. Heaps of discarded limpet shells are often found on high rocks well away from the waterline, in places where the gulls have roosted. GM

Nest Builders

Kelp gulls normally nest in colonies but will sometimes nest as individual pairs. Each pair builds a substantial nest from whatever vegetation and materials are available. Both parents contribute to the building, but males do most of the work. Colonies are near water on sloping ground or rocky ledges.

Winged Beggars

Young kelp gulls begin to grow juvenile plumage very soon after hatching, and they take their first flight at about 45 to 60 days. The parents continue to feed them for another couple of weeks, and the juveniles harass their parents in the air, chasing them and begging for food.

Terns

ORDER Charadriiformes

FAMILY Sternidae

Like gulls, terns are also found all over the world. All the Antarctic terns belong to the genus *Sterna*, with the exception of the brown noddy (*Anous stolidusis*). Although brown noddies are mostly tropical breeders, there is a thriving population on Tristan da Cunha in the south Atlantic.

True migratory birds, many species move between cooler breeding grounds and warmer waters with more abundant food resources. Terns are generally small to medium in size, and have white or light gray under parts, darker gray upper parts, and a black cap when in breeding plumage. They have long, forked tails and slender wings, making them more maneuverable and acrobatic flyers than the gulls. Terns forage by swooping down to pick food from the surface or plummeting into the water. They hover briefly, slam into the water, then surface and take off, with a crustacean or small fish in their bill. They steal food from each other and other species, but often lose food to larger gulls or skuas.

Like gulls, terns nest in colonies and may occupy the same colony site for many years. They protect their nesting sites by attacking intruders; in some colonies, thousands of birds will soar into the air and then swoop on a rat—or a human—who ventures too close.

Antarctic Tern

Sterna vittata

OTHER NAME Wreathed tern

LENGTH 14–16 in/35–40 cm

WINGSPAN 29–31 in/74–79 cm

STATUS About 50,000 birds. IUCN: Not listed. CITES: Not listed

This species, which is the most common tern in the Antarctic region, nests on most of the sub-Antarctic islands, as well as along the Antarctic Peninsula and New Zealand coasts, and some areas of South America and South Africa. Smoky gray on its upper parts with a white rump, it has a black cap on its gray head and a red bill. Its underparts are mostly white, and sometimes have streaks of gray.

They breed in November to December, nesting in colonies in rocky areas near the shore or just inland. Around the Antarctic Peninsula they usually establish their colonies in flat gravel areas away from beaches. If left undisturbed, they will return to the same location year after year, but they can be flexible: on Tristan da Cunha, they once nested on sandy beaches, but shifted their colonies to ledges on small offshore outcrops after rodents were introduced. During the non-breeding season, they move to the ice edges to forage; they may migrate northward as far as the South American coast, but—unlike the Arctic terns—they do not migrate to the Northern Hemisphere. Their favorite prey is small fish and krill, which they obtain by plunge-diving.

ANTARCTIC TERN

0°

Tristan da Cunha
Gough I
South Georgia
Bouvetøya
Prince Edward Is
Sth Sandwich Is
Is Crozet
Falkland Is
Sth Orkney Is
Is Kerguelen
Heard I
I Amsterdam
90°W
South Pole
90°E

Balleny Is
Polar Front

Campbell I
Macquarie I
Antipodes Is
Auckland Is
Bounty Is
Chatham Is

Distribution
Breeding sites

180°

Tireless Flyers

Found mainly along the Antarctic Peninsula and the sub-Antarctic islands near the continent, Antarctic terns are tireless flyers, although they do alight to rest sometimes. They feed during the day on crustaceans and small fish, which they mainly take by plunging into the water from a few feet above the surface. Occasionally, they will take prey from the surface by dipping while on the wing. They never land on the surface to take prey.

Classifying Gulls and Terns

Gulls and terns are closely related, terns probably deriving from a primitive gull-like ancestor. Recently, however, DNA studies suggest they should be grouped with skuas (Stercorariidae) and skimmers (Rynchopidae) in the Laridae family, each current family retaining its distinction but at the subfamily level. Gulls are an enigmatic group and, even with the powerful modern classification techniques, hybridization is a persistent source of taxonomic confusion, and the new classification is not universally recognized. The classic separation of gulls, terns, and skuas into separate families is used here.

Arctic Tern

Sterna paradisaea

LENGTH 13–14 in/33–36 cm

WINGSPAN 30–33½ in/76–85 cm

STATUS About 500,000 breeding pairs. IUCN: Not listed. CITES: Not listed

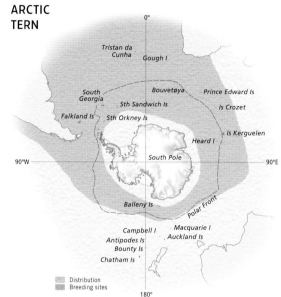

ARCTIC TERN

Distribution
Breeding sites

Winter Quarters

Arctic terns travel huge distances on their migration. They breed in the Arctic in the northern summer, but during their non-breeding season migrate far into the Southern Ocean—and even reach the Antarctic continent around the Antarctic Peninsula. There they occupy coastal and ice habitats like those of the high Arctic.

Kerguelen Tern

Sterna virgata

LENGTH 13 in/33 cm

WINGSPAN 28 in/71 cm

STATUS 2,400 breeding pairs. IUCN: Lower Risk—near threatened. CITES: Not listed

One of the most endangered of the terns, this species lives and breeds only on three island groups: on Iles Kerguelen, Iles Crozet, and Marion and Prince Edward islands in the southern Indian Ocean. The largest of the populations—about 2,000 pairs—is on Iles Kerguelen, where its survivial is threatened by introduced feral cats. The Kerguelen tern resembles the Antarctic tern, but it is a little smaller and duskier. Its gray underparts are a little darker than those of the Antarctic tern, and its bill is also darker. During the breeding season, the tern feeds largely on insects it finds in the uplands, but for the rest of the year it reverts to the typical tern prey of small aquatic animals.

The Arctic tern is a common black-capped tern that breeds exclusively in the high latitudes of the Northern Hemisphere. It resembles the Antarctic tern, but has longer wings, a longer, more slender bill, and a more delicate appearance. The underparts are lighter, with more white in the cheeks. In the northern winter, it flies halfway round the world to feed in the Antarctic pack ice or around the islands of the Antarctic Peninsula—one of the longest bird migrations. Having bred in the Northern Hemisphere summer, the Arctic tern has acquired its non-breeding winter plumage by the time it reaches Antarctic waters. This means that it is easy to distinguish between Arctic and Antarctic terns during November and December because Antarctic terns are wearing their breeding plumage. Later in the southern summer, the juvenile Antarctic terns are easily confused with winter-plumage Arctic terns. GM

KERGUELEN TERN

Distribution
Breeding sites

Intruder Alert

Most terns, including the Antarctic tern (right), nest in dense colonies on flat ground. When an intruder appears, the entire colony rises up to mob it. They are agile and swift flyers, and will dive at the intruder and peck at it aggressively.

Penguins

ORDER Sphenisciformes
FAMILY Spheniscidae

King Courtship
Mating for the king penguin (above) is a long process, which is usually instigated by the male after much displaying and parading.

With their sway-backed upright posture, smart black-and-white coloration, and their endearing waddling gait, penguins are the most distinctive of birds. Many people think of penguins as being strictly Antarctic, but only four species—the emperor, the Adélie, the chinstrap, and the gentoo—breed on the Antarctic continent proper, while seven others live and breed south of the Polar Front that defines the far southern oceans. Because penguins are flightless and must feed in cool water, they cannot travel across the warm waters of the tropics to inhabit the Northern Hemisphere. Only a few groups of Galapagos penguins (*Spheniscus mendiculus*) breed north of the Equator.

Grouped scientifically into 17 species, penguins are highly specialized, non-flying, marine birds, ranging in size from the little penguin, at 2¼ lb (1 kg) in weight and 16 in (40 cm) in height, to the emperor penguin, which weighs up to 84 lb (38 kg) and reaches 50 in (130 cm) in height. All have a dark back and a white front, and each species has a unique pattern, in some cases including orange or yellow, around the head and chest. Possibly their basic dark and pale coloration, repeated with variations, provides both protection from predators and camouflage while hunting: from above, penguins on the ocean surface look dark against the dark water, but when viewed from beneath, they appear light against a light sky.

Male penguins are generally heavier than females and have larger, more powerful bills and longer flippers. In most species, the difference between the sexes is hard to distinguish by eye, but behavioral observations do help. Scientists can sex macaroni penguins, but find distinguishing male and female gentoo and chinstrap penguins more difficult, although they have achieved a 95 percent success rate. Adélies are even more difficult to sex by eye, and scientists succeed only 85 percent of the time. Interestingly, penguins from high and low latitudes are more difficult to sex than the mid-latitude species, suggesting there may be some environmental factor that contributes to the presence (or absence) of sexual dimorphism. One reason may be the need for males to compete for nesting space and females; larger

Penguin Provender
Teeming colonies of penguins (below) were an invaluable source of protein and fats for the first Antarctic explorers: Nordenskjöld's land party of 1901–04 stockpiled 1,000 Adelies for winter.

males would be more successful, winning more battles. Another theory is that if the two sexes are different sizes they can effectively exploit different food resources, but this has never been shown to occur regularly.

Keeping Warm—and Cool

Unlike most seabirds, penguins have lost the ability to fly, so that they do not need to minimize their weight. On the contrary, they have developed heavy blubber reserves as insulation and for long-term energy storage, an adaptation usually seen only in mammals. Their feathers have also modified from flying equipment into waterproof insulation, which thickens the birds up and shapes them to hold air against the body. Penguins may not be able to take to the air, but they do "fly" through water, using their legs as rudders and rapidly beating their wings for propulsion. Swimming compresses their feathers and gradually forces out the insulating layer of air, so swimming penguins leave fine trails of bubbles.

Polar penguins are well insulated, and can remain in the water almost indefinitely without losing too much heat. But they are designed to retain body heat: when conditions are too warm, they have extreme difficulty cooling down. They shed heat through the undersides of their flippers and through their legs and feet, as well as by fluffing their feathers and panting. On a warm day, penguins may lie on their stomachs, flippers raised, feet waving in the air. In the heat of summer, bigger chicks that are still too young to go to sea and dissipate heat into the water have a very uncomfortable time.

Fishing the Southern Oceans

All the penguins are carnivores, but the various species prefer different prey, which range from small plankton to large fish and cephalopods. These prey types are all fairly active, so they are difficult to catch and must be swallowed whole before they can escape. Bird beaks do not have teeth, but penguin mouths have "teeth" made from keratin. Small hooks, like an extremely rough cat's tongue, the teeth project backwards inside their mouth and prevent prey from darting to freedom.

It is estimated there are at least 23 million breeding pairs of the 11 species of Antarctic penguins living and fishing in the region south of the Polar Front. Scientists have calculated that together, every year penguins must consume at least 3,900,000 tons (3.6 million tonnes) of fish, 594,000 tons (539,000 tonnes) of squid, and 15,300,000 tons (13.9 million tonnes) of crustaceans—mostly krill—every year.

The great bulk of this fishing takes place near the South Sandwich Islands and South Georgia. This is an immensely productive area. It supports 24 percent of all the breeding penguins that live within the Polar Front and it is the principal location for the consumption of marine resources by penguins. South Georgia alone provides 23 percent of all the food taken by penguins within the southern region. This massive feeding frenzy is due largely to macaroni penguins, which make up 94 percent of the penguins in the area. Another 19 percent of the prey consumed in the southern oceans is taken in the South Shetland Islands, in this case almost entirely by chinstrap penguins.

The Way Home

Most penguins seem to navigate by the sun, and are equipped with an internal clock to adjust for the time of day. Experiments have shown that if penguins are captured then released far from their colony, the weather is crucial to the operation of their homing instinct. In sunny conditions they will immediately head in exactly the right direction, but in overcast conditions it takes them longer to orient themselves and begin the trek to their colony. If it is very cloudy, they simply cannot work out what direction to take, and their homeward journey is erratic. They will wander in the wrong direction for hours, and then orient themselves correctly during brief clear spells but lose their way again when it clouds over. Evidently, penguins know which direction they want to take, but can find it only by the position of the sun in the sky.

New Feathers for Old

A penguin's feathers get damaged by the wear and tear of daily life, and must be replaced approximately once a year. Most birds lose and replace their feathers one by one, molting gradually over a period of time. The story is different for penguins. Each year, penguins undergo a rapid and complete molt, in which all their feathers are replaced at once. First, new feathers grow, pushing out the old feathers, which remain attached to the tips of the new ones. Eventually the old feathers are shed in batches, creating a veritable snowstorm of feathers around the penguin colonies. The new feathers then grow further, and they thicken up before the penguin is ready to go to sea again. The process usually takes about three weeks, and during that time every bit of the penguin's energy is focused on renewing its feathers.

First Molt

By its first winter, a king penguin chick can weigh almost as much as an adult. Over the winter, the chick loses weight, but when spring returns and fishing improves it regains weight. It also loses its stringy chick down, molting into immature feathers, which are similar to adult feathers but not so bright.

uropygial gland with their bill, then run the oil through their feathers. To get it on their head and neck—the hard-to-reach places—they put the oil on the edge of one of their flippers and rub their head over the flipper area. In some species, partners will preen each other's head and neck; this is called allopreening. Although allopreening may help to spread oil, it is considered to be mainly a nurturing or cleaning behavior: emperor penguins do not seem to have external parasites and they have not been observed allopreening. Adélies, chinstraps, and gentoos—the brush-tailed penguins—are exclusively Antarctic species, and they, too, do not allopreen, whereas the king penguins, which breed in warmer conditions, do. Allopreening is more common in warmer conditions when there are more parasites.

How Fast?

Despite some enthusiastic estimates of speeds of up to 30 to 37 mph (50 to 60 kph), small to medium-sized penguins, such as these Adélies, actually swim at about 3 to 6 mph (5 to 10 kph), or up to twice that in short bursts.

Polar Stoics

Emperor penguins (below) are no strangers to blizzards: they spend their lives in the pack ice and they breed on the permanent ice of the Antarctic continent and nearby islands.

Without their feathers, penguins have no insulation or waterproofing, so molting birds cannot go to sea to feed—they would freeze to death. From the time they begin to molt until their new feathers have grown in, penguins are very exposed to the elements and are at risk of freezing or starving, or both. In order to conserve energy, they move around very little while molting; they simply stand on shore sheltering from the weather, and use their blubber reserves to grow new feathers. Even so, the molting phase results in massive weight loss—a molting penguin can lose more than half its body weight in just over three weeks.

Preening, in which birds use their bills to stroke their feathers from base to tip, is essential for penguins. Penguins coat their feathers with an oil secreted by the uropygial gland, located at the base of the tail, which waterproofs the feathers. They squeeze oil out of their

Sharing Resources

Good breeding locations on the sub-Antarctic islands and along the Antarctic shoreline are rare, and during summer, when all penguins come ashore to reproduce, they gather in very large numbers. During the summer season, several species often share a general area, and its food supply, until they can reproduce, molt, and disperse throughout the southern oceans again. To accomplish this sharing of limited resources, penguins employ different feeding strategies and target different prey. Usually, one species fishes inshore, very close to land, while other species go farther out to sea to feed. If there are three or more species in the area, offshore feeders specialize in different types of food and partition their prey by both size and type. Foraging depth may be one of the deciding factors in the separation of resources, with the larger species of penguins, which are capable of longer, deeper dives, taking larger prey items.

Natural Fish Hooks

Penguins have no teeth, and must catch and swallow their food whole, using only their beak. This is not quite as difficult as it would seem because they have backward-pointing spines inside their mouths that can grip and hold onto their prey.

Breeding in Colonies

Most penguin species prefer large colonies of thousands, or even hundreds of thousands. To attract partners from among these vast throngs, the penguins put on attention-seeking courtship displays, including calling and highly ritualized posturing. Penguins are highly imitative birds, and when one couple begins a courtship ritual, all the neighboring birds tend to copy them. Experiments playing courtship sounds to pairs of royal and little penguins have found that it increased the rate of courtship and mating by shortening the time between pairing and egg laying. The cacophony of colony sounds was also recorded and played back to the whole colony and the result was an increase in synchronicity of egg laying. This may partially explain why larger penguin colonies tend to be more successful. The acoustic background of a colony influences its seasonal reproductive rate and synchrony, with larger colonies more

synchronous than smaller ones, and more successful. Synchronized egg laying produces many chicks all at the same time, so imitative mating and laying behavior is an effective survival technique, as it floods an area with a large number of chicks all hatched at the same time. This gives each individual chick a better chance of surviving the local predators: the shorter the period with chicks vulnerable to predators, the fewer chicks can be taken.

Penguins must provide all the protection they can for their eggs. They incubate the eggs in nests of rocks, tail feathers, bones, and sometimes plant material, but these are not good insulators, and do not provide much protection. To protect their eggs and keep them warm, penguins tuck them into a brood patch, which is a bare patch of skin surrounded by thick feathers low between their legs. At incubation, this patch becomes engorged with blood vessels located at the surface of the brood patch. The vessels are heated by the adult's body core and keep the egg warm. When a penguin stands up or swims between incubating bouts, the stomach feathers close over the patch to prevent too much heat loss.   LW

Disputing Right of Way

Tensions are high during the breeding season, and every small territorial incursion is taken as an invitation to battle. Breeding penguins will attack any other penguin that enters the small space surrounding their nest, and may even chase them a small distance to see them off, but they seldom inflict any serious physical damage.

Wingless Divers

Wingless divers, genus *Aptenodytes*, are the largest living penguin species. There are only two members in this group, the king penguin and the emperor penguin. These stately birds stand at an impressive height of more than 35 in (90 cm). The emperor penguin has a bold splash of golden yellow on the sides of the head, neck, and chest, and the king penguin has similar coloring but is brighter—marked with a rich orange color.

The two species are quite closely related, but they live in vastly different conditions. Emperors, the most southerly of all penguins, live and breed in the harshest of environments, the fast ice of the Antarctic continent, whereas king penguins inhabit the comparatively mild sub-Antarctic islands north of the Polar Front. Neither species builds a nest, even though king penguins live in a region where nest-building materials are readily available. Emperor penguins, on the other hand, raise their young deep on the southern ice, where there is not even a scrap of rock to build a nest.

Bringing Up Baby

King penguin chicks hatch in summer and reach their parents' body weight in around two months. However, when winter sets in and prey becomes scarce, they are not fed regularly and so lose weight. The next spring, when the prey returns, the parents can feed the chicks again, and their weight recovers to about 80 percent of the adult weight before they fledge, at approximately 13 months old. Emperor penguins hatch in mid-winter, four months later than kings, and mature steadily over five months. They fledge in December with a beak size and weight 40 to 60 percent that of adults—but with feet 80 percent of adult size. Presumably, they need large feet to walk from their birthplace in the fast ice to the open ocean, where they begin hunting independently.

Each strategy has its payoffs. King penguins' two-stage rearing means that energy is not wasted on weak youngsters that do not survive winter, and the chicks that do make it through have good survival prospects. The emperors' shorter rearing period means that they can breed annually and thus produce more offspring, balancing out the fact that many chicks perish long before reaching the ocean.

Rushing to Mate

King penguins are usually four years old when they first begin to breed, and after successfully raising a chick, the female is often ready to mate again a month sooner than the male. This means that successful pairs don't usually re-mate.

Survival Success

King penguin chicks (left) may have to wait several weeks for their parents to return with food. If it hatched in January, the chick has a much greater chance of survival than if it were born later in the breeding cycle.

Recognizing Their Own

Parents returning from fishing expeditions (right) locate their chicks by sound. King chicks are taught to recognize the characteristic rise and fall of their parents' voices from a very young age. The chicks respond with a high-pitched, three-noted whistle. This calling makes a king colony very noisy.

Emperor Penguin

Aptenodytes forsteri

LENGTH 40–50 in/100–130 cm

WEIGHT 84 lb/38 kg

STATUS At least 500,000; population thought to be
stable. IUCN: Not listed. CITES: Not listed

Emperor penguins have blue-gray backs shading to a
black tail, and their underparts are white flushed with
yellow, which strengthens to deeper yellow curving ear
patches. Uniquely among penguins, the chicks have
beautiful, pale gray body down and a black head with
a white eye mask.

Emperor penguins live in about 40 known colonies
south of latitude 65°S, and there are three regional
populations, in East Antarctica, the Ross Sea, and the
Weddell Sea area. They spend their lives in the pack
ice, avoiding open water beyond, and breed only on
the permanent ice attached to the southern continent
and nearby islands. Emperors are the only penguins
that breed on ice and snow rather than exposed land.
In order to feed, they must visit open water, and if open
water is not available they dive into the breathing holes
maintained by seals and into narrow cracks in the
continental ice. They feed on fish. Emperors have been
recorded diving for as long as 18 minutes to depths of
more than 1,300 ft (400 m). On most dives, emperor
penguins spend two-and-a-half to nine minutes holding
their breath under water.

Emperor penguins have many ways of dealing with
their rigorously cold environment. Large bodies retain
heat more efficiently than small ones do because their

Begging for Food

When a young emperor penguin chick
is hungry, it stretches as tall as it can,
waves its head from side to side, and
begs for food. After much pleading, the
parent regurgitates an oily mess of fish,
squid, and krill into the chick's throat.

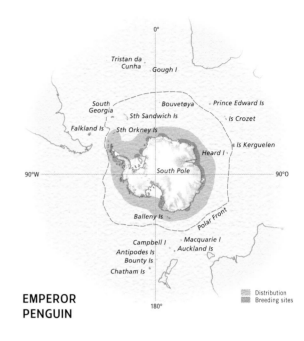

EMPEROR PENGUIN

Distribution
Breeding sites

surface area is small in relation to their weight, and the
emperor is the largest of all modern penguins; it is twice
as heavy as the king, its warmer-climate relative, even
though there is little difference in their heights.

All the emperor's extremities are reduced in size,
which limits the heat that can be lost through them. Its
head is small in relation to its body (although it has a
long beak), and its flippers and bill are proportionally
25 percent smaller than those of other penguins. In
addition, it has a highly developed blood counter-
current mechanism—a heat exchange system designed
to retain warmth in the body; it is twice as developed as
the king's system, and three times as effective as the
Adélie's. The emperor's body is designed to withstand
temperatures of 14°F (–10°C) before it must use body
energy to keep warm; king penguins must draw on body
energy at a temperature of only 23°F (–5°C). Like all the
penguins, emperors have a certain amount of insulating
blubber; however, their main form of insulation is their
feathers, which are dense and tightly packed, and lock
together to trap air in the layer of down against the skin.

Family Life

Living as far south as any animal in the world, emperor
penguins reproduce in indescribably savage conditions.
They have little time nor energy to spare on courtship,
and the pair bonds form quickly. The male lays his head
on his chest and calls out, sometimes repeatedly, and a
female responds by standing in front of him. Both slowly
raise their heads, stretching tall, and freeze into immobility
for a few minutes. Then they relax, and either pair or part.

Unlike many penguin species, emperors do not go
fishing to stockpile energy before laying their eggs. After
pairing for six weeks, the female lays a single egg. The
male immediately transfers it from her feet to his, and
covers it with his brood patch to keep it warm. The egg
is sheltered in this position by a fold of his skin, and the
male rocks back onto his heels to minimize the amount
of ice touching his feet. Having done her part for the
time being, the female heads to the sea to fish.

The males incubate the eggs through the winter in
constant darkness, surviving by huddling together; the

Going into a Huddle

Emperor penguins are the only adult penguins that huddle, and forming these tight groups
is essential for survival during the cold winter. It is believed that huddling reduces the loss
of body weight by 25 to 50 percent. In large colonies, up to 5,000 birds will form one huddle,
grouping together in as little space as possible. The huddle moves as a whole, with all the birds
keeping their backs to the wind. In a fantastically complex group dance, the windward birds
move down the flanks of the huddle, and then into the middle. Eventually, they are pushed out
to the edge and into the worst of the weather, before moving down the flanks and back into the
middle again. This circulating pattern ensures that all the birds get some time in the warmth, and
that no bird should freeze to death.

temperature often drops to –76°F (–60°C), with winds blowing at up to 110 mph (180 kph). After nine weeks, the chick hatches and the father gives it a small feed. Shortly afterward, the females come back and set about persuading the males to give up their chicks; they give in eventually and hand the chicks over to their mates. If a chick falls to the ice, another penguin may steal it, or, if it is left alone, it will most likely freeze to death in the extreme Antarctic cold.

Now, at last, the males are free to make for the sea to fish. They will have been fasting for four to six months and they will have lost about half their body weight by the time the mothers return to claim their offspring, but still they usually have to walk about 60 miles (100 km) across the ice to reach open water and much-needed food. Males never die of starvation while they are caring for their chicks. If the female is late returning, the male simply abandons the egg or the chick and makes for the sea to feed before it is too late for him to survive. Abandoning the chick may seem harsh and, indeed, even counterproductive—but the male will live to breed again, an older, wiser, and presumably more effective parent the next time round.

After the chicks have begun to hatch, two lines of emperor penguins can be seen snaking between the colony and the sea—one coming and one going. The chicks need constant feeding, and the parents take turns providing meals for about seven weeks. After that point, the growing chick must have more food than a single parent can provide. The chicks huddle together for warmth while both parents go off to the sea at the same time in search of fish.

A Warm Place
The temperature inside the brood patch, the bare patch of skin on the parent's belly (above) keeping the precious egg and then the young chick warm, is snug at over 86°F (30°C), even though the air outside is a chilly –22°F (–30°C).

Chicknapping
So strong is the brooding instinct in emperors (below) that chickless adults will try to steal any chick that leaves its parent for even a few seconds. Up to a dozen adults may try to claim a chick by pushing it under their belly feathers.

Identifying Colors

Most of the penguin species are easily identified by the coloration around the head and neck, and the brightly painted king penguin is no exception. Even the mandibular plate, part of the lower beak, is pink or orange in the adult.

King Penguin

Aptenodytes patagonicus

LENGTH 37 in/95 cm

WEIGHT 29 lb/13 kg

STATUS Approximately 2 million; most populations increasing. IUCN: Not listed. CITES: Not listed

The king penguin is the second largest living penguin species. Adults weigh around 29 lb (13 kg) and average 37 in (95 cm) in height. Females look identical but are a little smaller and slighter than the males. Unlike most of the penguins, kings have big splashes of color in their feathering. They are one of the more colorful penguins, with a beautiful spoon-shaped patch of bright orange on each side of their neck, fading into a paler orange upper chest. Their lower chest is white, and the rest of the penguin is a dark blue-gray with a silvery sheen, shading to black lower down the body. There are two king subspecies, *Aptenodytes patagonicus patagonicus*, which lives on South Georgia and the Falklands, and *A. patagonicus halli*, from Iles Crozet and Kerguelen, and Prince Edward, Heard, and Macquarie islands. The Latin name *patagonicus* is misleading because the king penguin is not found in Patagonia. Kings were hunted for oil in the nineteenth century, and several colonies were reduced or destroyed. The birds are increasing in numbers and expanding their range.

King penguins range widely in the southern oceans but stay clear of the pack ice, which they leave to their closest relative, the emperor penguin. Five of the seven breeding populations occur on islands just south of the Polar Front, but they prefer to fish in waters just north of the Polar Front, where the surface temperature is around 40°F (4.5°C). They are also often found fishing over the slopes of the continental shelf. They stay close to their breeding colonies in summer, but are somewhat pelagic, particularly in winter, and juveniles are regularly spotted several hundred miles from the nearest colony.

Wooing and Wedding

King penguins acquire their adult feathers soon after their second birthday. They spend the next two seasons practising courtship behavior, and mate for the first time at four years old. A male ready to mate returns from feeding at sea and advertises his availability by calling and stretching to his greatest height. If he succeeds in attracting a female, the two shake their heads together— perhaps a displacement activity to dispel any potential for conflict—and they occasionally call. Then one of the pair, usually the male, struts off in a head-swinging "attraction" walk, and the new potential partner follows behind in a similar fashion. Swinging their heads from side to side shows off the orange neck patches, and is believed to encourage following behavior. Sometimes a third penguin gets in the way of the forming pair during this ritual walk, which may lead to aggressive behavior, including wing flapping, mock beak stabbing, and, on occasions, even to fighting.

The next stage is "high-pointing," where the couple stand face to face, stretch slowly to their full height, and freeze for five or ten seconds before relaxing. The birds then sing antiphonally before bowing, and finally mating. After laying, they take it in turns to incubate the egg and brood the chick, strengthening their pair bond with each changeover by repeating the joint calling and bowing.

Once formed, a pair remains monogamous while raising the chick. However, long-term pair bonds are uncommon: only about 4 percent of king penguins mate with the same partner in the following breeding season. Possibly re-bonding is infrequent because females stop feeding their chicks, molt, feed, and are ready to mate again a full month earlier than the males. King penguins do try to get back to the same breeding place: one-third return to the exact location of their previous territory, and almost 95 percent return to within a few yards of where they incubated their last chick.

King penguins gather in vast numbers to reproduce in a staggered three-year cycle, gathering on the plains and shelving beaches of their sub-Antarctic island homes. Pairs first reproduce early in the southern spring, laying one large egg in November and incubating it by balancing it on their feet, protected by a fold of skin. They do not build nests; instead, each pair's territory is defined by, and limited to, pecking distance, although territories do shift a little as the penguins shuffle slowly along with the egg resting on their feet. The parents take turns incubating, in two- to three-week periods. The egg hatches in late January.

Public Baths

Like most penguins, kings like to scrub down after spending time on shore, where they are likely to get covered in guano and mud. When they first go to sea, they will spend some time rolling and splashing in the water, cleansing their feathers of any gathered grime, before leaving for their hunting areas.

KING PENGUIN

The chick does not fledge until the next November at the earliest—but first it must survive the long winter. Young chicks are at risk both from the elements and from predatory giant petrels and skuas, so at first one parent stays with the young while the partner goes off fishing. As in the incubation period, the parents take turns at fishing and guard duty until the chicks are so big that they need two parents to hunt and bring back enough food. At this stage, the chick joins a group of chicks known as a crèche. The crèche provides both protection from predators and some warmth through huddling together. Parents returning from the sea find their own offspring by calling and recognizing its voice. By the summer's end, a healthy chick will have reached a weight of around 24 to 26 lb (11 to 12 kg). As the winter approaches and food becomes harder to find, the parents can feed the chick only sporadically. The chicks lose about a third of their body weight over the lean winter period, but in spring, when prey becomes plentiful once more, the young penguins regain most of their lost weight before molting into juvenile plumage.

The Three-year Breeding Cycle

Breeding success has a price for king penguins. Healthy chicks that hatched early in the year will not fledge until November, so that its parents are not able to produce another egg until much later in the following breeding season. The second offspring usually hatches in March, and the parents feed it as much as possible before prey becomes scarce; however, it is inevitable that this later-born chick weighs less at the beginning of winter than a youngster hatched in January, so it is less likely to survive. If a late-born youngster does live, its parents feed it again the following spring and it fledges very late in the season. Any youngsters born after a successful late-born chick are always too young to survive winter, and the parents breed on an early cycle the next year.

Recovery of a Population

King penguins are large birds that gather in great numbers, and in the past they were ruthlessly hunted for their oil. Their numbers were seriously reduced, and some colonies were completely exterminated. However, with the end of sealing and whaling, most king populations seem to be growing at a yearly rate of 5 to 15 percent, and recently vacant areas are rapidly being colonized by nearby populations. There are now large, secure colonies at most breeding locations. Future risks to king penguins are largely limited to those related to the impact of humans on their small islands. An accidentally introduced disease, pest, or predator could inflict great harm on a large, closely packed colony in a very short period, and with human activity around the sub-Antarctic islands increasing, the risk of such introduction is increasing.

Swimming and Diving

Penguin bodies are designed for efficiency in water. Both emperors and kings average speeds of slightly less than 6 mph (9.5 kph) while fishing, although kings can reach 7 mph (12 kph). Kings with young chicks spend only a fifth of their time swimming at 6 mph, whereas parents feeding voracious larger chicks must work this hard for over a third of their time.

King penguins generally dive to depths of around 160 ft (50 m), but dives to 1,000 ft (300 m) have been recorded. Emperor penguins dive to an average depth of about 1,000 ft but can often reach close to 1,300 ft (400 m). These are not great depths compared with those plumbed by the sperm whale, which regularly dives to 1,000 to 2,500 ft (300 to 750 m) and can go as deep as 10,000 ft (3,000 m). However, for their size, penguins can perform very impressive dives. Emperor penguins have a probable aerobic dive limit of about four minutes, but two single dives have been recorded of approximately 22 minutes—which is more than five times that limit—to depths of between 130 and 200 ft (40 and 60 m). In both these cases, the penguins were breaking long fasts.   LW

Walking Home

King penguins seem to avoid climbing, preferring large level areas on plains and in valley mouths, especially where there is protection from the wind. While their colonies may be close to shore, many penguins still have a long way to walk if they have left a mate or chick at the rear of a large colony.

Brush-tailed Penguins

L ike the emperor penguin, the three brush-tailed penguins of the genus *Pygoscelis*— chinstrap, Adélie, and gentoo—live and breed in Antarctica proper. They are fairly large birds, standing taller than all but the emperors and the kings, and their feathers are monochromatic black and white.

The name of their genus comes from Greek words, meaning roughly "elbow-legs"—a reference to the shape of their leg bones, which in fact applies to all penguins— but the description "brush-tailed" may derive from their way of cocking their tail while swimming on the surface, so that it stands up from the water like a brush.

All three of these penguins nest on the land around the Antarctic continent and its adjacent islands. They colonize headlands and hills that are exposed as the snow retreats with the coming of summer. The amount of exposed land is very limited, but is essential to the success of a colony. The penguin builds its nest from small rocks uncovered as the snow melts. Although it is not much of a nest, it is better than nothing: meltwater drains away below it, and the rocks provide shelter from the wind. To shape the nest, one partner lies down and kicks backward with one foot, scraping a cup-shaped hollow out among the stones. The females lay two eggs, which the parents incubate in turns, lying along their rocky nest with the eggs tucked into their insulating brood patch. Given good weather and fishing, both of the chicks will fledge, but normally only one survives.

Gentoo Feeding
This penguin chick is getting a meal of krill from its parent, who has just returned from fishing with a stomach full of food, most of which will go to the chick. Gentoos prefer to hunt close to shore—their diet is mostly krill from south of the Polar Front, but is often more varied farther north.

Ancient Sites
This penguin rookery on the Windmill Islands is about 3,300 years old. Some Ross Sea rookeries date back 13,000 years. The oldest Antarctic Peninsula colony known is about 650 years old.

Precious Stones

B rush-tailed penguins value stones so much that when they return from the sea they often bring one to their partner. They carry stones for great distances, and often steal them from the nests of their rivals. They choose a stone with great care, and may replace one with another that apparently is better. They often pick up stones from outside the nest rim and move them to a more central location to "tidy" around the nest. Neighboring couples respond in kind, and boundary disputes may follow.

Adélie Penguin

Pygoscelis adeliae

LENGTH 28 in/71 cm

WEIGHT 11 lb/5 kg

STATUS At least 2.6 million pairs; 10 million immature animals. IUCN: Not listed. CITES: Not listed

Adélie penguins, named for the wife of French explorer Jules Sébastien Dumont d'Urville, are everybody's idea of the prototypical penguin. They have a simple and clearly demarcated but strong black-and-white coloring, with a blue-black back and head, and a white chest and throat. Their sole touch of color is a small patch of dark orange on the base of their bill, which is partially feathered, the black feathers making it look shorter than it really is. Around their eyes, Adélies have vivid white rings, which the young develop long before they grow their adult plumage. This arresting feature is a vital part of the penguin's armory: eyes down and crest erect, it exposes even more white in the sclerae of its eyes while it performs its three main displays of aggression—the direct stare, the fixed and alternate one-sided stare, and the crouch. The startling contrast of black and white constitutes an unmistakable threat.

Adélie penguins are rarely found north of 60°S, and always remain entirely south of the Polar Front, living on the Antarctic continent and nearby islands. They prefer to live within the pack ice, but small numbers colonize the South Shetland, South Orkney, and South Sandwich islands. Adélie rookeries will form anywhere that the penguins can reach, as long as it is free of ice: they will colonize ridges, hillsides, scree slopes, knolls, beaches, peninsulas, and moraines, provided that they are ice-free. Sometimes the route to the rookery is very steep, but this does not deter the penguins as long as there are no large steps, which the birds cannot negotiate with their short legs.

Adélie colonies may be very large, with populations of 20,000 to 30,000 birds common. There are seldom more than 100,000 penguins in one colony, although occasionally a single large colony may be home to more than 1 million birds.

Long Treks

Adélies, despite their small size, will travel long distances to colony sites. They create paths by trekking across long, snow-covered slopes toward exposed ground. Steep paths, rocks, and even cliff faces seem no obstacle, but rough ground, where their short legs become a disadvantage, take a toll.

Adélie penguins are shallow divers, and, like all other penguin species, they feed by pursuit diving, pecking out their food as they swerve from side to side under water. Adélies prefer to eat euphausiids, but will also take fish, amphipods, and cephalopods. Their diet varies according to what is available and where they are; for example, the diet of the Adélies around Princess Elizabeth Land and Wilkes Land is largely fish, while that of the Ross Sea populations is about half crustaceans, and that of the Terre Adélie penguins may include a higher proportion of cephalopods. Evidence suggests that Adélie penguins breed most successfully where krill is locally abundant.

ADÉLIE PENGUIN

- Distribution
- Breeding sites

All in the Timing

Timing is crucial for successful Adélie breeding, and the level of pair-bonding seems to depend on the latitude of the colony: the farther south the colony, the more important timing is to reproductive success and the less likely a penguin is to wait around for last year's partner to turn up. At Casey Station (about 66°S), 80 percent of Adélies rejoin their former partners, whereas at Cape Bird, 11 degrees farther south, only 56 percent reunite with the previous year's partner. Adélies at Cape Bird have a shorter breeding season, so they cannot spare the time to wait for last year's partner to return. Instead, they begin breeding as soon as possible, with a new mate.

Feathered Elegance
The Adélie penguin (above), replendent in black and white, presents a "dinner-suited gentleman" image.

Consummate Divers
Penguins are heavier than other birds and not at all suited to flying as this Adélie seem to be attempting (left): their bones do not have air spaces, their air sacs are reduced, and they have short feathers.

The Gathering of the Clans

Adélies are gregarious, and often gather together to molt on floes and bergs in the pack ice. Large gatherings of the penguins can often be found on the lee side of a hummock or a pressure ridge, sheltering until they can take to the water and fish again. Before going to sea, they congregate at the water's edge, calling back and forth to birds that are already in the water, until finally one penguin jumps. Many of the foremost birds follow this lead immediately, and then the birds at the back of the group move to the front and start calling to those already in the water—perhaps avoiding the land–sea interface where they are most vulnerable to the risk of being taken by leopard seals. When an Adélie comes in from the sea, it surfaces about 65 ft (20 m) from the beach, looks around, and then swims in and launches itself up to 7 ft (2 m) out of the water, gripping the ice or rock with long toenails to prevent falling back in.

At the onset of winter, Adélies disperse northward from the pack ice, the younger penguins leaving first. Many depart before molting, moving into the ice close to krill supplies. They may return to their breeding sites if storms open pathways through the ice. But the birds usually spend autumn and early winter in pack ice up to 400 miles (650 km) north of the Antarctic continent, and juveniles may winter even farther north than adults.

Homeward Bound

In spring—October in the southernmost locations, but earlier in warmer, more northerly sites—Adélies go back to their breeding colonies. The first to arrive are older, more successful individuals, up to eight years old; the last are the two-year-olds, about to breed for the first time. The home-coming time is governed by the ice, and first-time breeders may be very late when the pack ice is heavier than usual. Males arrive about four days before females. The more successful breeders have

places in the middle of the colony, where reproduction is usually highly synchronized: the pairs rebuild their nests in exactly the same location year after year.

Adélies cannot lay eggs until the surfaces of their colonies are clear of snow: once it has found a suitable location, a colony stays put. After the female has laid, males take the first incubation shift of two weeks, then parents swap. Either parent may be on the nest when the eggs hatch. After hatching, parents usually brood alternately, taking two-day shifts each for three weeks until the chick is large enough to be left on its own. Then both parents go to sea in order to satisfy their offspring's voracious appetite.

Swim Power
Adélie penguins are superbly adapted to aquatic life: a short, stiff tail and rearward feet together act as a rudder, steering a stubby, inflexible body. The "flipper" wings are powerful propellers on either side of the body.

Specialized Feathers

The feathers of penguins are highly specialized for insulation. The very short, broad, and flat rachis (center shaft) reduces the overall length of the feather and allows it to lie flat against its neighbor, and the closely spaced barbs, especially near the rachis and tip, provide good interlocking. Small, downy afterfeathers form a second layer of insulation.

Harsh Natural System

Penguin colonies afford easy pickings for leopard seals and orcas. Birds that escape with injuries may recover—or, like this doomed chinstrap, may become food for predators and scavengers, such as giant petrels and skuas.

Chinstrap Penguin

Pygoscelis antarctica

OTHER NAMES Ringed penguin, bearded penguin, Antarctic penguin, stone-cracker penguin

LENGTH 28 in/72 cm

WEIGHT 8½ lb/3.8 kg

STATUS: 6.5–7.5 million pairs; some populations are increasing rapidly. IUCN: Not listed. CITES: Not listed

Chinstraps have a white throat, chest, and underparts, and a black body and head. Their bill is black, and a distinctive thin black line runs from ear to ear through the white of their throat. It is this band that gives them their common name—chinstrap—but these penguins were frequently seen by early Antarctic explorers, and were also named for their habitat (hence "Antarctic") and their habits (hence "stone-cracker").

There are eight regional populations of chinstrap penguins, all based inside the Polar Front; 99 percent of all chinstraps live within the southern Atlantic Ocean. They inhabit both the sub-Antarctic and the Antarctic proper, preferring light pack ice. They breed on ice-free land on the Antarctic Peninsula and the islands south of the Polar Front, largely within the Scotia Arc. Chinstraps are very agile climbers, and their favorite nesting sites seem to be rocky slopes, headlands, rough foreshores, and high cliff edges.

Chinstrap penguins are specialist feeders and eat mostly crustaceans, especially krill. Chinstraps dive to less than 165 ft (50 m) most of the time, spending 40 percent of their time almost on the surface at a depth of less than 33 ft (10 m), and 90 percent of their time at less than 130 ft (40 m). Their maximum dive depth seems to be 230 ft (70 m), but they probably reach this depth only when they are desperately trying to find food for their demanding chicks. Chinstrap dives have been timed: their shallow-feeding dives last about 90 seconds only each, with a little over 30 seconds between dives. They generally feed for about two-and-a-half hours, covering only 3 miles (5 km) over a five-hour trip. One reason for the current research interest in chinstraps' eating habits is that these penguins seem to have been very successful recently. Their range has expanded and their numbers are rising, perhaps because there are

more krill in the southern oceans since there are reduced numbers of krill-eating whales, or possibly because of the reduction in winter ice in the Antarctic waters where the chinstraps hunt.

Noisy Breeders

Chinstraps are extremely noisy, and it is surprising just how cacophonous a colony can be. All their calls are louder than those of the other brush-tailed species, and they seem to call more frequently than other penguins. Away from the nest, chinstraps are gregarious and very inquisitive. Of all the penguin species, they are the least likely to back away from an encounter and the most renowned for attacking humans who encroach on their territory—which is how they got their reputation for being feisty and aggressive.

While they are breeding, chinstraps stay close to their home colonies, traveling a maximum of 60 miles (100 km) during the chick-rearing phase, and perhaps slightly farther when the eggs are incubating. During this period, they concentrate hunting energies around nearby ice floes, presumably in search of the krill that is likely to be found nearby. After reproduction, when chinstraps molt, some choose to wait it out just inland of their colony, whereas others seem to prefer to molt at a greater distance away.

Retreating from the Ice

In the winter, as the temperature falls and the pack ice starts to expand northward, chinstrap penguins move into the open waters north of the pack ice. They leave the Antarctic Peninsula and nearby King George Island

CHINSTRAP PENGUIN

Tristan da Cunha · Gough I

South Georgia · Bouvetøya · Prince Edward Is

Sth Sandwich Is · Is Crozet

Falkland Is · Sth Orkney Is

Heard I · Is Kerguelen

90°W · South Pole · 90°E

Peter I Øy

Balleny Is

Polar Front

Campbell I · Macquarie I

Antipodes Is · Auckland Is

Bounty Is

Chatham Is

Distribution
Breeding sites

180°

in early April and do not return until spring (late October). But they stay together, and over the winter large groups congregate in the open water north of the pack ice. Little is known about where chinstraps spend the non-breeding season, but they probably attempt to follow the krill and other organisms on which they feed.

Chinstraps have been spotted a staggering 2,000 miles (3,200 km) from their breeding site; however, this figure may be based on individuals that are extending the range of the species, rather than on the movements of the main population.

Studying Stress

The study of penguins is fascinating, but the disturbance of breeding penguins by human activity is the subject of constant concern and research. Initial investigations looked at obvious signs of distress, such as fleeing the nest and abandoning young at the approach of humans, or at the success or failure of reproduction. Currently, less obvious marks of distress, such as increased heart rates, are being examined and compared with the effects of normal stresses—for example, the disturbance shown by a nesting bird at the approach of a skua or other scavenger—but the effects and their implications are difficult to separate from normal fluctuations in reproductive success. One ingenious way of measuring stress is a heart-rate monitor in the shape of a fake penguin egg, which is placed under penguins.

Chasing Food
When chinstrap parents (above left) return from sea, they are immediately harassed by their offspring for food. They often lead older chicks on a chase through the colony, before they finally stop to feed them. This possibly serves to separate the weaker chicks from stronger ones when there is only enough food for one chick.

Southern Migration
The Antarctic Peninsula is currently warming, and this seems to be one of the factors that is encouraging chinstraps, such as these ones, to extend their range farther south. Colonies found at the southern end of their peninsula distribution limit did not exist 50 years ago.

Gentoo Penguin

Pygoscelis papua

OTHER NAME Johnny penguin

LENGTH 30 in/75 cm

WEIGHT 12 lb/5.5 kg

STATUS More than 300,000 pairs. IUCN: Lower Risk—near threatened. CITES: Not listed

GENTOO PENGUIN

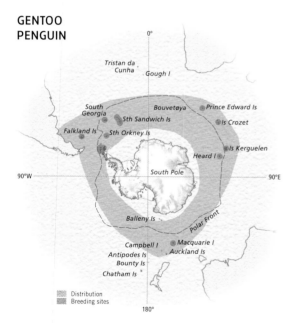

Distribution
Breeding sites

Gentoos have a triangular white flash above each eye, white flecks on the crown of their head, and bright orange bill and feet. They live slightly farther north than other brush-tailed species, remaining close to their breeding islands all year in sub-Antarctic and Antarctic regions right around the South Pole, but they move no farther south than 65°S, shunning the most southerly pack ice and the southern continent itself, except near the Antarctic Peninsula. Most breed within the Southern Ocean, but about 25 percent breed north of the Polar Front. Taxonomists identify two gentoo subspecies: the

southern group, *P. papua ellsworthii*, is fully migratory, following a clear path each year, and has smaller feet, flippers, and bill than the northern *P. papua papua*, which disperses from its breeding areas but does not migrate in the true sense.

When not breeding, gentoos mostly stay around the winter pack ice. The farther south they breed in summer, the more likely that the whole colony will move northward for winter, but generally they remain near their colonies, some staying as far south as possible in nearby open water. Many may leave their breeding site after molting to feed and regain their weight, but they never totally abandon a

What's in a Name?

Both the common and the scientific names of the gentoo penguin are mysteries. Gentoo, its common name, means "pagan" in Hindustani—a word believed to be of Anglo-Indian/Portuguese origin and somehow related to stories about dancing girls, but its etymology is unclear. As for the gentoo's scientific name, *papua*, the first stuffed specimens to reach England for describing were thought to be from the island of New Guinea, and were named *papua*, which was the Malay word for "curly." Nothing about this penguin merits the description "curly," so it may be a reference to the geography of its supposed place of origin.

Setting Up

Gentoos will nest with chinstraps or Adélies, but generally prefer lower ground than the other two brush-tailed species when nesting in the same area. On the Antarctic Peninsula, they usually locate their colonies less than about 300 ft (90 m) from the shore.

Being Resourceful

Suitable materials for nest building are often very scarce, so supplies will be obtained from any source. If small stones are not available, gentoo penguins may use coal, small bits of wood, and other debris found near some of the Antarctic Peninsula stations.

colony, and populations build up again after a few months. However, whole colonies have been known to move on from time to time, perhaps escaping parasite infestation: large numbers of ticks have been observed on gentoos. They peck constantly at each other and often squabble with their gentoo neighbors over nesting materials, but they do not come to blows with other penguin species.

Finding Food

Most gentoos feed in the day and return to their colony in the evening. They generally feed inshore, and find shallow-water prey when breeding. They go to sea en masse early in the day, then later they take to the water individually. They tumble in, swim briefly, then surface and clean themselves of the accumulated muck of the colony by rolling about and wiping their bodies all over with their flippers. When they come in to land after fishing, they will pause to inspect the landing site, and then surf in on a wave. They leap onto ice or rocks, and then shake their head and preen before moving back to their nest. Gentoo penguins living near Antarctica mostly consume euphausiids, but those breeding north of the Polar Front consume more fish. If need be, they will dive deeper than 540 ft (165 m) in search of their prey, but their diving is often less than 65 ft (20 m).

Breeding Behavior

Gentoos breed on ice-free land on the sub-Antarctic and Antarctic islands, and the Antarctic Peninsula, and may form colonies some distance inland or directly on the coast. They occupy slopes, headlands, flats, terraces, valleys, ridges, and cliff tops. Some scientists suggest underwater landforms dictate gentoo penguins' breeding sites so that they can always breed close to the shallow waters that are their most profitable hunting areas.

Around 90 percent of gentoo couples will locate their former mate in the next breeding season, but, unlike

Taking Over

Gentoo penguins inhabit one of Pléneau Island's many rocky promontories at the southern end of the Lemaire Channel in the Antarctic Peninsula region. Over the past century, more gentoos have been breeding here, largely displacing the Adélie penguins, who prefer the cooler climate of the continent.

Adélies, they often move to new nest sites, presumably because they have a longer breeding season, and therefore it pays them to wait for a successful partner to return before beginning to breed. After the pairing, the male collects stones, moss, and dirt—whatever is available—devoting his attention to foraging for good nesting materials, and leaving the actual building to his mate. Once the eggs are laid, the individual incubation shifts last from around 24 hours to four days. At the northern end of their range, gentoos start to lay in the winter or very early spring; on South Georgia they do not begin until the end of October, and on the Antarctic Peninsula not until mid-November or later. These times vary, depending on snow cover, with the season delayed if there is heavy snow, and starting early if the seasonal snowfall is unusually light.

Gentoos mostly choose elevated nesting sites at a safe distance from breeding grounds of the elephant seals that share their regions. Most of them will lay two eggs. In good years both of the chicks survive, but this is rare, except on the Antarctic Peninsula, where usually either both of the chicks survive or neither do. Survival rates for the chicks can almost always be traced directly to food availability during the breeding period. Gentoos feed close to their nest sites, about once every 24 hours when they are mating, laying, and incubating, far more frequently than the other brush-tailed species—and this daily feeding may be a factor in gentoo penguins being less likely to abandon their chicks.

Young gentoos first venture from the nest at about 20 days of age, and by 29 days most have joined other youngsters in a crèche, leaving both parents free to go to sea to fish. The youngsters fledge and go to sea at around 80 to 100 days, but at first often spend more time on the beach than in the water.   LW

True Divers

Penguins of the genus *Eudyptes* are sometimes called "crested" penguins, a reference to their jaunty yellow "eyebrow" plumes. They occupy various habitats in the Southern Hemisphere, though none is truly Antarctic, and some live well within the sub-Antarctic region. They range from 18 to 28 in (45 to 70 cm) in height, and all have red or red-brown eyes. All species except the Fiordland penguin spend up to five months of the year at sea.

Eudyptes penguins differ from the other species in several ways. Allopreening, where one penguin preens another, is common in this group of penguins, although it is rare in most other groups. Some eudyptids live in small groups, or even in single pairs, hidden away from their neighbors, whereas other penguin species are extremely gregarious. They always lay two eggs, about four days apart: the second is as much as 70 percent heavier, and the first is almost invariably discarded.

A Speedy Molt
Most penguins molt quickly during the short summer: non-breeders first in the middle of the summer; failed breeders may begin molting as soon as they lose their egg or chick; and successful breeders molt last, after their chicks have fledged and they have replenished their energy reserves at sea.

Rockhopping
True to their name, rockhoppers (left) nest among ledges, rocks, and grass, and prefer to come ashore amidst large rocks and rough seas. They scramble from the sea, hopping ahead of the incoming waves with both feet together. This preference for hopping is not shared by other members of the genus.

Macaroni Haven
South Georgia is home to millions of macaroni penguins (right), but unlike the island's king penguins, most of the macaronis live and breed far from easy human access. The island provides many good colony locations in tall tussock grass and on steep rocky hillsides, and the surrounding waters supply plentiful food.

Rockhopper Penguin

Eudyptes chrysocome

OTHER NAMES Crested penguin, tufted penguin, rocky
penguin
LENGTH 18–23 in/45–58 cm
WEIGHT 5–7½ lb/2.3–3.4 kg
STATUS 1.1 million pairs and declining.
IUCN: Vulnerable. CITES: Not listed

The rockhopper is the smallest of the true divers, and
the third smallest of all the penguins. It has the classic
black-and-white coloration, with a white chest, a black
head and back, and an orange bill and feet. Its crest
is a narrow band of yellow feathers above its eyes,
thickening to a bushy dangling crest behind its eyes.

Rockhoppers are agile climbers: keeping both feet
together, they scale steep cliffs quickly, gripping the rocky
surface with their strong, sharp toenails. They hop on
more level ground as well, and are surprisingly quick—
probably the fastest of the smaller penguins, especially
on steep or broken ground. They may have been given
the name rockhopper for their habit of landing on rocks,
and have been known to spurn perfectly good beaches
with smooth landing grounds, choosing instead to leap
out onto rocky, wave-swept cliffs. They are much more
likely than other penguins to jump into the water feet
first, and they also seem to enjoy bouncing down hills,
apparently without inflicting damage on themselves.

Rockhoppers are found in the sub-Antarctic regions
around the Pole, but only about 12 percent of them—
at five different locations—breed within the Southern
Ocean. Most of their population base inhabits the area

Natural Surfers

Swimming rockhoppers often gather
just beyond the breakers near their
colony—like a group of wet-suited
surfers waiting for the waves. They
usually come to shore together on one
good wave, and once in the shallows
there can be an undignified scramble
to get past the point where they might
get swept back in.

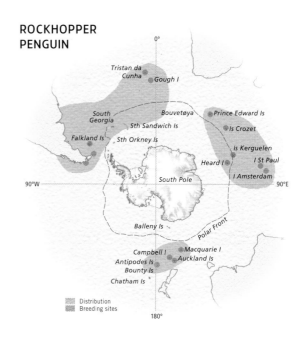

**ROCKHOPPER
PENGUIN**

north of the Polar Front, and in the winter, during the
colder non-breeding period, they disperse northward
into the ocean. They have been known to stay at sea
for long periods; there are tales of their feet becoming
encrusted with barnacles. Vagrants have been found
on the southern shores of all the southern continents.

These penguins are not large birds, so they do not
consume large prey. Their diet regularly consists of
the smaller crustaceans—amphipods, copepods, and
isopods—and in some areas small squid. Because they
are so small and their population in the Southern Ocean
is small, they account for only 1 percent of the energy
consumed by all the penguins in that ocean.

Rockhoppers nest close to water, in among rocks and on ledges. They line their nests with rocks, or grass when they can find it, and there have even been reports of them digging burrows on Ile Amsterdam.

Unusually for penguins that are living so far north of the Antarctic continent, rockhoppers have a remarkably precise breeding timetable—birds from the same site return to their nests on virtually the same day each year. Looser schedules, more attuned to each year's particular conditions, would be expected. Because rockhoppers live so much farther north than many other penguins, they often share their rugged rookeries, which may be in very unlikely and exposed spots, with albatrosses and shags. They have a strong attachment to their original nest sites, returning over and over to the identical spot, whereas in other species, loyalty to the partner is usually stronger than the site bond.

In southern rookeries, rockhoppers arrive between mid-October and early November, and lay two eggs between late October and mid-November. The first of the eggs is always smaller than the second and, if it has not already been jettisoned before it has had a chance to develop, it will be lost during squabbles that seem to pose much less of a hazard to the second egg. The eggs are incubated for 32 to 34 days and hatch from mid-December through January. The chick is brooded for 19 to 25 days, and the young fledge in February or March, after which the adults go to sea to feed for about five weeks, then return to molt, before finally heading back to sea, usually in April or May.

Like other penguins, rockhopper couples alternate hunting expeditions with nest duty. When it comes time to change shifts, both partners stretch their bills toward each other and trumpet loudly. Neighboring penguins often join in, directing calls toward the returning partner. As the relieving bird reaches the nest, both the partners trumpet vertically, with open bills pointed upward, lifting their flippers with the calls.

Despite their small size, rockhoppers are aggressive birds, and will attack whenever they feel threatened. They signal mild aggression by turning their head and bobbing it up and down, with flippers raised. If this fails to intimidate the opposition, one penguin will jab its open bill toward its opponent, and eventually the two birds lock bills. The conflict will end only when one of the penguins manages to seize the other by the nape of the neck. The victorious rockhopper may even drag its vanquished opponent along behind.

Golden Haired

The rockhopper's scientific name, *chrysocome*, means "golden haired," and so is equally suited to any of the penguins in this group. The golden crest is smaller in the rockhopper than in the macaroni or royal, but the small tufts, which often extend sideways, are distinctive. Rockhoppers have powerful bills for their size, and clear red eyes that can be quite disconcerting.

Close Encounters

Scientists differ about the classification of rockhoppers. Some believe that there are three geographically defined subspecies: *E. chrysocome chrysocome*, found in the Falkland Islands and off the coast of Chile; *E. chrysocome moseleyi*, found at Tristan da Cunha, Gough Island, Ile St Paul, and Ile Amsterdam; and *E. chrysocome filholi*, from the region around Iles Kerguelen, Prince Edward, Heard, and Macquarie islands, and the Auckland and Antipodes groups. Others contend that there are only two types: a northern race and a southern race, distinguished by differences in behavior, song, and bill length.

In any case, the relationships between the different *Eudyptes* species seem to be very close. Rockhoppers occasionally interbreed in the wild with macaroni, royal, and erect-crested penguins, and these matings have produced hybrid offspring, some of which have successfully reproduced.

Macaroni Penguin
Eudyptes chrysolophus
LENGTH 28 in/71 cm
WEIGHT 11–13 lb/5–6 kg
STATUS About 12 million pairs; some populations increasing, some in decline. IUCN: Vulnerable. CITES: Not listed

MACARONI PENGUIN

Distribution
Breeding sites

Macaroni penguins are the largest of the *Eudyptes*. They are black and white, with an orange bill and feet, and a bright crest. The crest is dark yellow or orange, growing back along the brow line from the center of the forehead—drooping over the edge behind the eyes, and hanging rather long at the back of the penguin's head. The birds got their common name because of their resemblance to eighteenth-century London dandies who went in for foppish dress and hairstyles associated with Italian culture.

Most macaronis live south of the Polar Front, in sub-Antarctic and warmer Antarctic waters. They occupy colonies on South Georgia and along the Scotia Arc, and a few pairs nest with other species in the South Sandwich Islands. They often drift for long distances north and south of their colonies, but always remain north of the pack ice.

There are about 12 million breeding macaroni pairs within the Southern Ocean, making up more than 50 percent of all breeding penguins south of the Polar Front. This is the dominant penguin in the Southern Ocean, in terms of both biomass and the consumption of resources, accounting for millions of tons of crustaceans and fish from the Southern Ocean each year.

Macaronis are noisy, pugnacious birds. They like to establish their rookeries on gently sloping ground, not too close to the water, with a pathway up to it along smooth, easily traveled stream beds. Smaller colonies often share rookeries with chinstraps or gentoos.

Breeding macaronis generally reach their colony between September and November. Older birds come ashore earlier than younger ones, and stay longer. The males arrive first and set about selecting or repairing a nest site, and the females come in eight days later. Then the birds pair, laying their eggs between October and November.

Like the other *Eudyptes* species, macaroni penguins lay a very small first egg, which is usually lost after the second egg is laid four to six days later. The female will take the first incubation shift and the male the second, both about 12 to 14 days long. Both parents preside over hatching, which takes place some time between mid-November and early January. The male broods the chick for the first two to three weeks while the female makes several hunting trips to the sea to feed herself and the chick. The chicks fledge after about 65 days, and then both parents go to sea to feed for a few weeks, returning to the beaches in March to molt.

Inter-species Mating
Macaronis are the most common penguins in the world, but even with around 12 million macaronis of the opposite sex to pick from, some birds choose to mate with penguins from other species, including rockhoppers.

Enough to Go Round
Penguin colonies make high demands on local waters for food. On the Prince Edward Islands, four species rely on the local food supply when they are breeding. Kings, rockhoppers, and macaronis eat 70 percent pelagic fish, whereas gentoos hunt out benthic shrimp and fish.

Royal Penguin

Eudyptes schlegeli

LENGTH 27½ in/70 cm
WEIGHT 11–13 lb/5–6 kg
STATUS 850,000 pairs. IUCN: Vulnerable.
  CITES: Not listed

Many scientists believe that the royal penguin is a subspecies of the macaroni, the only difference being the coloration of the face, chin, and throat. Royals have a white or pale gray face, and there may be some black in the throat, but they have the massive head plumes and the very large bill characteristic of the macaronis. Royals are largely, but not entirely, limited to Macquarie Island, south of Australia.

Royals are extremely robust. Like most penguins, they walk by alternating their feet, but occasionally they hop along with both their feet together, in the manner of their rockhopper relatives.

Even though both sexes have very large bills, royals are one of the easiest penguin species to sex when seen in pairs: the males have larger bills, and the females are seldom as white-cheeked as the males—in fact, they are more often gray.

Whereas most penguins are breeding by three to six years of age, royals have a prolonged adolescence, and it may be as much as 11 years before they breed. They breed in the same pattern as most *Eudyptes* penguins,

in areas where vegetation is available. They build their nests by excavation—from sand or stones—and carry stones to their nesting site. In sandy areas, they also scoop sand into their bill and bring it to the nest site.

When they are not breeding, royal penguins either migrate or disperse to the sea, where they spend much of their time, but little is known about their sea-going behavior or their sea range.   LW

Vulnerable Species

Royal penguins are isolated to breeding in a few large colonies on Macquarie Island and nearby islets. Because they breed only in this one small area, if anything should disturb the population, all royals would be at risk.

ROYAL PENGUIN

Tristan da Cunha
Gough I
0°
South Georgia
Bouvetøya
Prince Edward Is
Sth Sandwich Is
Is Crozet
Falkland Is
Sth Orkney Is
Is Kerguelen
Heard I
90°W
South Pole
90°E
Balleny Is
Polar Front
Campbell I
Macquarie I
Antipodes Is
Auckland Is
Bounty Is
Chatham Is
Breeding sites
180°

Other Penguins

Some of the penguin species rarely venture into the Antarctic or sub-Antarctic regions. Several of the crested penguins prefer to stay in warmer climates around New Zealand, as do the yellow-eyed penguin and the little penguin (the latter also living in Australia), and the four *Spheniscus* species, which are found in the warmer waters along the South American and South African coasts.

Noisy Neighbors

A pair of little penguins (left) on their nest peer out from their home in a small cave. Little penguins commonly nest in burrows, caves, crevices, or even under human-made structures along coasts. Many New Zealanders can testify to just how noisy a pair of penguins living under a coastal house can be.

Eudyptes Species

Three species of crested penguin occupy New Zealand waters. Their crests are paler and less obvious than those of the macaroni. Unusually for the penguins, the small Fiordland penguin (*E. pachyrhynchus*) breeds in winter and is faithful to both nest site and to partner. It is timid, and lives in small colonies or independent nests it builds out of vegetable matter and stones in temperate rainforest close to the sea. The Fiordland penguin feeds in groups during daylight.

The erect-crested penguin (*E. sclateri*) inhabits the islands south of New Zealand, and the Snares crested penguin (*E. robustus*) is found only on Snares Island. Uniquely, the erect-crested penguin can raise and lower its crest. Little is known of the movements of these birds at sea, but they are gregarious on land. They breed in large colonies in rocky breeding grounds.

Yellow-eyed Penguin

This penguin (*Megadyptes antipodes*) is restricted to southern New Zealand waters. It has pale yellow irises, a yellow forehead and crown, and a long, narrow bill. In August, it nests in isolation on grassy cliffs or in forests, feeding during the day and returning to land at dusk. It does not stray far from its breeding sites.

Little Penguin

Eudyptula minor, also known as the fairy, blue, or little blue penguin, is found in Australia and New Zealand. By far the smallest of all the penguins, at a mere 16 in (40 cm) high, the bird has a unique coloration—a pale, metallic blue-gray plumage with a white throat and chest. Completely sedentary, it is neither dispersing nor migrating, usually feeding during the daytime and then returning to its nest at dusk.

Spheniscus Species

The four *Spheniscus* penguins are fairly large black-and-white birds with no crest and a striped chest and neck pattern. These are the least Antarctic of all the penguins, and they range from the equatorial Pacific Ocean region to South Africa and to South America, although the magellanic penguin does venture as far south as the Falkland Islands. They nest in burrows or on rocky ledges and crevices, and their braying call has given them the sobriquet "jackass penguin."

The jackass or African penguin (*Spheniscus demersus*) breeds along the coasts of southern Africa

A Rare Penguin

The Fiordland penguin, unlike other crested species, breeds in single pairs or small groups scattered along long stretches of inhospitable coastline in southern New Zealand. Add to that a shy nature, and it is no surprise that, beyond knowing they are rare, we do not have a clear idea of population size.

Swimming with Penguins

Penguins are so well adapted for a swimming lifestyle that it comes as a surprise to sometimes see them bumping into one another underwater—and not just when they are rushing, even when they are merely milling around.

and is the only truly African penguin. It rarely strays far from land. It has a wide band of pink skin at the base of its upper bill and around its eyes, with a broad white band running from above its eyes around its cheek to its breast, and a black band across its breast and down the side of its white underparts.

The Galapagos penguin (*S. mendiculus*) is unique in being equatorial and, unlike other penguins, it molts more than once a year—before it begins breeding as well as after. It gathers in small family groups in the waters around the Galapagos Islands. Most of its head is black, with a thin white line curving from its eyes around its cheeks and onto its breast, where two thin black bands run across at flipper height.

With the characteristic *Spheniscus* black-and-white bands across its face, neck, and chest, the magellanic penguin (*S. magellanicus*) is medium-sized and has a black band around its white underparts, and another strong black band demarcating its head from its breast. This bird ranges through the Pacific and Atlantic coasts of southern South America and the Falkland Islands, occasionally appearing on South Georgia. It returns to its colonies around September, fledging its chicks and dispersing in March and April.

The Peruvian or Humboldt penguin (*S. humboldti*) lives off the Pacific coast of South America, from Peru to Chile. It has a dark head, a pink patch at the base of its lower bill, and a single black stripe across its neck

and breast. Peruvian penguins are timid on land but gregarious in small groups at sea. They lay their eggs year-round and neither migrate nor disperse, although the youngsters may drift in both northerly and southerly directions before returning to breed. Peruvian penguins nest in caves or burrows dug from sand and dirt a short distance from the ocean. LW

Squid for Science

Scientists occasionally use magellanic (seen here), gentoo, and rockhopper penguins as squid-catchers. About half the food penguins catch are squid types not caught by commercial or research trawlers fishing the same areas.

Antarctic Exploration

First Speculations

Continents Take Shape

Drawn in 1570 by Ortelius, this map records discoveries of Diaz, da Gama, and later Portuguese navigators of the coasts of Africa, India, east Asia, and Japan; of Columbus and Vespucci of America's east coast; and of Magellan of South America and the Pacific Ocean. Terra Australis, seen as a separate continent, includes Australia, New Zealand, and Tierra del Fuego.

Early Outlines

One of the earliest maps of the world, as it was then imagined, drawn in about 240 BCE by the astronomer and mathematician Eratosthenes. The lands around the Mediterranean are easily recognizable; the remoter regions of India and Scythia (China) are less familiar. This map also shows Parmenides's five climatic zones.

Antarctica is the only continent that, from the perspective of human thought, began as a sophisticated concept emerging from a series of deductions. In the sixth century BCE, Pythagoras, the Greek philosopher and mathematician, calculated that the Earth was round. About a century later, Parmenides divided the world into five climatic zones not unlike those that we know empirically today. He postulated frigid zones at the poles, a torrid zone at the Equator,

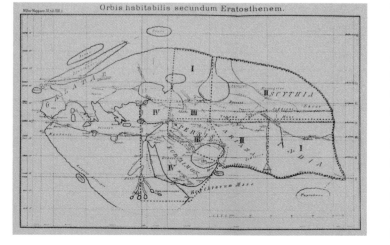

Orbis habitabilis secundum Eratosthenem.

and temperate zones separating these uninhabitable extremes of heat and cold. In the fourth century BCE, Aristotle suggested that the landmass of the Northern Hemisphere must be balanced by a large landmass in the south; later this became known as Terra Australis Incognita, "The Unknown South Land." Aristotle gave it a name: the north lay under the constellation Arktos (the Bear) at the time, so he called the other end of the world Antarktikos (meaning "opposite to the north").

Aristarchus, a third-century astronomer, was the first to expound that the Earth rotated around the sun.

In 240 BCE, Eratosthenes of Cyrene (now Egypt's Aswan)—who coined the word "geographica"—calculated the Earth's circumference by comparing shadow angles in two distant locations at the summer solstice. It was a simple, ingenious, and remarkably accurate method. Although the units he used were imprecise, he only overstated the circumference by about 15 percent. By CE 200, Pomponius Mela, the Roman geographer, had postulated the existence

Ptolemy's Rounded View
This 1482 edition of Ptolemy's world map of the second century CE (above) gives a detailed picture of Europe and the Mediterranean and Atlantic coasts of North Africa, with the Black and Caspian seas clearly recognizable. Unknown realms shown encircling the south make the Indian and Atlantic oceans mighty inland seas. This view of the world replaced the flat-Earth beliefs of the Dark Ages.

The South in 1600
Chica sive patagonice et australis terra (below), a 1600 copy of a 1599 map by Cornelis Wytfliet, is one of the first maps of South America that includes the Strait of Magellan, Tierra del Fuego, and the speculative Terrae Australis.

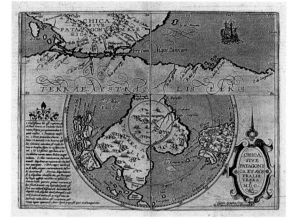

of a cold continent at the southern pole of a globe roughly the size that we now know it to be, spinning around the sun. It was the nearest to the truth that anyone would be for 1,500 years.

Claudius Ptolemy, the influential Alexandrian geographer of the second century CE, added greatly to cartography with his refinement of grid lines of latitude and longitude to give every location fixed coordinates, northern orientation for maps, and various scales for different uses. However, his mistakes had an equally great effect. He agreed that there must be a southern land, but thought that it linked the known continents and was fertile and inhabited. He also ignored the work of Eratosthenes, preferring that of the Greek astronomer Poseidonius, whose theories yielded a calculation of the Earth at about 75 percent of its true size. Because of Ptolemy's errors, Christopher Columbus expected to find Japan where he encountered America, and James Cook set out to find the riches of the supposedly fertile Great South Land.

The Dark Ages
After Ptolemy and long before Columbus, Magellan, and Cook, an age of ignorance set in. In CE 391, the great libraries of Alexandria were destroyed. An era of dogma based on the literal interpretation of the Bible began. Theological objections to the concept of an inhabited southern land proliferated.

The Book of Isaiah states: "It is He that sitteth upon the circle of the earth … that stretcheth out the heavens as a curtain, and spreadeth them out as a tent to dwell in." It was an utterance that gave rise to a perception of the Earth as a flat disk. The Gospel of St. Mark records God's command to Christ's disciples: "Go ye into all the world, and preach the gospel to every creature." The disciples did not go to the South Land—and therefore it could not exist.

The Age of Exploration
When the Christian city of Constantinople (now Istanbul) was occupied by the Turks at the end of the fourteenth century, fleeing scholars brought Ptolemy's works back to Italy. His *Geographia* was translated into Latin in 1405 and hundreds of copies had been printed by the end of the century. Europe started to look outward. The Portuguese Prince Henry the Navigator (1394–1460) is regarded as the initiator of the great age of exploration; although he himself did not travel much, he was the patron of many voyages in Portuguese caravels which ventured far down the west coast of Africa. Henry had ambitions to reach India by sea for missionary and trade purposes, but the farthest his vessels reached was to the coast of Sierra Leone. It was left to his compatriot Bartolomeu Diaz (c.1450–1500) to launch the age of Antarctic exploration. DM

Royal Patronage
Although Henry the Navigator took no part in the voyages he sponsored between 1419 and 1460, he established a famous school of navigation, built an observatory, and was the driving force behind the Portuguese discovery of new coastlines and sea routes.

Early Navigators

Vasco da Gama
It took the Portuguese navigator Vasco da Gama 23 days to sail from the Cape of Good Hope to Calicut, on the west coast of India, in 1497.

The Earth is Round
Ferdinand Magellan's circumnavigation of 1519–22 finally corrected Ptolemy's drastic underestimation of the size of the globe. This 1589 map detail shows Magellan's ship, the *Victoria*, forging bravely westward across the vast Pacific Ocean.

M ost early exploration of Antarctica was a process of whittling down the fabled Great South Land as empirical knowledge replaced conjecture. The Portuguese explorer Bartolomeu Diaz took the first significant step when he sailed down the west coast of Africa in 1487. At the time, it was still credible that Ptolemy was right and that Terra Incognita filled the bottom of the world with a coast across the temperate zone. In January 1488, Diaz was alongside the coast of what is now South Africa when storms drove him out to sea and he sailed south for a few days. When he turned east, he found no coast, and made landfall only by sailing north again. He had rounded the southern tip of Africa. He realized he had discovered a sea route to India, and Africa, at least, was not joined to the Great South Land.

When he returned, there was talk of his leading a voyage to India, but it was another Portuguese explorer, Vasco da Gama (*c*.1469–1525), who sailed to India in 1497, thus proving that the Atlantic and Indian oceans were not landlocked. In March 1500, the Portuguese navigator Pedro Cabral led a fleet of 13 vessels to India, but to avoid the becalming waters of the Gulf of Guinea they sailed southwest far enough to see (and claim) the land now known as Brazil.

Next came Amerigo Vespucci, born in Florence in 1451 and employed by Spain, then by Portugal, and finally by Spain again. In 1501 he led a Portuguese expedition that reached Brazil in January 1502 and proceeded south to the River Plate. There is some dispute about the authenticity of Vespucci's papers, but it is likely that he sailed south along Argentina's Patagonian coast. He concluded he had discovered not a new route to Asia but a whole new world. It was to become known as America, a name first used in 1507. Clearly, the world was much larger than Ptolemy had calculated, and this was confirmed when, on September 25, 1513, Spanish explorer Vasco Núñez

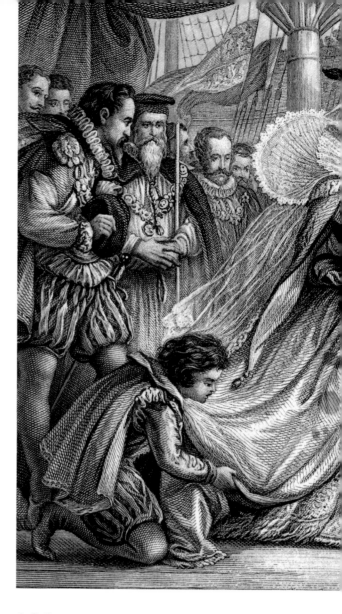

de Balboa crossed the Isthmus of Panama and became the first European to see the vast Pacific Ocean.

Circumnavigation

Born in Portugal in 1480, Ferdinand Magellan traveled east as far as Malacca (in what is now Malaysia) with the Portuguese fleet before seeking service elsewhere. He persuaded the king of Spain to send him on a voyage westward to prove that the Spice Islands—the Moluccas, now part of present-day Indonesia—could be considered part of Spain's western hemisphere of influence. To accomplish this, he needed to find a passage around the southern tip of South America.

Magellan's five ships left Spain in 1519. From Rio de Janeiro, he sailed southward, continually seeking a westward passage and continually being baffled. Finally, he rounded the Cape of the Virgins (Cabo Vírgenes) at 52°50′S and entered the wild strait that now bears his name.

The Strait of Magellan threads a tortuous and treacherous 330 miles (525 km) between the island of Tierra del Fuego and the South American continent. It took Magellan 38 days to reach the ocean that he named the "Pacific," because it appeared to be so peaceful. The fleet discovered the ocean was much larger than they expected: they had 14 weeks of near starvation before struggling into their next landfall at Guam. Magellan was killed on April 27, 1521, in the Philippines, but his surviving ships sailed on to the Moluccas (and, in 1522, returned to Spain), thus proving Magellan's thesis that the Spice Islands could be reached by sailing either east or west, and proving conclusively that the Earth was round.

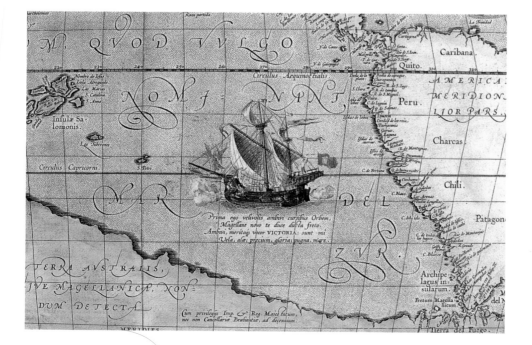

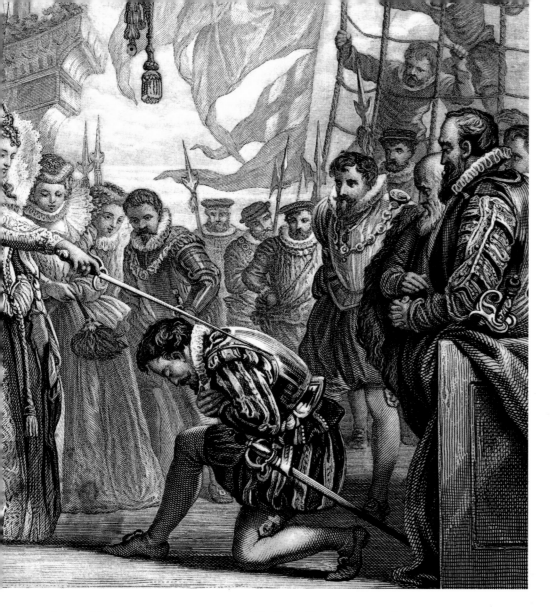

Atop the Cordillera
After pushing through the thick jungle and crossing the rivers and swamps of the Isthmus of Panama, in 1513 Vasco Núñez de Balboa and his Spanish troops (below) gazed upon a mighty body of water that became known as the Pacific Ocean after Magellan's expedition.

Arise, Sir Francis
Francis Drake returned in glory from circumnavigating the globe in 1580—only the second time this feat had been achieved—and was knighted on the deck of the *Golden Hind* by Elizabeth I.

Filling in the Gaps
The 1650 hemisphere map *Polus Antarcticus* by Johannes Jansson was re-engraved from the plate of a 1639 map by Henricus Hondius. Jansson's was one of the first printed maps to incorporate some of the information Abel Tasman brought back from his circumnavigation of Australia and New Zealand in 1642.

Sailing "in that sea"

In 1572, the brilliant English navigator, Francis Drake (*c.*1540–96), led an expedition to raid the Spanish possessions in Panama. There he had his first glimpse of the Pacific and formed the intention "to sail once in an English ship in that sea." In December 1577, he sailed with five vessels, with the goal of mapping the coast of the South Land that was believed to lie below the Pacific Ocean. He reached the entrance to the Strait of Magellan in August 1578. Drake's chaplain, Francis Fletcher, reported that the vessels had sailed through a "crooked strete," with mountainous land on either side "with such tops and spires into the aire and so rare a height as they may well be accounted amongst the wonders of the world …"

When he arrived at the Pacific Ocean in September, Drake did not experience the same benign conditions as Magellan. Rather, an "intolerable tempest" drove the *Golden Hind* southeast as far as 57°S. So, by chance, and traveling backward, Drake discovered the place where the Atlantic and Pacific oceans "meete in most large and free scope," in what is now the Drake Passage below Tierra del Fuego.

The Dutch

The year 1579, when the Low Countries broke away from Spanish rule, marked the birth of a new seafaring nation. The Dutch East India Company, founded in 1602,

was to have a profound influence on exploration and trade for the next 200 years. It became illegal for ships other than the company's to use the Strait of Magellan, so when the wealthy Amsterdam merchant Isaac le Maire organized a private expedition, he decided to seek out the southern seaway of which Drake spoke.

In 1615 his two vessels, the *Eendracht* and the *Hoorn*, commanded by Wilhelm and Jan Schouten, set sail for Patagonia under the overall command of Isaac's son, Jacob le Maire. Fire destroyed the *Hoorn* during the course of the voyage but, on January 29, 1616, the *Eendracht* became the first ship to round Cape Horn, which Wilhelm Schouten named Kaap Hoorn after his birthplace in northern Holland.

By 1642, the Dutch colony of Batavia was well established and ships of the Dutch East India Company had explored some of the west coast of Australia. But no one knew if Australia was part of the legendary Great South Land, so Abel Janszoon Tasman was sent from Batavia with instructions to explore the Indian and Pacific oceans. Over ten months, he reached about 49°S, then proceeded north to discover Tasmania. He continued to New Zealand, which he assumed was the western shore of the southern continent, and returned to Batavia, having circumnavigated the continent of Australia without ever seeing it. The size of the Great South Land had shrunk again.

Over the following century, various expeditions discovered land they thought was the great southern continent but which always turned out to be small sub-Antarctic islands. It was left to James Cook, of Britain's Royal Navy, to undertake the next major exploration of Antarctic waters, in 1772–75. DM

James Cook
1772–75

Cook in the Antarctic

James Cook made the first Antarctic circumnavigation during the summers of 1772–75, crossing into the Antarctic Circle three times. Returning to England, he declared, "no man will ever venture farther than I have done, and … the lands which may lie to the South will never be explored."

W hen the Royal Society asked the British Admiralty in 1768 to commission a voyage to the Pacific Ocean to observe the transit of Venus across the sun, James Cook was their rather surprising choice of leader. Alexander Dalrymple—who optimistically expected the Great South Land to begin south of the tropics and be populated with 50 million inhabitants—had been lobbying to lead a voyage south.

Cook, born in Yorkshire on October 27, 1728, was an experienced seaman and a self-taught navigator and mathematician, but was largely unknown. He sailed from Plymouth in HMB *Endeavour* on August 26, 1768, and returned on July 12, 1771. Having observed the transit, he followed further orders to find Terra Australis Incognita. Cook claimed New Zealand and the east coast of Australia for Britain, and mapped a great deal of the Pacific Ocean.

Cook's first voyage, successful though it was, did not discover the fabled southern continent. The naturalist

Joseph Banks, who accompanied Cook on the voyage, had deduced from his observations: "That a Southern Continent exists I firmly believe," while Cook concluded that "as to a Southern Continent I do not believe any such thing exists unless in a high latitude." But he was determined to resolve the issue once and for all, and with his second venture a question that was more than two millennia old would be answered.

The Second Voyage

In July 1772, Cook, now a commander, set out with HMS *Resolution* and *Adventure* with instructions from the Admiralty for "prosecuting [his] discoveries as near to the South Pole as possible." First, though, he was to look for Cape Circumcision, mapped by French explorer Jean-Baptiste Bouvet de Lozier in 1739 and potentially the tip of the southern continent. The expedition was equipped with provisions for 18 months and the most up-to-date chronometer. Banks had fallen out with Cook and the Admiralty over such matters as accommodation for himself and his party of 17, which included two French horn players; in Banks's place, Cook had the erudite, but demanding and difficult John Reinhold Forster, later described by a Cook biographer as "one of the Admiralty's vast mistakes."

Cook sailed south from Cape Town but could not find Cape Circumcision. Exploring south and east, he realized that it could not have been part of a continent, and wrote dismissively: "I am of the opinion that what M. Bouvet took for land … was nothing but mountains of ice surrounded by field ice." On January 17, 1773, the *Resolution* and the *Adventure* became the first ships to cross the Antarctic Circle. Cook ventured only a short distance inside the Circle before encountering heavy ice and retreating to the northeast, little realizing that he was only 80 miles (130 km) from the Antarctic continent. The summer was over so he turned sail and headed directly to New Zealand.

Cook's two ships had lost touch in a storm off New Zealand's Cape Palliser, but Cook took the *Resolution* south again in November 1773. On December 20, he crossed the Antarctic Circle at about longitude 148°W. He had seen the first iceberg, but continued eastward through heavy ice. He sailed north to explore ocean not covered on his previous voyage, but turned south again at 122°W—to the disappointment of his crew, who thought they were on their way to Cape Horn and home. Instead, they crossed the Circle again on January 26, 1774. Over the next few days, they dodged north and south through thick fog and heavy ice. On January 30, they encountered "field ice" dotted with "Ice Hills or Mountains, many of them vastly large." Cook wrote: "it was indeed my opinion that this ice extends quite to the Pole, or perhaps joins to some land to which it has been fixed since creation." With some uncharacteristic pride, but with typical caution, he went on: "I will not say it

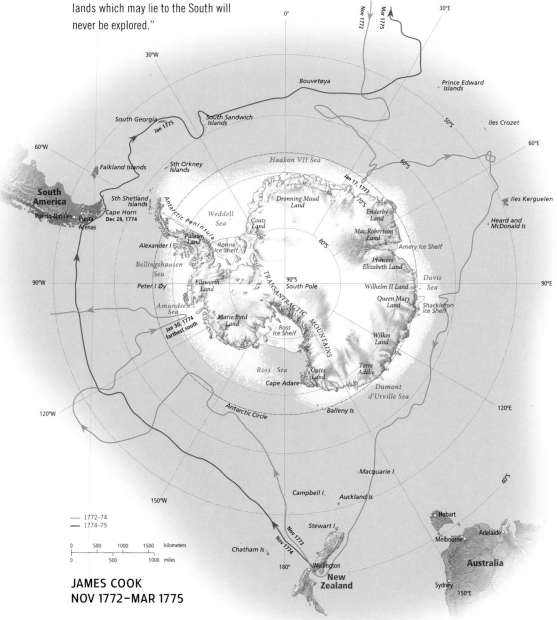

JAMES COOK
NOV 1772–MAR 1775

— 1772–74
— 1774–75

0   500   1000   1500   kilometers
0   500   1000   miles

Timeless Peaks
Heading back to England in January 1775, Cook discovered the island of South Georgia, which he named for the king and claimed for Britain. The rearing, jagged mountain scene has changed little since it was sighted by the crew of the *Resolution*.

Images of Icebergs
The painting *Ice Islands with Ice Blink* (below) is attributed to William Hodges. Some of Hodges's paintings of James Cook's second voyage are displayed in London's National Maritime Museum.

Islands of Ice
William Hodges's paintings provided the first glimpse of Antarctica's picturesque but hostile environment. Here, beside a huge, castellated "ice island," men in the ship's boats collect ice for water and shoot seabirds for food, while the *Resolution* stands by.

was impossible anywhere to get in among this ice, but I will assert that the bare attempting of it would be a very dangerous enterprise … I whose ambition leads me … as far as I think possible for man to go, was not sorry at meeting with this interruption …" He turned north, having reached a latitude of 71°10′S at longitude 106°54′W—an achievement that was not to be equaled again for 50 years, and then not in this most difficult part of the Southern Ocean.

Over the next few days Cook sailed eastward on both sides of the Antarctic Circle before winter (and pack ice) drove him back to warmer regions. Faced

Ne plus ultra

The contesting boasts of two men who were on the *Resolution* when she reached Cook's southernmost latitude on January 30, 1774, must have entertained the crew on the long voyage home. English explorer George Vancouver waited until the ship was ready to tack about, and then climbed to the end of the bowsprit to exclaim, "*Ne plus ultra*" ["no further is possible"]. But Swedish doctor and naturalist Anders Sparrman had a rival claim: "I went below [to his stern cabin] … to watch … the boundless expanses of Polar ice. Thus … I went a trifle farther south than any of the others … because a ship … always has a little stern way before she can make way on a fresh tack."

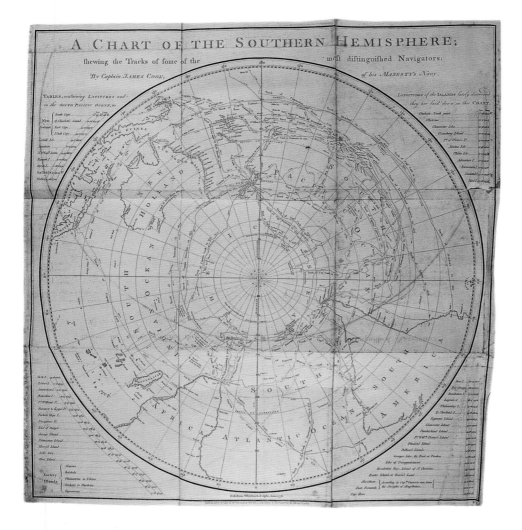

Cook's Second Voyage

George Forster, son of the expedition's naturalist, published this map in 1777. It shows the route of the expedition, and details individual sightings by Cook in the *Resolution* and by Furneaux in the *Adventure*. Cook's second voyage finally established the extent of Terra Australis Incognita—it had to lie south of the Antarctic Circle.

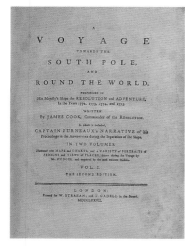

Momentous Work

The narrative of Cook's second voyage, published in 1777, was an epic tale that established the framework for virtually all subsequent exploration of the Antarctic region.

with impenetrable ice, Cook and his crew willingly retreated, weary of the "dangers and hardships, inseparable with the navigation of the Southern Polar regions." He sailed south again from New Zealand in November 1774, heading for the polar spring thaw. For five weeks, the *Resolution* sailed east toward Cape Horn. On December 28, the ship rounded the Horn and continued eastward through the south Atlantic in search of the land reportedly sighted by the London-born merchant Antoine de la Roche in 1675. Cook found it—far from where it had been placed by de la Roche. It had been considered a potential promontory of the southern continent, but at its southwestern tip Cook realized that it was an island. He named the point Cape Disappointment, and the island South Georgia.

From there, Cook followed the 60° latitude through fog, but in a sea clear of ice. On January 27, 1775, he encountered an iceberg, the harbinger of a lot of sea ice. He sighted some islands and, tantalisingly, some mountain peaks, but in the fog and sea ice he could not discern whether they were located on islands or on the point of a large landmass. He optimistically named it Sandwich Land, after the first lord of the Admiralty. Later he wrote of this land and ice: "I firmly believe that there is a tract of land near the Pole, which is the source of most of the ice which is spread over this vast Southern Ocean … I can be bold to say, that no man will ever venture farther than I have done and that the lands which may lie to the South will never be explored. Thick fogs, snow storms, intense cold and every other thing that can render navigation dangerous one has to encounter." He went on: "these difficulties are greatly

heightened by the inexpressible horrid aspect of the country, a country doomed by nature never once to feel the warmth of the sun's rays, but to lie forever buried under everlasting snow and ice."

Cook returned to England on July 30, 1775, after a voyage that had lasted three years and 18 days, "In which time," he noted, "I lost but four men and only one of them by sickness."

The first circumnavigation of Antarctica had been achieved, and the quest for the mythical bounty of the fertile great southern continent was over: it was known that it did not exist. Cook wrote: "I had now made the circuit of the Southern Ocean in a high latitude … in such a manner as to leave not the least room for the possibility of there being a continent, unless near the Pole and out of the reach of navigation …"

The Third Voyage

James Cook's third Pacific voyage once more took him into Antarctic waters. He sailed on the *Resolution* again, this time accompanied by the *Discovery*. The stated

purpose of the expedition was to return a Tahitian native to his home, but the true scope of this voyage was much wider: Cook was instructed to find another geographical myth, the Northwest Passage—the long-sought northern shortcut from Europe to Asia.

He sailed via Cape Town to inspect some islands discovered by the French captain Yves-Joseph de Kerguelen-Trémarec in 1772 in the southern Indian Ocean. Today they are called Iles Kerguelen, but Cook thought Desolation Islands was a more apt name for them. He proceeded through the Pacific Ocean, along the west coast of America, and through the Bering Strait to cross the Arctic Circle.

On the voyage to the north Cook had discovered the Hawaiian islands (which he named the Sandwich Islands), and when the Arctic winter set in he returned there. At first the native Hawaiians were friendly, but on February 14, 1779, at Kealakekua Bay on Hawaii, Cook intervened when one of his boats was stolen. A fight ensued, and he was killed by islanders. It was a tragic end to a remarkable life.

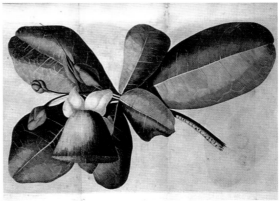

By happy chance, William Wales, astronomer on the *Resolution*, went on to teach at the Mathematical School at Christ's Hospital, London. Samuel Taylor Coleridge was one of his pupils in the 1780s, and the schoolmaster's account of ice and albatrosses perhaps inspired the poet's *The Rime of the Ancient Mariner*, which vividly, and remarkably accurately, evokes an environment he experienced only via hearsay. DM

Resident Year-round
Although king penguins now preen undisturbed on South Georgia (above), Cook's reports of the island's abundant sea mammals unleashed a century of exploitation that virtually exterminated fur and elephant seals, and later whales, from these shores and waters.

Pacific Flower
John Reinhold Forster replaced Joseph Banks as naturalist on James Cook's second voyage. He was assisted by his young son, George Forster, a fine natural history artist who completed 420 botanical and 271 zoological paintings on the voyage.

Thaddeus von Bellingshausen
1819–21

James Cook died in 1779. The world had to wait for 40 years for another explorer to match his achievements. In May 1819, the Estonian-born explorer Thaddeus Thaddevitch von Bellingshausen took command of a Russian expedition to Antarctica. The historian of the Heroic Age of Antarctic exploration Hugh Robert Mill described Bellingshausen's voyage as "one of the greatest Antarctic expeditions," which was "well worthy of being placed beside that of Cook," and a masterly continuation of Cook, supplementing it in every particular, competing with it in none."

Born in 1778 and a naval cadet from the age of ten, Bellingshausen had served as fifth lieutenant on the first Russian voyage around the world, led by Admiral Adam Johann, Baron von Krusenstern, from 1803 to 1806.

For the Russian Antarctic expedition, planned as part of a burgeoning nationalism, Bellingshausen was chosen to command the two ships, the *Vostok* and the smaller and slower *Mirnyi*, with a total of 189 officers and crew members. He was allowed very little time to prepare for this monumental undertaking: indeed, the expedition sailed in late July 1819, only about two months after Bellingshausen was recalled to St. Petersburg from survey work in the Black Sea. The instructions he was issued were succinct: to build on the explorations of Cook, whom the Russians greatly admired, and "to approach as closely as possible to the South Pole, searching for as yet unknown land, and only abandoning the undertaking in the face of insurmountable obstacles." The quest to reach the South Pole had begun.

A Modest Hero

Few visitors realize the significance of the name of the stretch of the southeast Pacific Ocean to the west of the Antarctic Peninsula: the Bellingshausen Sea. Bellingshausen's expedition was often overlooked in Antarctic history because his logs were lost and his unassuming personal record was available only in Russian until 1902, when a German translation appeared; there was no English version until 1945.

On November 20, 1819, his two ships left Rio de Janeiro for South Georgia, where Bellingshausen completed Cook's survey by mapping the island's southern coast. He then followed Cook's route to the South Sandwich Islands and, in better conditions than Cook had experienced, found that they were just more small islands, and that there were more of them than Cook had observed. Continuing eastward, he crossed the Antarctic Circle on January 15, 1820—his was the second expedition to do so—but did not even note this first crossing in his narrative. But while Cook sailed 24 degrees of longitude within the Circle, Bellingshausen was to cover more than 42 degrees.

On January 16, he would have seen the continent if the weather had been fine, but he was sailing through snow when he observed "a solid stretch of ice running from east through south to west." It is likely that he was viewing an ice shelf at the base of the Haakon VII Sea, which may have extended far out to sea at that time. Because geographers regard continental ice as part of the landmass—otherwise the Antarctic Peninsula would be classified as an archipelago—the first sighting of "Antarctica" may be credited to Bellingshausen. Some historians have contended that what he saw was pack ice, but his use of the Russian term for "continental ice" works against that argument.

Land in Sight!

Bellingshausen's information for February 5 is more definite. On a day with good visibility, he wrote: "The ice to the SSW is attached to cliff-like, firmly standing ice: its edges were perpendicular and formed bays, and the surface rose in a slope towards the south, over a distance whose limits we could not see from the cross-trees." He must have been looking at the Lazarev Ice Shelf. But he needed supplies, so he headed north to

Icy Fringe Sighted
Bellingshausen came closer than Cook did to Antarctica. In February 1820, he sighted what was probably the Lazarev Ice Shelf, and discovered Peter I Øy in January 1821.

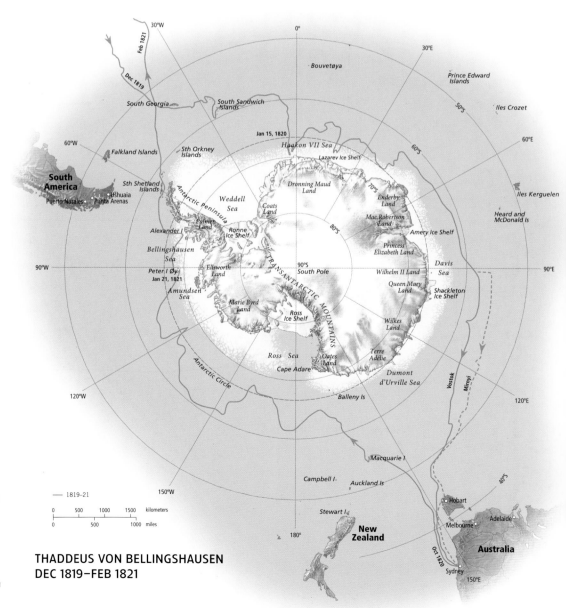

**THADDEUS VON BELLINGSHAUSEN
DEC 1819–FEB 1821**

Grand Harbor

Deception Island in the South Shetlands group is a drowned caldera and one of the finest natural harbors in Antarctica. When Bellingshausen sailed through its narrow entrance (on the far left of this photograph) in January 1821, he found 18 British and American sealing ships at anchor, even though the island had only been discovered the previous year.

Bellingshausen's Ships

Although here they appear identical, the *Mirnyi* was not only a much smaller ship but was also much slower, which frustrated Bellingshausen in the *Vostok*. The creator of this image had probably not been south himself—unlike William Hodges, who painted scenes of Cook's voyages from personal experience.

Icy Impediment

Heavy sea ice near Cape Adare makes navigation in the area difficult. In a tribute to early Antarctic explorers such as Bellingshausen, the experienced twentieth-century sea captain John King Davis noted: "with our auxiliary steam power and specially strengthened hull, [we] had an immense advantage over our gallant predecessors."

Island Profiles

The navigator on Bellingshausen's expedition drew these careful sketches of the islands that make up James Cook's "Sandwich Land." They can be recognized from the same points of view today. Cook and Thule islands are remnants of the rim of a large volcanic crater, and the caldera between them, though open, is of similar formation to Deception Island.

Bristol Island, 7 miles distant

Thule Island   Cook Island, 15 miles distant

Port Jackson (now Sydney). On October 31, he sailed south again to complete his circumnavigation of the South Pole. Like Cook before him and Shackleton to come, he encountered unseasonably heavy ice as he approached the continent. Thus he missed the Ross Sea and could not cross the Antarctic Circle until six weeks later, on December 14.

On January 21, 1821, he reached the southern-most point of his voyage—69°59′S—and discovered Peter I Øy, the first non-ice land ever seen within the Antarctic Circle. He also discovered Alexander I Land (now Alexander Island), stating: "I call this discovery 'land' because its southern extent disappeared beyond the range of our vision." It is in fact a large island, which is separated from the Antarctic continent by a narrow channel but linked to it by ice. Bellingshausen sailed on to the South Shetland Islands, which he mapped, and on Teille Island (now Deception Island) he encountered American and British sealing ships working the island group. Then on January 25, a young American captain,

Nathaniel Palmer, came aboard; Palmer was later to make a dubious claim to have been the first person to see the Antarctic continent.

One of Bellingshausen's ships, the *Vostok*, had been shipping water since leaving Port Jackson, and at the end of January, as winter approached, Bellingshausen decided to turn north to Rio de Janeiro, arriving there on March 27. Over the next month, he overhauled the ships and then sailed back to Kronstadt, via Lisbon, completing the voyage on August 4, 1821. The end of his narrative is typically brief and factual: "We had been absent for 751 days. During that time we had been at anchor in different places 224 days and had been under sail 527 days. Altogether we had covered 57,073½ miles [91,830 km] … During the course of our voyage we had discovered twenty-nine islands: two of these were in the Antarctic, eight in the South Temperate Zone, and nineteen in the Tropics."

He did not mention that only three men had died throughout the voyage—a record James Cook would have admired. And he had achieved the remarkable feat of circumnavigating Antarctica closer to the coast than had Cook—so close, in fact, that he was the first to see emperor penguins, the largest and southernmost-dwelling of all penguins. But the voyage had suggested few commercial possibilities and was largely ignored in Russia; and, because Bellingshausen's charts and logs were not accessible to non-Russian readers, the voyage was overlooked by other nations. Bellingshausen rose to the rank of admiral during the next 30 years of his naval career, and then was appointed governor of Kronstadt, the role he held when he died in 1852. DM

The First Sight and the First Step

The South Shetland Islands, off the west coast of the Antarctic Peninsula, are central to the question of who was the first to "see Antarctica." The South Shetlands were discovered in February 1819 by sealer William Smith, whose report of his find triggered the island sealing boom. The next summer, Smith was employed by the British Admiralty to survey the islands under the command of Edward Bransfield.

The First Sight

On January 30, 1820, Bransfield and Smith saw and charted part of the Antarctic Peninsula, and named it Trinity Land (now Trinity Peninsula). Exactly two weeks earlier, Bellingshausen had probably seen the icy fringe of Antarctica, far to the west. However, he had glimpsed ice and the British duo sighted rocky mountains. In November 1820, Nathaniel Palmer, the captain of the sealer, the *Hero*, sailed through the narrow entrance of Neptune's Bellows and into the spectacular caldera known as Port Foster in Deception Island. He may have been the first to do so; Bransfield and Smith had seen the island on January 29, 1820, but did not investigate in the thick fog. Palmer met Bellingshausen in January 1821, and later reported that "I informed [him] of … the discovery of land … and it was him that named it Palmers Land." However, Palmer's only likely sighting was ten months after Bransfield and Smith.

The First Step

The honor of the first to take a step onto the Antarctic continent probably belongs to the crew of a boat from the *Cecilia*, an American vessel skippered by John Davis. On February 7, 1821, he recorded "… open

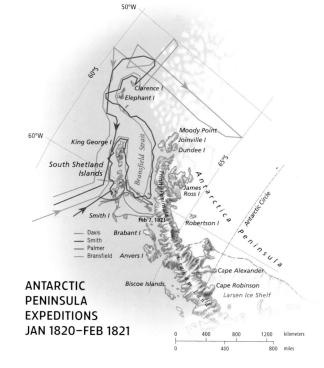

ANTARCTIC PENINSULA EXPEDITIONS JAN 1820–FEB 1821

cloudy weather and light winds a standing for a large body of land in that direction SE at 10 AM close in with our boat and sent her on shore … I think this southern land to be a continent."

The landing probably took place at Hughes Bay (64°13′S, 61°20′W), and significantly pre-dates the other documented claim, that of the Norwegian businessman Henryk Johann Bull, who led a whaling expedition to the Ross Sea region in 1895. Hughes Bay is located in what is now known as the Davis Coast, but the names of the crew who made the historic one-hour landing are unknown.   DM

View of the Peninsula
From Deception Island (foreground), the mountains of the Antarctic Peninsula can be seen across Bransfield Strait. British mariners Smith and Bransfield charted part of the peninsula in January 1820. Ten months later, the American sealer Nathaniel Palmer explored Deception Island, and from there spied these icy peaks.

James Weddell
1822–24

Travelers to Antarctica still speak of James Weddell with awe. Although the Weddell Sea, one of the two great indentations in the Antarctic continent, is a spawning ground of polar ice, in 1823 Weddell sailed farther south than was conceivable at the time. His record stood for 18 years, until James Clark Ross ventured into the waters on the other side of Antarctica. Steam had ousted sail before Weddell's achievement was replicated in the same location.

Weddell was British, but he was born in Ostend, in the Netherlands, in 1787. He joined the Royal Navy at the age of nine, and then alternated between the Royal Navy and the Merchant Navy, before taking command of the 160-ton (145-tonne) sealer, the *Jane*, in 1819. He returned to England in 1821 with enough seal skins to purchase the smaller, 65-ton (59-tonne) *Beaufoy* to partner the *Jane*.

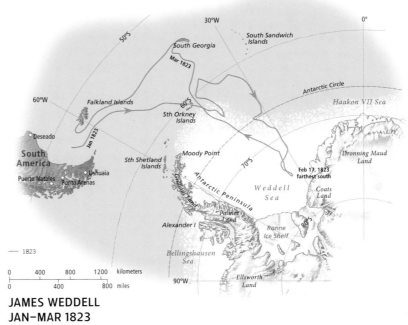

**JAMES WEDDELL
JAN–MAR 1823**

An Inauspicious Start

The voyage that set Weddell's mark on Antarctic history sailed down the River Thames on Friday, September 13, 1822. Directly after the inauspicious date, while still in English waters, there was a collision and the *Jane* was damaged. The vessels sailed to the South Orkneys area but found few seals there—and they were unlike any that Weddell had seen before. (He took the skins and skulls back to Robert Jameson of Edinburgh University, who declared it was a new species: the Weddell seal, *Leptonychotes weddelli*.)

But the purpose of the voyage was seal prospecting: the immediate task was to find more seals, or more land where they might live. Weddell offered a reward of ten pounds—almost as much as an able seaman's annual wage—for the first sighting of land. By February 1823, he had concluded that such a discovery would only be

made to the south. He pushed on, into a prevailing wind through intense cold and a sea jagged with ice. It was a miserable voyage, but Weddell's judgment was partially vindicated on February 16 by a change in the weather: the wind shifted to the west, the ice disappeared, and flocks of seabirds were reflected in the calm sea. The next day they reached 71°34′S, 30°12′W. No ship had ever been so far south.

A Stroke of Luck

In these ideal conditions, Weddell pressed on south. On February 19, he wrote a paragraph that those who know the area still read with wonder: "In the evening we had many whales about the ship, and the sea was literally covered with birds of the blue peterel kind. NOT A PARTICLE OF ICE OF ANY DESCRIPTION WAS TO BE SEEN … had it not been for the reflection that probably we should have

Fearless Sealer

In 1819 James Weddell left the Royal Navy to become a sealing skipper. On his first, very successful, voyage to the South Shetlands, he discovered the South Orkney Islands. In February 1823, seeking new seal-hunting grounds, he pushed farther south than anyone had before him, into the icy sea that bears his name. Weddell's feat was not to be matched until 1912.

Skill and Daring

In uncharacteristically ice-free seas, Weddell's flagship, the brig *Jane*, followed by her smaller companion, the cutter *Beaufoy*, reached their farthest point south on February 20, 1823. Although Weddell made the most of his opportunities, the voyage failed in its primary objective of finding new sealing grounds; nor did he discover new lands.

As Seen by Weddell

Weddell's drawing of a "Sea Leopard" clearly shows the mottled fur, the twinned rear flippers, and horizontal posture of a true seal. Weddell would have been more familiar with the more upright posture and separated rear flippers of the fur seal.

obstacles to contend with in our passage northward, through the ice, our situation might have been envied." The capital letters are his.

On February 20, a rising south wind forced Weddell to make a decision. At noon, the ship was at 74°15′S, 34°16′W, and only three icebergs were visible, but the winter was closing in, and there was ice to be crossed in sailing north. Caution won the day, and Weddell fired the canon, raised the colors, and issued the men an extra allowance of rum. He named his discovery the George IV Sea; in 1900 it was renamed the Weddell Sea.

Weddell was fortunate in his easy passage into the Weddell Sea—many who followed did not fare so well, even in larger, purpose-built ships. His feat was so difficult to replicate that skeptics doubted his veracity,

and he had three of his crew swear under oath that the ship's logs were correct. But James Clark Ross, the next to hold the title of "farthest south," wrote generously that Weddell "was favored by an unusually fine season, and we may rejoice that there was a brave man and daring seaman on the spot to profit by the opportunity."

The *Jane* and the *Beaufoy* reached England in July 1824. In *A Voyage Towards The South Pole*, published in 1825, Weddell wrote: "I have only done that which every man would endeavor to accomplish, who in the pursuit of wealth, is at the same time zealous enough in the cause of science to lose no opportunity of collecting information for the benefit of mankind." The *Jane* had to be scrapped in 1829. Five years later, Weddell died in poverty in London, aged 47. DM

Weddell Seal at Rest

On the South Orkneys, James Weddell discovered a new species of seal, which was named *Leptonychotes weddelli.* The encounter proved more significant to science than to commerce. Compared with the valuable fur seal populations found farther north, Weddell seals have much poorer, thinner coats, which were of little economic worth.

Sealers and Whalers

It has been said that few did more to endanger Antarctic wildlife than James Cook, who, after his second southern voyage, told of islands teeming with seals and the Southern Ocean filled with whales. But Cook was not the first to note this profusion; earlier explorers, including Francis Drake, had also reported it.

Since at least the twelfth century, peoples of the Arctic had used seal and whale oil for lighting and seal furs for clothing, and southern seals had long provided oil and clothing for the Patagonians. The wholesale slaughter of seals and whales by European profit-takers, however, was a direct result of the demands of the Industrial Revolution with its improved technologies.

In Quest of Bounty

European sealing probably began with Spaniard Juan de Solfo, who sailed to South America in 1515; he was killed and eaten by the natives, but his crew took seal skins back to Seville. In the eighteenth century, Chinese furriers discovered how to remove the coarse outer hair of seal pelts, leaving the soft inner fur intact. Sealing soon became a maritime gold rush, beginning in 1764 on the Falkland Islands, and rapidly spreading across to South Georgia and the coast of South America.

In the late seventeenth century, piratical English navigator William Dampier called at the Juan Fernandez Islands off the coast of Chile and wrote: "seals swarm around … there is not a bay nor rock … but it is full of them." In 1797, throngs of sealers arrived, and by 1807 few seals remained. A single vessel took off 100,000 skins, and the total harvest was estimated at 3 million. Forty thousand skins were shipped to London from the Falkland Islands in 1788, the year when the first British sealers reached South Georgia. In 1800, a New York sealer took 57,000 skins from the island, and in 1825 James Weddell calculated that 1.2 million fur seals had been killed on South Georgia. Within this short period, more than 120,000 seal skins had also been plundered from Macquarie Island.

Biscoe and the Enderby Brothers

As their quarry approached extinction, sealers sought new hunting grounds. During those voyages, important discoveries were made. Nationalism had instigated the circumnavigations of Antarctica by James Cook and Thaddeus von Bellingshausen. However, the third was an uneasy mixture of exploration and sealing.

John Biscoe, at 36, was an ex-Royal Navy seaman employed by the Enderby Brothers, the firm that had sailed into Boston Harbor with a cargo of tea in 1773, and thus triggered the "Boston Tea Party" and the American War of Independence. The company needed new ventures, so it switched to whaling and sealing. The Enderby Brothers tradition combined exploration with exploitation, and captains with education and naval experience were recruited to collect flora and fauna. This was a policy that eventually bankrupted the company, but it left an enduring legacy of Antarctic exploration. Biscoe was to seek out new sealing grounds, but also to make discoveries in high southern latitudes. His vessels, the *Tula* and the much smaller *Lively*, were sparsely equipped and provisioned, with a crew of 29. They sailed from England in July 1830 and, often at an agonizingly slow 3 knots or less, crossed the Antarctic Circle in January 1831. On January 28, they arrived at their farthest point south: 69°S at 10°43′E.

Biscoe ventured even farther south than Thaddeus von Bellingshausen, and in one way he had better luck. He definitely first sighted land on February 28, 1831, at 66°S, 47°20′E, when "several hummocks" had resolved themselves into "the black tops of mountains … through the snow." After two days of trying to land, Biscoe gave up. He named his discovery Enderby Land. On March 3, his ships were separated by a storm that drove the *Tula* north. The indefatigable Biscoe turned the damaged *Tula* south again, hoping that the ice had blown away from the shore. It had not, and with some crew injured and others sick with scurvy, he continued on north. When the vessel reached Hobart, only Biscoe and four others could stand, and two had died.

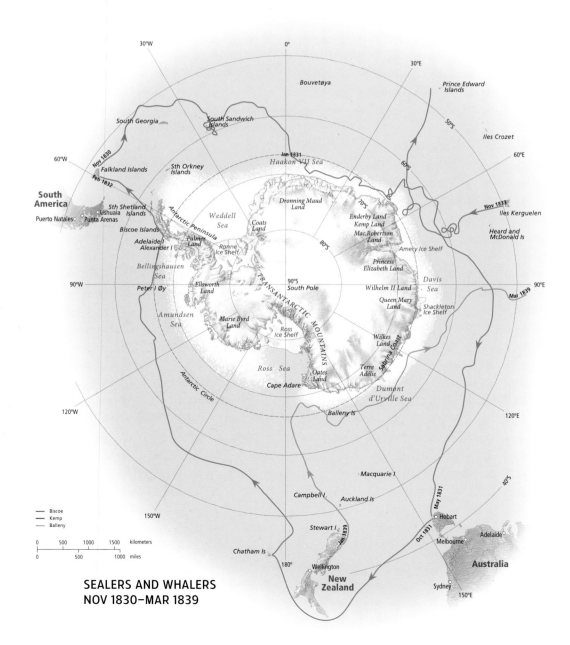

SEALERS AND WHALERS
NOV 1830–MAR 1839

No Escape

Fur seals are agile rock climbers, but sealers pursued them remorselessly. The bodies were quickly stripped and the skins thrown down to the waiting boat, then taken to the ship, where they were salted to preserve them. This lithograph shows a gang at work on Beauchene Island, in the Falkland Islands, in the 1830s.

On September 3, Biscoe and the *Tula* headed to the south again, but as they left port they met the *Lively*. The ship's crew had a harrowing tale to tell. Seven of her crew of ten were dead, and the survivors had been marooned in Port Phillip Bay (now Melbourne) when the ship had drifted out to sea while they were foraging ashore; it had taken weeks to recover and refloat her.

The two ships headed southeast in October. After some unrewarding sealing in New Zealand waters, they continued slightly south of east to cross Cook's path, and discovered an island at 67°15'S, 68°30'W; Biscoe named it Adelaide Island for England's queen. In late February, they arrived in the South Shetland Islands, having circumnavigated Antarctica and finally established there was a large continent at the heart of the ice. Biscoe wrote that "I am firmly of the opinion that this is a large continent as I saw to an extent of 300 miles [500 km]."

Biscoe needed to find seals before winter set in, but he failed, and the *Tula*'s rudder was damaged in a late storm. He retreated to the Falkland Islands, where the *Lively* was wrecked. He wanted another season in Patagonia to make the voyage a financial success, but one by one his hard-pressed crew deserted.

On February 8, 1833, the bedraggled *Tula* sailed into London with just 30 seal skins to show for a voyage of 30 months. Biscoe received a gold medal from the Royal Geographical Society, went on to skipper craft in the West Indies and around Australia, and sailed south again as far as 63°S on a whaling venture in 1838. In 1842, a public appeal in Tasmania raised money to send him back to England, but he died during the voyage.

William Dampier

In 1684, the buccaneer–naturalist William Dampier spent time in the Juan Fernandez Islands watching the seals and sea lions. He wrote best-selling books about his voyages and about his scientific observations, books which were highly valuable to later explorers.

Return from the Brink

This fur seal pup at Cooper Bay, South Georgia, represents the gradual revival of fur seals after their virtual extinction in the early nineteenth century. James Weddell calculated that by 1825 about 1.2 million fur seals on South Georgia had been slaughtered.

Profitable Work

This watercolor by Augustus Earle shows a sealer on one of the islands of Tristan da Cunha, stripping the skin from an elephant seal in the process known as flinching, or flensing. The tools of the sealer's trade were a sharp spear and a whetted knife.

Enduring Legacy

In 1833, the *Magnet*, under the command of British sealer Peter Kemp, left London for Antarctica in quest of seals. Little is known of Kemp—even the date of his birth is uncertain—but he went to Iles Kerguelen and on to the Antarctic Circle, where he saw land to the south at about 58°E on Boxing Day. He called it Kemp Land, and then turned north for more sealing at Iles Kerguelen. He fell overboard and drowned on the return voyage. The Kemp Coast, as it is now known, was not seen again until Australian explorer Douglas Mawson visited it almost a century later.

Even less is known about the life of another British sealer and whaler, John Balleny, but his name lives on in a group of five islands rising sheer from the water and lying across the Antarctic Circle south of New Zealand. They were named for him by the British Admiralty, in a tribute to a significant voyage.

In July 1838, the *Eliza Scott*, under Balleny, and the much smaller *Sabrina*, under Thomas Freeman, sailed from England via the Cape of Good Hope and Iles Amsterdam to New Zealand. On February 1, 1839, they reached 69°02′S at 172°11′E; there the way south was blocked by ice. But it showed that there was a possible route along this longitude: the way into the Ross Sea had opened a crack. On February 9, Balleny saw what are now the islands that bear his name, and he briefly landed there on February 12, taking a sample of rock

from an iceberg; this was the first landing below the Antarctic Circle. He followed a course some 300 miles (500 km) south of Cook's voyage on the *Resolution*, and on March 2 he saw land to the south, which he named the Sabrina Coast. It became a memorial to his ship: in a violent storm on March 24, a blue distress flare was the last that was seen of the *Sabrina* and her crew. The *Eliza Scott* returned to England via Madagascar and St. Helena, bearing just 200 seal skins. Balleny's log and charts were paraded before the Royal Geographical Society and were useful to James Clark Ross, but the man himself faded into obscurity.

While early sealing brought wildlife populations to the verge of extinction, early whaling was less successful. The whaling industry's time for the mass destruction of species other than southern right whales had to wait for the invention of the explosive harpoon head and the factory ship in the mid-nineteenth century. But in the course of their bloody trade, the early whalers and sealers, hardy men who pursued unexplored lands as enthusiastically as they hunted quarry, left an enduring legacy in Antarctic exploration. DM

Spoils for Oil

Weddell reported that at the time of his visit to South Georgia, the "sea-elephants were nearly extinct, not less than 20,000 tons [18,145 tonnes] of elephant-oil having been shipped to the London markets alone ... since the reports by Captain Cook."

Peril Versus Profit

This scene, painted in about 1820 by Thomas Buttersworth, vividly shows the dangers of hunting whales from rowing boats in iceberg-strewn waters. These primitive techniques limited the whalers to hunting only right whales, which floated when killed and so could be hauled back to the ship.

Prey No Longer

Even the aggressive appearance of male southern elephant seals failed to protect the species from being savagely hunted to near extinction. Now the species is protected, and any approach by humans is for peaceable scientific study only.

Jules-Sébastien Dumont d'Urville
1837-40

Respected Leader
Dumont d'Urville's voyage to the Southern Ocean was his third major expedition for the French Navy. It was on this voyage that he made his most famous Antarctic discoveries, which earned him promotion to rear admiral. His efforts led to the French claim to a slice of Antarctica.

It is due to Jules-Sébastien Dumont d'Urville that there is a small French territorial claim amidst the giant Australian claim to Antarctica. It is called Terre Adélie after his wife. The Adélie penguin (*Pygoscelis adeliae*) is named for the land he discovered.

Dumont d'Urville was born on May 25, 1790, in Normandy, to an aristocratic but poor family. He was an entomologist, a classicist, a botanist, a brilliant linguist, and a respected leader, although somewhat aloof. Dumont d'Urville joined the French Navy in 1807, was second in command of a Pacific voyage in 1822, and leader of another in 1826. Both those voyages were on the same vessel: for the second, the *Coquille* was renamed the *Astrolabe*. He returned to the Pacific in 1829, and wrote a famous account of the voyage.

When Dumont d'Urville proposed another Pacific voyage in 1835 he was 45—old by Antarctic leadership standards. His request was granted. He was ordered to go via the Southern Ocean, with ambitious instructions to extend his explorations "towards the Pole as far as the polar ice will permit." There was to be a bonus paid to the crew if they reached 75°S—farther south than Weddell's British voyage of 1823. Dumont d'Urville took two ships, the *Astrolabe* and the smaller *Zélée*, with a complement of 183 officers and crew. He sailed from Toulon on September 7, 1837, and was off the coast of Tierra del Fuego by January 1838.

Ice and Storms
Unhappily for Dumont d'Urville, and for his crew, they struck unseasonably heavy sea ice at 59°30′S and had to dodge icebergs at 62°S. On January 21, they tried to find a way through the pack ice, but by the twenty-fifth they had been forced to retreat northward toward the South Orkney Islands. A storm stopped them landing for fresh supplies of seal and penguin meat, and they sailed south again. When they encountered pack ice in early February, Dumont d'Urville bravely elected to push into it. They stopped in a stretch of open water and the crew celebrated, but Dumont d'Urville, who had been ill for weeks, wrote: "I went to bed and could hear their rowdy celebrations … The rashness and imprudence of this move unfolded before my eyes … The only way out was by the way we had come in … was there not reason to fear … ice would completely block our exit?"

He was right. Before midnight they were ice-bound, and it took five days of hard labor to break into the open water. They then continued westward, surveying and mapping around Graham Land and the South Shetlands before the ships retreated to Chile. Dumont d'Urville wrote: "Either Weddell struck an exceptionally favorable season or he played on the credulity of his readers … As soon as just one other captain has penetrated only five or six degrees farther than us in the same area will my doubts be allayed and Weddell right in my eyes."

Another Try
When they reached Valparaíso on May 25, two of the crew were dead and seven others deserted. But a week later, they set out across the Pacific, landing at many islands along the way, and arrived in Australia in March 1839. Early in 1840, after further Pacific exploration, they sailed south from Hobart. They found tabular icebergs at 64°S on January 18. Their erratic compass suggested that they were close to the South Magnetic Pole. The next afternoon an officer climbed the mast and reported land ahead. After a frustrating few days becalmed, the ships reached the land on January 21 and launched a boat from each ship. On a tiny islet just off the mainland and barely within the Antarctic Circle, the excited boat crews landed, unfurled the French tricolor, and collected some granite chips and a few unlucky penguins.

Dumont d'Urville named the area Terre Adélie, "to perpetuate … my deep and lasting gratitude to my devoted wife." An officer reported they drank a bottle of Bordeaux wine that someone had brought along "to the glory of France." They continued westward along the coast for another couple of days, but a bad storm blew up and both ships were nearly lost.

JULES-SÉBASTIEN DUMONT D'URVILLE JAN 1838-FEB 1840

Tantalizing Horizon

Despite infuriatingly calm weather that slowed their approach, on January 20, 1840, Dumont d'Urville's ships came in sight of a high, snow-covered land stretching from southeast to northwest. The next day they landed on a small islet just offshore of the continent itself, before heading north again.

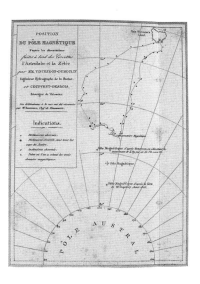

Plotting the Pole

Dumont d'Urville's journals about the expedition were published in 1841 and included maps of his coastal sightings of Terre Adélie. This inset detail shows the expedition's plotting of the South Magnetic Pole.

Something to Celebrate

On January 19, 1840, Dumont d'Urville's expedition finally crossed the Antarctic Circle. To mark the occasion, the crew of the *Astrolabe* paraded around the decks in bizarre costumes (including a penguin with an outsize beak, center right), and "Father Antarctic" visited to welcome Dumont d'Urville to his frozen domain. Similar rituals are still enacted when novices cross the Circle.

It was time to return to Hobart, but there was one more unexpected adventure. On January 29, a ship loomed out of the fog in the east—it was the *Porpoise* from the American expedition led by Charles Wilkes. The astonished French maneuvered to meet the ship, which changed course and departed without contact. Later, the secretive Wilkes claimed that the Frenchman had turned and run; Dumont d'Urville replied: "I would have been glad to give our co-explorers the results of our researches … it seems the Americans were far from sharing those feelings."

Dumont d'Urville sailed north on February 1, reached Tasmania on February 18, and returned to Toulon on November 6, after exploring more of the south seas. The voyage had aged him far beyond his 50 years, but his efforts were rewarded. He was promoted to rear admiral and given the gold medallion of the Société de Géographie—its highest honor. But he had little time to enjoy his fame. In May 1842 he, his wife, and his son perished in a train accident—a tragic end to a family whose names would forever be associated with the frozen south. DM

Charles Wilkes
1838–42

Charles Wilkes

Charles Wilkes (1798–1877) led the first United States expedition to Antarctica. Six ships set off in 1838, but only two survived the expedition. None reached the continent, although lookouts on three of the ships reported sightings. Wilkes's claim to have seen land just three days before Dumont d'Urville is regarded as highly suspect.

The United States Exploring Expedition, nicknamed Ex Ex, left Chesapeake Bay on August 18, 1838, commanded by the determined but irascible and largely inexperienced 40-year-old lieutenant, Charles Wilkes. The fleet was led by the *Vincennes*, a 860-ton (780-tonne) war sloop and the first United States Navy ship to circumnavigate the globe; the others were the *Peacock*, *Porpoise*, *Sea Gull*, *Flying Fish* (the baby of the fleet, a 106-ton [96-tonne] New York pilot boat), and the *Relief*, a capricious supply ship. More than 400 men were on board, including nine scientists. At Orange Harbor, Tierra del Fuego, Ex Ex split into three groups: the *Vincennes* and the *Relief* were to survey the southern tip of South America; Wilkes, on the *Porpoise* with the *Sea Gull*, would try to outdo James Weddell; the *Peacock* and the *Flying Fish* would sail southwest in an attempt to pass Cook's southernmost point.

Perilous Ventures

Winter was approaching—it was already February 26, 1839—and Wilkes gave up a mere ten days later, well short of the Antarctic Circle, defeated by impenetrable ice and inadequate clothing. On their retreat to South America, the *Peacock* and the *Flying Fish* were driven apart by a gale. The valiant little *Flying Fish* had a

particularly bad time; she disintegrated further in every storm, and the sailors were constantly wet. When the thermometers broke, they suspended a tin cup of water and resolved to travel south until it froze. This they did, and on March 22 they reached 70°14′S—still a degree short of Cook—before turning north.

A depleted expedition regrouped in Valparaíso at the end of June: the *Sea Gull* had been lost in a storm off Chile, and the cumbersome *Relief* was sent home. The remaining ships sailed to the Pacific. Scientifically, this leg was more successful than the Ex Ex's first Antarctic venture, but the relationship between Wilkes and his scientists and officers deteriorated. In Sydney, in December 1839, officials were aghast at the terrible condition of the ships and their lack of equipment, heating, and waterproofing for their next sojourn in the icy south. The question of Wilkes's sanity—or at least his judgment—also became a matter of widespread concern. When Wilkes offered the naturalists on the expedition the chance to stay behind and meet the ship in New Zealand in March 1840, they all accepted.

Claims and Counterclaims

For his second Antarctic exploration, Wilkes had been ordered to sail south from Tasmania as far as he could, to explore the continent westward to longitude 45°E, and then to proceed to Iles Kerguelen. He substantially modified these instructions, sailing only as far as 105°E and then heading for New Zealand's Bay of Islands. He also entreated his captains to "avoid a separation as the lives of … the squadron may be jeopard[iz]ed by it." Within ten days, the *Flying Fish* and then the *Peacock* were separated from the others and from each other. The *Flying Fish* reached the ice but wisely turned toward New Zealand in early February. On January 11, 1840, the *Porpoise* and the *Vincennes* reached the "icy barrier" at 64°11′S, 164°30′E, and, proceeding to the west, fortuitously met up with the *Peacock* five days later. That same day, the lookouts on all three ships reported seeing land—but nothing was entered in any

CHARLES WILKES
FEB 1839–MAR 1840

Iceberg Landing
Wilkes claimed several sightings of the Antarctic coast during January and February 1840, but the closest he came to a landing was on this large iceberg, where the crew enjoyed themselves as they collected ice for water. The snowy hills behind may or may not have been the continent—many areas Wilkes marked on his chart as "land" were sailed over by later explorers.

of the ships' logs. Visibility was good, but it remains doubtful whether they were seeing snowy land or a glacier, or merely trapped icebergs. Wilkes later wrote in his narrative for January 19 that he had been "fully satisfied that it was certainly land," and named it Cape Hudson after the skipper of the *Peacock*. His chart showed land 40 miles (65 km) away, but later voyages established that the closest land lies 120 miles (190 km) from his position that day. The first sighting of land that was entered in the ship's log was on January 28.

On January 24, the *Peacock* damaged its rudder in ice and limped back to Sydney. The *Porpoise* and the *Vincennes* pressed on to the west. Six days later, *Porpoise* sighted Dumont d'Urville's *Astrolabe* through the fog. According to the commander of the *Porpoise*, Cadwalader Ringgold, he intended to pass under the stern of the other vessel but when he saw it putting on more sail, "without a moment's delay, I hauled down my colors and bore up on my course before the wind." The *Porpoise* continued sailing westward before setting course for New Zealand on February 14.

The *Vincennes* continued its coastal survey, skirting the edge of the pack ice under an increasing cloud of paranoia. Wilkes wrote resentfully: "I cannot help feeling how disgusting it is to be with such a set of officers (one or two I must except) who are endeavoring to do all in their powers to make my exertions go for nothing."

On February 21, the *Vincennes* reached what is now known as the Shackleton Ice Shelf, which extends almost 100 miles (160 km) out to sea. Wilkes sailed for Sydney to report his success. He had followed the pack ice for more than 1,700 miles (2,700 km). He had also seen enough to confidently proclaim that the landmass was a continent—and he was the first explorer to do so.

Wilkes was certain he was about to announce the first discovery by a national expedition, rather than by whalers, of an Antarctic landmass south of the Pacific and Indian oceans. But word arrived from Hobart that Dumont d'Urville had seen land on the afternoon of January 19. Wilkes countered that he had seen land

on the morning of the same day, and later that he and others had first seen land on January 16. Wilkes's cause was not helped when James Clark Ross returned from Antarctica to report that he had sailed across an area marked as solid land on a chart Wilkes had given him.

Home at Last
It was another two years before the expedition returned home. The *Vincennes* docked in New York on June 10, 1842, and by July 25 Wilkes was facing a court martial. He was found guilty on just one charge—that of illegal punishment of seamen by ordering more than 12 strokes of the lash—and he was reprimanded. Although this did not seem to hinder Wilkes's naval career, his talent for controversy persisted: in 1861, during the American Civil War, he boarded a British mail ship in order to remove some Confederate officers and nearly dragged Britain into the war. In 1866, Charles Wilkes retired from the United States Navy with the rank of rear admiral, and he died in February 1877. DM

Providential Escape
In the first foray into Antarctic waters, one ship sank, a second retired hurt, and in March 1839, Wilkes's own ship, the *Porpoise*, narrowly avoided being wrecked on the fog-shrouded shores of Elephant Island while returning from the Weddell Sea. The *Porpoise* rejoined the remaining ships at Valparaíso. The much depleted fleet then sailed via Sydney to the other side of Antarctica.

James Clark Ross
1839–43

When James Clark Ross sailed from England on October 5, 1839, both Charles Wilkes and Jean-Sébastien Dumont d'Urville had already ventured into the Antarctic region and had retreated to the north to wait out the polar winter. There was fierce competition among the Antarctic expeditions of the late 1830s. Ross was spurred on by patriotism and personal ambition to surpass others' achievements.

Ross was already an experienced polar navigator. Born on April 15, 1800, he had joined the Royal Navy before his twelfth birthday. His most significant Arctic achievement was locating the North Magnetic Pole in 1831. This—and his Byronic appearance—had won him great popular acclaim in Britain. He was described by a fellow officer as "the finest officer I have met with … He is perfectly idolized by everyone." When British scientists pressed for an expedition to find the South Magnetic Pole—the "Magnetic Crusade"—James Clark Ross was the obvious choice to lead it.

His ships were the 410-ton (372-tonne) *Erebus* and the smaller *Terror*—bomb ships, built to withstand the recoil of mortars. These bleak names were not reflected in conditions on board; Ross ensured that they were

waterproof, warm, and well provisioned. The *Erebus* was under Ross's command and the *Terror*'s captain was Francis Crozier, who had sailed the Arctic with Ross. Also on board was 22-year-old naturalist Joseph Hooker, who was to bring London's Kew Gardens to their full Victorian glory. Just before the ships departed, the sealer and whaler John Balleny brought news of land below the Antarctic Circle and a possible break in the pack ice along the 170° meridian.

Arriving in Hobart in August 1840, Ross received permission from the governor of Tasmania to build the observatory "Rossbank" on a hill above the town. But there was bad news as well as good: Ross's national pride was pricked to learn that Dumont d'Urville and Wilkes had both spent their second Antarctic seasons seeking the Magnetic Pole. Ross set sail southward on November 12 and crossed the Antarctic Circle on New Year's Day, 1841, near the area that John Balleny had suggested. He pushed south, sometimes through heavy ice, to break into the open sea on January 9. He had discovered the Ross Sea, the best ocean access to the geographic South Pole. Ross's was the first expedition equipped with vessels suitable for ice work—but they

Polar Experience
James Clark Ross had already made six expeditions to the Arctic, and had located the North Magnetic Pole, when he sailed south in 1839 in search of its southern equivalent.

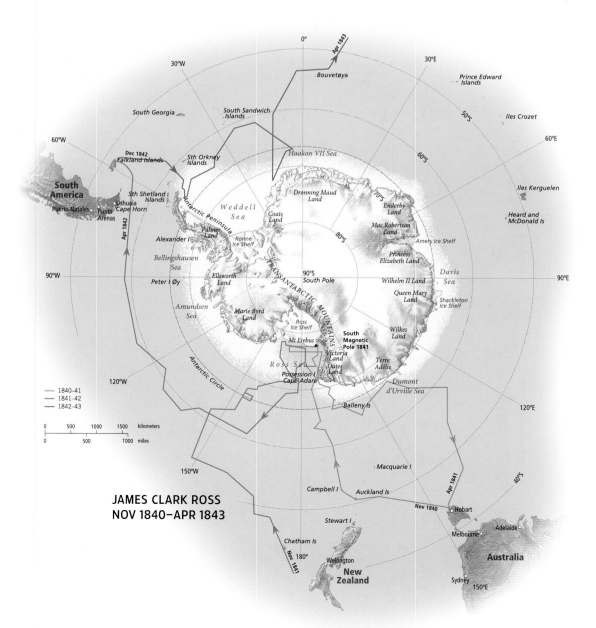

JAMES CLARK ROSS
NOV 1840–APR 1843

— 1840–41
— 1841–42
— 1842–43

were still sailing ships. Decades later, the Norwegian explorer Roald Amundsen noted that: "These men sailed right into the heart of the pack … It is not merely difficult to grasp this; it is simply impossible—to us, who with a motion of the hand can set the screw going, and wriggle out of the first difficulty we encounter. These men were heroes."

Ross was orientated toward the quest for the South Magnetic Pole and was disappointed to encounter land instead of open water. He calculated that the Magnetic Pole was only 500 miles (800 km) away, but clearly there was no direct sea route. But the discovery of new land was worthy of ceremony, so on January 12, 1841, Ross and Crozier landed on an island they later named

Possession (71°52′S, 171°12′E) and claimed and named Victoria Land for the young Queen Victoria.

Fire and Ice

Proceeding southeast, with the fading hope of finding a seaway to the Magnetic Pole, they passed Cook's farthest point south on January 11 and Weddell's 74°15′S on January 22. When they reached 74°20′S, they celebrated with a double ration of rum. Six days later, they sighted what Hooker described as "a fine volcano spouting fire and smoke." Ross named it Mount Erebus and its smaller neighbor Mount Terror. Seeing an active volcano 12,450 ft (3,795 m) high and covered in snow was remarkable enough. As they approached

Forged in Fire
A volcano spurting a "dark cloud of smoke, tinged with flame" was astonishing in this land of ice and snow. Ross named it Erebus for his flagship, and called its smaller neighbor Mount Terror.

No Way Through
As soon as he sighted the Ross Ice Shelf, which he named the Victoria Barrier, Ross realized that "we might with equal chance of success try to sail through the Cliffs of Dover, as to penetrate such a mass." He described these icy white cliffs as "extending from its western extreme point as far as the eye could discern to the eastward."

a low white line filled the horizon. Ross wrote that it was "a perpendicular cliff of ice between one hundred and fifty feet and two hundred feet [45–60 m] above … the sea, perfectly flat and level on top, and without any fissures or promontories on even its seaward face." Ross marked it on his chart as a "Barrier," as it killed their last hope of sailing to the Magnetic Pole; it is now the Ross Ice Shelf. Sailing 390 nautical miles (720 km) eastward, they finally found a bay in the Barrier that a crewman described as the "most rare and magnificent sight that ever the human eye witnessed."

Ross's party could congratulate themselves. They had collected rocks and plants, taken the first Antarctic sea soundings (and had established that the edge of the Barrier ice was afloat), found the Ross Sea, and, by rigorous measurement, established that the South Magnetic Pole was much farther south than predicted. The two ships tied up by the Rossbank observatory in Tasmania on April 6, 1841.

Southward Again

The voyage south the following summer is better known for its tribulations and celebrations than for any ground-breaking achievements. The two ships sailed via Sydney and New Zealand's Bay of Islands and crossed the Polar Front on December 13, 1841. Coincidentally, they again crossed the Antarctic Circle on New Year's Day, but this

time the ships were locked in pack ice. An ice floe between the two ships became a dance floor. Ross dressed up as a woman, and, as one officer later wrote to his sister, "Captain Crozier and Miss Ross opened the ball with a quadrille … Ices and refreshments were handed round … You would have laughed to see the whole of us, with thick overall boots on, dancing, waltzing and slipping about … Ladies fainting with cigars in their mouths—to cure which the gentlemen would politely thrust a piece of ice down her back … a 'lady' burnt the back of my hand with a cigar."

Forcing Through Ice
The *Erebus* and the *Terror* battled through the icebergs to reach the Barrier at 78°10'S on February 23, 1842—a farthest south record that would not be surpassed for 50 years. Both ships were lost during the search for the Northwest Passage in 1845–48.

Claimed for the Queen
Ross sighted the coast of Victoria Land soon after entering the open sea now named for him. He landed on an offshore island, and there, ankle-deep in guano from its raucous rookery, took possession of all he could see. The most prominent mainland peak Ross named for Captain Edward Sabine, the leading authority on terrestrial magnetism.

But this was the last light moment for weeks. The ships remained trapped as the ice drifted north, and on January 19, 1842, a gale ground the ice against the ships, smashing their rudders. Nevertheless, on February 23, they were at the Barrier at 78°10′S—their farthest south, a record that would not be surpassed for 50 years. They planned to winter on the Falkland Islands, and the next day set out on the long voyage around Cape Horn.

At first they saw few icebergs, but some emerged through the fog and snow of March 12. The conditions worsened around midnight. Ross ordered the *Erebus* to heave to for the night. But even as sails were being furled, an iceberg loomed dead ahead. Ross described how "the ship was immediately hauled to the wind on the port tack ... But just at this moment the *Terror* was observed running down upon us, under her top-sails and foresail; and as it was impossible for her to clear both the berg and the *Erebus*, collision was inevitable ... Our bowsprit, foretopmast, and other smaller spars, were carried away, and the ships hanging together, entangled by their rigging, and dashing against each other with fearful violence, were falling down upon the weather face of the lofty berg under our lee, against which the waves were breaking and foaming to near the summit of its perpendicular cliffs." Half-naked seamen roused from sleep struggled desperately, while Ross, according to various officers, "calmly gave the order to loose the sail ... as if he were steering into any harbor."

In two strokes of luck, the ships parted and Crozier found a gap through what had seemed a single berg. In its lee, he was able to inspect the damage to the *Terror*. The *Erebus* was almost completely incapacitated and destruction against the iceberg seemed inevitable. Their only chance was the intricate maneuver "stern board," in which a square rigger sails backward—this in a storm with the rigging and sails tied in knots. Few could hear Ross's calm commands in the chaos, but his do-or-die tactic was successful and the *Erebus* joined the *Terror* in the shelter of the iceberg. Amazingly, two days later, the ships were able to continue north to the Falklands, arriving on April 6.

The Final Venture
Christmas 1842 was the ships' third one surrounded by ice as they approached the Antarctic Peninsula on their last southward voyage. Ross wanted to extend Dumont d'Urville's exploration and perhaps follow Weddell's southward path, although he knew that no discovery could be expected along that track. Apart from some surveying around the eastern side of the tip of the Antarctic Peninsula, their main discovery was ice in all its permutations, cutting every access to the south. After six frustrating weeks, finally forced to give up, they resumed an eastward course.

At 40°W, the point where Weddell had found open water, they found solid pack ice. However, the farther east they sailed, the farther south lay the edge of the pack ice. They crossed the Antarctic Circle for the third time to reach 71°30′S, 14°51′W, on March 5, 1843.

By March 11, they were north of the Circle, and on September 2 they saw England again after almost four-and-a-half years. DM

Sunlit Glacier
Ross sailed along the "Great Ice Barrier" for two successive years, unable to find a landing place. The Barne Glacier, which descends from the western slopes of Erebus, was named by Scott's expedition.

The *Challenger* Expedition
1872–76

Wyville Thomson
Professor of natural history at the University of Edinburgh, Thomson (1830–82) encouraged the Royal Society to fund a global expedition to undertake oceanographic research. He sailed as chief scientist.

For many years, scientists believed that the deep oceans were "lightless and lifeless" but, between 1868 and 1870, dredging off the Scottish coast disproved this belief, and the science of oceanography was born. The first voyage to explore the possibilities of the fledgling science was effectively a joint venture between Britain's Royal Navy and the Royal Society. HMS *Challenger* sailed from Sheerness, Scotland, under the command of George Nares (1831–1915) on December 7, 1872, and returned to Portsmouth three-and-a-half years later, on May 24, 1876, with such a wealth of scientific data that it took from 1885 to 1895 to compile it into 50 volumes: they contained some 30,000 pages of information, which would take decades to analyze.

The expedition's goals were global and ambitious. Among a number of other tasks, it was to "investigate the physical conditions of the Deep Sea, in the great Ocean basins … in regard to Depth, Temperature, Circulation, Specific Gravity, and Penetration of Light."

The *Challenger*, a corvette with a displacement of 2,593 tons (2,352 tonnes), had been adapted for survey and scientific purposes, and carried six civilian scientists, as well as 20 officers and 200 crew. The southward voyage was remarkably slow because they established the painstaking and time-consuming procedures for scientific dredging and sampling.

On February 16, 1874, the *Challenger* became the first steamship to cross the Antarctic Circle, and she went on to reach 66°40′S at 78°30′E. However, the much-quoted steam-powered vessel claim is rather misleading, since the greater part of the long voyage, both within and beyond the Antarctic Circle, was conducted under sail. As sub-lieutenant Lord George Campbell wrote, steam was used only "in a prolonged calm when, if we had coal to spare, we steamed slowly on our way." The *Challenger* was not ice-strengthened, and even using a combination of sail and screw, the ship barely survived her brief time among the icebergs.

Nevertheless, whether she was powered by the 16,000 sq ft (1,485 sq m) of sails on the three masts or by the 1,200-horsepower engine, the *Challenger* completed a remarkable scientific voyage. She sailed across the Atlantic to Brazil, then returned eastward,

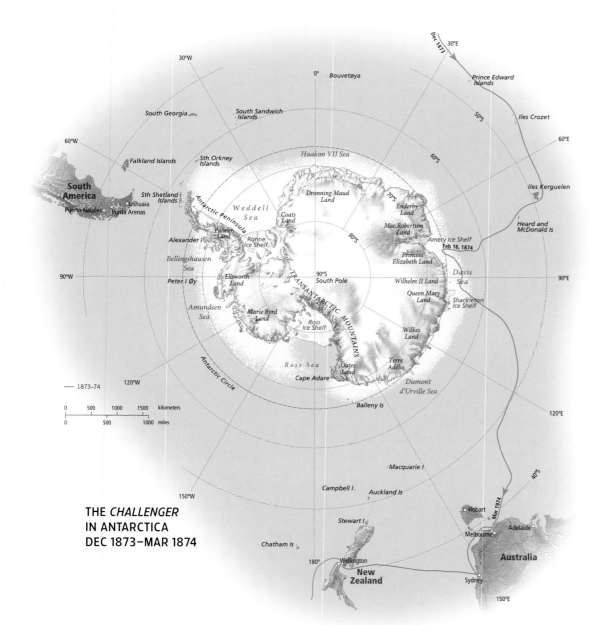

**THE *CHALLENGER*
IN ANTARCTICA
DEC 1873–MAR 1874**

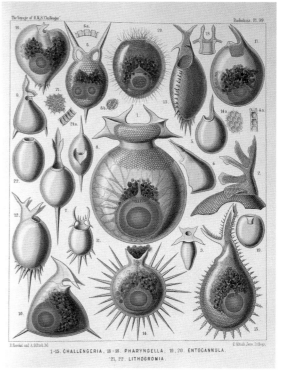

1–15. CHALLENGERIA, 16–18. PHARYNGELLA, 19, 20. ENTOCANNULA. '21, 22. LITHOGROMIA.

Relevant Records
The scientists aboard the *Challenger* wrote meticulous reports and made very accurate illustrations that are still used by marine biologists. This series of radiolarians—tiny protozoans with complex skeletons, usually composed of silica—are named after *Challenger* and the expedition's scientists.

Exploring the Depths

The *Challenger* was by far the largest ship yet to venture into Antarctic waters, and the first steam-powered vessel to do so. Although it was in the ice for less than three weeks of its long circumnavigation, the expedition made some very significant biological and geological discoveries, using dredging equipment never before employed in these latitudes.

Unfamiliar Marine Creature

The giant sea star was one of about 400 animal species brought to the surface by the *Challenger*'s dredges. Such creatures had never before been seen and were of great interest to scientists.

calling at Tristan da Cunha and Iles Kerguelen—where they surveyed the coast around the islands—and at Heard Island. Only days after leaving Iles Kerguelen, the crew saw their first iceberg and many more soon followed. The *Challenger* remained in the ice for less than three weeks before she continued northward to Melbourne, and then to Japan and the north Pacific. The ship did not go back to the Southern Ocean, but returned to Britain by way of Tahiti, Chile, the Strait of Magellan, and Montevideo.

Extending the Frontiers

Chief scientist Professor Wyville Thomson reported the *Challenger* had made 362 observation stations during the voyage, "at intervals as nearly uniform as possible." Although the expedition's sojourn in the south was so short, it made a great contribution to Antarctic science, including establishing that there was a permanent cell of high atmospheric pressure over Antarctica.

Earlier expeditions to southern ice had observed the rocks and debris trapped in icebergs drifting northward, but it was left to the *Challenger* to dredge the sea floor of the Southern Ocean for the rocks released from these melting icebergs. The rocks the ship laboriously hauled to the surface with her 18-horsepower winch were quite

different from the rocks of the sub-Antarctic islands. "The *Challenger* had dredged up fragments of mica schists, quartzites, sandstones, compact limestones, and earthy shales, which leave little doubt that within the Antarctic Circle there is a mass of continental land quite similar in structure to other continents," reported the naturalist John Murray. Thus, the expedition had obtained evidence of a distinctive southern continent. South of 43°S, her dredging haul included about 400 animal species, more than 80 percent of which had never been seen before.

When the *Challenger* steamed into Portsmouth in May 1876, she had covered 68,900 nautical miles (127,600 km). Captain Nares was no longer on board, having been recalled from Hong Kong to lead the British Arctic Expedition of 1875–76 into the Canadian Arctic.

But while Nares was seeking the North Pole and discovering the Challenger Mountains, the *Challenger*'s scientists and crew were methodically continuing their scientific work. They took daily magnetic observations while at sea. They collected flora, fauna, and geological samples on a range of sub-Antarctic islands, and they acquired a wealth of information about ocean currents and water temperatures, meteorology, and about the sea floors of the basins of the Southern, Atlantic, and Pacific oceans. Their researches included the discovery of a great number of previously unknown marine species.

The *Challenger* expedition had been a very long, slow voyage because the scientific program required the ship to stop for most of the day every 200 miles (320 km) to dredge and take soundings. It was extremely productive, providing information about the oceans, and Antarctica, that is of value to this day. DM

Commercial Ventures
1892–95

Captain of Commerce
After several voyages to the Weddell Sea—he discovered petrified wood on Seymour Island in 1892 and the Larsen Ice Shelf in 1893—the Norwegian Carl Anton Larsen (1860–1924) went on to found South Georgia's lucrative whaling industry. He captained the *Antarctic* on Otto Nordenskjöld's ill-fated Swedish expedition of 1901–04.

The powerful Australian interest in the Antarctic continent can be traced back to Henryk Johan Bull, a Norwegian who emigrated to Melbourne in 1885. In *The Cruise of the Antarctic*, Bull reported his disappointment when an attempt by Australian scientists to raise an Antarctic expedition failed, and his belief that "an expedition on commercial lines would possibly find supporters."

Whaling in Antarctic waters went back to the early nineteenth century, but at that time there were enough whales in the Northern Hemisphere to discourage the longer and more hazardous voyage south. The northern whale populations began to fall and alternative products such as mineral oils emerged, but demand for baleen from right whales—mainly for the "whalebone" used in corsets—kept the quest alive. In 1873, a German expedition explored the Antarctic Peninsula but found only rorqual whales. In the 1892–93 summer, a Scottish expedition explored the northwestern Weddell Sea but found no right whales among the many it saw.

Foyn and Antarctica
A Norwegian expedition under Carl Anton Larsen was working in the same area that summer, but with little success in terms of whaling. When Larsen returned the following summer, he had better luck with seals than with whales. His 1893–94 voyage was notable because he found petrified wood at Cape Seymour—the first fossils to be discovered on Antarctica. Larsen named Foyn Land (now the Foyn Coast) in 1893 after Svend Foyn, the Norwegian inventor of the explosive harpoon

head; Foyn mounted the new device on the steam-driven catchers which replaced rowing boats, innovations that revolutionized the whaling industry and made him rich.

In 1893, after Henryk Bull failed to raise any finance in Melbourne, he went to Norway to see Foyn. Within 15 minutes, Foyn had promised him a ship—a 249-ton (226-tonne) steam whaler called the *Kap Nor*, which was renamed the *Antarctic* for the voyage south. The vessel ran aground on a preliminary whaling expedition to Campbell Island but she sailed from Melbourne on September 26, 1894, called in at Hobart, and then set off for Antarctica on October 13. The ship's captain was Leonard Kristensen, and Henryk Bull was on board in the loosely defined role of manager.

Although the expedition posed little threat to the whale population of Antarctica, it added greatly to the controversy about who had been the first to set foot on the Antarctic continent: if John Davis had not in fact landed on the Antarctic Peninsula in 1821, as he had claimed, then Bull's party was the first recorded landing on the continent. The date was January 24, 1895. Bull wrote: "Cape Adare was made at midnight. The weather was now favorable for a landing, and at 1 AM a party, including the Captain, second mate, Mr. Borchgrevink, and the writer, set off, landing on a pebbly beach of easy access after an hour's rowing through loose ice, negotiated without difficulty … The sensation of being the first men who had set foot on the real Antarctic mainland was both strange and pleasurable, although Mr Foyn would no doubt have preferred to exchange

**CARL LARSEN
NOV 1893–MAR 1894**

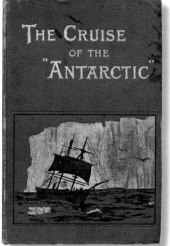

Ross Sea Voyage
Henryk Bull's book, *The Cruise of
the Antarctic*, fired enthusiasm for
commercial whaling in the Ross Sea
region and also sparked off the Heroic
Age of exploration.

Whaling Innovator
Svend Foyn's contribution to the
commercial whaling industry—both
as inventor and as sponsor of Bull's
exploratory whaling expedition—is
commemorated in the naming of the
Foyn Coast and Foyn Island.

this pleasing sensation on our part for a right whale even of small dimensions." Bull left the question of who was the first to step ashore open.

Rival Claims

Others were not so modest. Carsten Borchgrevink, who was born in Oslo and migrated to Australia when he was 24, was on board as a "generally useful hand." He had claimed to have been a student at Christiania (Oslo) University, and Bull took him on because he could also fill the role of scientist. Bull acknowledged having an ulterior motive: "His family was known to me ... and the prospect of a real companion was so cheering that I promised to do my best for him." But Borchgrevink's credentials were not rated highly; Bull added: "It is true we intended to build an extra cabin for ... colonial men of science ... but the qualifications and letters of introduction brought by Mr Borchgrevink were not such as to warrant ... building extra cabins for him." Borchgrevink was in the landing party, and subsequently (and repeatedly) claimed that he was first ashore. In one account he stated: "I do not know whether it was the desire to catch the jellyfish (seen in the shallows), or from a strong desire to be the first man to put foot on this terra incognita, but as soon as the order was given to stop pulling the oars, I jumped over the side of the boat."

His claim was disputed by at least two others. In his journal, Kristensen wrote: "I was sitting foremost in the boat, and jumped ashore as the boat struck, saying 'I have then the honor of being the first man who has ever put foot on South Victoria Land'." But the third claimant has considerable merit: in view of the water temperature and shipboard protocol, there is reason to believe the New Zealand crewman A. H. F. von Tunzelman, who maintained that he was in the bow and jumped out first to steady the boat so that the captain could disembark.

This rather childish dispute aside, the *Antarctic* expedition was an important development in the exploration of Antarctica because the voyage opened the way for the land explorations of the Heroic Age. In *The Cruise of the Antarctic*, Bull wrote: "We have proved that landing on Antarctica proper is not so difficult as it was hitherto considered, and that a wintering party have every chance of spending a safe and pleasant twelvemonth at Cape Adare, with a fair chance of penetrating to, or nearly to, the magnetic pole by the aid of sledges and Norwegian skis."

The immediate commercial returns from the voyage were few: it had found no baleen whales, and while the *Antarctic* had been exploring, the owner, Svend Foyn, had died, so that soon after the return to Melbourne on March 12, 1895, the ship was instructed to return to Norway. But the expedition is remembered in scientific circles, and there is a Bull Island in the Possession Islands group. This was where the crew of the *Antarctic* landed and found lichen—the first vegetation to be discovered in Antarctica, an environment previously believed too bleak to allow plant life to survive.   DM

Adrien de Gerlache
1897–99

New Challenges
Whether by accident or design, the scientific expedition led by Belgian naval lieutenant Adrien de Gerlache (1866–1934) was the first to face the rigors of an Antarctic winter.

At the Sixth International Geographical Congress in London, in July 1895, the English geographer Clements Markham said: "The exploration of the Antarctic regions is the greatest piece of geographical exploration yet to be undertaken."

Meanwhile, the Brussels Geographical Society was assembling a scientific expedition initiated and led by a young naval lieutenant, Adrien Victor Joseph, Baron de Gerlache de Gomery. In Norway, de Gerlache bought the *Patric*, a 275-ton (250-tonne) whaler, and renamed it the *Belgica*. His second-in-command was another Belgian, Lieutenant George Lecointe. The crew included a Rumanian chief scientist, Emile Racovitza; the Pole Henryk Arçtowski as geologist; a 25-year-old Norwegian, Roald Amundsen, as first mate; and as ship's doctor, 32-year-old American Frederick Cook, who had been in Greenland with Robert Peary (both were later to claim they had reached the North Pole).

The *Belgica* left Antwerp on August 16, 1897, with new cabins and laboratory—additions that overloaded the vessel and made it painfully slow. The plan was to sail along the eastern side of the Antarctic Peninsula, to winter in Melbourne, and to visit Victoria Land the following summer. They reached Punta Arenas early in December, but de Gerlache decided to explore Tierra del Fuego, and so they did not arrive off the Antarctic Peninsula until January 20, almost the end of the polar summer.

Ice-bound
They explored the islands that line what is now Gerlache Strait over the next few weeks. They crossed the Antarctic Circle on February 15 and encountered pack ice, but pushed on before becoming trapped in ice. Amundsen wrote that they "faced the prospect of a winter in the Antarctic with no winter clothing … without adequate provisions … and even without lamps enough … It was a truly dreadful prospect."

Pack ice can move as fast as 10 miles (16 km) per day, and it carried the *Belgica* westward. De Gerlache had sailed well southwest of the Antarctic Peninsula, and the *Belgica* became trapped at 71°30′S, 85°16′W, on March 2, 1898; she would not be free until March 14, 1899, at 70°30′S, 103°W.

It was a harrowing time. The sun sank below the horizon on May 17, not to reappear for 70 days. In *Through the First Antarctic Night*, Frederick Cook outlined the daily routine: "Rise at 7.30 am; coffee at 8; 9 to 10, open air exercise; 10 to 12, scientific work … for the officers, and for the marines, bringing in the snow, melting snow for water, replenishing the ship's stores, repairing the ship, building new quarters, making new instruments, and doing anything which pertains to the regular work of the expedition; 12 to 2 pm dinner, and rest or recreation, 2 to 4, official work … 6 to 7, supper; 7 to 10, card-playing, music, mending, and on moonlight nights, excursions. At 10 o'clock we went to sleep."

Cook was a tireless source of strength, boosting morale and providing medical help when symptoms of distress, even insanity, surfaced. As Amundsen wrote: "He … was the one man of unfaltering courage, unfailing hope, endless cheerfulness, and unwearied kindness,"

Cold Storage
Cook and Amundsen knew the importance of fresh meat in preventing scurvy, and had stored seal and penguin carcasses in the ice before wildlife fled the advancing winter. But, as Amundsen related, "The commander … developed an aversion to the flesh of both … He was not content only to refuse to eat it himself, but he forbade any of the ship's company to indulge in it." De Gerlache and Lecointe took to their beds with scurvy, and Cook and Amundsen took over. Cook described penguin meat as tasting like "a piece of beef, odiferous cod fish and a canvas-backed duck roasted together in a pot, with blood and cod-liver oil for sauce." He persuaded de Gerlache to take it as a medicine, and soon they all started to recover.

Cook described the momentous occasion when the bright top of the sun reappeared: "For several minutes my companions did not speak … we could not … have found words with which to express the buoyant feeling of relief, and the emotion of the new life which was sent coursing through our arteries." But their optimism was premature: summer progressed, but no way out of the ice presented itself. They faced the prospect of another winter trapped in the ice with little hope of survival. On New Year's Eve, they saw open water 2,100 ft (640 m)

Blue Sky, White Peaks
Towering, snowy peaks rise high above Cape Renard, which separates the Danco and Graham coasts on the west side of the Antarctic Peninsula. They were first sighted in 1898 by the men on board the *Belgica*.

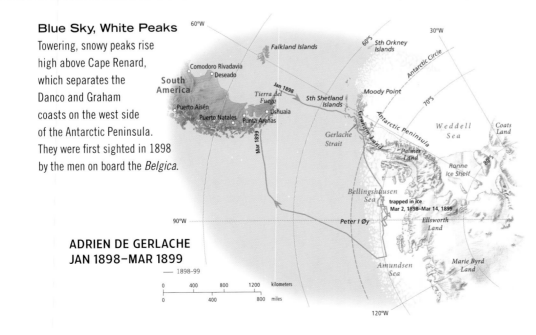

**ADRIEN DE GERLACHE
JAN 1898–MAR 1899**

— 1898–99

0   400   800   1200   kilometers
0   400   800   miles

First Photographic Records
Watched by a curious emperor penguin, in March 1898, the *Belgica* is trapped in ice in the Bellingshausen Sea. It was 377 days before the ship was freed.

Geological Theory
Geologist Henryk Arçtowski made some significant discoveries. He was the first to postulate that the Antarctic Peninsula was a continuation of the Andes chain through the islands of the Scotia Arc. This was an important piece in assembling the jigsaw puzzle of plate tectonics.

away. Cook suggested cutting a passage to freedom, although they had only three 4-ft (1.2-m) saws. It was only a small basin of water they were making for, and they were relying on Cook's faith that this was where any break to open water would appear. By the end of January they had hacked through over 2,000 ft (600 m) of ice. Then one morning, Amundsen wrote, they woke, "to discover that the pressure from the surrounding ice pack had driven the banks of our channel together, and we were locked in as fast as ever."

Escape
Not surprisingly, depression set in. But on February 15, the ice relented and the passage they had cut reopened and extended right up to the ship. They took the *Belgica* to the basin under power, but they were still 7 miles (11 km) from freedom and winter was approaching. Then, on March 14, "the miracle happened—exactly what Cook had predicted," Amundsen wrote. "The ice opened and the lane to the sea ran directly through our basin! Joy restored our energy, and with all speed we made our way to the open sea and safety."

The *Belgica* returned to Antwerp on November 5. Despite its problems, the expedition gathered much valuable scientific data. Arçtowski believed, correctly, that his bathymetric measurements suggested that the Antarctic Peninsula was an extension of the Andes—not directly across the Drake Passage but through the Scotia Arc. In addition, this was the first expedition to collect meteorological readings below the Antarctic Circle through a winter. The lowest temperature they recorded was –45°F (–43°C).

The participants had varied futures. Amundsen became the first to reach the South Pole (among many achievements) and de Gerlache completed several Arctic journeys. Cook subsequently claimed to have been the first to climb Mount McKinley, North America's highest mountain, and to be first to the North Pole, but in both cases he was found to have falsified his records. He was imprisoned for fraudulently using the United States mail in 1923, released in 1929, was pardoned in 1940, and died the same year. DM

The First Antarctic Night
Two days after the *Belgica* reached the Antarctic Peninsula, a young sailor, Auguste-Karl Wiencke, fell overboard; he held onto a line for some time, but sank before he could be saved. During the long winter, paranoia and fierce arguments developed. Some men took to their bunks for days at a time, and geophysicist Emile Danco died of cold.

Carsten Borchgrevink
1898–1900

In 1894–95, Carsten Borchgrevink accompanied Henryk Bull on the *Antarctic*, and when he returned he claimed that most of the achievements of that voyage were his and his alone. It was not an endearing performance. However, as Hugh Robert Mill noted: "No one liked Borchgrevink very much but he had a dynamic quality and a set purpose to get out again to the unknown South that struck some of us as boding well for exploration."

British publisher Sir George Newnes had made his fortune publishing penny magazines, such as *Tit-Bits*, *Country Life*, and *The Strand Magazine*, for the newly educated masses. Newnes put up 40,000 pounds for Borchgrevink's British Antarctic Expedition. The *Pollux*, a whaler of 267.7-ton (243-tonne) net, was bought and renamed the *Southern Cross*. The expedition sailed from London in August 1898.

The aim was to spend a winter at Cape Adare and undertake some land exploration. This was the first land-based Antarctic expedition, and they took more than 70 dogs from Siberia and Greenland, handled by two Laplanders, to haul sledges. The ship's master was Bernhard Jensen, who had been second mate on the *Antarctic*. The scientists were William Colbeck, of the

Royal Naval Reserve, as cartographer, navigator, and magnetic observer; the Norwegian zoologist, Nikolai Hanson, assisted by Hugh Blackwall Evans; and Louis Bernacchi, a physicist from Tasmania.

They sailed from Hobart on December 17, 1898, and spotted their first ice on December 30 at 61°56′S. It took 43 days to get through the pack ice; fortunately the *Southern Cross* had been ice-strengthened. They sighted the Antarctic continent on February 15, 1899, and landed at a large black triangle of flat gravel on the western side of Cape Adare two days later. During ferocious storms, it took 12 days to land the wintering party's supplies. Six men were stranded ashore at one point, and survived only because they were kept warm by dogs in the tent the Laplanders had brought. Soon afterward, they erected two prefabricated huts made of Norwegian pine on the shore; this was named Ridley Camp, after Borchgrevink's mother and son.

Stir-crazy

On March 1, the *Southern Cross* departed for New Zealand, leaving ten men behind to brave the Antarctic winter. On May 15 the sun set, not to rise again until July 27—72 days later. The small party retreated to the

Tireless Researcher
Physicist Louis Bernacchi carried out a full program of work, even when ice-bound. He returned to Antarctica with Scott's *Discovery* expedition.

Bleak Ramparts
Through the sea ice near Cape Adare, Mount Minto looms out of the clouds. Borchgrevink chose desolate Ridley Beach nearby as the site of his base. Although it was a good spot for the camp itself, the inaccessible terrain behind prevented the expedition from undertaking any significant inland journeys of exploration.

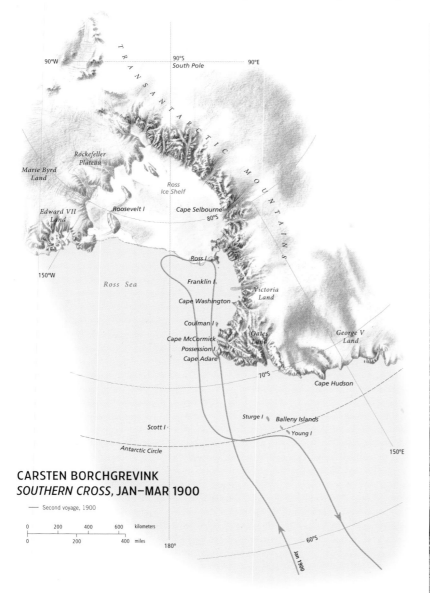

CARSTEN BORCHGREVINK
***SOUTHERN CROSS*, JAN–MAR 1900**

— Second voyage, 1900

First Winter Expedition

After landing at Cape Adare with Bull in 1895, Borchgrevink (1864–1934) returned in February 1899, determined to winter on the Antarctic continent. Although lacking leadership skills, he held his team together sufficiently to prove it was possible, a breakthrough that paved the way for the major explorations of the Heroic Age.

Close Quarters

Borchgrevink's huts of prefabricated Norwegian pine have withstood more than a century of gales and blizzards due to the ingenious system used for interlocking the logs into a robust and rigid structure. The nearby huts of Scott's northern party, ten years younger, have virtually disintegrated. Though cramped and dark inside, the Norwegian huts provided adequate shelter, and were a model for later expeditions wintering farther south.

huts and cut outdoor work to a minimum. However, the confinement indoors had its hazards.

Relations between Borchgrevink and the other men worsened; a petulant document he produced stated that "the following things would be considered mutiny: to oppose C. E. B. [Borchgrevink] or induce others to do so, to speak ill of C. E. B., to ridicule Mr. C. E. B. or his work, to try and force C. E. B. to alter contracts." There were other, more tangible problems: on July 24, the huts nearly burned down when Colbeck left a candle burning and set fire to his bunk, and on September 1 three of the party were almost asphyxiated when a wind shift filled the cabin with coal fumes. Hanson had been sick on the voyage and for much of the winter, and on October 14 he died, probably from beri-beri; he never saw the thousands of Adélie penguins that crossed the ice toward the camp site and their breeding grounds only two days later. He was buried at the summit of Cape Adare.

During the year at Cape Adare, the land party made several survey trips deep into Robertson Bay, penetrated a short way into the interior, and carried out a full program of scientific measurements.

Much of our knowledge of the expedition comes from Louis Bernacchi's diary: for January 1, 1900, he noted "A new year and the last of the nineteenth century! A year spent entirely within the Antarctic Circle, and the ice and snow visible each day, for 365 days." The *Southern Cross* was expected to return in January, and the anxiety mounted as the days passed. The entry for January 28 runs: "At about 8 o'clock in the morning when all

were asleep in their bunks, a voice in the room calling 'Post!' broke upon our slumbers. It was the voice of Capt Jensen. With what hysterical joy we tumbled out of our bunks to welcome him!"

They left Cape Adare on February 2. As Bernacchi wrote, "We are not sorry to leave this gelid, desolate spot, our place of abode for so many dreary months!" Over the next few weeks, they sailed into the Ross Sea, landing at Possession, Coulman, and Ross islands, and traveling southward across the Ross Ice Shelf to an estimated 78°50′S—the farthest south yet. They turned north on February 22, crossed the Antarctic Circle on February 28, and reached New Zealand's Stewart Island on March 31.

Belated Recognition

The news that it was possible to survive a winter ashore in Antarctica was noteworthy, but the expedition's return attracted little interest. The Boer War had intervened and Robert Falcon Scott's National Antarctic Expedition was preparing for departure, amid great excitement. The public was interested to learn that there were no large land animals in Antarctica—which meant that expeditions would not need guns against polar bears!

Although Borchgrevink was made a fellow of the Royal Geographical Society and knighted by the king of Norway, he did not receive the recognition that he considered his due. His account, *First on the Antarctic Continent*, was badly written and did not further his cause. In 1930, when the Royal Geographical Society belatedly awarded him the Patron's Medal, it stated: "When the *Southern Cross* returned, this Society was engaged in fitting out Captain Scott to the same region … and the magnitude of the difficulties overcome by Mr Borchgrevink were underestimated … It was only after … that we were able to realize the improbability that any explorer could do more in the Cape Adare district than Mr. Borchgrevink had accomplished." The explorer died in Norway on April 21, 1934.

Despite their eagerness in fleeing Antarctica after that grim winter, Bernacchi and Colbeck were soon heading south again—both returned to Cape Adare, Bernacchi with Scott on the *Discovery* and Colbeck on the *Morning* to rescue Scott's expedition. DM

Robert Falcon Scott
British National Antarctic Expedition: 1901–04

It was fitting that Robert Falcon Scott (1868–1912) had named his son Peter Markham Scott, after the president of the Royal Geographical Society, Sir Clements Markham, who took an unknown Royal Navy officer and transformed him into the legendary Scott of the Antarctic. Recent analysis has shown that Scott was neither a fortunate leader nor a good planner. But he documented some of the definitive events of the Heroic Age of Antarctic exploration as well as achieving much for Antarctic science.

Markham had presided over the Sixth International Geographical Congress, held in London in 1895. Three important scientific expeditions that resulted from that congress were neatly dividing the frozen continent: the German expedition led by Erich von Drygalski went to the region south of the Indian Ocean; Otto Nordenskjöld traveled to the Antarctic Peninsula; and Britain focused its attention on the Ross Sea region.

The British National Antarctic Expedition

Markham had recognized Scott's potential when he first met the 18-year-old midshipman in 1887. Twelve years later, when Scott volunteered to lead the British National Antarctic Expedition, Markham accepted his offer. The expedition combined the might of three of the powerful institutions of Britain—the Royal Geographical Society, the Royal Society, and the Royal Navy—and its aims combined exploration and science. In keeping with these goals, the expeditioners were a mix of Royal Navy and Merchant Navy personnel, as well as an array of scientists of various disciplines.

Their vessel was the *Discovery*—the first ship to be designed and built in Britain for scientific exploration. She was constructed in Dundee of oak strengthened with internal beams and clad in a steel bow. The king, Edward VII, came to the ship at the Isle of Wight to wish them well, and the *Discovery* sailed on August 6, 1901. The king's blessing did not prevent the slow but sturdy ship from pitching and rolling at sea, nor did it stop a persistent leak. The expedition called in at Macquarie Island and Auckland Island on its way to Lyttelton, New Zealand. It had an enthusiastic send-off when it sailed from there on December 21, although one unfortunate seaman fell to his death during the celebrations.

The ship crossed the Antarctic Circle on January 3, 1902, and reached Cape Adare on January 9. Scientist Louis Bernacchi was the only member of the expedition who had previously visited Antarctica—two years earlier as a member of the Carsten Borchgrevink expedition. Continuing into the Ross Sea, the *Discovery* followed the edge of the Ross Ice Shelf eastward, searching for the land reported by Ross, and eventually sighted the mountains of what is now Edward VII Land. At Balloon Bight (now the Bay of Whales) on February 4, Scott rose in a tethered, army-supplied hydrogen balloon and became the first person to fly over Antarctica. He was followed up by Sub Lieutenant Ernest Shackleton, who become Antarctica's first aerial photographer.

The expedition established its base at Hut Point on Ross Island and began to make some excursions. On a badly managed trip to Cape Crozier, seaman George Vince died when he slipped over a cliff, and Frank Wild displayed Antarctic ingenuity when he hammered nails into the soles of his boots to provide extra traction on the ice. The Adélie penguins around the base were a source of endless fascination. Scott recorded that the emperor penguins began to head south at the start of winter—this was the first recorded observation of their remarkable breeding pattern.

On April 23, the sun disappeared below the horizon, not to return until August 22. The officers and the men continued to live on the *Discovery* and remained busy throughout the winter. For diversion, there was a series of theatrical performances and the monthly *South Polar Times*, which was edited by Ernest Shackleton and printed on the press that had been included with the expedition's equipment. The officers and men messed apart but they ate the same food. There were regular inspections and morning prayers were said on the mess deck. Of the inspections, Frank Wild wrote that they were: "Somewhat unnecessary in the circumstances, but as one of the sailors remarked 'Oh well, it pleases him and it doesn't worry us'."

In early October, Charles Royds led a party to Cape Crozier where, on October 12, Reginald Skelton found emperor penguins with well-developed chicks—this was a clear indication that the birds had hatched during the bitter Antarctic winter.

Bird's-eye View
Scott was the first balloonist in the Antarctic, ascending to 800 ft (245 m) in the basket below a hydrogen balloon. Shackleton, the only other one in the party to go up, took aerial photographs. Their height of ascent was limited by the weight of the heavy tethering rope.

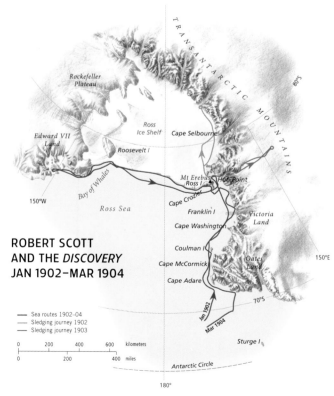

ROBERT SCOTT AND THE *DISCOVERY* JAN 1902–MAR 1904

— Sea routes 1902–04
— Sledging journey 1902
— Sledging journey 1903

Stark Reminder

Looking back from the tip of Hut Point across the *Discovery* hut to McMurdo Station, the cross in the foreground is a memorial to George Vince, who died near this spot on March 11, 1902. On the first sledging expedition, Vince slid off an ice cliff into the sea. His body was never found.

A Sledging Expedition

The first-ever attempt to sledge to the South Pole left the *Discovery* on November 2, 1902. The three-man party comprised Scott, Ernest Shackleton, and Edward Wilson, with 19 dogs. Twelve men had left earlier to deposit supply caches for the polar party on their return journey. The groups soon met up and traveled together until November 15, when the polar party pressed on alone. Scott was confident at first, but their progress was slow; the dogs were underfed and overloaded, and the men mostly had to haul the sledges themselves. Scott and Shackleton began to annoy each other. Years later, the peacemaker Wilson recalled an event on the ice cap. After breakfast, while he and Shackleton were loading the sledges, Scott shouted: "Come here, you bloody fools." Wilson asked if Scott was addressing him, and Scott replied "No." "Then it must have been me," said Shackleton, and when Scott was silent, continued: "Right, you're the worst bloody fool of the lot, and

every time you dare to speak to me like that, you'll get it back." It was an odd exchange at the bottom of the world, where each relied on the others for survival.

Scott's attitude to dog sledging reveals what later proved to be a fatal flaw in his Antarctic philosophy. "In my mind," he wrote, "no journey ever made with dogs can approach the height of that fine conception which is realized when a party of men go forth to face hardships, dangers, and difficulties with their own unaided efforts … Surely in this case the conquest is more nobly and splendidly won."

But they did not win. In fact, although they set a new farthest south of 82°16′ on December 30, 1902, they did not travel beyond the Ross Ice Shelf. By then all were suffering from scurvy, and the last few dogs died on the return journey. Shackleton, the largest man, suffered most from scurvy and lack of food, and even had to ride briefly on the sledge at the end of January. By February 3, 1903, they were back at Hut Point.

Their Names Live On

Many of the expedition's officers and scientific staff are now immortalized in place names in Antarctica. On the stern of the *Discovery* in 1901 are (from left): Wilson, Shackleton, Armitage, Barne, Koettlitz, Skelton, Scott, Royds, Bernacchi, Ferrar, and Hodgson.

The Second Winter

A relief ship, the *Morning*, had reached Ross Island on January 23 under the command of William Colbeck, who had discovered what is now Scott Island on the voyage down the Ross Sea. However, Colbeck could not sail within 8 miles (13 km) of the *Discovery* because the sea ice had not fully melted over summer. It never did melt that summer, so the *Discovery* remained set in ice for the coming winter. But at least the expedition had fresh supplies. Scott decided to send back "one or two

Wide Waters

Even at the coldest time of year, there is a polynya (a large area of open water) at McMurdo Sound and Ross Island. The nearest part of Ross Island is Cape Bird, the site of large Adélie penguin colonies in summer. The high points of Ross Island are Mount Terror (center) and Mount Erebus (right).

Place of Refuge

Inside the *Discovery* hut, which some of Shackleton's Ross Sea party later relied on for their survival. The interior of the hut is stained black by smoke from the blubber that was their fuel for heating, cooking, and lighting.

Midwinter Festival

On June 23, 1902, Scott wrote: "The mess-deck was gaily decorated with designs in colored papers and festooned with chains and ropes of the same material, the tables loaded with plum pudding, mince pies, and cakes ... we left the men to enjoy their Christmas fare with an extra tot of grog."

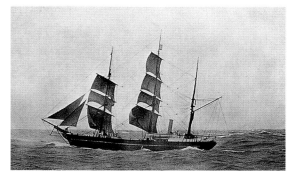

Low in the Water
The *Morning* was sent as the relief ship. James Doorly wrote: "There are few oceans so tempestuous as that globe-encircling expanse to the southward ... the great seas broke continually over the ship, and one night during a gale one of the quarter boats was dragged out of its tackle and swept away."

Antarctica's Big Bird
As Scott wrote: "various devices were resorted to in our endeavors to capture birds for our collection, and sooner or later most of the species were brought on board. The larger albatrosses were caught by towing a small metal triangle, well baited ..."

undesirables" and, with the approval of the surgeon Reginald Koettlitz—but not that of Albert Armitage, his second-in-command—included Shackleton, claiming that he was not fit enough to continue. The real reason may have been that Shackleton was charismatic and popular, whereas Scott had to rely on his rank to keep control. Reportedly, Shackleton wept as he sailed away.

There was a series of excursions over the winter of 1903 and into the following summer. Bernacchi took over the editorship of the *South Polar Times*. Armitage had led a sledging trip into the mountains of southern Victoria Land on the other side of McMurdo Sound in November 1902, and that group became the first to stand on the polar ice cap. Scott took a party into the same area on October 26, 1903, and journeyed well beyond the point Armitage had reached. When they returned on Christmas Day, after having been away for 59 days and covering 725 miles (1,170 km), Scott had formed a strong bond with the two seamen he had traveled with: Edgar Evans and William Lashly. Both of them were to accompany him on his next voyage.

Homeward Bound
The next summer, the British government sent the *Morning* and the *Terra Nova* back to Ross Island to evacuate the party. They arrived on January 5, 1904, with instructions that all the expeditioners were to leave by the end of summer, even if it meant abandoning the *Discovery*. Luckily, the ice broke up in February (with the help of dynamite) and the relief ships were able to reach the *Discovery* on February 14—two days later the ship was free. The two relief vessels immediately sailed north. Scott took the *Discovery* on a surveying voyage along the coast of Victoria Land.

They were back in New Zealand by the beginning of April, and arrived off Spithead, near Portsmouth, on September 10, 1904. Scott was promoted to the rank of captain on that day. His two-volume account of the expedition, *The Voyage of the Discovery*, has become a classic of Antarctic literature. Six volumes of scientific reports were published by the British Museum. DM

Inhospitable Shore
In *The Voyages of the Morning*, junior officer James Doorly wrote of the "remarkable cone-shaped islet close off [Scott Island]," now named Haggits Pillar. He related how a party "after some difficulty" landed, claimed Scott Island, and left a record of its visit.

Erich von Drygalski
German Antarctic Expedition: 1901–03

Pioneer Glaciologist
Erich von Drygalski (1865–1949) was born in Prussia, in what is now Kaliningrad. He undertook fieldwork in glaciology in Greenland, before expeditioning to Antarctica and later to Spitsbergen in the Arctic region of Norway. On returning from Antarctica, he spent many years preparing his expedition research for publication.

After the International Geographical Congress, held in London in 1895, there was widespread interest, especially from geographers and other scientists, in taking up the challenge of discovery in Antarctica, the most remote and least known region of the planet. There was an enthusiastic response in several European countries, and in 1898 the German South Polar Commission was formed to mount the first German Antarctic Expedition.

Erich von Drygalski, professor of geography and geophysics at the University of Berlin, was appointed as the leader of the expedition. He had already led two polar expeditions, to Greenland in 1891 and 1893 to study glacial ice, and his scientific qualifications were impeccable. The generous state funding enabled him to commission a purpose-built ship, the *Gauss*, and to recruit a crew of 32, which included five scientists. Scientific research and geographic discovery were the two primary goals of the expedition, and so Drygalski selected as his field of operations an unexplored area of the East Antarctica coast.

Icebound
The *Gauss* left Kiel in August 1901 and reached Iles Kerguelen five months later, then headed south at the end of January 1902. Ice conditions were difficult, but on February 21, at about 90°E, Drygalski recorded the appearance of "coherent, uniform white contours … ending abruptly at the water's edge forming cliffs 40 to 50 meters [130 to 165 ft] high … the area behind rose gently to about 3,000 meters [10,000 ft]." Drygalski named this new land for Kaiser Wilhelm II. Attempting to approach more closely, the *Gauss* became trapped between two ice ridges and, despite desperate attempts to free the ship, by early March it was clear that there was no escape from the ice.

Pressure Ridges
When the *Gauss* became trapped in pack ice in the Davis Sea, Drygalski continued his glacialogical studies. He later published a 20-volume work about the expedition and its findings.

Drygalski had ensured that the ship was provisioned and the crew prepared to winter in the ice. Life on board ship quickly developed a pleasant routine: "Sundays were beer nights, Wednesdays were lecture nights, Saturday nights were best of all … with a glass of grog, united in games or conversation." There were also card clubs, a smoking club, and a band.

But the scientific work was not neglected, and the expeditioners maintained continuous meteorological records and they undertook wildlife surveys. From mid-March, Drygalski organized dog-sledging expeditions to the coast, and about 55 miles (90 km) from the ship, he discovered an extinct volcano and named it Gaussberg. The explorers collected geological samples and made magnetic observations, but winter put an end to further excursions when one sledging party nearly perished in a prolonged snowstorm at the end of April.

A hot air balloon, the first to be used in Antarctica, took Drygalski to a height of nearly 1,600 ft (490 m), from which, as he reported by telephone to the ship's deck, "the sight was grandiose. I could see the newly discovered Gaussberg … the only ice-free landmark in the surrounding area."

On the basis of this aerial reconnaissance, an inland expedition was planned to 72°S—much farther south than Gaussberg (66°40'S). However, when the longer days and better weather of spring necessary for this

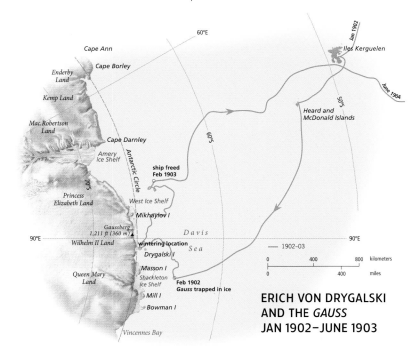

ERICH VON DRYGALSKI AND THE *GAUSS* JAN 1902–JUNE 1903

Remember Thy Name

The Drygalski Ice Tongue, 12 miles (20 km) at its widest and 43 miles (70 km) long, extends far off the coast of Victoria Land into the Ross Sea. It was named by Robert Scott for his contemporary, the renowned German explorer and scientist. Drygalski Island, near where the *Gauss* wintered over, is another feature named after him.

journey eventually arrived, Drygalski decided that his priority was to free the ship from the ice, so the idea was regretfully abandoned.

Making their Escape

The crew set to with saws, pickaxes, and explosives in an attempt to dig, cut, and blast their way out of the ice toward open water. But it was slow going until one day Drygalski noticed that, where soot from the ship's funnel had stained the ice, the melting was quite rapid. So all hands were called to lay a trail of garbage and any other dark, disposable matter up to the ice edge. "Within a month we had a long water channel almost two meters [7 ft] deep. Although there were still four to five meters [13 to 17 ft] of ice underneath, the channel widened constantly." Even so, the men's patience was sorely tried; it was not until early February 1903, almost 11 months after the ship was trapped, that they "felt two sharp jolts in rapid succession … it was like a revelation," as the ice finally broke up.

Heading west, the expedition explored the coastline for another two months in dangerous ice conditions that finally forced Drygalski to abandon the attempt and head north at the end of March. They finally returned to Kiel in November 1903. Although the expedition discovered little new territory, it produced 20 volumes of scientific reports under Drygalski's expert editorship. LC

Full to the Wind

Drygalski took this photograph of the *Gauss* under full sail. The ship was a three-masted barquentine (with an auxiliary engine) of similar design to Amundsen's *Fram*.

Otto Nordenskjöld

Swedish South Polar Expedition: 1901–04

Leader from Academia

Otto Nordenskjöld (1869–1928) was a university lecturer in geology and mineralogy at the Uppsala University, Sweden. Before going to Antarctica, he had led a successful expedition to research glacial geology in Patagonia.

The plan for the Swedish expedition led by Otto Nordenskjöld was simple: the *Antarctic*, Henryk Bull's old sealing ship, under the command of Carl Anton Larsen, would leave Nordenskjöld and a party of scientists to explore the eastern side of the Antarctic Peninsula for a year while others on the ship conducted scientific work elsewhere; the two parties would be reunited the following summer.

Parting Company

The *Antarctic* left Gothenburg on October 16, 1901, and arrived in the South Shetland Islands in January 1902. In February, Nordenskjöld's scientific party of six landed at Snow Hill Island, where they erected a 13½ by 21 ft (4 by 6.4 m) prefabricated hut. With two others, Nordenskjöld made a dog-sledge journey of 375 miles (600 km) south to the King Oscar II Land coast in early October. In December, Nordenskjöld discovered the fossil remains of a giant penguin—the largest ever found—on nearby Seymour Island. Ice conditions remained severe all through the summer, and when the sea froze on February 18, 1903, they knew they were trapped for another winter.

The *Antarctic*, which had spent the winter in the sub-Antarctic islands, turned south on November 5, 1902, to collect the party at Snow Hill Island. After surveying a section of the west coast of the Antarctic Peninsula, Larsen headed for the east coast but found the way blocked by heavy sea ice. Gunnar Andersson, Toralf Grunden, and Samuel Duse were put ashore at Hope Bay on December 29 to establish a depot and try to reach the scientists at Snow Hill Island over land.

Andersson and Larsen had a contingency plan. If the *Antarctic* arrived at Snow Hill Island and the land party had not arrived by January 25, they would assume the overland attempt had failed, and the ship would return to get them at Hope Bay between February 25 and March 10. If the *Antarctic* had not reached Snow Hill Island by February 10, the overlanders and the scientists would return to Hope Bay to await the ship.

Andersson, Grunden, and Duse set off southward. Their only map was an inaccurate chart by James Clark Ross from 1843. They crossed the ice to Vega Island, but open water prevented them from reaching Snow Hill Island. They returned to Hope Bay by January 13 and waited for the *Antarctic*.

Across Wild Waters

Carl Larsen guided five of the stranded crew of the *Antarctic* across the waters of the Erebus and Terror Gulf in a small whaleboat during a 14-day trek across ice and water to Snow Hill Island.

Past Drifting Icebergs

The expedition sailed between the tip of the Antarctic Peninsula and Joinville Island through a passage Nordenskjöld named Antarctic Strait (now Antarctic Sound). Vessels passing through often find the sound blocked with icebergs pushed out from the Weddell Sea.

Unkempt but Fit and Well

Toralf Grunden, Gunnar Andersson, and Samuel Duse (left to right), who were trapped at Hope Bay for nine months, supplemented their meager supplies with any penguin, seal, and fish they could catch. Their unshaven faces, blackened by blubber and soot, show the state they were in when reunited with the others at Snow Hill Island.

Shipwreck

The *Antarctic* had pushed deep into the ice of Erebus and Terror Gulf, and on January 10, 1903, the ship underwent the first of several violent squeezes from the ice pack. The damaged and sinking vessel (which was holding a large collection of scientific samples) was abandoned on February 12, and the 20 men—and the ship's cat—crossed 25 miles (40 km) of rough sea ice and patches of open water to Paulet Island. Arriving there on February 28, they built a stone hut, 23 by 33 ft (7 by 10 m). They had salvaged a ton of food from the ship's stores, and they killed 1,100 penguins as a food supply. Over the long, dark winter there was little to do, so meals and taking temperature readings became the highlights of each day.

Those at Snow Hill Island made several excursions and recorded hourly meteorological observations (which differed significantly from those of the first winter), but the three stranded at Hope Bay had little but survival to occupy them. They built a primitive stone hut with a tarpaulin roof, which they could warm with a blubber stove to a relatively balmy few degrees below freezing. They took turns at cooking, and developed an elaborate ritual of formally thanking the cook after each meal. Assuming the *Antarctic* was lost and their whereabouts unknown, on September 29, 1903, they started off toward Snow Hill Island and hope of rescue.

That same day, at Snow Hill, Nordenskjöld and Ole Jonassen set out on an excursion; they soon had to turn back, but began again on October 4. By October 12, they had explored the whole west coast of James Ross Island and were traveling along the north coast of Vega Island. Far out on the ice they saw what they took for penguins—and then realized were men. It was the three from Hope Bay, and the two parties met with "delirious eagerness" at a point they renamed Cape Well-met.

Meanwhile, several search and rescue expeditions were being prepared, including one on an Argentinian corvette, the *Uruguay*, under the command of Lieutenant Julian Irizar, who was Argentina's naval attaché to Britain. Arriving at Seymour Island on November 7, Irizar's crew found a boathook that Nordenskjöld had left on a signal cairn less than two weeks earlier. When they investigated further, they spotted the nearby camp of two of Nordenskjöld's men, who led Irizar to the Snow Hill Island hut. The two marooned land parties were saved, but it seemed the *Antarctic* and her crew were lost.

At about 10:30 that night the dogs started barking. They were signaling the arrival of Larsen and five others, who reported that all but one of the *Antarctic*'s crew had survived and were on Paulet Island. Nordenskjöld wrote: "No pen can describe the joy of this first moment … when I saw amongst us these men, on whom I had only a few minutes before been thinking with feelings of the greatest despondency."

They had come the long way round. When the ice broke up in October, Larsen's party had rowed to Hope Bay in a whaleboat, only to find the three stranded there had left for Snow Hill five weeks earlier. With only a tent pole for a mast and the tarpaulin from Hope Bay as shelter, they crossed the treacherous Erebus and Terror Gulf in their tiny craft, but had to walk the last 15 miles (24 km) across solid sea ice.

Reunion

The *Uruguay* solved Captain Larsen's final problem: how to rescue the 14 men left on Paulet Island. On the evening of November 11, 1903, when the botanist Carl Skottsberg heard a sound, he thought: "I must have been dreaming. The sound is repeated. It must be so … The boat is here! … I thump at the sleepers beside me: can't you hear it is the boat … The shouts are so deafening that the penguins awake and join in the cries; the cat, quite out of her wits, runs round and round the walls of the room."

The Swedish expeditioners were taken to Buenos Aires. On the way, they called in at Hope Bay to pick up the rock samples and paleobotanical fossils that Andersson had collected. This was typical of an expedition that carried out a full scientific program throughout a remarkable, harrowing adventure. DM

**OTTO NORDENSKJÖLD
AND THE *ANTARCTIC*
JAN 1902–FEB 1903**

William Spiers Bruce
Scottish National Expedition: 1902–04

Scientific Know-how
Science was the motivation for the *Scotia* expedition of 1902–04. William Bruce was a keen oceanographer and he had with him a meteorologist, botanist, zoologist, and taxidermist.

Marine Sample
This large cuttlefish taken in Scotia Bay at Signy Island was one of many marine invertebrate specimens that the *Scotia* brought back from Antarctica. It was a collection that remained unequaled for several decades.

William Bruce, leader of the Scottish National Expedition, first visited Antarctica at the age of 25 in the summer of 1892–93 as doctor on a whaling expedition from Dundee to the Falkland Islands and beyond. During the voyage, he also carried out oceanographic research, and this experience fired Bruce (1867–1921) with the idea of returning with his own scientific expedition.

This ambition led to his refusal of Scott's invitation to join the *Discovery* expedition as its naturalist, and to his eventual success in persuading the Coats brothers, prosperous Glasgow industrialists, to finance a private Scottish expedition after the British government had refused to help.

Bruce invited Thomas Robertson, the commander of his earlier whaling expedition, to captain his ship, a Norwegian whaler renamed the *Scotia*; and as well as a crew of 25, he recruited seven scientists, including a bagpipe-playing laboratory assistant. This exceptionally large contingent demonstrated that science, not pole-seeking, was the expedition's goal, with the focus on hydrographic research of the Weddell Sea and a survey of wildlife in the South Orkney islands.

Working through the Winter
After a two-month voyage from Scotland, the expedition reached the Falkland Islands in January 1903, and the South Orkneys a month later. Encountering heavy ice, the *Scotia* pushed on to the South Sandwich Islands but became beset not much further south, at 70°25′S.

She broke free a week later, but further progress south was impossible, and only with great difficulty did Captain Robertson work the ship northward, toward the

South Orkneys, where he found a winter anchorage in a sheltered bay on the south coast of Laurie Island.

The *Scotia* was soon frozen in. Bruce established a shore base, with a stone hut as living quarters and magnetic and meteorological observatories and other structures for the scientific work. Fieldwork continued all winter, including detailed surveys of the coastline, with soundings being taken through holes cut in the sea ice, and botanical studies of the stunted vegetation of the island. In the spring, returning penguins were serenaded on the bagpipes by the laboratory assistant before being killed to provide much needed fresh meat; their skins were preserved by the expedition's zoologist.

It was not until late November 1903 that the *Scotia* was freed from the ice, by the combined efforts of a timely northeasterly gale and the crew with ice saws and explosives. Bruce left six men ashore, including the biologists to study the penguin colonies, and he sailed north to refit the ship in Buenos Aires. He persuaded the Argentinian government to continue meteorological observations on Laurie Island and brought back three Argentine scientists with him to take over. The Scottish meteorologist and the cook were left behind to assist the Argentinians when the *Scotia* headed southward again in late February 1904.

An Important Discovery
This time the ice was less troublesome, and the ship reached 72°18′S quite quickly, before being forced to a halt. A sounding showed they were in comparatively shallow water. On climbing to the crow's nest, Captain Robertson was delighted to see an ice plateau to the southeast. After breaking out of the ice, the ship moved along a fairly clear lead parallel with this coast, which Bruce named Coats Land in honor of his patrons, and followed it for 150 miles (240 km) toward the southwest, reaching as far as 74°S.

Bruce deduced that this long coastline must be an extension of Enderby Land, and therefore part of the continent rather than just an island—a discovery that was as significant as Ross's of McMurdo Sound on the other side of the continent. They were trapped in the ice by a northeasterly gale and fearful of being frozen in for the winter (as would happen to Shackleton's *Endurance* 12 years later in almost the same area), but luckily a southwesterly gale re-opened the pack and the *Scotia* fled northward, reaching Northern Ireland in July 1904.

The four scientists and the cook left on Laurie Island continued their work throughout the winter, and were relieved the following summer by the Argentinian ship the *Uruguay*, which had rescued Otto Nordenskjöld's expedition the previous summer. Another Argentinian party took over observations for 1905, thus initiating an unbroken series of expeditions that endures to the present and makes Laurie Island the oldest continually operating base in the Antarctic region. LC

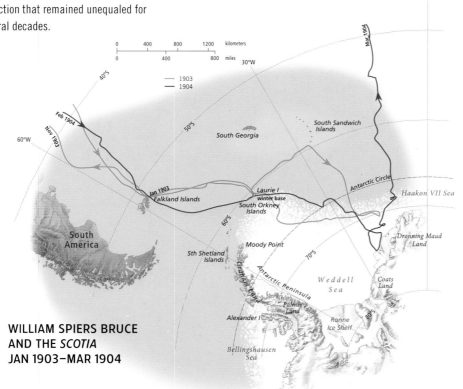

**WILLIAM SPIERS BRUCE AND THE *SCOTIA*
JAN 1903–MAR 1904**

Skinned and Roasted

After being serenaded with bagpipes, emperor penguins (left) on Laurie Island were killed and eaten. But the emphasis of the Bruce expedition was science, so their skins were preserved.

Captive Listener

It may seem that the emperor penguin (above) admires piping—until one notices the string tying it to Kerr, the laboratory assistant. The expedition doctor noted that "these lethargic, phlegmatic birds" responded with complete indifference to a wide range of bagpiped tunes.

Through Fog and Ice

Laurie Island narrows to an isthmus at the modern Orcadas Station, where Bruce established his meteorological observations base in 1902, then handed it over to the Argentinian government. This view from the shore of Scotia Bay looks westward across the base to Jessie Bay, solidly packed with ice.

Jean-Baptiste Charcot
French Antarctic Expeditions: 1903–05, 1908–10

Jean-Baptiste Charcot's father was Jean-Martin Charcot, the founder of modern neurology and an important influence on Sigmund Freud. Although young Jean followed his father into medicine, he was fascinated by the sea. When he inherited a fortune, he invested much of it in the construction of the *Français*, a three-masted polar exploration vessel of 270 tons (245 tonnes) displacement.

The First Expedition: the *Français*

When news of the disappearance of Otto Nordenskjöld's expedition reached Europe, Charcot resolved to rush to his rescue. He wrote to his friend, industrialist Paul Pléneau, asking if he would like to participate. Pléneau's reply was a telegram: "Where you like. When you like. For as long as you like." He went as the expedition's photographer. The *Français* expedition attracted strong official and public support, and left Le Havre on August 27, 1903. The destination was the western Antarctic Peninsula, where an ambitious program of surveying and science was planned as soon as the *Antarctic* had been located. However, in Buenos Aires they met the rescued Nordenskjöld expeditioners, and so the French party continued its own explorations. The *Français* left Ushuaia on January 24, 1904, and was in the ice of the South Shetlands by the beginning of February.

Despite being hampered by a temperamental ship's engine, the men of the *Français* managed to discover and name the excellent anchorage of Port Lockroy and they passed through the Lemaire Channel as far as the Biscoe Islands, before retreating to the northern part of what is now Booth Island in the face of the advancing winter ice. Here the crew moored the ship in a bay that Charcot named Port Charcot after his father, and sat out the winter of 1904. Despite a large library of books and musical recordings, and expansive provisioning which extended to daily fresh bread and ample rations of wine and spirits, the men's morale fell dramatically during the enforced inactivity of winter.

Exploration began again the following summer, but on January 15, 1905, the ship hit a rock off Alexander Island and began to sink. The crew made a massive effort to stem the leaks and the *Français* managed to limp back to Argentina. Charcot sold her to the Argentinian government for use as supply ship to its new South Orkney Islands station.

The expeditioners returned to France and were greeted as polar heroes. The publication of 18 volumes of scientific reports amply justified that standing.

Ship Aground

When the *Français* struck a submerged rock near Cape Tuxen, some of the crew tried to stem the leak, while others hand-pumped for 23 hours a day. Even as the ship finally sailed to safety, the crew held to their original goal and completed their mapping of the Palmer Archipelago on the Antarctic Peninsula.

For Marguerite

Named by Charcot for his new wife, Marguerite Bay (left) is a major embayment on the western Antarctic Peninsula, where Graham Land to the north meets Palmer Land. Over the mountains is the Larsen Ice Shelf.

Rocky Memorial

This cairn on Megalestris Hill behind Port Circumcision (above) was erected by Charcot's expedition to commemorate sailing the *Pourquoi-Pas?* into the port on New Year's Day, 1909.

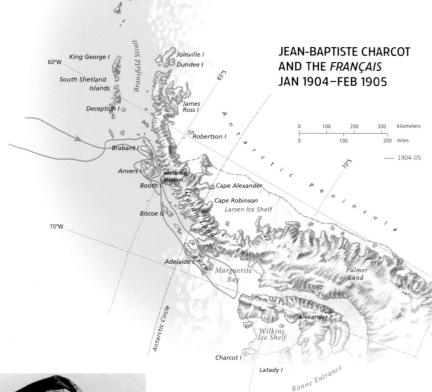

JEAN-BAPTISTE CHARCOT
AND THE *FRANÇAIS*
JAN 1904–FEB 1905

Attracted to the Cold
Jean-Baptiste Charcot (1867–1936) was born to wealth, but chose to spend much of his working life in harsh polar conditions. He always pointed out he had named Charcot Land and other features after his famous father, not after himself.

The *Pourquoi-Pas?* Expedition

As a young man Charcot had owned a yacht called *Pourquoi-Pas?* ("Why Not?"). Buoyed by his earlier success and quite generous government funding, he designed another polar vessel, and he gave her that name. The strong, well-equipped, and comfortable vessel weighed 880 tons (800 tonnes), and was fitted with a purpose-built 550-horsepower engine. The ship left Le Havre on August 15, 1908, with 30 scientists, officers, and crew on board, as well as Charcot's new wife, Marguerite, who left the ship in Punta Arenas.

Just before Christmas, they reached Deception Island. On January 1, 1909, they found a sheltered small anchorage at Petermann Island, and named it Port Circumcision after the feast day on which it was discovered. Only three days later, things started to go wrong. With Lieutenant E. Godfroy, who specialized in tides and atmospherics, and E. Gourdon, a geologist and glaciologist, Charcot took a boat south to Cape Tuxen and the Berthelot Islands. They planned to be away for the day, so took no extra supplies, shelter, or clothing. Soon after finishing their last food, the ice closed in and it began to snow. The three were lucky to survive long enough to be rescued by the *Pourquoi-Pas?* three days later. Just a few hours after the rescue, the *Pourquoi-Pas?* ran onto rocks near Cape Tuxen. Eventually, the crew managed to shift enough cargo so that the ship could pull free of the rock at the top of the tide 24 hours later. They sailed back to Petermann Island for makeshift repairs.

The rest of the summer of 1909 was spent charting south of the Antarctic Circle, farther south than Charcot had ventured previously. As the winter approached, the ice forced the *Pourquoi-Pas?* back to Port Circumcision, where the crew securely moored the ship and ran steel wires across the mouth of the cove to keep icebergs out. The four huts built on shore were lit by electricity generated by the ship. Despite all their precautions, a storm in June drove several icebergs into the bay, and one of these damaged the ship's rudder so severely that

it could be repaired only with difficulty. At the end of the winter the men sailed north to Deception Island to pick up fresh provisions. There a diver told Charcot that the ship's hull was gravely damaged, and that the vessel would sink if there was another incident.

Nevertheless, Charcot resolved to turn south again to conduct his summer program. Beyond Marguerite Bay, which he had named for his new wife the previous summer, he discovered new land at 70°S—later named Charcot Land (although it is now known to be an island) for his famous father. Proceeding west along the edge of the ice, the *Pourquoi-Pas?* turned her bow north only on January 22, 1910, when she had reached 124°W. When the crew saw Peter I Øy in the distance, they were the first to do so since Bellingshausen in 1821.

A Victorious Return

Charcot and his expeditioners were greeted by a crowd of enthusiastic residents and by floods of congratulatory telegrams when the *Pourquoi-Pas?* arrived at Punta Arenas. After the ship underwent substantial repairs in Montevideo, she sailed to Rouen, arriving on June 5, 1910, to a tumultuous reception.

The fuss was warranted. The expedition had charted 1,240 miles (2,000 km) of coastline, had given form to the whole western side of the Antarctic Peninsula, and had collected so much scientific data that it would take a decade and 28 volumes for it all to be published. Roald Amundsen, who had described Charcot as "the French savant and yachtsman," credited him with "opening up a large extent of the unknown continent … and the scientific results were extraordinarily rich."

Charcot was showered with honors. He went on to win a Distinguished Service Cross while he was serving with the British Navy in World War I. He never returned to Antarctica, but he continued his research in the cold waters of the north Atlantic. On September 16, 1936, the *Pourquoi-Pas?* was wrecked off Iceland; there was only one survivor. Charcot went down with his ship. DM

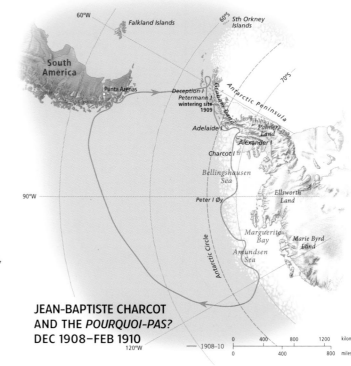

**JEAN-BAPTISTE CHARCOT
AND THE *POURQUOI-PAS?*
DEC 1908–FEB 1910**

Ernest Shackleton
British Antarctic Expedition: 1907–09

When Ernest Shackleton (1874–1922) arrived back in England on June 12, 1903, he found that Robert Scott's 1901–04 expedition, from which he had been virtually sacked, was a controversial subject amid accusations of incompetence. Shackleton, by contrast, was regarded as a polar hero. He was soon declared fit for more Antarctic service, and he used his fame to raise his social standing. After a short period as a journalist, he was appointed the secretary of the Royal Scottish Geographical Society, fought a losing election for a parliamentary seat, and set about organizing his own Antarctic expedition.

On February 8, 1907, Shackleton's main sponsor, William Beardmore, promised a large loan. Shackleton began assembling his British Antarctic Expedition just three days later. However, he soon learned that Scott was also planning to sail south again, and regarded not just the *Discovery* base but the whole of Ross Island as rightfully his; he insisted that Shackleton should stay away from that area. Under pressure from their mutual colleague Edward Wilson, Shackleton agreed to Scott's outrageous constraint.

Establishing Winter Quarters
The expedition left England on August 7, 1907, aboard the *Nimrod*. The ship also carried an Arrol-Johnston motor car made at a Scottish factory that was owned by Beardmore. Shackleton, who left later, met the ship in Lyttelton; because the *Nimrod* could only make 6 knots, catching up was not difficult. The 330-ton (300-tonne) ship was 40 years old and battered, but Shackleton could not afford his first choice, the *Bjørn*. To save coal, the *Nimrod* was towed from New Zealand into the ice by the larger *Koonya*, owned by New Zealand's Union Steamship Company and captained by Frederick Evans. The ships left Lyttelton on New Year's Day, 1908. On January 15, after two weeks of life-threatening storms, the *Koonya* turned north just past the Antarctic Circle. Shackleton noted that the *Koonya* was the first steel vessel to cross the Antarctic Circle.

At 9:30 AM on January 23, they sighted the Ross Barrier (now the Ross Ice Shelf) and sailed along it, looking for the bay in the ice that Scott had named Balloon Bight after he and Shackleton went aloft there. They passed the bay where Borchgrevink had landed in 1900, "but it had greatly changed." Then they came to a large harbor, which Shackleton named the Bay of Whales because "it was a veritable playground for these monsters," before he fled the heavy ice that was driving into the bay. When Shackleton realized that this was in fact where he had ballooned, but that it had changed dramatically because a large section of the ice shelf had broken away, he resolved to make his winter base on land rather than on ice. When pack ice prevented him from continuing east, he broke his promise to Scott and turned west toward Ross Island. But even there sea ice

stopped the *Nimrod* from approaching the old *Discovery* hut at Hut Point. They erected their prefabricated hut at Cape Royds. The *Nimrod* left a wintering party of 15 when she sailed for Lyttelton on February 22.

Overland Excursions
Their first major land excursion was the first ascent of Mount Erebus. Edgeworth David (the leader), Douglas Mawson, and Alistair Mackay, with a support party of three, left on March 5, and struggled through savage blizzards to the summit (12,450 ft/3,795 m) on March 10; they were back two days later.

On October 29, Shackleton, Frank Wild, Jameson Boyd Adams, and Eric Marshall set out for the South Pole. Their plan was to walk and use ponies, even though the advantages of skis and dogs for polar travel were by then widely recognized. The car, Antarctica's first, was virtually useless. By November 26, Shackleton could again claim a record for the farthest south—farther than he had penetrated with Scott—but he was worried that they were already rationing food.

In early December, they became the first men to reach the southern extent of the Ross Ice Shelf and to climb the vast Beardmore Glacier onto the polar ice

Lionized by the Press
After the *Nimrod* expedition, London's *Daily Telegraph* gushed that: "In his photographs Mr. Shackleton is nothing more than an intelligent but ordinary naval officer. In reality he radiates the fascination of an indefinable force ... If he has the face of a fighter, he has the look of a poet; one must be both a fighter and poet to accomplish what he has done."

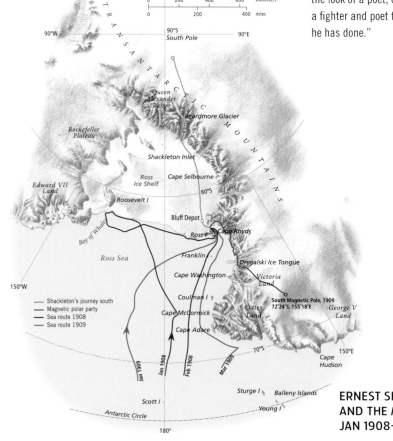

**ERNEST SHACKLETON AND THE *NIMROD*
JAN 1908–MAR 1909**

cap. The climb was a nightmare, and on December 7 their last pony fell into a crevasse and died. Just after Christmas, the slope became easier and they were on the Polar Plateau, but they had only one month's food to reach the Pole and return to their supply dump—a journey of almost 500 miles (800 km). They were all suffering from altitude sickness. Shackleton described the symptoms: "My head is very bad … as though the nerves were being twisted up with a corkscrew and then pulled out." He was torn between his ambition to reach the Pole and his sense of responsibility for his men. By January 2, he was writing: "I must … consider the lives of those who are with me. I feel that if we go on too far it will be impossible to get back … man can only do his best … time is going on and food is going also." On January 9, they left everything behind and made a one-day push to the south—and came within 100 miles (160 km) of the Pole— and back. At 88°23′S, they hoisted the Union Jack, took some photographs, and left a commemorative brass cylinder. Looking southward, Shackleton noted: "We feel sure that the goal we have failed to reach lies on this plain … we stayed

only a few minutes … Whatever regrets may be, we have done our best." By the time they returned to the camp that day, they had traveled 41 miles (66 km).

On the return journey through the Transantarctic Mountains, they hoisted a sail on their sledge and quickly reached the top of the Beardmore Glacier, but the descent to their food dump while weak and short of food was extremely draining: "I cannot describe adequately the mental and physical strain of the last 48 hours," Shackleton wrote of January 26 and 27.

Smoking Mount

The Erebus ascent group consisted of Edgeworth David, Douglas Mawson, and Alistair Mackay, plus a support group of three. The party reached the summit through savage blizzards to achieve the first ascent of the mountain.

Slippery Ascent

On one of the few surviving recordings of Shackleton's voice, he recounts the first ascent of Mount Erebus. He took this photograph of the climbers setting out on the expedition's first sledging journey at about 9 AM on March 5, 1908. He reported that they had trouble keeping on their feet in the soft snow.

Time Capsule

The interior of the hut was restored by a New Zealand team to look as it did in 1908. Shackleton wrote that the cooking range was designed "to burn anthracite coal continuously day and night and to heat a large superficial area of outer plate, so that there might be plenty of warmth given off in the hut."

A Narrow Escape

They were near death yet, on January 31, Shackleton surreptitiously gave Frank Wild his breakfast biscuit. Wild wrote: "I do not suppose that anyone else in the world can thoroughly realize how much generosity and sympathy was shown by this; I do by God I shall never forget it." They all were suffering from acute dysentery and were "appallingly hungry."

On February 21, Shackleton wrote: "Our need is extreme and we must keep going … food lies ahead and death stalks us from behind." Two days later, they arrived at Bluff Depot and food supplies, but Marshall was so ill that Shackleton and Wild left him with Adams and pressed on to arrange rescue. They found that the building at Hut Point was empty, but there was a note saying that the ship was nearby, so they set fire to a hut

to attract attention. It worked; by March 1, both were on board the *Nimrod*. After "a good feed of bacon and fried bread," the indomitable Shackleton set off with three companions on the 18½-hour trek to rescue Adams and Marshall. Shackleton wrote: "We were all safe on board at 1 AM on March 4." The polar party had walked 1,700 miles (2,740 km) in 128 days.

To the South Magnetic Pole

In the meantime, the three who had climbed Mount Erebus were attempting to reach the South Magnetic Pole. The northern sledging party of the two Australian geologists, Edgeworth David and Douglas Mawson, and the naval surgeon Alistair Mackay, tried using the motor car to transport supplies across the sea ice, but soon abandoned it in favor of man-hauling—a challenging task, as one of the two sledges weighed 606 lb (275 kg) and they had some 1,000 miles (1,600 km) to cover.

They had left Cape Royds on October 5, 1908, while supply depots were still being set up for the geographic polar bid. The northern party walked northward on the sea ice along the west edge of Victoria Land (which they claimed for Britain on October 17), with Mawson as the pathfinder. The heavy loads and soft snow slowed them down, so they cached some of their rations and relied on seal and penguin meat, cooked on an improvised stove. They discovered that their intended route up the Drygalski Ice Tongue proved impractical.

David recalled that, by December 20, "We had not yet climbed more than 100 feet [30 m] or so above sea level … We knew that we had to travel at least 480 to 500 miles [770 to 800 km] to the Magnetic Pole and back to our depot, and there remained only six weeks to accomplish this journey." They pushed on with a single sledge weighing 670 lb (305 kg), and climbed a route they named Backstairs Passage.

Standing Still

The hut at Cape Royds has changed little, but there has been considerable restoration—in 1947 visitors found the boards off the roof of the hut and the garage had already collapsed. The hut is now in the care of the New Zealand Antarctic Heritage Trust.

Proud Moment
The northern sledging party (from left: Mackay, David, Mawson) reached what they thought was the South Magnetic Pole, and claimed the area, at 3:30 PM on January 16, 1909. Mackay and David set up the Union Jack while Mawson positioned the camera. David recalled that he pulled the string to trigger the camera: "Then we gave three cheers for His Majesty the King."

Pony Power
Fridtjof Nansen, the greatest Arctic explorer of his time, had advised Shackleton against using horses for drawing sledges in Antarctica. But, through the Hong Kong and Shanghai Bank, Shackleton ordered a dozen Manchurian ponies for the expedition.

By January 15, 1909, they were up on the Polar Plateau and more than 7,000 ft (2,135 m) above sea level. Their compass was only 15 minutes off vertical, but that did not mean the three had only 15 nautical miles (equivalent to about 28 km) more to travel: they had to allow for the daily movement of the Magnetic Pole, and they might even have to chase it. According to David, "Mawson considered that we were now practically at the Magnetic Pole, and that if we were to wait for twenty-four hours taking constant observations at this spot the Pole would, probably … come vertically beneath us." But rather than wait for it to come to them, they decided to go to the approximate mean position of the Pole. Like Shackleton had done a week earlier, on January 16, they left everything and made a one-day rush for their goal. When they got there, they hoisted the Union Jack, claimed the area for the British empire, gave three cheers for the king, and took a photograph. Several years later, calculations made from Mawson's observations showed the party had been close to the area of polar oscillation but had not penetrated it.

A Fortunate Rescue

To meet the *Nimrod* at the depot on the Drygalski Ice Tongue, the northern party had to average 17 miles (27 km) for 15 days. They approached the depot on the morning of February 4; only a day later they heard a rocket, and rushed from their tent to see the *Nimrod*. Their journey of 1,260 miles (2,030 km) over 122 days ended with their first baths in four months.

Only later did they learn how lucky they were to be found. The *Nimrod* had to search a daunting 200 miles (320 km) of coast for them. John King Davis, the first officer, was on duty from 4 AM to 8 AM on February 3. When Captain Evans came on deck and confided that he thought there was little chance of finding the land party, he asked Davis if his watch had examined the

entire coast. Davis admitted that a small section had been obscured by icebergs. They decided to steam back for four hours. Sailing into a narrow fault in the ice behind the obstructing icebergs, they saw that "upon the crest of a little knoll of ice was a green conical tent." The expeditioners had been farther up the glacier when the *Nimrod* had first passed this spot. If the northern party had been left to make their own way back to Cape Royds, their chances of surviving would have been very slight because that late in the summer much of the sea ice had broken up.

The *Nimrod* sailed back to Cape Royds and then to Hut Point, where they found Shackleton and his party. On the way back to New Zealand, Shackleton sailed along the coastline beyond Cape Adare to map the Terre Adélie coast as far west as possible, and reached 166°14′E—beyond any previous effort. They reached Lyttelton on March 25 and England in June.

Shackleton was knighted for his achievements, but he thought he had failed, even though he had paved the way to the Pole. As Roald Amundsen was later to write of Shackleton: "Seldom has a man enjoyed a greater triumph; seldom has a man deserved it better … I admire in the highest degree what [Shackleton] and his companions achieved with the equipment they had. Bravery, determination, strength they did not lack. A little more experience … would have crowned their success." Amundsen added what could well be the defining judgment: "Sir Ernest Shackleton's name will for evermore be engraved with letters of fire in the history of Antarctic exploration. Courage and willpower can make miracles. I know of no better example than what that man has accomplished." DM

Polar Footwear
Shackleton brought ski boots and finnesko on the expedition with him. Finnesko are boots of reindeer fur filled with sennegrass, which adds insulation and absorbs moisture so feet stay warm and dry.

Roald Amundsen
1910–11

Roald Amundsen has been called a consummate professional in an age of amateurs. The story of his journey to the South Pole certainly reveals organizational skills that his competitors lacked, but there were also controversial elements, even errors, in the Norwegian polar conquest.

Amundsen's chronicle of his expedition reveals a warm-hearted, generous personality, who was far from the cunning manipulator that the supporters of Scott painted him at the time. And he strenuously rejected the title of adventurer: "Adventure is only an unwanted interruption ... [The explorer] does not seek titillation, rather previously unknown facts," he wrote. "... an adventure is simply an error in his calculations ... a very unhappy proof that no one can take into consideration all eventualities ... Any explorer experiences adventures. They excite him and he looks back on them with pleasure, but he never seeks them out."

A Change of Direction
Amundsen had been first to conquer the Northwest Passage, and had spent a winter in Antarctica with Adrien de Gerlache on the *Belgica*. Although he was a meticulous planner, in 1909, when he was aiming to be the first to the North Pole and had borrowed Fridtjof Nansen's *Fram* for the purpose, events forestalled him: in September of that year, both Robert Peary and Frederick Cook claimed the North Pole. As Amundsen recalled: "Just as rapidly as the message had traveled over the cables I decided on my change of front—to turn to the right-about, and face to the South." With a secrecy that later attracted criticism, he told only his brother, and later Lieutenant Thorvald Nilsen, captain of the *Fram*. The crew found out in Madeira; until then, they believed they were sailing around Cape Horn and up to the Bering Strait to take advantage of the Arctic drift. They all agreed to stay for the new destination.

On June 7, 1910, the *Fram* left Christiania (now Oslo) and sailed from Madeira on September 9, with 19 men and 97 dogs—the number of dogs had risen to 116 by January 1911. Amundsen telegraphed his intentions to Scott, stating: "Beg leave to inform you *Fram* proceeding Antarctic. Amundsen." He notified his sponsors and Nansen in the same way. But secrecy was not Amundsen's only odd trait: he would not take a doctor on expeditions, mainly because he thought that, being well educated, a doctor might encourage dissent.

Having studied the reports of previous expeditions, Amundsen decided to establish his base on the ice at the Bay of Whales because it was closer to the Pole than Ross Island. The *Fram* arrived there on January 14, and the crew erected their prefabricated hut at "Framheim." They killed seals and penguins for food: Amundsen wrote: "An emperor penguin just came on a visit—soup-kettle." On February 4, Scott's ship, the *Terra Nova*, visited for a day, but Scott was already on

Ross Island. Amundsen's men established supply depots as far as 82°S. The *Fram* departed northward before the Bay of Whales froze over.

Surprisingly, on an expedition so well planned, there was one key blunder: they had the *Almanac* for 1911, but had forgotten the 1912 edition, which contained essential information for fixing the position of the South Pole. This meant that if Amundsen were to claim the Pole, he had to be there before the end of 1911.

When the sun appeared on August 24 after four months of darkness, the sledges were already packed for the polar run. Amundsen set off for the South Pole in September—very early in spring—but had to return because conditions were too cold for survival. The dogs,

One-shack Town
Framheim in February 1911 was an orderly place. The *Fram* was about to depart and the first of the depot-laying parties was ready to set out. Some 900 cases had been unloaded and the hut was built. Its walls were tarred and the roof covered with tarred paper.

High Achiever
It seems likely that Roald Amundsen (1872–1928) was not only the first to walk to the South Pole, but may also have reached the North Pole first (in an airship), as well as being the first to sail through the Northwest Passage.

Skin-side Inside
Amundsen, dressed here as he went to the South Pole, learned from the Arctic peoples. He wrote, "no one had yet wintered on the Barrier so we had to be prepared for anything ... with the richest assortment of reindeer-skin clothing ... after the pattern of the Netchelli Eskimo."

Strategic Positioning

Much of Amundsen's success came from his choice of sites. The Bay of Whales, seen here in early morning, was 1 degree closer to the Pole than McMurdo Sound. Shackleton had turned away from there, fearing the Ross Ice Shelf could calve and leave his whole camp adrift.

Taking an Observation

The *Fram*, captained by Lieutenant Thorvald Nilsen, sailed the expedition members to Framheim in the Bay of Whales, spent the winter in Buenos Aires, then returned to collect the men in the 1911–12 summer.

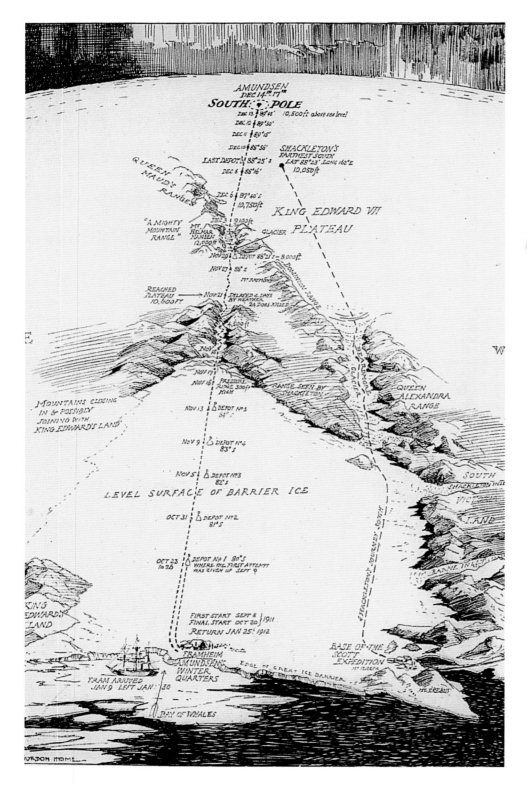

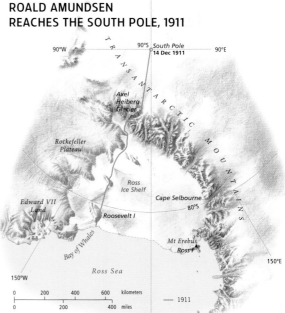

Inland Mapping
In the early twentieth century, overland expeditions into the Antarctic continent began to make mapping of the interior possible. Nevertheless, early records were really just sketch maps of the journeys. Shown here are Amundsen's 1911 route to the Pole (left) and Shackleton's 1908–09 route (right).

after six months of leisure, did not "understand that a new era of toil had begun." On October 19, five men started for the South Pole on skis and sledges, and by November 7 they were far enough across the Ross Ice Shelf to see mountains to the south.

At the base of the climb up to the polar ice cap, Amundsen outlined his plan: "The distance … from this spot to the Pole and back, was 683 miles [1,100 km] … we decided to take provisions and equipment for sixty days on the sledges, and to leave the remaining supplies … and outfit in depot … We now had forty-two dogs. Our plan was to take all forty-two to the plateau; there twenty-four of them were to be slaughtered, and the journey continued with three sledges and eighteen dogs. Of the last eighteen, it would be necessary, in our opinion, to slaughter six in order to bring the other twelve back to this point. As the number of dogs grew

less, the sledges would become lighter and lighter, and when the time came for reducing their number to twelve, we should only have two sledges left."

His calculations were almost perfect: it took eight days less than the time allowed, and they returned with 12 dogs. They climbed the Axel Heiberg Glacier from November 17 to 21, and, with some sadness, because a "trusty servant lost his life each time," they shot the dogs at a camp they called the Butcher's Shop at the top of the glacier.

On December 8, they passed Shackleton's 88°23′S. It was a milestone for the 39-year-old Amundsen, who wrote: "No other moment of the whole trip affected me like this. The tears forced their way to my eyes …" They laid down their final depot at the spot, and made a push for the South Pole.

Arriving at the South Pole
Of December 14, 1911, Amundsen wrote: "At three in the afternoon a simultaneous 'Halt!' rang out from the drivers. They had carefully examined their sledge-meters and they all showed the full distance—our Pole by reckoning. The goal was reached, our journey ended … I had better be honest and admit straight out that I have never known any man to be placed in such a diametrically opposed position to the goal of his desires as I was at that moment. The regions around the North Pole—well, yes, the North Pole itself—had attracted me from childhood, and here I was at the South Pole. Can anything more topsy-turvy be imagined?

"… Pride and affection shone in the five pairs of eyes that gazed upon the flag, as it unfurled itself with a sharp crack, and waved over the Pole. I had determined that the act of planting it—the historic event—should be equally divided among us all. It was not for one man to do this; it was for all who had staked their lives in the struggle, and held together through thick and thin.

"… Everyday life began again at once … Of course, there was a festivity in the tent that evening—not that champagne corks were popping and wine flowing—no, we contented ourselves with a little piece of seal meat each, and it tasted well and did us good."

Four-legged Friends
Voyaging south, Rønne the sailmaker (above) danced with one of the dogs— "in the absence of lady partners"—to a tune on the fiddle. Amundsen reported that kenneling was adjusted to house dogs that were friendly with one another side by side, and that these dogs would be in the same teams.

Pole Position
On December 14, 1911, Amundsen, with Helmer Julius Hanssen, Olaf Bjaaland, Oscar Wisting, and Helge Sverre Hassel, reached the South Pole. They spent three days there, making sure they had reached the correct position. Today, the Pole is marked by the permanent structures of Amundsen–Scott Base.

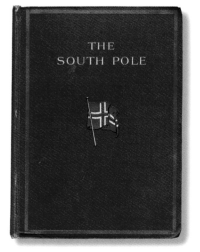

Traveler's Tale
Roald Amundsen's book, *The South Pole*, reveals a charming man of self-deprecating humor.

With the exactness that characterized the whole expedition, the polar party spent the next few days surveying around their camp until they were certain they had reached the Pole by any calculations. On December 17, at the point they concluded was the Pole, they erected a tent with a Norwegian flag on top and "inside the tent, in a little bag, I left a letter, addressed to HM the King, giving information of what we had accomplished. The way home was a long one, and so many things might happen …" They also left some clothes and a sextant, before lacing the tent and turning to the north.

The Secret of Success
By January 6, they were back on the Ross Ice Shelf, and at 4 AM on January 25 they reached Framheim, after a 99-day journey. They were probably fitter and

healthier than when they left. The science of nutrition was in its infancy, but at Framheim the Norwegians ate a healthy mixture of wholemeal bread, berry preserves, and undercooked seal meat. Whereas Shackleton's party barely struggled back to rescue and Scott's party did not come back at all, because their rations resulted in starvation and nutritional deficiencies, Amundsen's men ate well, and they had so much food on hand that he could write: "We are bringing the purveyors of our sledging samples of their goods that have made the journey to the South Pole and back in gratitude for the kind assistance they afforded us." Shackleton sacrificed a biscuit; Amundsen brought back food as souvenirs.

Five days after they arrived back at their coastal base, they were sailing back to Hobart on the *Fram*. On March 7, 1912, Amundsen telegraphed news of his success to his brother, who told the world. When Amundsen heard Mawson needed dogs for his Australasian Antarctic Expedition, he donated 21, keeping only puppies and the survivors of the South Pole conquest. DM

Custom Designed
The *Fram* (meaning "Forward") was built with a rounded hull so that the pressure of ice would cause her to rise, rather than be crushed. The ship was launched in 1892, and had already performed well in the Arctic before Nansen lent her to Amundsen for his polar expedition.

Robert Falcon Scott
The Last Expedition: 1910–12

Robert Scott arrived in Melbourne on October 12, 1910, on his way to Antarctica to unwelcome news: Roald Amundsen was also bound for the South Pole. The race had begun. Scott was annoyed that the Norwegians were aiming for the Pole that he considered his by right. Science was important to his second polar expedition, but so was claiming the Pole for Britain. From the start, Scott doubted that he would win the race. He wrote: "I don't know what to think of Amundsen's chances. If he gets to the Pole, it must be before we do, as he is bound to travel fast with dogs and pretty certain to start early."

Scott had announced his Antarctic expedition in September 1909, but money was slow to come in, despite widespread endorsement. The *Terra Nova*, which had come to his support on the previous expedition, was to be the expedition ship. Edward "Teddy" Evans was to be his second-in-command. Lawrence "Titus" Oates and Apsley Cherry-Garrard each contributed 1,000 pounds, and were recruited as an officer and assistant zoologist, respectively. Oates wrote to his mother: "Points in favor of going. It will help me professionally as in the army if they want a man to wash labels off bottles they would sooner employ a man who has been to the North Pole [*sic*] than one who has only got as far as the Mile End Road. The job is most suitable to my tastes. Scott is almost certain to get to the Pole and it is something to say you were with the first party. The climate is very healthy although inclined to be cold."

On the Way to the Pole

The *Terra Nova* left Britain at the beginning of June 1910, and Scott joined the ship in South Africa. The expedition photographer, Herbert Ponting, boarded in Lyttelton with 19 Siberian ponies and 33 sledge dogs (and one collie bitch) shipped from Russia. On January 4, 1911, after three weeks in the ice, the ship was off Ross Island. Because Hut Point and Cape Crozier were frozen in, they erected their hut at Cape Evans. It had a wardroom for the 16 officers and scientists, and a separate mess deck for the nine crewmen.

Before winter, a western party led by Griffith Taylor went to the Dry Valleys. A northern party sailed the *Terra Nova* toward King Edward VII Land looking for a suitable wintering-over site, and soon returned to tell of meeting Amundsen and the *Fram* at Framheim at the Bay of Whales; this land party changed plans and were taken to Cape Adare (and, consequently, became known as the eastern party). Scott was surprised to hear that the Norwegians were so close; he had assumed that they would start from the Weddell Sea side of the continent. "There is no doubt that Amundsen's plan is a very serious threat to ours," Scott wrote. "He has a shorter distance to the Pole by 60 miles [100 km]—I never thought he could have got so many dogs safely to the ice. His plan for running them seems excellent. But above and beyond all he can start his journey early in the season—an impossible condition with ponies."

Scott knew the limitations of his ponies. He had led the southern party, intending to set up a supply depot

Author of Merit
Robert Falcon Scott entered the Royal Navy at the age of 13. He died on the ice aged 43. His legacy was his writing, in which he captured the spirit of the Heroic Age of Antarctic exploration.

Doomed to Die
"Titus" Oates and these ponies (far left) were setting out to their deaths. Oates, an expert horseman, had no part in selecting the Manchurian ponies. They had been bought in Harbin and shipped from Vladivostok to New Zealand, where Oates first saw them, and he was not impressed with the animals.

Familiar Vessel
The *Terra Nova* (left) was the 770-ton (700-tonne) Dundee whaler that had come to Scott's rescue on the *Discovery* expedition. Apsley Cherry-Garrard reported that "there was not a square inch of the hold and between-decks which was not crammed almost to bursting … Officers and men could hardly move in their living quarters."

A Home at Cape Evans

Beyond the Cape Evans hut is Mount Erebus. The hut's interior appearance now owes as much to Shackleton's later Ross Sea party, which was marooned here and lived off Scott's excess rations, as it does to the Scott expedition.

Tribute to Heroic Failure

A 9-ft (2.7-m) memorial cross stands on the summit of Observation Hill. It lists the names of the doomed polar party, and the lines: "Who died on their return from the Pole. March 1912. To strive, to seek, to find, and not to yield." Lord Tennyson's words were Cherry-Garrard's suggestion.

at 80°S but, although the 26 dogs performed well, the eight ponies kept sinking in the soft snow. Even when the party switched to marching at night, when the snow was firmer, progress was slow. Finally, One Ton Camp was established 36 miles (58 km—equivalent to about 31½ nautical miles) short of their goal—a shortfall that would prove to be fatal.

Scott insisted on a regular routine through the long, dark winter months. The most remarkable exploit was the

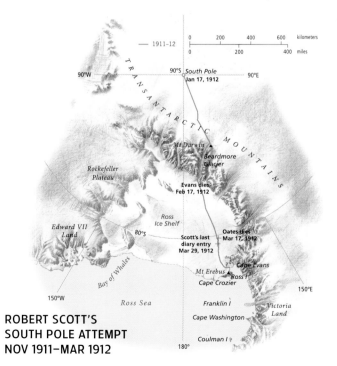

**ROBERT SCOTT'S
SOUTH POLE ATTEMPT
NOV 1911–MAR 1912**

bleak excursion recounted in Apsley Cherry-Garrard's *The Worst Journey in the World*. With Edward Wilson and "Birdie" Bowers, Cherry-Garrard set out in the dark, bitter cold for Cape Crozier, 60 miles (100 km) away, to collect emperor penguin eggs because Wilson believed that they might provide a clue to the evolution of reptiles into birds. They returned five weeks later, with just three eggs, and ravaged faces. (The eggs were analyzed years later, with inconclusive results.) Cherry-Garrard's masterpiece of polar literature begins: "Polar exploration is at once the cleanest and most isolated way of having a bad time which has been devised," and ends: "if you have the desire for know-ledge go out and explore … If you march your Winter Journeys you will have your reward, as long as all you want is a penguin's egg."

The Assault on the Pole

On October 24, 1911, the motor sledges set off to lay depots, and on November 1 the main polar party, led by Scott, left Cape Evans. The two groups comprised 16 men, 10 ponies, 33 dogs, and 15 sledges, including the two motorized ones. By November 6, the polar party had passed the abandoned motor sledges—their crew had continued south, but were now man-hauling. On November 21, the two groups met, and on December 9 they reached the bottom of the Beardmore Glacier: Scott was literally following in Shackleton's footsteps. Here they shot the surviving ponies, and men and dogs carried on up the glacier. It was hard work, especially

Precision Instrument

Lieutenant "Teddy" Evans surveying with the 4-in (10-cm) theodolite that was used to locate the South Pole. Evans had been considering leading his own expedition to Antarctica to explore what is now the Edward VII Peninsula, but Scott adopted the plan as his own and offered Evans the role of his second-in-command.

Cape Evans in Winter

A painting by the expedition's chief scientist, Edward Wilson. A naturalist and doctor, Wilson instigated "the worst journey" to Cape Crozier during the winter of 1911, and perished on the journey back from the Pole.

as they could not ski, and even Scott admitted that "it would be impossible to drag sledges on foot." Two days later, the dogs were sent back and 12 men continued, hauling three heavily laden sledges. On December 14, the day that Amundsen reached the South Pole, Scott was still on the Beardmore Glacier, and writing: "We got bogged again and again, and, do what we would, the sledge dragged like lead … Considering all things, we are getting better on skis." By December 21, Amundsen was on his way home and Scott's party was at Upper Glacier Depot. One of the supporting teams turned back, and eight men continued with two sledges.

On January 3, Scott made a fatal decision. He told his last support team to turn back the next day, but that one of them, Bowers, would come with him to the Pole. The one sledge was already fully laden with supplies for four and Bowers had no skis, so the polar team would be limited to his pace. Tom Crean, "Teddy" Evans, and William Lashly headed back, while Scott, Wilson, Oates,

Bowers, and Petty Officer Edgar Evans carried on. They were never to be seen alive again.

At first things went well. On January 9 they passed Shackleton's farthest south point (88°23′S), and Scott wrote: "All is new ahead." But, by January 11, he was expressing doubts: "We ought to do the trick, but oh! for a better surface." On January 16, they saw a cairn of Amundsen's; Scott's frustration was clear: "The worst has happened … sledge tracks and ski tracks going and coming and the clear trace of dog's paws … The Norwegians have forestalled us and are first at the Pole … All the day dreams must go; it will be a wearisome return. Certainly we are descending in altitude—certainly also the Norwegians found an easy way up."

They reached what they thought was the Pole the next day, and found the Norwegians' tent the day after. Scott wrote despairingly: "The Pole. Yes, but under very different circumstances from those expected. We have had a horrible day … Great God! this is an awful place and terrible enough for us to have labored to it without the reward of priority … Now for the run home and a desperate struggle."

The Way Back

The dispirited party started back but, by January 23, Scott reported that Evans, the strongest of them, "is a good deal run down," and that "Oates gets cold feet." They reached the Upper Glacier Depot on February 7. Although they had mislaid or misjudged a day's supply of biscuits, and although Scott had noticed that Evans was "going steadily downhill," they then spent two days adding 35 lb (16 kg) of mineral specimens collected at Mount Darwin to their sledging load. Scott noted: "The morraine was obviously so interesting that … I decided to camp and spend the rest of the day geologizing. It has been extremely interesting." The fossils that the rocks contained later proved invaluable in assessing the geological history of Antarctica, but the time spent collecting them and the effort spent in carrying them contributed to the polar party's deaths.

Near the foot of the Beardmore Glacier, Edgar Evans lost his life on Saturday, February 17. He had suffered several bad falls, and was straggling behind, and then disappeared. Found partly undressed and delirious, he sank into a coma and died that night.

By early March, they were all in desperate condition, particularly Oates, who had severe frostbite in his feet. On March 10, Scott commented that: "Things steadily downhill. Oates' foot worse. He has rare pluck and must know that he can never get through. He asked Wilson if he had a chance this morning and of course Bill had to say he didn't know. In point of fact he has none. Apart from him, if he went under now, I doubt whether we could get through. With great care we might have a dog's chance, but no more."

A week later, Scott wrote in his journal: "Friday March 16 or Saturday 17—Lost track of dates but think the last correct … At lunch the day before yesterday, poor Titus Oates said he couldn't go on: he proposed we should leave him in his sleeping bag. That we could not do." They struggled on. The same entry records, "He slept through the night before last, hoping not to wake;

Scott's Sanctum
Scott, often uncomfortable in his dealings with people, was a skilled writer with a versatile mind. In the Cape Evans hut, he appeared to be happiest reading and writing at his desk in his den, or presiding over serious discussions on a wide range of topics at the large oak table.

but he woke in the morning—yesterday. It was blowing a blizzard. He said 'I am just going outside and may be some time.' He went out into the blizzard and we have not seen him since … I take this opportunity of saying that we have stuck to our sick companions to the last."

On March 19, a blizzard held them down in their tent just 11 miles (18 km) short of One Ton Depot, and their fuel ran out three days later. Scott's diary entry for March 29 records his last words: "Outside the door of the tent it remains a scene of whirling drift. I do not think we can hope for any better things now. We shall stick it out to the end, but we are getting weaker … and the end cannot be far.

"It seems a pity but I do not think I can write more—

"R. Scott

"Last entry

"For God's sake look after our people."

Aftermath
Eight months later, on November 12, 1912, expedition survivors found the tent and their bodies. The tent was closed and they erected a cairn of ice above it, with skis forming a rough cross on top. They marched 20 miles (32 km) south to look for the body of Oates, but found the area covered in snow, and so built a cairn in his memory. Finally, they erected a huge wooden cross on the summit of Observation Hill on Ross Island, with a dedication that concluded: "to strive, to seek, to find, and not to yield."

Judgment of the British Antarctic explorers of the Heroic Age has shifted in prevailing attitudes. Scott was cast as a hero for most of the twentieth century, but recent critics have suggested that his only achievements were to write well and to die well. Shackleton has even been called unpatriotic because, in 1909, he had enough provisions to reach the Pole but turned back. He and his companions could have claimed the Pole for Britain and died on their way back. DM

Scott's Other Expeditioners

The eastern party of six, led by Victor Campbell, landed at Cape Adare from the *Terra Nova* on February 18, 1911 and erected a hut near the Borchgrevink's expedition's base. But bad conditions limited their explorations before they were picked up by the ship on January 3, 1912. They then decided to explore along the west coast of Victoria Land, starting at Terra Nova Bay on January 8. They had ample supplies for their intended stay of about six weeks. However, the ship could not get back, so they dug a snow cave on Inexpressible Island and spend the winter there. Leaving at the end of September 1912, they traveled along the sea ice to Ross Island, and survived only because they found some food caches along the way. They reached Hut Point on November 6, and found a note reporting the polar party had not returned by the start of winter. They were recuperating at Cape Evans on November 25 when the search party returned with news that the bodies of Scott, Wilson, and Bowers had been found.

The *Terra Nova's* last stop in Antarctica before sailing to New Zealand and England was at the snow cave—the crew concluded that "No cell prisoners ever lived through such discomfort."

Irrepressible Island
The eastern party first named the 7-mile (12-km) long island Southern Foothills, but later called it Inexpressible Island after living in the tiny snow cave.

Wilhelm Filchner

Second German Antarctic Expedition: 1911–12

Before he set out for Antarctica in 1911, Wilhelm Filchner (1877–1957) had made expeditions to Russia, central Asia, and Tibet. A Swiss by birth, he was a Prussian military officer and surveyer by profession. He prepared for his Antarctic expedition with military efficiency. He bought a Norwegian ship that had been purpose-built for polar work, renamed it the *Deutschland*, and took six members of his team to Spitsbergen in Arctic Norway for training. His original goal was to cross the Antarctic continent, launching two expeditions from the Ross and Weddell seas, but limited funds curtailed his plans, and only the *Deutschland* sailed for the Weddell Sea in May 1911, leaving South Georgia in December. Fighting its way south, at the end of January 1912 the expedition sighted an ice shelf 130 ft (40 m) high—now named after Filchner—and steamed south along it to a bay at 77°45′S, which they named after the ship's captain, Richard Vahsel.

Here they started to build their base on ice, and it was almost finished when, on February 18, tons of ice broke away from the ice shelf and took their base and some men with it. The *Deutschland* set off in pursuit.

In an extremely hazardous exercise, over the next few days, the men managed to salvage most of the materials and animals, including dogs and horses, and ferried them to the ship by lifeboat. Undeterred, Filchner went on to establish two depots at points on the ice shelf that he hoped were more stable.

A Sighting Disproved

Within days, the ship was icebound. The crew settled into their winter quarters, setting up tents for scientific observations, carrying out oceanographic research into the movement of pack ice, and exercising the dogs and the horses whenever the weather permitted. In June, Filchner led a dangerous expedition to try to locate land sighted by the American sealer, Benjamin Morrell in 1823, in a position only 45 miles (70 km) east of the ship. At 37 miles (60 km), they saw nothing, and found no bottom at 4,900 ft (1,500 m), thus disproving the existence of Morrell Land. Over six months, the ship drifted northward through 10 degrees of latitude. The *Deutschland* finally broke out of the ice in October and reached South Georgia two months later. LC

Spectacular Cleft
Drygalski Fiord lies on the southeast coast of South Georgia. It was charted by the Filchner expedition of 1911–12 and named for the leader of the earlier German polar expedition of 1901–03.

Nobu Shirase
Japanese South Polar Expedition: 1910–12

The first Japanese South Polar Expedition venture originated with the ambition and determination of an obscure lieutenant in the Japanese Army. Nobu Shirase (1861–1946) battled popular ridicule and government snubs, and eventually sought support from a former premier of Japan, Count Shigenobu Okuma, whose prestige attracted sufficient public subscriptions to fund the expedition.

Shirase sailed from Tokyo in December 1910 in an antiquated whaling vessel, the *Kainan Maru*. His expedition was dogged by vile weather, hostility, and outright racist publicity. When the *Kainan Maru* put in at Wellington in February 1911, the New Zealand press ridiculed its "crew of gorillas." The polar regions, they claimed, were no place for "such beasts of the forests."

Rough seas and thick ice slowed the expedition's progress south to the Ross Sea. They sighted the coast of Victoria Land, but could not land, and finally they turned back at Coulman Island. When they reached Sydney in May, the crew were refused accommodation ashore, and lived like beggars until Edgeworth David, who had been a member of Ernest Shackleton's 1907 expedition, intervened on their behalf.

Raising the Japanese Flag
Shirase's first objective had been the South Pole, but it was now clear that both Amundsen and Scott would be way ahead of him, so on his second foray south he concentrated on the exploration of Edward VII Land. In much better conditions than previously, the expedition reached the Ross Ice Shelf in January 1912 and landed at Kainan Bay at the eastern end. Heading west, they were startled to see a ship ahead—it was the *Fram*, waiting for Amundsen's party to return from the Pole.

Eventually, they reached the Ross Ice Shelf by climbing a cliff 330 ft (100 m) high. Shirase and four companions set off southward, on what has became known as the "Dash Patrol," with dog sledges. Though inexperienced and hampered by frequent blizzards, they covered nearly 185 miles (300 km) before raising their flag at 80°5′S and burying a cylinder containing a record of their achievement.

The party returned to the Bay of Whales and the *Kainan Maru*, and in June 1912 the expedition returned to Japan, where they were welcomed by a large crowd and acclaimed as heroes. LC

First Voyage Limit
Coulman Island marked the southern limit of the *Kainan Maru*'s first season in Antarctica. The Japanese reached the island on March 14, 1911, but here, at 73°28′S, the ice was too thick for them to proceed farther, so, with winter approaching, they returned north.

Patient Progress
Members of the underfunded Japanese expedition showed the same resolve as many better known expeditions. They spent 60 hours cutting a path to the top of the Ross Ice Shelf before beginning a journey inland.

Adapted for Antarctica
The *Kainan Maru* (meaning "Southern Pioneer") was a three-masted wooden fishing boat from Shibaura, Tokyo. Its displacement was 220 tons (200 tonnes), and an 18-horsepower auxiliary steam engine was added. The ship was clad in ¼-in (6-mm) sheet steel.

Douglas Mawson
Australasian Antarctic Expedition: 1911–14

Complete Commitment
Douglas Mawson (1882–1958) was explorer, survivor, and geologist. His Antarctic experience won him world renown. New Zealander Eric Webb described him as "an intellectual leader with utter motivation and selfless dedication to his objective."

The *Aurora*, 1914
Mawson described the *Aurora* as "by no means young [but] still in good condition and capable of buffeting with the pack for many a year ... The hull was made of stout oak planks ... The bow ... was a mass of solid wood, armored with steel plates."

After Ernest Shackleton's *Nimrod* expedition, Edgeworth David said of Douglas Mawson: "[he] was the real leader and was the soul of our party to the magnetic pole. We really have in him an Australian Nansen, of infinite resource, splendid physique, astonishing indifference to frost." Scott and Shackleton both asked the Australian explorer and geologist to join their next expeditions. Mawson turned Scott down when Scott rejected his proposal to explore the coast west of Cape Adare. Shackleton and Mawson developed an extensive Antarctic exploration proposal, but it came to nothing.

So Mawson promulgated his own Australasian Antarctic Expedition. Unlike other Antarctic explorers, he was not aiming for the South Pole; instead, he was most interested in the part of Antarctica lying below Australia, in what he called the Australian quadrant. Initially he envisaged three land parties, but modified this to two: the main party of 18, under Mawson, was based in Terre Adélie, and the western party of eight, led by the experienced Antarctic expeditioner Frank Wild, was based in Queen Mary Land.

Mawson's ship was the *Aurora*, an old Dundee whaler of 675 tons (612 tonnes), captained by John King Davis, with a complement of four officers and 19 crew. Mawson also bought a Vickers REP monoplane; it had crashed in October 1911 during a fund-raising display in Adelaide, but he took the wingless fuselage to Antarctica to use as a motor sledge.

On December 2, 1911, the expedition sailed from Hobart and they established a five-man radio station at Macquarie Island before continuing southward. On January 8, 1912, Mawson's main party landed on the Antarctic mainland and the *Aurora* sailed west with Wild's party on January 19.

"The Home of the Blizzard"
The main party had landed in what seemed an ideal natural harbor, which Mawson named Commonwealth Bay. He called the point where they erected their hut Cape Denison. Too late, he found that they were at the

base of a funnel for strong winds that came down from the polar ice cap; he wrote: "The climate [was] little more than one continuous blizzard ..." It was a trying winter, but the expeditioners recorded meteorological observations, and erected aerials so they could send Morse messages to Macquarie Island: this was the first use of radio in Antarctica—although it was only toward the end that they received replies.

In August, in the ice of a slope about 5 miles (8 km) south of the hut, they excavated and provisioned a comfortable cave. They named it Aladdin's Cave, and it was to prove a lifesaver. Because of spring blizzards, it was November before the five planned sledging expeditions set out. All left between November 8 and 10, knowing that they must be back to meet the *Aurora* by January 15, 1913. One party, aiming for the South Magnetic Pole, got to within 50 miles (80 km) of their goal, at an altitude of 5,900 ft (1,798 m), on December 21, before a lack of time and low food supplies forced them to turn back. They returned to the Cape Denison hut on January 11.

A second party planned to use the crippled aircraft to explore the hinterland of Terre Adélie. Once they even managed to drive it up a steep slope, but on the second day several pistons broke, and they abandoned it. On December 5 they found a tiny black meteorite—the first ever discovered in Antarctica. The party turned back on December 26 and after a difficult journey reached the Cape Denison base on January 17.

The third and fourth parties traveled together for 46 miles (75 km), and then the near-east group turned back to map the coast between the glacier named after the expedition's Swiss skiing expert and mountaineer, Xavier Mertz, and Cape Denison. In one small stretch they found an archipelago of 154 little islands.

The eastern coastal party had continued eastward, making its difficult way across the rough Mertz Glacial Tongue, then sea ice, then the Ninnis Glacial Tongue, and then sea ice again, as far as a towering coastal promontory of columnar lava that they named Horn Bluff—an expanse of spectacular "organ pipes," with outcrops of sandstone and coal. They started back on December 21, but struck bad weather and soft snow, and passed several days with virtually nothing to eat before reaching one of the food depots. They were back in the comparative safety and warmth of the hut at Cape Denison on January 17.

A Beleaguered Party
The fifth party, however, was overdue. Mawson had left on November 10, with Mertz and the ex-Royal Fusilier Belgrave Ninnis. With three sledges and 16 dogs, they had made good time at first. But disaster struck on December 14, not long after they had abandoned one of the sledges. Mertz was leading on skis, Mawson was behind with the first sledge, and Ninnis was in the rear

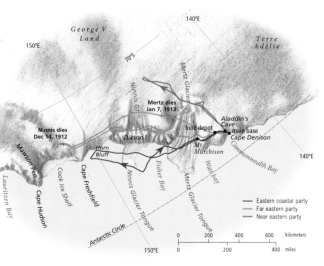

**DOUGLAS MAWSON
EASTERN PARTIES
NOV 1912–FEB 1913**

"The Home of the Blizzard"
The stores and equipment were landed at Cape Denison, the head of the boat harbor, to form the main base station. In the distance, men sledge materials to where the hut is to be built.

Their own situation was desperate indeed. Ninnis, in the supposedly safer rear position and with the best dog team, had been carrying most of the food, the main tent, and other vital supplies. "May God help us," wrote Mawson, as he contemplated their return journey. They started the return journey on a more southerly route than they had taken outward in order to avoid the most dangerous crevasses, and survived by killing and eating the six remaining dogs. It was December 28 when they dispatched Ginger, the last of the dogs.

By January 3, 1913, Mertz was suffering badly from dysentery and severe frostbite. Three days later he was riding on the sledge; the next day he had several fits, and struggled to get out of his sleeping bag. Mawson woke a couple of hours later to find that his companion was "stiff in death." There have been insinuations that, in the desperate situation he was in, Mawson practiced cannibalism, but he had enough food, and his actions in burying Mertz under blocks of snow and marking the position argue against this. To lighten his load, he cut his sledge in half with a penknife, sewed a makeshift sail, and carried on.

Change of Purpose
When Mawson's wireless station was built on Macquaire Island, sealers had huts at the southern end of the peninsula. Dr A. L. McLean wrote, "everywhere [was] covered with the bones, bleached skeletons and putrid carcasses of sea elephants." Today, the isthmus is a peaceful place of scientific inquiry.

with the other one. Mawson wrote: "My sledge crossed a crevasse obliquely and I called back to Ninnis … to watch it, then went on, not thinking to look back again …" But when he did look back, Ninnis had disappeared through a hole in the ice bridge spanning the crevasse. Mawson and Mertz peered over the edge and saw two dogs, and a tent and food bag caught on a ledge; there was no sign of Ninnis. Lacking a rope that was long enough to climb down, they called fruitlessly for four hours, then read the burial service.

and a message left only six hours earlier by three other members of his expedition: it reported that Amundsen's Norwegian expedition had reached the South Pole, and that the rest of his own men were safe.

Mawson's problems were not over. The snow along his route turned to slippery ice, and he had discarded his crampons. He fell frequently, "until I expected to see my bones burst through my clothes," and he made new crampons from his wooden theodolite case and from nails and screws.

He arrived at Aladdin's Cave on February 1, only to be pinned down by a blizzard for a week. On February 8, he made his way to the Cape Denison hut, where he saw the *Aurora* sailing into the distance, and "my hopes went down." Then three men waved from the harbor and ran to meet him: six men had stayed in case any of Mawson's party returned.

The *Aurora* was immediately recalled by radio, but the weather conditions made landing impossible, so she resumed her course to pick up Wild's western party and returned to Hobart on March 15, 1913. The ship returned to Commonwealth Bay on December 13 to collect Mawson's party, carried out marine research along the Antarctic coast for two months, and docked in Adelaide on February 26, 1914.

Douglas Mawson's scientific achievements had preceded him, and he was treated as a hero when he arrived home. On June 29, 1914 (the day the Austrian Archduke Ferdinand was assassinated in Sarejevo, triggering World War I), he was knighted in London.

The Western Party

While Mawson's party endured their ordeals, the *Aurora* had taken Frank Wild's party to the Shackleton Ice Shelf of Queen Mary Land, 1,500 miles (2,400 km) to the

Fieldwork

At New Year, the eastern coastal party set up camp at Penguin Point where, as Cecil Madigan wrote, "a rock wall three hundred feet [90 m] high formed the sea-face, jutting out from beneath the ice slopes of the buried land." Besides seals and Adélie penguins, they found some new birds' eggs and mites living in the moss there.

The Last Lap

Mawson recorded that the skin had peeled away from their groins, and on January 11 he wrote: "My feet felt curiously lumpy and sore." The soles of his feet had come off, and had to be strapped on with bandages. But sudden death was more immediately threatening than frostbite. On January 17, he fell through an ice bridge; suspended on a thin rope, he was precariously anchored by his half-sledge. He contemplated giving up by releasing himself from the rope, but hauled up to the surface, where he "cooked and ate dog meat enough to give me a regular orgy," and wove a rope ladder that he tied to himself and the sledge—and subsequently used on several occasions. By January 29, he was close to the Cape Denison hut. That day he found a cache of food

Restored Hut

Mawson noted that, soon after the main base hut at Cape Denison was completed in February 1912, it had become "so buried in packed snow that ever afterwards little beyond the roof was to be seen." The expedition's hut has since been restored.

Wireless Contact

Mawson's expedition was the first to have radio contact with the outside world. Walter Hannam sent messages that were received at Macquarie Island but the replies did not come through. Only in February 1913, after Hannam had left, was the first two-way communication established, relaying through the radio base at Macquarie.

Roving Camera

Frank Hurley, a Sydney postcard photographer, had been accepted into the expedition after inveigling his way into Mawson's overnight railway carriage, and went on to make a remarkable film of the expedition. Two weeks after he returned to Australia with Mawson, he was filming a motoring trip into outback Australia when an Aboriginal tracker arrived with a cabled invitation to join Shackleton's *Endurance* expedition. Two months later he sailed to Buenos Aires to meet the ship, and was soon trapped with it in the ice.

Marking Time with Music

Winter evenings passed slowly in the hut, but books and a gramophone helped to entertain the expeditioners (from left to right, standing): Mawson, Madigan, Ninnis, and Correll; (sitting round the table): Stillwell, Close, McLean, Hunter, Hannam, Hodgeman, Murphy, Laseron, Mertz, and Bage.

west. Vast quantities of supplies were dragged by aerial cable to the ice shelf before the ship sailed away. Their hut on the ice was habitable by February 28, 1912, but the radio masts they had erected were blown down by the very first blizzard. On clear days they could see the mainland, 17 miles (27 km) to the south; they visited it in mid-March and left a cache of supplies.

Wild wrote: "The most bigoted teetotaler could not call us an intemperate party. On each Saturday night, one drink per man was served out, the popular toast being 'Sweethearts and Wives.' The only other convivial meetings of our small symposium were on the birthdays of each member, Midwinter's Day and King's Birthday."

In August, just before they planned on beginning the spring skiing program, their hut nearly burned down while the acetylene generators were being charged. The first journey was to the east; six men and three dogs left on August 22 and returned on September 15. There were two expeditions during the summer months: one to the eastern coast under Wild, and one to the western coast under Evan Jones. One man was left at the hut.

Wild's group found the going heavy, particularly across the crevasses of the Denman Glacier. At the end of their outward journey, on December 22, they climbed to the peak of Mount Barr-Smith and were back at the hut on January 6, 1913. Jones's party had found vast colonies of penguins and other nesting birds on their way. They spent Christmas at the small, bare, extinct volcano that Erich von Drygalski had named Gaussberg when the Germans sledged there ten years before, and the German cairns were still visible on the summit. They were 215 miles (345 km) from their hut, but were back there on January 21.

The *Aurora* had not returned by the end of January, and the party began to stockpile seal meat and blubber in case they were marooned on the ice shelf for another winter. Saturday, February 23, was the anniversary of the departure of the *Aurora*, and in a howling blizzard they faced the possibility that the ship would not return. But the *Aurora* arrived the following day with most of the other expeditioners, and they were quickly boarded and sailed back to Australia. DM

Ernest Shackleton

Imperial Transantarctic Expedition: 1914–17

When the news broke that Roald Amundsen had reached the Pole, Shackleton wrote: "The discovery of the South Pole will not be an end to Antarctic exploration. The next work is a transcontinental journey from sea to sea, crossing the Pole." He turned his ambitions toward achieving that goal himself. The project was not without its critics—including Clements Markham, then aged 84 but still very influential. Shackleton also faced the embarrassment of his brother Frank being jailed for fraud, and his own finances were always uncertain. But, with characteristic determination and his Irish charm, he pulled it all together.

Shackleton's first plan was to take one ship, disembark at Vahsel Bay, at the head of the Weddell Sea, and travel 1,800 miles (2,900 km) overland while the ship sailed around to meet them. If he were late and the ship had to depart, a small whaler would be left behind and he could have a chance at another of his ambitions: an open-boat journey across Antarctic waters. It was a foolhardy plan, and he was persuaded instead to use two ships, with a party on the Ross Sea establishing food depots for him on the ice shelf. Shackleton was now 40, but he optimistically predicted that he could complete the journey in 100 days, a rate faster than Amundsen's, yet over completely unknown terrain.

The ship bought for the Weddell Sea sector was the Norwegian-built *Polaris*, captained by the New Zealander Frank Worsley and renamed the *Endurance*, from Shackleton's family motto: "By endurance we conquer." For the Ross Sea party, he had Mawson's *Aurora*. On August 1, 1914, the *Endurance* sailed from London after a good luck visit from Queen Alexandra. Four days later, England declared war on Germany. Shackleton offered his ship and men to the war effort, but the Admiralty told him to continue south. Three days out from Buenos Aires, a 19-year-old stowaway, Perce Blackborrow, was found. Shackleton castigated him—and then took him on as a member of the crew, bringing the numbers to 28 expeditioners and crew.

Shackleton made for the whaling settlement of Grytviken, about halfway along the north coast of South Georgia. He arrived in November 1914 to hear of the worst ice conditions ever reported in the Weddell Sea. They sailed south on December 5, 1914, and were in solid pack ice by December 11, with another 1,000 miles (1,600 km) to cover to Vahsel Bay.

Shackleton had already selected his polar party: surgeon Alexander Macklin; the Australian photographer Frank Hurley; second-in-command Frank Wild; the polar veteran Tom Crean; and George Marston, the artist from the *Nimrod* expedition.

The Ice Closes In

They saw land on January 10, 1915, and by January 18 they were only 80 miles (130 km) from their goal. But next day the ice closed around the *Endurance*; she was frozen in "like an almond in toffee," as one man noted, and began to drift gradually north with the pack ice. So began ten months of entombment in the drifting ice. They could not even take their base equipment across to the coast they could see, because that would mean lugging a hut and vast quantities of supplies and equipment across broken ice.

It was at this point Shackleton's leadership qualities became apparent. They might break free eventually—or the ship might be crushed in the ice. Meanwhile, the *Aurora* half of the expedition was laying supply depots on the Ross Ice Shelf that would never be used. Yet Shackleton betrayed no mental anguish. As Macklin noted: "We could see our base, maddening, tantalizing, Shackleton at this time showed one of his sparks of real greatness. He did not rage at all, or show outwardly the slightest sign of disappointment; he told us simply and calmly that we must winter in the Pack, explained its dangers and possibilities; never lost his optimism, and prepared for winter." They built small igloos—"dogloos" —on the ice for the huskies, and settled into a strict routine of shipboard life. Some parts of the *Endurance* were better insulated than others; Shackleton arranged for warmer areas to be created by repositioning cargo. Scientific experiments, entertainments, and amateur theatricals passed the time, led by Shackleton, whose boundless good spirits kept up their optimism.

Slowly they drifted northward. About the time the sun reappeared, there were moments of terror as the

A Place in History

Ernest Shackleton (1874–1922) faced his greatest challenge on the *Endurance* expedition. Biographer Margot Morrell stated, "Shackleton never planted a flag at the South Pole, he never made any of his goals, and he never earned all the money he wanted. Yet he was doing what he wanted to do, and he did it well enough to earn a place in history."

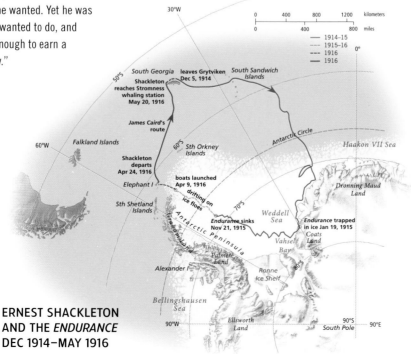

ERNEST SHACKLETON AND THE *ENDURANCE* DEC 1914–MAY 1916

At Patience Camp

The strange device between Frank Hurley (left) and Shackleton is a blubber-fuel stove that Hurley, who had trained as a metal worker, made out of the *Endurance*'s ash chute and old oil drums. Although resourceful, Hurley could be difficult, so the canny Shackleton picked him as a tent-mate to diffuse tension.

Ice Press

Trapped in the ice of the Weddell Sea in January 1915, the *Endurance* was still upright and under full sail, giving hope that she would eventually break free. Ten months later, she sank after drifting 570 miles (920 km) locked in the ice that crushed her.

Cutting the Ice

The sight of a clean wake behind the *Endurance* (right) as it pushed deep into the ice of the Weddell Sea was rare. On December 14, 1914, Shackleton wrote, "The young ice did not present difficulties to the *Endurance*, which was able to smash a way through, but the lumps of older ice were more formidable obstacles." Soon she was trapped.

pack ice formed pressure ridges and squeezed the *Endurance* until her timbers groaned and bent. The pattern continued for three months, and on October 18, 1915, tilted the ship at an angle of 30 degrees before releasing it, now with substantial leaks. Worsley suggested that they might "have to get out and walk," and Shackleton decided that they must be ready to abandon ship at any time. The following day, they retreated to five tents on the ice, amid "the groaning and cracking of splintering timbers." After their first night on the ice, Shackleton announced: "Ship and stores have gone—so now we'll go home." His matter-of-fact optimism was an inspiration.

It was another month before the *Endurance* was finally crushed; she sank on November 21. They had all been forbidden to go back on board the ship, but Hurley wanted to salvage his film and photographic plates, under mushy ice but in sealed tins in an ice chest. He stripped to the waist, hacked through the

ice chest's walls, and saved his film. The glass plate negatives were too heavy to carry so, in what has been described as one of history's great editing tasks, Hurley selected 120 plates and handed the rejected 400 to Shackleton, who smashed them on the ice. Hurley, ever the photographer, kept an album of prints, his movie film, and a box Brownie camera and three rolls of film, with which he continued to record the expedition.

Camping on the Ice

After the *Endurance* sank, they could only wait for the ice where they were camped to drift north and break up. Toward Christmas, they moved 10 miles (16 km) from Ocean Camp to Patience Camp, planning to take to the three boats salvaged from the *Endurance* in an attempt to reach land.

The men lived on the ice for five months, surviving toward the end largely on fresh seal and penguin meat. They shot most of the dogs in January 1916 and the remaining two teams at the end of March. Then the ice began to break up, so on April 9 they took to the boats and the numbingly cold, stormy seas. They drifted with the ice for 2,000 miles (3,220 km), past the tip of the Antarctic Peninsula. Their only course was to steer for Elephant Island, a desolate lump of rock; but at least it was not an ice floe.

They were in the boats for seven days and nights, sometimes camping on ice floes, before landing at Elephant Island's Cape Valentine. They could not stay on the exposed shore. They found a more sheltered spot 9 miles (15 km) down the north coast, with a gravely beach; it was still desolate, but there was plenty of wildlife, which they would need for food. The night they moved there and named it Cape Wild, a gale shredded their remaining tents. Morale was at its

Beginning of the End

On October 24: "The *Endurance* groaned and quivered as her starboard quarter was forced against the floe, twisting the stern-post and starting the heads and ends of planking. The ice had lateral as well as forward movement, and the ship was twisted and actually bent by the stresses."

The End

It took several weeks for the *Endurance* to sink (right). Shackleton noted: "The ship presented a painful spectacle of chaos and wreck" and "to a sailor [a ship] is more than a floating home … in the *Endurance* I had centered ambitions, hopes, and desires… she is slowly giving up her sentient life."

lowest: "Dejected men were dragged from their sleeping bags and set to work," one man recorded in his diary.

Shackleton decided he and five of the fittest men would sail to South Georgia for help in the largest of the boats, the *James Caird*—over 800 miles (1,290 km) through some of the wildest seas in the world. Harry McNeish, the ship's carpenter, raised the gunwales of the two-masted, 22-ft (7-m) whale boat and covered the decking with canvas. They launched the boat before ballasting it with a ton of rocks. On April 24 they set out, with Worsley as captain and navigator; Shackleton, Crean, and McCarthy went too, along with John Vincent and McNeish, who may have been included to remove potential troublemakers from Elephant Island.

Unable to take more than four sun sightings over a fortnight, and with the setting of his chronometer being uncertain, Worsley steered by dead reckoning and gut instinct, knowing that if he were wrong they would sail past South Georgia into the south Atlantic, and the men on Elephant Island would never be found. Of the 17 days of their journey, ten were in full gales.

South Georgia

When they sighted South Georgia on May 8, after 15 days at sea, they were on the western, windward side of the island. They landed at King Haakon Bay on the evening of May 10 after several near disasters along the coast. The Norwegian whaling stations were on the other side of the island, over a wall of high mountains. The island had never been crossed before. Only three were fit to attempt the crossing: Shackleton, Worsley, and Crean. They made three attempts to find a pass through the mountains before finding a way over just as daylight was failing. If they stayed at high elevation overnight they would freeze to death. Beneath them, disappearing into the mist and darkness, lay a long, steep precipice of snow. "It's a devil of a risk but we've got to take it. We'll slide," Shackleton said. Coiling their climbing rope beneath them, they sat down one behind the other, each locking arms and legs around the man in front, and pushed off into the darkness.

"We seemed to shoot into space," wrote Worsley. "For a moment my hair fairly stood on end. Then quite suddenly I felt a glow, and knew that I was grinning! I was actually enjoying it … I yelled with excitement and found that Shackleton and Crean were yelling too." They descended 1,500 ft (460 m) in seconds. They stumbled on, falling, slipping, and barely conscious. At dawn, Shackleton recognized a rock formation at Stromness Bay. At 6:30 AM, they heard the welcome sound of the steam whistle blast that awakened the Stromness workers. They were saved.

They had to clamber down a glacier and struggle through a waterfall still, and it was 3 PM, after 36 hours without rest, when they walked into the outskirts of the Stromness whaling station. Filthy, and with their faces blackened by blubber smoke, salt-matted hair and long beards, they were a fearsome sight; two children—their first human contact—ran from them in fright. The station foreman eventually led the three men to the house of the manager, Thoralf Sørlle, who had been on board the *Endurance* two years earlier.

"Well?" said Sørlle.

"Don't you know me?" asked Shackleton.

"I know your voice," he replied doubtfully. "You're the mate of the *Daisy*."

"My name is Shackleton."

The amazed whalers received the castaways with great warmth. A ship was sent around to the east coast to collect the other three crew of the *James Caird* and

the little boat itself. As Worsley wrote: "Every man on the place claimed the honor of helping to haul her up to the wharf." But it was to be another five months before the ice conditions made possible the rescue of the men marooned on Elephant Island.

Shackleton made three increasingly desperate bids to rescue them before he finally managed to get through the pack ice in the *Yelcho*, a little steel-hulled tug the Chilean government had lent him, under the command of Luis Pardo. Astonishingly, Shackleton had not lost a single man during the two-year ordeal.

Shackleton's task was not over: the Ross Sea party had to be rescued, and money raised for the *Aurora* to go south again under the command of John King Davis. Shackleton sailed directly from South America to New Zealand to join the expedition. Davis found Shackleton "somewhat changed," and allowed him to sign on only as a supernumerary, but he recorded that when at sea Shackleton's "old greatness of spirit shone out." DM

Sleeping Quarters
The 22 left at Elephant Island all slept underneath the two remaining boats upturned on rock walls. Hurley wrote: "The 'Snuggery' grows more grimy day by day: everything is an oleaginous sooty black … Oil mixed with reindeer hair, bits of meat, senegrass, and penguin feathers form a conglomeration which cement the stones together."

Waving Farewell
In their books, Shackleton and Hurley both said this photograph depicted the rescue from Elephant Island. However, in *The Endurance*, Caroline Alexander noted that Worsley's book said it was the *James Caird* leaving for South Georgia. Hurley modified the negative to provide a fitting final image for the expedition.

No Way to Cross
The night before they left to cross the South Georgia, Shackleton was unable to sleep: "My mind was busy with the task of the following day … No man had ever penetrated a mile from the coast of South Georgia at any point, and the whalers I knew regarded the country as inaccessible."

Ernest Shackleton
The Ross Sea Party: 1915–17

The other part of the expedition, the Ross Sea party on the *Aurora*, experienced as many perils as did those on the *Endurance*; indeed, while Shackleton lost not a man under his direct command, three died on the other side of Antarctica.

The *Aurora* was to moor off Ross Island over winter as a base for a sledging party to lay food depots across the Ross Ice Shelf and onto the Beardmore Glacier for Shackleton's homeward journey. The ship left Hobart on Christmas Day, 1914, and arrived at Ross Island on January 9, 1915, captained by Aeneas Mackintosh, who had lost an eye on Shackleton's *Nimrod* expedition. Also on board was experienced polar veteran, Ernest Joyce. Shackleton was vague about which of the two was in command, which led to friction and disaster.

Mackintosh decided to use Scott's Cape Evans hut, and on January 24 the first depot-laying expedition set out. Mackintosh was eager to start in case Shackleton arrived at the end of this first summer, and—against Joyce's advice—he pushed the dogs too hard; all but five died, and much of the task remained uncompleted.

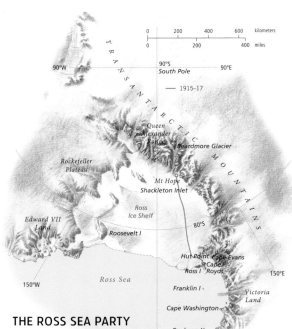

**THE ROSS SEA PARTY
JAN 1915–JAN 1917**

The Ship Vanishes

The *Aurora* was to be a supply and accommodation base until spring, when the laying of supply depots would begin, and by mid-March she was secured at Cape Evans. The ten men who would lay the depots were ashore with minimal equipment. But on the night of May 6 there was a violent storm, and by 3 AM the *Aurora* had vanished with its crew—its mooring had blown away, leaving only bent anchors and snapped hawsers. It had happened before. They expected the *Aurora* to return, but the ship was as firmly gripped in ice as was the *Endurance* on the other side of the continent; it was almost a year before she broke free.

The group knew they were marooned for the winter. They frantically searched to find what Scott had left behind at the Cape Evans hut. They found tins of jam, flour, cake, chocolate, oatmeal, and some pemmican, but no soap, tobacco, or medical supplies. Kerosene and two Primus stoves were life-saving. Their only clothing was what they were wearing and some spare underwear. They killed seals for meat and fuel—and continued to take scientific observations.

Rescue might be two years away, but they still had to lay supply depots for Shackleton in the spring or he would certainly die on the ice shelf. They made plans to carry 4,000 lb (1,800 kg) of supplies onto the ice shelf and establish the farthest depot at Mount Hope, at the foot of the Beardmore Glacier, at 83°40′S. Although Joyce was deferring to Mackintosh, relations between them were strained. In mid-August, Mackintosh and chief scientist Fred Stevens reached Shackleton's hut at Cape Royds and found cigars, tobacco, food, and soap.

During September, they hauled supplies across to Hut Point, the starting point for sledging south. Joyce

wrote: "Most of us wore the canvas trousers made from Scott's old tent, and they froze on us like boards." On October 26, they found a sledge left as a marker by Cherry-Garrard, a note for Scott, and six boxes of dog biscuits impregnated with cod liver il. On January 3, 1916, one Primus was burning its own metal, and the four depending on it had to turn back. The men were failing too: by 82°S, Mackintosh was weakening and Arnold Spencer-Smith, the chaplain, had scurvy. The others left him in a tent and pushed on to establish the southernmost supply depot at Mount Hope on January 26. When they returned, Spencer-Smith could not walk; they carried him on the sledge until he died on March 8, then buried him in the ice. By now they were all suffering from scurvy. A blizzard stopped them close to where Scott and his companions had perished, and they seemed set for the same fate.

But they reached Hut Point on March 11, having been away for more than six months. The four who had

Uncertain Fate

When the survivors were reunited at Cape Evans, those who crossed from Hut Point were sad but not surprised that Mackintosh and Hayward had not survived. Dick Richards recalled, with considerable understatement, that, "From July 1916 until January 1917, when the rescue ship arrived, it was a bit of a struggle to get by."

Cloud Break
For the Ross Sea party, spectacular nacreous ("mother of pearl") clouds, which are seen in the polar region at the end of the coldest months, would have broken the monotony after the long, dark Antarctic winter.

Deep Freeze
The desiccated remains of one of the Ross Sea party's dogs in its collar by the Cape Evans hut. The expedition's work was made infinitely harder by Captain Mackintosh pushing the dogs too hard too soon, so that most had died before the main task of laying supplies was completed.

turned back earlier had reached Cape Evans. Now the survivors at Hut Point could eat seal meat and recover their strength. Joyce calculated that they must wait four months for the surface to be stable enough to cross to Cape Evans, but by May 8, after a blizzard had cleared much of the sea ice, Mackintosh announced he and Hayward would walk across the 12 miles (20 km) that remained. It was a suicidal decision: a blizzard arose, and the two were never seen again. On the night of July 15, the remaining three, Joyce, Dick Richards, and Ernest Wild, crossed to Cape Evans and rejoined the other four. The seven men had a long wait for rescue.

Meanwhile, the *Aurora* and her crew had drifted 700 miles (1,130 km) in the ice. North of the Antarctic Circle, as summer approached, they hoped that the ice would open up and release the ship, but the looser ice did even more damage; when the *Aurora* was finally freed in March 1916, she was leaking and relied on a makeshift rudder to cover the 2,000 miles (3,200 km) to New Zealand. On April 3, 1916, she was taken under tow to Port Chalmers, near Dunedin.

Shackleton Returns
Rescue came on January 10, 1917. Dick Richards was outside looking for seals when he saw a ship's smoke. He walked back into the hut and said casually: "There's a ship out there." The ship was the *Aurora*. The ultimate miracle was when the men realized that one of those

A Mirage
One day when Ross Sea party member Dick Richards stepped outside in search of seals, he saw a ship's smoke. He calmly reported to the others what he had seen, but it took some time to realize it was the *Aurora* with Shackleton on board.

approaching was Shackleton—not lost in the south as they had thought. John King Davis, commanding the *Aurora*, wrote of the encounter: "They were just about the wildest looking gang of men I have ever seen … But the marks of their great physical and mental privations went deeper than their appearance … factors had combined to render these hapless individuals as unlike ordinary human beings as any I have ever met … They were dazed by it; we were excited."

The *Aurora* docked in Wellington on February 9. Joyce claimed their expedition was "without parallel in the annals of polar sledging," and it was justified. The *Discovery* sledging party of Scott, Shackleton, and Wilson was out on the ice for 93 days; Shackleton's "farthest south" took 120 days of sledging; and Scott's tragic final venture lasted 150 days. The Ross Sea party was on the ice for 199 days, without adequate clothing or support, and burdened with food for the failed polar party—an incredible achievement. DM

The *Quest* Expedition
1920–22

By 1920, Ernest Shackleton was again planning to visit the Antarctic region. His friend John Rowett put up money for the new venture, and a dilapidated Norwegian sealing boat was acquired and renamed the *Quest*, at the suggestion of Shackleton's wife, Emily. Frank Worsley was captain again, Frank Wild was Shackleton's deputy, and Alexander Macklin was one of three doctors. The expedition's aims were vague: oceanography; looking for sub-Antarctic islands; mapping uncharted coast; searching for Captain Kidd's treasure in Trinidad and for pearls in the Pacific.

The *Quest* left London on September 17, 1921, and sailed from Rio de Janeiro on December 17. Shackleton was obviously unwell, but on January 2, 1922, he wrote: "At 1 PM we passed our first berg. The old familiar sight aroused in me memories … Ah me! the years that have gone since in the pride of young manhood I first went forth to the fight. I grow old and tired but must always lead on." On January 4 they sighted South Georgia, and Wild recalled: "I recognized once more the old buoyant, optimistic Boss." After a cheerful evening in Grytviken they returned to the ship. Macklin told Shackleton to take it easy. "You are always wanting me to give up something," replied Shackleton. "What do you want me to give up now?" The old expeditioner then suffered a massive heart attack. He died at 2:50 AM on January 5. He was 47 years old.

After holding a service and erecting a cairn to his memory, the *Quest* continued its voyage, but achieved little. Shackleton's body was taken to Montevideo, but his widow requested that he be buried at South Georgia, where he was interred on 5 March. Symbolically, his is the only grave with its head facing to the south, toward Antarctica, rather than to the east. The *Quest* returned to England on September 16, 1922.

A Companion's Tribute

Frank Worsley wrote: "I knew that I should never look upon his like again. He was not only a great explorer: he was a great man … by his genius for leadership he had kept us all in good health … by sheer force of his personality he had kept our spirits up … by his magnificent example, he had enabled us to win through … His was a proud and dauntless spirit … We would have gone anywhere without question just on his order. What more glowing tribute could any man wish for?"

Shackleton's death marked the end of the Heroic Age of Antarctic exploration. DM

Hero's Grave

The reverse of the granite headstone bears lines by one of the explorer's favorite poets, Robert Browning: "I hold that a man should strive to the uttermost for his life's prize."

A Memorial Cairn

This simple cairn, built by Shackleton's comrades from the *Quest*, is at Hope Point, on the northern shore of King Edward Cove. Shackleton was buried facing south, toward the scene of his greatest triumph: his epic struggle across the ice, over wild polar seas and glaciers, which made his name synonymous with the Heroic Age of Antarctic exploration.

Final Resting Place

Shackleton died in South Georgia on January 5, 1922. When the *Quest* returned in April, seven comrades from the *Endurance* paid their respects at his graveside. Frank Wild, his trusted lieutenant, is second from the left.

British Imperial Expedition
1920–22

Despite its grandiloquent name, there had never been a smaller expedition to spend a winter in Antarctica than the British Imperial Expedition. Essentially, it consisted of only two men: the 19-year-old geologist Thomas Bagshawe, and the 22-year-old Royal Naval Reserve Lieutenant Michael Lester, who had had some surveying experience. The two spent ten months, from March 1921 to January 1922, at Waterboat Point on the northwestern coast of Graham Land.

The expedition originally had two other members: its official leader, John Cope, who had been surgeon and biologist for the Ross Sea party of Shackleton's 1914–17 expedition, and the Australian aviator Hubert Wilkins, who had spent four years in the Arctic with Vilhjalmur Stefánsson's 1913–18 expedition. Cope wanted to fly 12 war-surplus planes to Edward VII Land and thence to the South Pole, but he could not secure the financial backing, and so he turned his sights to the Antarctic Peninsula, where whaling ships operated and transport was easier to arrange. He also gave up his idea of flying, proposing instead to extend the surveying work done by Otto Nordenskjöld's 1901–04 expedition with a dog-sledge journey southward from Snow Hill Island along the western and southern shores of the Weddell Sea.

The four met at Deception Island on Christmas Eve, 1920, hoping to find a whaler to take them to Hope Bay on the tip of the Antarctic Peninsula, and then to make their way southward along the eastern coast of Graham Land. But Hope Bay was icebound, and the whaler deposited them at Waterboat Point on the west coast instead. They still hoped to cross the peninsula, but there was no way over its mountainous spine.

Cope, his credibility in tatters, gave up, and so too did the frustrated Wilkins. They found a whaler to take them home, but Bagshawe and Lester "volunteered to stay for the winter and carry on … the scientific work …" Incredibly, they managed to do so, keeping a two-hourly meteorological log, an ice log, a natural history log, and, for a month, an hourly tidal log.

"Never were there such devoted scientists … these two young men possessed that quality so annoying to the great Napoleon, of not having the sense to know when they were defeated," according to Frank Debenham, the first director of the Scott Polar Research Institute. LC

Splendid Isolation
Today Waterboat Point is the site of the Chilean summer station Presidente Gabriel González Videla, near the ruins of Lester and Bagshawe's boat-hut. The cliffs and soaring mountains that frustrated the expedition's efforts to cross to the Weddell Sea provided a spectacular setting for their stay.

Messing About in a Boat

Thomas Bagshawe and Michael Lester made their home in the stranded boat, 30 ft long by 10 ft wide (9 m by 3 m), after which Waterboat Point was named, extending it with packing cases to provide a tiny "lounge" and kitchen, and sewing canvas food bags together to serve as windows. In these cramped, leaky quarters, they cooked on a stove made from an old oil drum, and lived almost entirely on seals and penguins. For Bagshawe, however, "visiting Antarctica was … sheer delight," and when they left both he and Lester "felt miserable as we watched the hut disappear from sight … it was to us a haven of comfort."

Discovery Investigations
1921–51

The modern whaling industry of Antarctica began in 1904 in the waters around South Georgia, the Falkland Islands, and the South Shetland Islands, an area of acknowledged British sovereignty. Catches—and profits—soared. In 1908 the British government established the Falkland Islands Dependencies (FID), with the aim of regulating the species and numbers of whales taken, and of licensing the shore flensing plants, where whales were processed. Part of each license fee was set aside for research, and in 1913 a whaling committee was established to consider whether whales needed protection from exploitation, and, if so, how this could be achieved. The committee's first report, published in 1915, had concluded that "on its present scale, and with its present wasteful and indiscriminate methods, whaling is an industry which, by destroying its own resources, must soon expire." The wholesale slaughter of whales was exacerbated by the huge wartime demand for whale oil, which was a valuable source of glycerine for explosives; it reached a peak in 1915–16, when nearly 12,000 whales were taken.

Saving the Whales

The committee reconvened after the war, and its 1920 report emphasized that the depletion of whale stocks could be avoided only by control of the whaling industry. However, effective control was impossible because not enough was known about the habits and ecology of the whales. Thus were born the Discovery Investigations, named after Robert Scott's ship, the *Discovery*, which became its first research vessel. These investigations were concentrated on "conducting research into the economic resources of the Antarctic with the particular object of providing a scientific foundation for the whaling industry." They continued until 1951, ranging in scope from "the greatest animals this world has ever seen down to the microscopic life of the sea." The final report of the committee was published in 1980.

The first director of research, Stanley Wells Kemp, was an experienced marine biologist. His responsibilities involved "planning, organizing and carrying out one of the most comprehensive schemes of oceanographical research ever undertaken by any country in the world."

Grytviken, Early Days
When this was a working whaling station, whales were hauled onto the open space between the jetties and stripped of their blubber, a process known as flensing. The blubber was fed into huge pressure cookers in the building with chimneys (center) and the oil pumped into tanks behind. The meat, bones, and guts were turned into fertilizer and animal feed.

Grytviken in 2000
Comparing the 1920s picture (below opposite) with that of 2000 (left) shows how the whaling station grew and then decayed. The pink-roofed house, once the manager's residence, is now a museum. Shackleton's grave stands in the small graveyard on the farther shore. Most of the buildings have since been demolished, leaving their inner workings exposed.

A Steam Whaler
Scientific research on whales and whaling was first proposed in 1913, and this led to the launch of the Discovery Investigations in 1924, a program devoted to examining Antarctica's economic resources, in particular, whaling.

The *Discovery* at Sea
In 1925 Scott's ship, the *Discovery*, was recommissioned as a whaling research ship. After her escape from the ice of McMurdo Sound in 1904, she had spent six years in the Canadian Arctic, five years as a wartime supply vessel in the Russian Arctic, and another four years laid up on the Thames.

Working Jetty
At its peak, during the 1915–16 season, nearly 12,000 whales were slaughtered. This abandoned jetty in South Georgia would have been bustling with whalers bringing in their catches, and ships loading whale oil and other products for the theater of war in Europe.

Kemp led the first expedition, which left England in September 1925 and reached South Georgia in February 1926. The captain was John Stenhouse, an experienced navigator and master of square-rigged sailing ships; he had taken command of the *Aurora* in the Ross Sea during Ernest Shackleton's 1914–17 Transantarctic Expedition. Also on board were eight biologists and oceanographers, including Alistair Hardy, who described spending hours "fishing for knowledge in this deep water world we cannot see," with dredges, nets, and water bottles at a series of "stations" across the Southern Ocean.

Scientists at Work
A shore laboratory sited in Cumberland Bay on South Georgia, close to the whaling station at Grytviken, had already been set up and the scientists continued their collecting, often working "on the platform among whales in all stages of dismemberment. Little figures busy with long handled knives like hockey sticks, look like flies as they work upon the huge carcasses, whose girth is not far short of a railway carriage …" From the data that they collected, the scientists gradually determined what were the breeding times, gestation periods, rates of growth, and ages at maturity for various species.

The following season, in 1927–28, the *Discovery* was joined by the smaller *William Scoresby*, which was equipped as a trawler and whale-catcher as well as a research ship. Between them, the two ships covered a wide area between the Falkland Islands and the South Shetlands, marking whales so they could track their migration routes, taking water samples, and collecting krill and plankton. In 1929, the *Discovery* was replaced by the *Discovery II*, a purpose-built 1,250-horsepower research ship, 230 ft (70 m) long. Between 1929 and 1951, the *Discovery II* made six voyages, each of two years, including two circumnavigations of Antarctica, and the *William Scoresby* made eight voyages in all.

On its fourth commission, in 1935–37, the ship was diverted to rescue the aviators Lincoln Ellsworth and Herbert Hollick-Kenyon when their plane crashed near the Bay of Whales. The same season, a survey party at King George Island was nearly lost when its launch broke down and was wrecked.

Most Discovery investigations were less eventful, but the research work undertaken on these expeditions laid the foundations for modern scientific conservation of whales, and for the establishment of the International Whaling Convention in 1936, and, eventually, to the effective regulation of the whaling industry.   LC

Hubert Wilkins
1928–29

Possibly the greatest adventurer and explorer that Australia has produced, Sir Hubert Wilkins was knighted in 1928 in recognition of his pioneering flight over polar ice from Alaska to Spitsbergen, in Arctic Norway. He had had a distinguished and hair-raising photographic and flying career during World War I, and thereafter regarded the airplane as the natural means of exploring the polar regions. Initially, he had been involved with the ill-managed British Imperial Expedition to Graham Land in 1920–22, until he lost patience with the venture's cheerful amateurism, but the expedition did take him to Deception Island. With the American aviator Carl Eielson, he then flew with great distinction in the Arctic from 1925 to 1928.

The fame that these Arctic flights attracted secured financial backing from American newspaper magnate, William Randolph Hearst. In return for news and radio rights, Hearst raised finance for an Antarctic expedition in the southern summer of 1928–29, after the Australian government had declined to finance Wilkins's dream of exploring and photographing the south polar regions and perhaps flying to the South Pole.

The First Flight

Early in November 1928, Wilkins's small expedition of just five men, with Eielson as the chief pilot, reached its base in the relative shelter of Deception Island, the doughnut-shaped caldera that lies 50 miles (80 km) northwest of the Antarctic Peninsula. The *Hektoria*, the Norwegian whaler that sailed the party there, carried two single-engined Lockheed Vega high-winged monoplanes, one of which Wilkins and Eielson had already used to good effect in the Arctic.

Twelve days later, on November 16, the two aviators made the first powered flight in the skies of Antarctica. They had originally planned to use skis for take-off and landing, but the inadequate ice in Deception's harbor forced them to use conventional wheels, albeit on a somewhat unconventional runway. They took off from a very rough and less-than-straight airstrip scraped out of the black sand beach. The first flight over Antarctica lasted a mere 20 minutes; after a few brief circuits of the island, and with weather conditions deteriorating, the plane was back on the ground. But the exploration of Antarctica by air had begun.

Pioneer Aviator
Hubert Wilkins (1888–1958), with Carl Eielson, was the first to fly an airplane in Antarctica. He was also the first to successfully fly across the Arctic Ocean, for which he was knighted in 1928. After his Antarctic exploits, he flew on the airship *Graf Zeppelin* on the first round-the-world flight in 1929.

Walkabout Rocks

Wilkins claimed the Vestfold Hills area (site of Davis Station) for Australia in the summer of 1938–39. This cairn is known as Walkabout Rocks because Wilkins protected his handwritten proclamation by wrapping it in a copy of the magazine *Walkabout*.

Memories and Relics

This black sand beach on Deception Island was used as an airstrip by Wilkins's Lockheed Vega in 1928. The hangar and derelict Otter aircraft (the plane has since been removed) are relics of the British Antarctic Survey base that operated on the island from 1943 until 1967, when a volcanic eruption covered the base in cinders, forcing a hurried evacuation.

Stepping Out

In 1929 Wilkins married an Australian actress, Suzanne Bennett, whom he had met at a reception in New York in honor of his successful trans-Arctic flight of the previous year. They were married for 29 years. They had no children.

Ten days later both aircraft flew, again locally over the island. However, on landing, one of them skidded from the ice into the water, where it lay suspended only by its wings. It was salvaged with considerable difficulty.

Finally, on December 20, after days of frustrations with the weather, Eielson and Wilkins made the historic flight of over 1,300 miles (2,100 km) that was to demonstrate the feasibility of exploring the Antarctic region by air, and achieve many of the expedition's aims. They flew for 11 hours from Deception Island and back, across to the mainland of the Antarctic Peninsula, and then south to the Antarctic Circle and beyond, flying over much territory that had never been reached by any previous explorers, let alone been seen.

Wilkins carefully recorded their journey extensively in photographs and sketches. Some of his observations and deductions proved to be inaccurate: he erroneously concluded, for example, that sections of the Antarctic Peninsula were part of an archipelago, rather than an extension of the mainland. Nevertheless, the value of the airplane in the exploration of Antarctica had been demonstrated beyond any doubt.

The Transcontinental Dream

In November of the following year, Wilkins returned to Deception Island to collect the two aircraft, which had been stored there for the winter. Again, he was financed largely by Hearst, but Eielson was not with him this time. Remembering the difficulties of taking off and landing on Deception Island's rugged volcanic lava, he had sailed south to the Antarctic Circle in search of easier flying bases. He found none; nevertheless, he did make a number of flights with the airplane rigged on floats, although none matched his epic journey of 1928.

For several years Wilkins continued with his dream of flying across the southern continent from the Weddell Sea to the Ross Sea, but his involvement with Antarctic aviation continued only at sea level. In a non-flying capacity, Wilkins assisted and advised the American millionaire, Lincoln Ellsworth, in his frequently frustrated attempts to make a major transcontinental flight during 1934–35, and he played the same roles for Ellsworth's final expedition, in 1938–39. PL

Richard Byrd
1928–41

American flying ace, Richard Byrd, claimed to have been the first to fly over the North Pole, on May 9, 1926, and in 1927 he was pipped at the post by Charles Lindbergh in the famous race to fly solo and nonstop across the Atlantic from New York to Paris. By the late 1920s, Byrd had one overriding ambition: he wanted to be the first to fly over the South Pole. It was clearly a more difficult proposition than the North Pole, but Byrd was backed by such great American financial names as Edsel Ford, who provided the Ford Trimotor aircraft that was to be the backbone of Byrd's operation, and the millionaire philanthropist John D. Rockefeller, who had extensive media interests.

Rather than seek flying bases in milder conditions farther north, at the end of December 1928, Byrd established a camp, which he called Little America, on the Ross Ice Shelf, a little way inland from the Bay of Whales. The expedition, which included more than 80 men, carried some sophisticated radio equipment for maintaining contact with the outside world—mainly North America—to recompense his media backers.

First Over the Pole: 1928–30

Byrd's first Antarctic flight took place on January 15, 1929, followed two days later by a more significant flight during which he discovered a mountain range that he promptly named after Rockefeller. For the next month, the expedition concentrated on the hitherto unexplored lands east of Little America, using the Ford Trimotor and

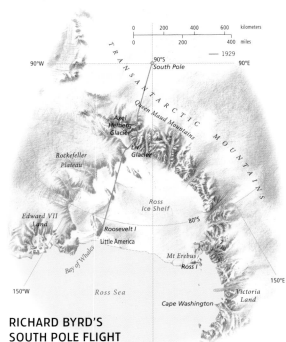

**RICHARD BYRD'S
SOUTH POLE FLIGHT
NOV 1929**

the other two expedition airplanes, a single-engined Fokker Universal and a Fairchild.

By March, however, flying conditions had become virtually impossible; the Fokker was destroyed on the ground in a blizzard, and the remaining two aircraft were deliberately buried in large pits to protect them for the winter. About half the crew wintered underground, in the most sophisticated Antarctic camp constructed up to that time, and it was not until early October that

Polar Competitor

Richard Byrd (1888–1957) was already an American hero for his success in flying over the North Pole in 1926 when he turned his sights to the South Pole. With the advice of Roald Amundsen, who led the only party to return from the Pole, and substantial corporate backing, he was sure of succeeding.

Christmas Fare

Byrd's third expedition established two bases, one on the Antarctic Peninsula, the other at the Bay of Whales. In charge of this West Base was Dr Paul Siple, a veteran of Byrd's two previous expeditions, and destined to follow him as leader of the post-war US Antarctic Program. He is seen here celebrating Christmas 1941 with other expedition members.

Fly Like a Byrd

The fuselage of the Ford Trimotor is unloaded onto the Ross Ice Shelf from the *City of New York* in December 1928. Byrd made his first flight in January 1929, two months after Hubert Wilkins's inaugural Antarctic flight. The Trimotor also made a successful polar flight the following November.

Fields of Ice

Flying over the vast, featureless Polar Plateau, with Bernt Balchen as pilot, Byrd as navigator, and two crew, the Ford Trimotor reached the South Pole on November 29, 1919. They did not land, and, after a little over nine minutes, turned the plane around and headed back to Little America on the edge of the Ross Ice Shelf.

Over the Mountains
In early October 1929, a geological party set out to survey the rugged Queen Maud Mountains, a major subdivision of the Transantarctica Mountains, extending for 500 miles (800 km) from the Ross Ice Shelf.

a land-based geological party was able to embark on an extended expedition to the Queen Maud Mountains, and final plans for the polar flight could be finalized. Byrd was doubtless spurred on by the news that Wilkins intended to return to Antarctica: the thought of being thwarted by the Australian in his quest to be the first to fly to the South Pole was a driving force.

The Ford Trimotor lacked the range to fly to the Pole and back nonstop. On November 19, mounted on skis, it first flew to the bottom of the Axel Heiberg Glacier to establish a fuel dump: the purpose was to reduce the plane's fuel dependency for the polar flight by some 435 miles (700 km). Possibly too much was deposited; on the way back, the aircraft ran out of fuel. It had to make an emergency landing 100 miles (160 km) from Little America. The plane came down safely, but the party had to wait three days for more fuel to be flown in.

The attempt on the Pole began on the afternoon of November 28 with a crew of four, including Bernt Balchen as pilot and Byrd as navigator. Encouragingly, the geological party at Queen Maud Mountains had reported clear weather there, but no one knew what the interior had in store. The principal barrier was the inland Polar Plateau, and the greatest drama of the flight unfolded as Balchen tried to fly up the Liv Glacier into the skies above the plateau. Severe turbulence hammered the plane, and it was unable to climb. They faced an unenviable choice: either turn back, or jettison weight in the form of vitally needed fuel or food that would be essential in the case of a forced landing. It was the food that went, and the aircraft finally gained enough height—just—to clear the top of the glacier.

The rest of the flight was largely uneventful, and they flew over the South Pole at 1:14 AM on November 29, nine hours and 56 minutes after departure. Little was visible of a fairly featureless landscape. Later, Byrd commented: "One gets there and that is all there is for the telling. It is the effort to get there that counts." With fuel such an important consideration, the aircraft turned around after just nine minutes and headed back to base. Refueling en route went smoothly, and the aircraft landed back at Little America just after 10 AM.

Back to the Pole: 1933–35
Byrd returned to America to great acclaim. The fame and fortune that he craved was his for the asking, in the form of honors and adulation. It also meant he was able to obtain further financial backing, so that when he turned his sights southward again, some four years later, he was able to enlist the support of many of the industrial giants of the United States. Edsel Ford was once again a backer, but the mainstay of Byrd's aerial fleet this time was a large, new Curtiss Condor biplane, supported by a Fokker, a Fairchild, and an autogiro. Like the previous venture, this expedition was large, and its base, reached in December 1933, was on the site of Little America, which had been left to the elements four years earlier. They built a new camp, restoring parts of the original and adding to it.

Rather than making record-breaking flights—if any were left to be made—the aim of this expedition was to explore and map the great uncharted terra nullius that extended far to the east. Scientific research was another major goal, particularly the study of the meteorology and geology of Antarctica.

From January 1934 until February 1935 (except for the very significant over-wintering time, when no flying was possible), about 400,000 sq miles (over 1 million sq km) were photographed and surveyed from the air, including vast tracts of Antarctic lands that previously were unexplored, indeed unseen. In the process, the autogiro was wrecked and the Fokker crashed in flying accidents, but no lives were lost. Ground parties also carried out extensive studies although, understandably, over a much smaller area.

Government Takes Over: 1939–41

Byrd's 1933–35 undertaking was among the last of the large, privately funded Antarctic expeditions: thereafter governments became involved. The German interest in Antarctica—through the Schwabenland Expedition of 1938–39—worried the United States, and in late 1939 Byrd went south again, this time leading a government-funded expedition. Once more, aircraft played a leading role; the expedition included two Curtiss Condors, a Beechcraft monoplane, and a twin-engined, all-metal Barley-Grow floatplane. By now, Antarctic ventures had become almost routine, and the showmanship of earlier expeditions had faded away. Byrd's third expedition conducted extensive survey work, scientific research, and evaluation from two bases until March 1941, when the party returned to the United States. PL

Flying South

New technology, especially aircraft, became the most potent force in the modern era of Antarctic exploration. Richard Byrd, one of the pioneers of polar aviation, added a sophisticated twin-engined seaplane, the Barley-Grow T8P1, to the fleet for his third expedition in 1939–41.

BANZARE
1929–31

T he 1920s saw a surge of interest in Antarctica by several nations. Norway was eager to extend its whaling grounds from the south Atlantic to the Ross Sea and East Antarctica. France wanted to protect its interests in Terre Adélie, based on Dumont d'Urville's voyage of 1840, as well as in Iles Kerguelen and other sub-Antarctic islands.

In response, in 1923 the British made a territorial claim to the Ross Dependency, to control and to profit from the expansion of whaling, and at the 1926 Imperial Conference, it articulated a policy to "paint Antarctica red," a reference to the practice at the time of mapping the British empire in red.

In particular, Britain had set its sights on the so-called Australian sector, which had been explored by Douglas Mawson's 1911–14 Australasian Antarctic Expedition. Mawson himself was also keen to extend his discoveries; he emphasized that "over half the circumference of the globe remains to be charted in high southern latitudes," and that, in particular, "the great Antarctic region lying to the southwards … is a

Lost Opportunity

Douglas Mawson was 47 when he returned to Antarctica as leader of BANZARE. Although committed to its program of scientific studies, the expedition offered Mawson little of the overland exploration at which he excelled, and he was disappointed with its results.

heritage for Australians … nearer to the Commonwealth than the distance between the east and west coasts of our own continent."

A Hard-working Ship

Mawson's persistence led to the British, Australian and New Zealand Antarctic Research Expedition (known as BANZARE), funded by the three governments. Its purpose was to chart the coast of Antarctica from 160°E to 85°E. The Falkland Islands lent Robert Scott's old ship, the *Discovery*, which was then being used for the Discovery Investigations, for two seasons. BANZARE was led by Mawson, and the captain for the first voyage was his skipper from 1911–14, John King Davis. The expedition left London in August 1929.

As well as survey work, the plans included a full program of biological and hydrographic research; 12 scientists were on board, along with the ship's crew of 17, plus a Gipsy Moth biplane for aerial reconnaissance and for mapping. The ship was not just overcrowded; Captain Davis was convinced that "a more unsuitable ship for oceanographical work than the *Discovery* could not have been built." She was seaworthy, but rolled so badly that it was like "a sea going rodeo," and was slow, under-powered, and lacking in coal-carrying capacity.

BANZARE sailed from Cape Town in October, first to Iles Kerguelen and to Heard Island, then southward to Enderby Land—which had not yet been claimed—to forestall the Norwegian expedition, under Riiser Larsen, that was already in the area. Several times Mawson sighted land and he made three flights over it, but was able to land only once, on Possession Island (53°37′E) on January 13, 1930. Next day, at 49°E heading west, he met Larsen in the *Norwegia*, heading east; by mutual agreement, and in accordance with their governments' respective policies, the two expeditions turned around, marking 45°E as the boundary between Norway's claim to the west and Britain's claim to the east. With her coal supplies running down, in mid-January, the *Discovery* turned north, arriving in Adelaide two months later.

Staking a National Claim

The *Discovery* sailed southward again from Hobart in November, once more led by Mawson, but with K. N. Mackenzie as captain and a similar complement of scientists and surveyors. Their route led south to Macquarie Island, where they found the derelict huts and wireless masts from the 1911–14 expedition, and then on to the Ross Sea, where the *Discovery* refueled from the huge whaling factory ship, the *Sir James Clark Ross*: it was a striking contrast between the old and the new. Following the coast westward, they encountered heavy pack ice and could not land until they reached Commonwealth Bay, the site of Mawson's old base—the old hut was filled with ice. A proclamation claiming the region for Britain was read there, and the expedition

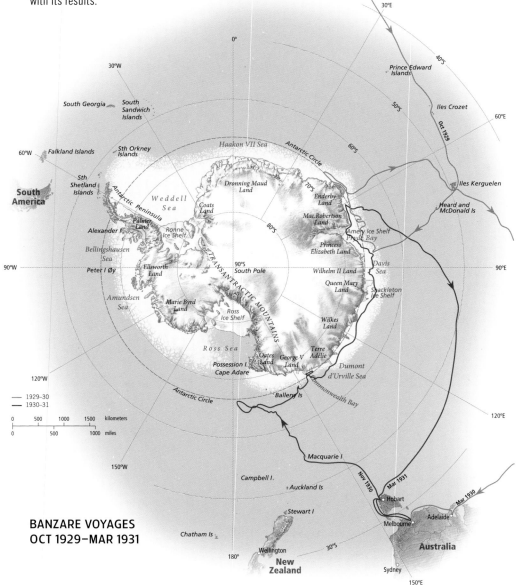

**BANZARE VOYAGES
OCT 1929–MAR 1931**

Home Sweet Home

The scientists went ashore for ten days on Heard Island, where they used a hut at Atlas Cove, built by Norwegians as a refuge for shipwrecked whalers. After he barricaded the door against marauding elephant seals, Frank Hurley praised his temporary home as "warm, dry, rent free, and no taxes."

pressed on, sailing over "land" reported by Charles Wilkes in the 1840s; Mawson reasoned that it must have been an ice tongue. They landed farther west and raised the flag at Scullin Monolith and Cape Bruce; they also sighted the Casey, David, and Masson ranges of Mac.Robertson Land, and, with the help of the Gipsy Moth, discovered the large indentation of Prydz Bay near the Amery Ice Shelf.

Bad weather and heavy ice hampered both of the BANZARE voyages, and Mawson was disappointed with the results; he was unable to undertake the overland exploration he would have liked, and instead had to rely on aerial reconnaissance. Even so, the expedition gathered a wealth of information about a previously unknown part of Antarctica, and resulted in the 1933 affirmation of Britain's claim to the southern continent between 160°E and 45°E; later, in 1936, this was to become the Australian Antarctic Territory. LC

Mawson at Red Dome

The BANZARE expedition spent time at Iles Kerguelen and made extensive surveys along the many fiords that pierce its coast. The scientists noted the contrasts between the stunted vegetation of the mainland, devastated by rabbits, and the luxuriant growth on the isolated smaller islands

Flying the Flag

At noon on January 5, 1931, at his old base at Cape Denison, the Union Jack was hoisted, the British anthem sung, and Mawson read a proclamation in the name of the monarch taking possession of all the area around Commonwealth Bay as King George V Land.

British Graham Land Expedition
1934–37

The British Graham Land Expedition represents the watershed between the heroic and the modern phases of Antarctic exploration, delimited by the outbreak of World War II. It was the last primarily private expedition for several decades, with the last mainly sail-powered ship, while it was the first to combine the traditional use of dogs and skis with modern tractors, motor boats, and aircraft. Its leader was John Rymill, a South Australian farmer and grazier with a lifelong passion for the polar regions.

He was a member of the first British Arctic Air Route Expedition to Greenland under Gino Watkins in 1930–31, and he led a second expedition (1932–33) after Watkins's death in an accident in 1932. Rymill resolved to realize their common ambition of organizing an Antarctic expedition. He managed to raise 20,000 pounds from the Royal Geographical Society and the Colonial Office, and used it to purchase a 112-ft (34-m) three-masted, topsail schooner (called the *Penola*, after his family's farm), a single-engine Fox Moth biplane, boats, dogs, sledges, a tractor, a hut, a hangar for the plane, and other equipment for two years in Antarctica.

An Economical Venture

Although the expedition was run on a shoestring, Rymill had no trouble recruiting four veteran companions of his Arctic expeditions and three young scientists, as well as a naval crew to sail the *Penola*. They set sail in September 1934 for the western coast of Graham Land, intent on determining whether it was a peninsula or an archipelago.

It was a question that was raised by Hubert Wilkins's pioneering flights south from Deception Island in 1929, on which he claimed to have identified ice-filled fiords cutting through from the Bellingshausen Sea on the west to the Weddell Sea on the east.

The *Penola*'s engines broke down just south of the Falklands, but they sailed on and coaxed the ship into an anchorage in the Argentine Islands, using the plane and the motor boat. There Rymill established his first base, and kept using this novel air–sea reconnaissance technique to lay depots farther south for winter sledging journeys. During a flight in February 1935, he sighted

The Expedition Ship
The *Penola* was a 30-year-old schooner, only capable of 4 knots under sail, and with an unreliable engine. But under her determined captain, "Red" Ryder, she made a major contribution to the success of the expedition. After winter spent in the Argentine Islands, the ship was sent to the Falklands for a refit before returning to retrieve the expeditioners.

A New Southern Base
The Fox Moth biplane was invaluable in reconnoitering a route that the *Penola* could follow from the Argentine Islands to a more southerly base for the second year's exploration. A site was found in the Debenham Islands, at Marguerite Bay, and stores and equipment landed.

Strait and Narrow

Penola Strait separates the Argentine Islands from the Antarctic Peninsula. The site of the expedition's first base on Winter Island became Base F of the post-war British network of stations set up to counteract Chilean and Argentine claims to the area. Base F later became Faraday Station and is now Ukraine's Vernadsky Station.

Rymill's Return

Rymill (at right) and the expedition's doctor, Edward Bingham, on their return from crossing Graham Land to the Weddell Sea. They spent a month exploring valleys and glaciers on the Weddell Sea coast, but found no evidence of Wilkins's channels cutting through the peninsula.

Marguerite Bay, south of Adelaide Island. This was the expedition's next target after a winter spent exploring islands and coast by tractor, dog sledges, and skis.

The voyage was difficult. Strong currents and the katabatic winds sweeping down from the mainland glaciers made for treacherous sea ice. In his biography of Rymill, John Bechervaise wrote that the conditions tested to the full Rymill's "almost uncanny judgement in respect of sea ice. He seemed to know by instinct the degree of risk, over sea ice and crevasses, which could sensibly be taken." Bechervaise also recorded a fellow expeditioner's comment that Rymill's "strong sense of responsibility not just to see the expedition through but for the party he had selected … made a remarkably happy expedition under a leader whom all of us liked and greatly respected."

Peninsula or Archipelago?

In January 1936, the *Penola* moved the party south to the Debenham Islands in Marguerite Bay, returning to South Georgia for a refit during the second winter while the land-based party was making its most significant discoveries. Reconnaissance flights in March revealed the complex archipelago between Adelaide Island and the mainland to the north, but cast doubt on Wilkins's postulated sea-level links with the Weddell Sea. Once the sea ice firmed, the ski-equipped plane could range more widely, and in August a narrow sound between Alexander I Land and the mainland was revealed.

In early September, a dog-sledge expedition went off to explore the sound, and penetrated more than 600 miles (960 km) along it. Here the sound opened out to the southwest, suggesting that Alexander I Land was an island (as it later proved to be), but they found no sea-level channels to the east. A second sledging trip of 235 miles (375 km) across the plateau also failed to uncover any evidence of links to the Weddell Sea.

Rymill headed north when the *Penola* returned to collect the land party in March 1937. Having followed the coast from 64°30′S to 72°30′S, the expedition had successfully demonstrated that Graham Land was not an archipelago but a peninsula. LC

Versatile Transport

The single-engined De Havilland Fox Moth aircraft could fly with either skis or floats, and was small enough to be towed by the *Stella*, the expedition's motor launch, as seen here. This versatility was enhanced by the use of dog- and motor-sledges.

Lincoln Ellsworth
1934–39

With the South Pole "conquered" by plane, the next aerial challenge was a trans-Antarctic flight. The ebullient American millionaire Lincoln Ellsworth had flown in Arctic regions, and had financed an unsuccessful attempt by Hubert Wilkins to reach the North Pole in 1931 by submarine. The two teamed up again when Ellsworth decided to organize an expedition to the south: Ellsworth to fly and Wilkins to provide the managerial expertise from sea level. The plan was to fly from the Ross Sea to the Weddell Sea and back. Although hardly a crossing of the continent at its widest point, this would still be a return flight of around 3,420 miles (5,500 km), and would cover substantially unexplored territory over much of the area between the flightpaths of Wilkins in 1928 and of Byrd in 1929.

A Chapter of Accidents

They acquired a ship, the *Wyatt Earp*; recruited Bernt Balchen, who had been Byrd's pilot to the South Pole in 1929; and took delivery of a newly designed, two-seater, single-engined Northrop Gamma metal monoplane.

Thereafter, success did not come readily. They reached the Bay of Whales in January 1934 but, after a brief test flight, the Northrop Gamma was damaged beyond any hope of on-site repair when the ice shelf broke apart beneath it. They retreated northward to rebuild the aircraft, and returned to Deception Island in November with a modified plan to fly one way, in the opposite direction, and finish at Little America II.

This time, a broken engine rod on the very first engine run—and no spare—forced the *Wyatt Earp* to go back to South America for a replacement. When she returned, the ice at Deception Island had melted (the aircraft was to use skis), and the expedition was forced to relocate 100 miles (160 km) farther south. Here the weather delayed them for more than a month; they tried to take off on their record-setting flight on January 3, 1935, but poor weather forced Balchen to turn back after an hour. Ellsworth abandoned the project for the summer, but it was still some time before the ship was able to sail north through the pack ice.

A Dream Achieved

For his third attempt, in November 1935, Ellsworth replaced Balchen with the English-born Herbert Hollick-Kenyon. They established a northern base on Dundee Island, just off the northeastern point of the Antarctic Peninsula, but on November 20 disaster struck again.

A mere couple of hours into the proposed flight of 2,300 miles (3,700 km)—which was the longest period that Ellsworth had achieved in Antarctic skies—a fuel gauge threatened to disintegrate, and the attempt was abandoned. The next day, after some five hours in the air, bad weather forced them to again turn back. Two days later, on November 23, they took off again, and finally achieved Ellsworth's dream, although hardly in the shortest time possible. Part way through the flight, their radio failed and communication with the *Wyatt*

Tortuous Trip
Ellsworth (right) with Herbert Hollick-Kenyon, co-pilot of the *Polar Star* on the successful flight from Dundee Island to near Byrd's Little America base. Anticipating a 14-hour flight, they finally reached Little America II on foot 22 days later.

The Rescue Party
The trans-Antarctic flight ended at Little America II, where they waited for their ship to arrive. On losing radio contact, however, the ship had raised the alarm and Australia had sent the *Discovery II* to rescue them.

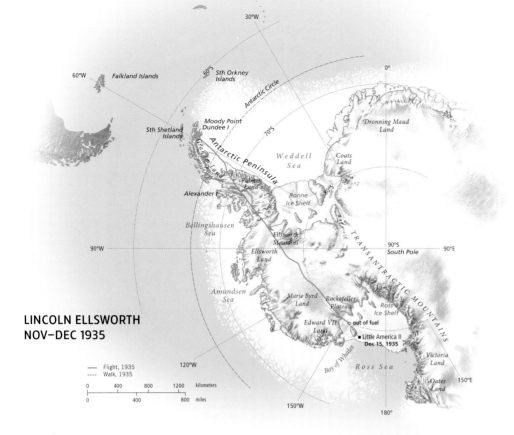

**LINCOLN ELLSWORTH
NOV–DEC 1935**

— Flight, 1935
···· Walk, 1935

0   400   800   1200 kilometers
0   400   800 miles

Historic Aircraft
Ellsworth's Northrop Gamma monoplane, which made the first trans-Antarctic flight. Despite running out of fuel only 25 miles (40 km) short of its destination, it was recovered by the *Wyatt Earp*.

High Hopes Dashed
Ellsworth (left) and co-pilot, Norwegian Bernt Balchen, in the cockpit of the *Polar Star*. They planned to fly across Antarctica from the Ross Sea to the Weddell Sea and back in January 1934. But the plane was damaged on the ice shelf, which delayed the flight a year.

Earp was cut off, but on this occasion they did not turn back. Some 14 hours into the flight—roughly two-thirds of the way—they landed, stopping for about 19 hours for Ellsworth to pursue territorial claims and stretch his legs. The weather deteriorated once they were back in the air, and they had to make two forced landings. The first delayed them for three days, the next for a blizzard-racked eight, before they took off for the final 500 miles (800 km) to the Bay of Whales—only to run out of fuel less than 30 minutes from their destination.

Eventually, on December 15, after a ten-day trek, they reached the Ross Sea coast on foot and found the remains of the Little America II base, which had been abandoned in February of that year. Ellsworth had left an emergency plan with Wilkins, including plans for a rescue, so they settled down to wait for the *Wyatt Earp*.

Help did arrive, but from an unexpected source. When the *Wyatt Earp* lost radio contact with Ellsworth back on November 23, Wilkins had raised the alarm.

The Australian government asked the British to provide a suitable rescue ship. The *Discovery II* was sent from Hobart to the Bay of Whales. On January 14, 1936, an Australian Airforce DH60 Gipsy Moth from the rescue ship flew over the stranded party and directed them to the *Discovery II*. Five days later, the *Wyatt Earp* arrived at the Bay of Whales, delayed by heavy ice and the need to investigate various other collection points that Ellsworth had pre-arranged in his emergency plan.

Another Try

Ellsworth made one more expedition, to the other side of the southern continent, in the summer of 1938–39. Again he was accompanied by Wilkins, but each was making, or reinforcing existing, territorial claims to the same area on behalf of their respective governments, Ellsworth from the air and Wilkins from the ground. Ellsworth may also have been stung by suggestions that his epic flight of 1935 could not truly be described as trans-Antarctic. At the back of his mind, therefore, was the possibility of a flight across the Antarctic continent to the South Pole, and then on to the Bay of Whales on the other, western, side.

Dreadful weather and heavy ice prolonged the voyage south from Cape Town; it took about two months to reach the Antarctic mainland. The *Wyatt Earp* spent much of the first two weeks of 1939 maneuvering to avoid icebergs and only three brief flights were made in an Aeronca aircraft. On January 11, a flight of greater interest took place in a single-engined Lockheed Delta. An inadequate ski runway meant that it could not carry enough fuel for a trans-Antarctic flight; instead, in three hours, Ellsworth and pilot, Canadian Jack Lymeburner, flew inland to lay claim from the air to an area previously claimed by Australia. (In the three days preceding this, Wilkins had landed three times to reinforce Australia's claim.) But just a few days later, a serious injury to one of the members of the ship's crew forced them to turn back to Australia, making another tempestuous crossing of the Southern Ocean. PL

The *Schwabenland* Expedition

The *Schwabenland* Expedition was one of the shortest Antarctic expeditions ever: only three weeks south of the Circle. Mounted by Hermann Göring in a bid to claim territory, it was led by Captain Alfred Ritscher of the German Navy, and was the first attempt by any nation at large-scale aerial mapping of Antarctica. It employed two Dornier Wal flying boats, which were launched by catapult from their mother ship, *Schwabenland*—all borrowed from the airline Lufthansa.

The expedition reached Dronning Maud Land in late January 1939. Over seven days, its aircraft made 16 flights and surveyed 232,000 sq miles (600,000 sq km)—about half of which were recorded in 11,000 photographs—ranging from 4°50´W to 16°30´E. The expeditioners ventured 370 miles (600 km) inland, and discovered several mountain ranges. The aircraft never landed on the continent, but along their flight path aluminum darts topped with swastikas were dropped to establish the German claim. The claim was disputed by Norway—the same area had been explored more thoroughly by Riiser Larsen's expeditions of 1927–30—which proclaimed sovereignty over the area between 20°W and 45°E on January 14, 1939, five days before the *Schwabenland* Expedition arrived, and was recognized by all other claimant nations of the time: Australia, Britain, France, and New Zealand. After World War II, Ritscher published the surviving aerial photographs, providing valuable information about a previously unknown region. LC

Operation Highjump
1946–47

An entirely new concept in Antarctic exploration, Operation Highjump was the first predominantly military Antarctic operation. Run by the United States Navy, it was the biggest Antarctic venture ever, with 13 ships, 23 aircraft, and over 4,700 personnel. The operation originated in the massive demobilization of the United States armed forces after World War II, the escalating cold war between the United States and the Union of Soviet Socialist Republics, and the realization that Antarctica was one area of the world in which this competition could be played out.

In August 1946, the United States Navy initiated the Antarctic Development Project to meet this challenge. Its main purpose was to build up personnel, material, and logistics capacity in polar regions, but it was also to consolidate United States interest in areas previously explored, find suitable base sites, and determine the feasibility of building and maintaining permanent bases, including air fields. At the time, scientific research was only a secondary concern.

The immediate objective was to establish a base on the Ross Ice Shelf, near Richard Byrd's Little America III of 1940–41, and to photograph as wide an area as was possible from the air. An ambitious program of aerial mapping of the entire coastline was also included.

Three-part Armada

The central group, which comprised an icebreaker, two supply ships, a submarine, and a command ship under Admiral Richard Cruzen, was to penetrate the Ross Sea pack ice and establish the base. But in one of the worst ice years on record, this was a difficult undertaking. The submarine proved to be a dangerous liability, and had to be escorted out of the ice by the icebreaker, the *Northwind*, leaving the thin-skinned supply ships to fend for themselves; one of them tried to blast its way through the pack ice using cannonfire to escape a menacing iceberg. Eventually, however, the *Northwind* brought the convoy safely to the Ross Ice Shelf.

Little America IV, a 200-tent camp with a runway hewn out of ice, was set up to receive the six R4D aircraft, which were brought to the edge of the pack ice by aircraft carrier. The ship's bridge was in the middle of the flight deck and the aircraft were too wide to get past it—they could use only half the length of the flight deck and had to be launched by JATO (Jet Assisted Take Off). But, with Richard Byrd on board, they landed safely on the ice on January 29–30, 1947. From there, Byrd made his second flight over the South Pole, and 100 miles (160 km) beyond. In over 200 hours of flying, sophisticated aerial mapping cameras recorded vast areas of new territory, defining the shape of the Ross Ice Shelf, revealing the nature of the Dry Valleys, and confirming the existence of the unbroken chain of the Transantarctic Mountains. At the end of February, Little America IV camp was evacuated; the

tents, aircraft, and equipment of this most expensive and technologically advanced Antarctic operation were abandoned, and were never to be used again.

Highjump's eastern group, comprising three Martin Mariner PBM seaplanes and a seaplane tender, and accompanied by a tanker and a destroyer under the command of Admiral George Dufek, was assigned the area from the Amundsen Sea to the Weddell Sea. This was the worst sector for weather and sea conditions, and there was an early setback when they lost three men and a seaplane, and had to mount a protracted rescue of the surviving crew members. Some useful mapping was done along the coast of Marie Byrd Land and south Graham Land, but none in the Weddell Sea. The overall results were disappointing.

The western group, with similar resources under Captain Charles Bond, worked west from the Balleny Islands off the coast of Oates Land. They encountered better weather, and on February 1, 1947, made the most significant discovery of the entire operation, when Commander David Bunger sighted the ice-free oasis in Queen Mary Land that now bears his name: Bunger Hills. This group mapped over a third of the coastline, from 160°E to 40°E, and in places they penetrated some 500 miles (800 km) inland.

A Limited Success

For all its massive scale, and all the claims made for it, Operation Highjump photo-mapped a mere 25 percent of its initial objective. More than half of the photographs that were taken were useless because of lack of ground control (although this was partially remedied by the use of helicopters in Operation Windmill the following year), and only 35 percent were incorporated into published maps. Nonetheless, this was a great advance on the existing maps of Antarctica—and Operation Highjump certainly succeeded in making the United States Navy a world leader in the militarization of Antarctic scientific research and exploration. LC

Nostalgic Lunch
Admiral Richard Byrd (right), officer in charge of Operation Highjump, and his scientific advisor, Dr Paul Siple, on their fourth expedition. Siple had led the Little America III venture in 1940–41 and he found his old base more than 2 miles (3.2 km) northwest of its original position, and 5 ft (1.5 m) under ice. He dug his way in for lunch with comrades from the Byrd expedition.

Short-lived Encampment
Supplies unloaded onto the sea ice in the Bay of Whales (USS *Yancey* in the foreground). All equipment had to be taken by tractor-drawn sledges to the site of Little America IV over 1 mile (1.6 km) away. After the three-month operation the base was abandoned, never to be used—or, indeed, even seen—again.

Ice Sculpture

A spectacular windscour in the coastal ice cliffs surrounding O'Brien Bay, near Casey Station in Wilkes Land. The bay is named after US Navy Lieutenant Clement O'Brien, who served as a radio communications officer with Highjump's 1947–48 follow-up, Operation Windmill.

Ground Control Site

First mapped from the air, Peterson Island in the Windmill Islands was used by Operation Windmill in 1948, which gave its name to the group of about 50 islands. A plaque commemorating an Operation Highjump landing is sited close to the fiberglass Apple hut now used by Australian scientists from nearby Casey Station.

The International Geophysical Year

1957–58

Old Comrades

Although nominally in command of Operation Deepfreeze, Admiral Byrd (left) was 68 when he visited Antarctica for the last time in 1955, leaving management to Admiral George Dufek. Paul Siple (right) again visited Little America with his old boss before taking charge of Amundsen–Scott Base.

The International Geophysical Year (IGY) focussed attention on science in and about Antarctica, and it was a stepping stone to the Antarctic Treaty of 1959. The idea originated in a discussion among British and American physicists in 1950, and it took its theme from previous International Polar Years, of 1882–83 and 1932–33. The second had been a period of minimum sunspot activity; because 1957–58 would be a period of maximum activity, findings would provide valuable comparisons. The IGY proposal was endorsed by the International Council of Scientific Unions, which set up a committee in 1952 to organize what became the IGY.

Its scientific program focussed on two areas, Antarctica and outer space, and ran for 18 months from July 1, 1957.

Twelve nations—Argentina, Chile, New Zealand, Australia, South Africa, Belgium, Britain, France, Norway, the United States, Japan, and the USSR—established 50 stations in Antarctica, with more than 5,000 personnel, to conduct a series of research programs on glaciology, meteorology, geology, geomagnetism, and upper atmosphere physics. The largest contributors were the USSR, with seven stations, as well as a temporary one at the Pole of Inaccessibility (the point on the continent farthest from all its coasts) occupied for 22 days; and America, also with seven stations, including Amundsen–Scott Station at the South Pole. (Britain had a total of 13 stations in the Antarctic region, but 11 were on the Antarctic Peninsula and adjacent sub-Antarctic islands and most pre-dated the IGY.)

Operation Deepfreeze

Deepfreeze, which was under the overall command of Admiral George Dufek, was the four-year program through which the United States met its IGY objectives. Dufek's task was to coordinate the United States Navy, Army, Air Force, Marines, and Coast Guard with those of other nations, particularly New Zealand, and to have all its seven bases operational, equipped, and staffed with civilian scientists by July 1957—a formidable challenge, even though staged over several seasons.

Deepfreeze I (1955–56) established the bases, McMurdo Station on Ross Island and Little America V on the Ross Ice Shelf. Eight US Air Force Globemaster aircraft made flights from Christchurch to an ice runway at McMurdo, and seven ships "converged on Antarctica and constructed Little America and McMurdo stations on schedule." In March 1956, after the aircraft and the ships departed, they left 93 men at McMurdo Station and 73 at Little America for the winter.

From this bridgehead, Deepfreeze II (1956–57) brought in 12 ships, 13 aircraft, and 4,000 men to undertake the far more difficult tasks of establishing Amundsen–Scott Base at the South Pole, and Byrd Base, 620 miles (1,000 km) inland from Little America. Army engineers dynamited through a crevasse field 8 miles (13 km) wide to build an "all-weather highway" for the main supply route to Byrd, where CBs (members of the Construction Battalion) constructed the base in the week between Christmas and New Year.

Dufek was only the eleventh person to stand at the South Pole when the first ski-equipped Navy Skytrain landed there on October 30, 1956, after a four-hour flight from McMurdo. It was accompanied by a wheeled Air Force Globemaster, which circled overhead ready to drop survival equipment in case the Skytrain crashed on landing or take-off. In fact, the Skytrain's skis froze to the ice during the 50-minute stopover, and 15 JATO (Jet Assisted Take Off) bottles were needed to get the plane airborne. In late November, the advance CB team were flown in to begin construction, and over the next month the Globemasters parachuted around 825 tons (750 tonnes) of equipment onto the South Pole. The base was finished by March 1957, and 17 men were left to winter over, led by Paul Siple, a veteran of Byrd's early expeditions and of Operation Highjump.

Meanwhile, a joint United States/New Zealand base was established at Cape Hallett on the western shore of the Ross Sea; Wilkes Base in Knox Land was built in 17 days; and two icebreakers battled through the Weddell Sea to set up Ellsworth Base on its southwestern shore.

During Deepfreeze III (1957–58), 350 men wintered over at the seven United States bases and conducted scientific research. Little America V was weather control

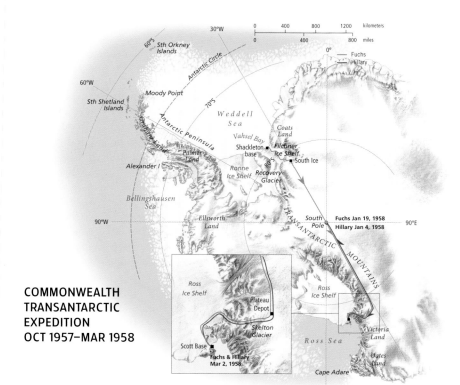

COMMONWEALTH TRANSANTARCTIC EXPEDITION
OCT 1957–MAR 1958

Delayed Welcome
Below Mount Erebus on Ross Island, the crew of one of the three surviving Snocats of Fuchs's triumphant crossing party are greeted as they reach Scott Base after a 99-day marathon.

Success Despite Setbacks
Sir Edmund Hillary, New Zealand conqueror of Everest (left) and Dr Vivian Fuchs, director of the British Antarctic Survey, were a formidable team to lead the first successful crossing of the Antarctic continent. They shared the honors: Fuchs completing the whole crossing and Hillary becoming the first to reach the Pole overland since Scott.

for the entire Antarctic meteorological network, and an international staff coordinated the daily observations. Seismic soundings at the South Pole revealed that the ice cap there was nearly 9,000 ft (2,745 m) thick, and glaciological traverses helped to delineate the contours of the bedrock far below.

Transantarctic Expedition
Ernest Shackleton's dream of 40 years earlier was also fulfilled during the IGY when a British Commonwealth team made the greatest traverse of all—crossing the Antarctic continent from coast to coast. As Shackleton had planned for his Imperial Transantarctic Expedition, so the Commonwealth Transantarctic Expedition used a two-pronged attack. The main expedition, led by Vivian Fuchs, started from the Weddell Sea. It was preceded by a two-year effort to establish a base camp, first in 1955 at Vahsel Bay, where eight men spent the winter of 1956. The following summer an advance base was established by air 275 miles (440 km) inland at South Ice, and three men wintered there. Thirteen others, including Fuchs, spent the winter at the Vahsel Bay base, preparing the vehicles and aircraft to be used in the crossing the following summer.

But bad weather delayed the departure of the route-finding party to South Ice until early October; even then, and with the help of dog teams scouting ahead, the journey took until mid-November. The main party did not reach South Ice until late December, and started for the Pole on Christmas Day. They were using the most advanced over-snow vehicles available— three Weasels, four Snocats, and a Muskeg tractor—with the route being pioneered by two dog teams. Even so, the steep ice ridges with frequent vertical drops and numerous crevasse fields took their toll. Three vehicles had to be abandoned, but the remainder finally reached the Pole on January 19, 1958.

Fuchs's party was beaten to the Pole, however, by the irrepressible New Zealander, Edmund Hillary, leader

Doomed Station

Wilkes Station (above), built in 1957, was run for the IGY by the US Navy and handed over to the Australian National Antarctic Research Expeditions (ANARE) in 1959. It soon became clear that the site—in a hollow—was prone to snow drift. It was largely abandoned in 1965.

Small Beginnings

New Zealand's Scott Base (above right), established by Edmund Hillary on the southern tip of Ross Island, is about one-tenth the size of its mighty US neighbor, McMurdo. The hut in which Hillary and his team spent the winter of 1957 is on the far left, in front of the main buildings. It is now a museum.

of the Ross Sea party. Hillary's designated role was to support the crossing party by laying depots from the Ross Sea toward the Pole. He had established Scott Base, a little south of McMurdo, on Ross Island in January 1957, and then spent the rest of the summer finding a route over the Skelton Glacier onto the Polar Plateau, and stocking advance depots from the air.

After wintering at Scott Base with his party of 22, Hillary and three others set out with two dog teams, three Massey Ferguson TE20 tractors fitted with rubber wheel tracks and towing a caravan for living in, and sledges loaded with fuel and cargo. With the help of air support, they established two more supply depots, the last only 500 miles (800 km) from the South Pole.

At this point, all their vehicles were in good shape and they had enough fuel, so Hillary decided to press on rather than wait for Fuchs, who was already behind schedule. The last lap, over the rough terrain of the Polar Plateau at an altitude of 10,000 ft (3,050 m), lowered fuel efficiency, and they arrived at the Pole on January 4, 1958, with less than 20 gallons (100 liters)

of fuel left. Hillary joked that "our tractor train was a bit of a laugh;" nonetheless, this was the first expedition to reach the Pole overland since Robert Falcon Scott's, nearly 46 years earlier.

Fuchs's party did complete the transcontinental crossing, joined by Hillary's group for the final stretch down the Skelton Glacier. It was plagued by frequent breakdowns, the Snocats opened crevasses that the lighter Massey Ferguson tractors had negotiated easily, and navigation was tricky due to the proximity of the South Magnetic Pole, but they finally reached Scott Base on March 2, 1958, after traveling 2,160 miles (3,475 km) in 99 days.

The journey was a superb achievement in itself, but the seismic survey and other scientific investigations that were undertaken en route provided unique data on a completely unknown area of the continent, and added substantially to the scientific value of the IGY. So much data was collected on the overland expedition that three world centers were established to evaluate it, thus further extending the spirit of international cooperation.

"A Continent for Peace and Science"

Perhaps the IGY's greatest success was that it focussed attention on the necessity for a permanent international regime to manage the Antarctic region, and to prevent it from falling prey to the cold war conflict between the superpowers. In December 1959, after 18 months of negotiations, all of the 12 nations that had participated in the IGY signed the Antarctic Treaty.

The treaty banned military activity, guaranteed free access to the region for scientific research, and solved the problem of territorial claims by stating that the treaty did not "endorse, support or deny any territorial claim," and by prohibiting any further claims. Today, Antarctica remains under this continuing regime. LC

Detaille Base

One of the British bases on the Antarctic Peninsula, on Detaille Island, had been established in 1956, prior to the IGY, but was abandoned three years later—it was impossible to reach regularly for resupply.

Research at McMurdo

Since 1956, the United States has had a full-time research station at McMurdo Station on Ross Island, and from that first year it has been the largest base on the continent. A wide range of scientific research is conducted here, with significant logistical support.

On Half Moon

Argentina, which began establishing bases on the Antarctic Peninsula in 1946–47, was a participant in the IGY and one of the original signatories to the Antarctic Treaty. Camara Base on Half Moon Island opened in 1953 and closed in 1997–98, although it has been used occasionally since then.

Antarctica
Today

Which Nations Claim What?

Friendly Flakes

Snowflakes fall on a station entry sign with its implicit promise of comfort. Except during the fiercest blizzards, travel in familiar Antarctic territory with the help of satellite navigational aids and modern radar is relatively easy, even in inclement weather.

The Antarctic "Pie"

The Antarctic region is divided into segments shaped like slices of a pie, each slice being under the jurisdiction of a nation that was a signatory to the Antarctic Treaty of December 1, 1959, which arose out of the International Geophysical Year of July 1957 to December 1958.

Uniquely among the continents of the world, Antarctica belongs to no country— the outcome of decades of negotiation and of dispute, both on the southern continent itself and through remote diplomatic channels. By the 1920s, there were seven claimants to territory in Antarctica. These claims were based on discovery and on "a sufficient display of authority" to show they occupied and controlled the land to which they laid claim.

Australia claimed the largest portion— 42 percent, which was originally claimed by Britain and then placed under Australian control. In 1924 France claimed a small slice (a 6-degree arc) within the Australian claim as Terre Adélie, because of the exploration and claim made by Dumont d'Urville in 1840. Since 1923, when Britain handed it over, New Zealand has claimed the Ross Dependency—effectively the 50°S sector below New Zealand that embraces the whole of the Ross Sea. Norway claimed Dronning Maud Land in 1939 through its whaling activity. However, while other nations claim a slice from the South Pole to latitude 60°S, Norway does not define the outer limit of its claim and is really interested only in a 65-degree arc of the coastal strip. There is a section of West Antarctica that is not claimed by any nation.

The problem sector centers around the Antarctic Peninsula, where, like a fan of playing cards, the claims of Chile (37-degree arc), Argentina (49-degree arc), and Britain (60-degree arc) overlap. However, in geopolitical terms, two absentees from the list of claimants were just as important as the claimants in the development of the unique solution that is represented by the Antarctic Treaty. In 1924, the United States adopted the Hughes

Doctrine (named for Charles Hughes, secretary of state from 1921 to 1925) which states that mere discovery is not sufficient for territorial claim: actual settlement is required. The United States could have based a claim on the explorations of Wilkes, Byrd, and Ellsworth, but it has not done so, nor has it recognized any other claims. When Norway claimed Dronning Maud Land in 1939, the then Soviet Union refused to recognize any claims to Antarctica, even though it could have based an all-encompassing claim on Bellingshausen's 1819–21 circumnavigation. Subsequently, Hitler ordered the annexation of the Norwegian claim by Germany, and sent two Dornier Wal flying boats to drop aluminum darts capped with swastikas right across the region.

Warfare Averted—Twice

During and immediately after World War II, Argentina, Chile, and Britain engaged in a flurry of base-building. When Britain suggested taking disputed claims to the International Court of Justice, Chile and Argentina rejected this solution; they wanted an international conference to resolve all the claims over the continent.

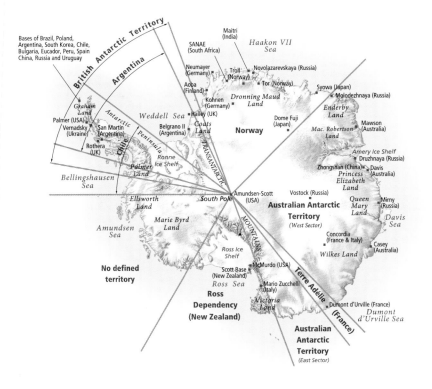

A Station on Ice
Wilkes Station was built by the United States in 1957, and in 1958 it was run jointly with Australia. After two years, the United States withdrew, and the ice took over eventually.

Life at the Pole
Antarctica is a vast and unique laboratory for the study of life on the Earth. The world's last wild continent is a place where settlement has been determined by science, not conquest, and where both science and tourism remain strictly controlled.

In 1948 both Argentina and Britain sent warships to the region, and armed conflict seemed likely.

The peace-making suggestion of the United States was that Antarctica should fall under the trusteeship of the United Nations, and this evolved into the concept of a condominium, or joint sovereignty, by the seven claimant nations plus the United States, but excluding the Soviet Union. Only Britain and New Zealand were in favor of this contentious idea. Meanwhile, a legal adviser to the Chilean government, Professor Julio Escudero, put forward the "modus vivendi" proposal that all national claims be effectively frozen so that scientific research programs could continue. This intelligent compromise was to become the heart of the Antarctic Treaty 11 years later.

However, the mere proposal was not the end of the matter. In January 1949, the Soviet Union announced that it would not accept the legitimacy of any Antarctic agreement to which it was not a party. The closest that Antarctica came to warfare was in February 1952, when Argentinian troops fired over the heads of members of a British party who were rebuilding the Hope Bay Station, which had been destroyed by fire four years earlier. (South Georgia and the Falklands lie outside the treaty area, so the 1982 conflict does not count.)

A Continent for the World
The remarkable spirit of scientific cooperation (at the height of the cold war) shown during the International Geophysical Year (IGY), which ran for 18 months from July 1957 to December 1958, showed that nations could work together in Antarctica. The 12 nations that participated in the IGY operated 60 field stations across Antarctica and its islands, and produced huge amounts of research data. From the IGY, the Antarctic Treaty was born, and on December 1, 1959, 12 nations signed the treaty: Argentina, Australia, Belgium, Britain, Chile, France, Japan, New Zealand, Norway, South Africa, the Soviet Union (now the Russian Federation), and the United States. There are 46 signatories to the Antarctic Treaty now. Strengthened by the Madrid Protocol of 1991, Antarctica belongs to no nation, and remains a nuclear-free continent for science and exploration. DM

Antarctic Law

Claiming Territory
This obelisk at Chile's Presidente Eduardo Frei Montalva Base (known simply as "Frei") symbolizes the cooperative nature of the Antarctic community. However, the base itself supports Chile's territorial claim over this slice of Antarctica. An obelisk in Patagonia marks the mid-point of Chile's claimed territory—from the Peruvian border to the South Pole.

Antarctica in the third millennium is a unique continent—a nature reserve that is consecrated to world peace and to scientific research. It is also unique in international law, not only because of its isolation and its climatic extremes but because it has no permanent human populations, so therefore there is no natural sovereignty over land, and it also contains the only significant unclaimed slice of territory on the Earth. This uncertain jurisdiction can lead to precarious circumstances: for example, the claims of Britain, Chile, and Argentina overlap across the Antarctic Peninsula. However, some other nations do not regard territorial claims on Antarctica as valid.

Antarctica is underpinned by a unique legal system. A group of 27 "consultative countries," with another 19 observing parties, uses the 1959 Antarctic Treaty as the foundation for a legal regime that applies to the whole area south of latitude 60°S, including, to some extent, the "high seas." The rules are formulated at the annual meetings of the 27 decision-making countries. Each country's legal system must endorse these decisions, and then they become binding on the citizens of each country, as well as on those of the 19 countries that do not participate directly but have agreed that they will abide by the treaty rules.

World Heritage
By popular consensus, Antarctica belongs to the world, and countries and individuals working there must obey rules that have been formulated over half a century to protect this remarkable part of the Earth. The members of the Antarctic Treaty are not regarded as owners of the Antarctic continent but as custodians and keepers of mutually sanctioned laws.

Participating countries have become "polar police," albeit armed only with diplomacy. They do have the power to scrutinize other countries' behavior, guided by the rules that they themselves have formulated, using environmental impact assessments, mandatory inspections, and exchange of information to monitor each other's compliance. Ultimate authority, however, rests with independent countries and their power to invoke their laws against their citizens. Australia, for example, claims 42 percent of the Antarctic continent as sovereign territory, and applies territorial law and ordinances to its Australian Antarctic Territory (AAT); however, an Australian citizen anywhere in the Antarctic region is subject to Australian law. It is not necessary for Australian sovereignty to be recognized by any other country for the government of Australia to make laws specifically for the AAT, or to apply other appropriate civil and criminal law generally. So, if an Australian citizen breaks any Antarctic law (even if outside the AAT), the full force of domestic law can be invoked.

The case is the same for all countries whose citizens are operating in the Antarctic. Problems arise only when one country tries to claim jurisdiction over someone, or something, that belongs to another country. A ship at sea, for instance, is under the jurisdiction of the country of its registration, but the owners, the masters, the tour operators, the charterers—even the passengers—may have to contend with many different laws in the event of violation. Each circumstance requires an individual approach, depending on the nature of the violation and the applicable law. Fundamentally, however, on land south of latitude 60°S, all persons are theoretically subject to the laws of their country of citizenship.

Peaceful Solutions
If a whole range of Antarctic Treaty countries were to be involved in a crime committed in Antarctica, which might lead to a dispute about jurisdiction, they could call upon the treaty's dispute resolution procedure, which clearly states that the participating countries should try to resolve the argument among themselves in the first instance. Because all countries active in the Antarctic are reasonably unanimous about its heritage values, any disputes are most likely to be resolved peacefully and to the mutual satisfaction of all; indeed, this has been the case so far. If the crime also involves a country outside the treaty group, they can attempt to summon international sentiment against the outside party and try to reach a peaceful accommodation through diplomacy, which is how the treaty asks its members to behave. Because the 46 treaty countries include those with a capacity to operate in this environment, diplomacy has so far proved adequate. JJ

Flying the Flag
The flags of the 12 original Antarctic Treaty signatories fly at the South Pole, outside the newly erected buildings of the Amundsen–Scott Station. The treaty was made in 1959 and it came into effect in 1961.

A Picture From the Past

Oscar Wisting, seen here at the South Pole with his dog team in 1911, was part of Roald Amundsen's expedition. The treaty parties decided that all dogs should be removed from Antarctica by 1994 for environmental reasons.

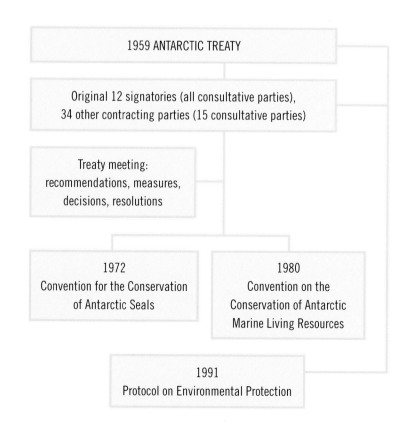

Treaty Activities

A flow chart illustrates how, from the original Antarctic Treaty of 1959 with its 12 signatories, membership of the Antarctic community has grown and measures have been adopted to protect the region's marine life and its unique environment.

Antarctic Treaty Parties

STATE	EFFECTIVE DATE	STATUS
Argentina	06.23.61	ATCP/OS/C
Australia	06.23.61	ATCP/OS/C
Austria	08.25.87	Acceding state
Belarus	12.27.06	Acceding state
Belgium	07.26.60	ATCP/OS
Brazil	05.16.75	ATCP 12.09.83
Bulgaria	09.11.78	ATCP 25.05.98
Canada	05.04.88	Acceding state
Chile	06.23.61	ATCP/OS/C
China	06.08.83	ATCP 07.10.85
Colombia	01.31.89	Acceding state
Cuba	08.16.84	Acceding state
Czech Republic	06.14.62	Succeeding state [1]
DPR Korea	01.21.87	Acceding state
Denmark	05.20.65	Acceding state
Ecuador	09.15.87	ATCP 19.11.90
Estonia	05.17.01	Acceding state
Finland	05.15.84	ATCP 09.10.89
France	09.16.60	ATCP/OS/C
Germany [2]	02.05.79	ATCP 03.03.81
Greece	01.08.87	Acceding state
Guatemala	07.31.91	Acceding state
Hungary	01.27.84	Acceding state
India	08.19.83	ATCP 12.09.83
Italy	03.18.81	ATCP 05.10.87
Japan	08.04.60	ATCP/OS
Netherlands	03.30.67	ATCP 19.11.90
New Zealand	11.01.60	ATCP/OS/C
Norway	08.24.60	ATCP/OS/C
Papua New Guinea	03.16.81	Acceding state
Peru	04.10.81	ATCP 09.10.89
Poland	06.08.61	ATCP 29.07.77
Republic of Korea	11.28.86	ATCP 09.10.89
Romania	09.15.71	Acceding state
Russian Federation	11.02.60	Succeeding state [3] ATCP/OS
Slovak Republic	06.14.62	Succeeding state [4]
South Africa	06.21.60	ATCP/OS
Spain	03.31.82	ATCP 21.09.88
Sweden	04.24.84	ATCP 21.09.88
Switzerland	11.15.11.90	Acceding state
Turkey	01.24.96	Acceding state
United Kingdom	05.31.60	ATCP/OS/C
United States of America	08.18.60	ATCP/OS
Uruguay	01.11.80	ATCP 07.10.85
Ukraine	10.28.92	ATCP/OS [5]
Venezuela	03.24.99	Acceding state

KEY

ATCP   Antarctic Treaty consultative party

OS   Original signatory (automatically ATCP)

C   Claimant state (automatically ATCP)

1. The Czech and Slovak republics inherited Czechoslovakia's obligations as a "contracting party" with effect from January 1, 1993, the date of their succession to the treaty.
2. The German Democratic Republic was united with the Federal Republic of Germany on October, 2, 1990. GDR acceded to the treaty on January 19, 1974, and was recognized as an ATCP on October 5, 1987.
3. Following the dissolution of the USSR, Russia assumed the rights and obligations of being a party to the treaty. The USSR had been an original signatory to the treaty.
4. See footnote 1 above.
5. Ukraine has asserted that it succeeded to the treaty following the dissolution of the USSR and thus should be entitled to ATCP status. Consultative party status was granted on May 27, 2004.

Tourism in Antarctica

Adventurous tourists have been visiting Antarctica since the 1950s. Lars-Eric Lindblad, of Lindblad Travel in New York started tourism on a regular basis in 1966 when he chartered an Argentine vessel, the *Lapataia*, to take some tourists from South America to the South Shetland Islands, a distance of 620 miles (1,000 km) across the stormy Drake Passage. They, like the tourists who have followed them, traveled south to find the last wilderness on the Earth, the final frontier on the planet, and to feel like their heroes: Shackleton, Mawson, Amundsen, and Scott.

Ship-based tourism is still the most dominant form of tourism in Antarctica, allowing travelers to view the southern continent from a floating hotel. The alternative for the average tourist is to participate in flights over the Ross Sea region. A few intrepid souls also participate in polar adventure tours and climbing trips to Vinson Massif (16,066 ft/4,897 m) and other high mountains, or even ski to the South Pole.

Antarctica is a unique place—there are no hotels ashore, no local markets to visit, and the shopping opportunities are limited to a few gift shops at the stations of some of the treaty party countries that administer Antarctic affairs through the complex web

known as the Antarctic Treaty System. Most tourist landings are carried out using inflatable boats called Zodiacs, and more than 90 percent of all tourism activities take place in the Antarctic Peninsula region and on the offshore islands to the west of the peninsula.

The season for visiting the Antarctic region runs from the middle of November to the middle of March. For the remainder of the year, the sea ice surrounding the continent makes tourist landings impossible.

Tourism Management

The management of Antarctic tourism is an interesting issue. When the Antarctic Treaty came into force in 1961, it did not specifically mention non-governmental activities, such as tourism. At that time, it was not apparent that significant numbers of tourists would ever want to visit a place that scientists on government-sponsored expeditions had claimed for themselves. As tourism increased, the treaty parties passed several recommendations that have affected the way tourism operations are conducted.

However, it was not until 1998, when the Protocol on Environmental Protection to the Antarctic Treaty (also known as the Madrid Protocol) came into force, that the

Safety First

Arranging a private expedition to Antarctica is a daunting exercise. These expeditioners crossing the Neumayer Glacier are bound by their nations' commitment to the Antarctic Treaty. Their government will approve the venture only with proof that all eventualities have been planned for.

Wildlife Watchers

Antarctica's wildlife is so abundant and approachable that anyone can develop a tourist mentality. From the sea ice, the crew of a United States icebreaker watch a minke whale taking advantage of the channel their vessel has created.

Antarctic continent had a comprehensive, systematic management regime addressing all human activities, including tourism issues. Under the protocol, all human activities are subject to an environmental impact assessment process, and all activities that have more than a transitory impact require a comprehensive environmental impact evaluation. The protocol is given effect through legislative measures in the various countries that are signatories to it. For example, in the United States, it is an offense to "harass" Antarctic wildlife (that includes frightening a flying seabird so that it departs its nest or approaching a penguin too closely). The penalties for infringement can include jail terms and fines of up to $US10,000.

To prevent tourists from having a negative impact on the environment (and to self-regulate their own activities), Antarctic tour operators established the International Association of Antarctica Tour Operators (IAATO) as early as 1991, and IAATO developed its own guidelines for the safe and responsible conduct of tourism. It established a code of conduct for tourists, which was subsequently modified by the treaty parties and formed the basis for Recommendation XVIII–I, which provides the "Guidance for Visitors to the Antarctic and the Guidance for Those Organizing and Conducting Tourism and Non-Governmental Activities in the Antarctic." Tourists aboard all IAATO member vessels must attend compulsory lectures to familiarize themselves with the code of conduct for shore visits. They learn the minimum distances to be kept between people and wildlife; that there are no toilet facilities at landing sites; and that taking food ashore is strictly prohibited, as is smoking during landings. The people who pay a lot of money to visit Antarctica do expect to find a pristine place. This puts the pressure on the tour operators and travelers to do the right thing and help to keep Antarctica clean.

Visitor numbers have been increasing steadily during the past decade. During the 2006–07 season, 24,470 passengers landed from ships, the operators of which were IAATO members, and 4,152 people traveling on four non-IAATO vessels made landings;

an additional 6,930 passengers participated in Antarctic cruises that did not land.

Antarctic tourism today is the best managed tourism in the world, and other destinations are encouraged to adopt the measures used in Antarctica to improve their own practices. A visit to Antarctica is often the highlight of people's lives—they come back changed, having experienced nature at the most fundamental level. They also return from the region with a new yardstick against which to measure environmental degradation at home.

The Future

Research suggests that the current pattern of ship-based tourism will continue. It is also expected that Russian-registered vessels will still play a significant role. There is, however, also the possibility that some day, large passenger aircraft will make the short flight across the Drake Passage to an improved airstrip on King George Island, from where people could be transported to waiting cruise ships. This would save four or five uncomfortable days of sailing across the stormy Drake Passage, and would probably open up Antarctica to new and less adventurous segments of the tourism market.

Should this scenario become a reality, tourism will certainly spread farther south along the west coast of the Antarctic Peninsula and will encroach on previously unvisited sites, thus increasing the potential for negative impacts. Strict regulation will be necessary (perhaps in the form of a quota system, such as those applied in other natural and cultural sites around the world) in order to safeguard the wildlife and scenic beauty of the most unspoiled wilderness on this planet.

The early part of the twenty-first century has also seen the appearance of large passenger vessels in Antarctic waters. Such vessels can carry up to 2,500 passengers and 1,200 crew but do not make landings. The sinking of the MV *Explorer* in November 2007 is a warning to the operators of all vessels, but especially to those who venture south in large ships that have not been ice-strengthened. TB

Deceptive Strength

The sea ice and bergs along the coast of the Antarctic Peninsula dampen wave action—a welcome change after the often tempestuous Drake Passage. Ice arches contain hundreds of tons of apparently solid ice, but they are unstable and can collapse at any time.

In the Master's Steps

Hikers on South Georgia are literally following in the footsteps of Ernest Shackleton as they descend into the ruins of Stromness whaling station, although the building is now too dangerous for them to follow his path through the station itself. Growing public interest in Shackleton and his men's amazing battle for survival during the *Endurance* expedition has resulted in some specialized tourist voyages.

Keeping a Balance

Antarctica was long protected from the impact of humans by the barrier presented by the Southern Ocean. Now, however, scientists and tourists are crossing that barrier in increasing numbers, and its isolation will no longer protect the environment of Antarctica. Human activity affects Antarctica at three levels: locally, as scientific and tourist visits; regionally, as resource exploitation, such as whaling and fishing; and globally, as industrially caused changes, such as ozone depletion and climate change.

First Ventures

John Rymill's British Graham Land Expedition was the first Antarctic venture to use aircraft and motorized tractors and boats, as well as dogs and skis. Inevitably, the vehicles and animals had adverse effects on the environment, as have the many humans who have visited the region.

Local Impacts

The types of environmental impacts caused by people in the Antarctic are much the same as elsewhere, with the important difference that extremely high standards of environmental stewardship apply in Antarctica. It has not always been so; in the early days of Antarctic exploration, few people realized that human activity could harm the environment. But attitudes gradually changed as science became the focus of Antarctic activities with the establishment of the Antarctic Treaty System (ATS) in 1959. Conservation was on the agenda of the first Antarctic Treaty Consultative Meeting in 1961. Since then, the environment has become the treaty's major concern.

The Protocol on Environmental Protection to the Antarctic Treaty (the Madrid Protocol) sets out to provide comprehensive protection of the Antarctic environment and its fauna and flora. The protocol also protects the wilderness and esthetic values of Antarctica, and its value as an area for science. The protocol recognizes that people will visit Antarctica and that some impacts are inevitable. Any building will change the landscape, and can destroy plant and animal habitats; stations generate human waste and exhaust fumes; even going for a walk damages the fragile soils. The protocol requires that activities should be planned on the basis of information that is sufficient to allow prior assessment of impacts, which involves research to ensure the consequences are known.

Tourists and scientists alike want to experience the unique wildlife of Antarctica at close quarters, but approaching animals too closely can cause stress and lower their chances of breeding successfully. Research into the response of Antarctic animals to human activities is being used as the basis for guidelines so that visitors may be able to enjoy the region's wildlife without endangering it.

Although other local environmental impacts are not inevitable, they do constitute a definite risk. All human activity in Antarctica necessarily relies on fuel, but conditions in this harshest of regions make the handling of fuel containers difficult at times, and small fuel spills are a common occurrence. More seriously, a number of shipping accidents have released large quantities of fuel, and there may well be similar large-scale spills in the future. The formulation of contingency plans to reduce damage when these spills occur is a top priority.

One of the most environmentally damaging events anywhere in the world is the introduction of species, whether deliberate or inadvertent, because once they are established they are practically impossible to eradicate. The ecologies of almost all the sub-Antarctic islands have been significantly changed by introduced plants and animals. Introduced pathogens could also

Not at Home Here

These reindeer seem perfectly at home at Stromness, on South Georgia, but reindeer do not belong in the Southern Hemisphere. Introduced species such as these can be the cause of serious harm to the delicate plant communities found on sub-Antarctic islands.

Animal Welfare

Antarctica's unique wildlife is the focus of extensive scientific attention. Research protocols are carefully set up and scrutinized by independent animal ethics committees, with the aim of ensuring that intrusive techniques do not generate unnecessary pain, disturbance, or stress for the animals being studied.

Disputed Territory

The British survey hut at Port Lockroy was built in the middle of a gentoo penguin colony, and the birds use the footpaths as their own. When staff arrive each spring, they must gently discourage a few penguin pairs from nesting on the wire-mesh steps to the hut's front door.

Alternative Technologies

Previously, Antarctic stations relied on fossil fuels, but burning fuel contributes to pollution and greenhouse gases, and the transportation of fuel oil carries the risk of oil spills. Today, alternative technologies, such as this wind generator at Casey Station and solar power, are being tested in Antarctica to reduce dependence on fossil fuels.

Recyling Rubbish

Waste generated at McMurdo Station is sorted for recycling before being shipped out of Antarctica. In the past it was standard practice simply to deposit the waste from Antarctic stations in open waste-disposal tips. Happily, this is no longer permitted, and most waste is now taken back to its country of origin.

cause havoc; there is no evidence that human activity has introduced disease to Antarctic wildlife so far, but it could happen. Currently, measures to reduce the risk are being formulated within the ATS.

Most visitors to the Antarctic are either members of one of the national Antarctic programs or tourists. Many more people visit Antarctica for recreation (14,000 in 1999–2000) than as part of a national program (4,000 for the same period), but the total number of person-days in Antarctica for national programs far exceeds the number for tourism. National programs also have a far greater infrastructure presence in Antarctica than tourist operations; there are as yet no permanent tourist facilities there. People visit Antarctica because of its special environment, whether as tourists to experience the unique wildlife and landscapes, or as scientists to study phenomena that are of global importance. It would be ironic if their visits were to threaten the very reasons for going there.

Hunting and Fishing

The earliest human impact on the Antarctic was the completely uncontrolled exploitation of the region's wildlife. The first regular visitors to Antarctica and the Southern Ocean went to hunt its wildlife, and very quickly they had an enormous, and adverse, impact on its seal populations. As early as 1830—just six decades after James Cook's first foray south of the Antarctic Circle—sealers had destroyed many of the once-teeming fur seal colonies of the sub-Antarctic islands. From seals, they shifted their attention to penguins as a source of oil, and then to the great whales. The sealing industry declined when it became economically unviable, but whaling, which began on a large scale in the Southern Ocean in the 1900s, has been regulated since 1949 by the International Whaling Commission in an attempt to maintain its viability. Nevertheless, many companies stopped their whaling activities only because of the falling catches and diminishing profits. In recent years, the commission has introduced more effective protective measures: by the 1960s blue and humpback whales were fully protected; fin and sei whales were protected by the 1970s; and in 1986 the commission suspended all commercial whaling. It is now limited to "scientific whaling."

The effects of early wildlife exploitation are still apparent today. Fur seal populations have exploded in many locations, probably as a result of ecological imbalances caused by the hunting of them and their competitors. King penguin populations in most localities are now growing enormously; however, the populations of long-lived great whales have not yet recovered.

Now that large-scale sealing and whaling have been prohibited, fishing is the only major resource extraction activity in the Antarctic Treaty area. Fishing can cause adverse impacts in a number of ways: by taking too many fish, by reducing the food available for natural predators of the targeted fish, by destruction of habitat, and by killing non-target species (such as albatrosses caught on longlines). Fisheries in the Southern Ocean are regulated by the Convention for the Conservation of Antarctic Marine Living Resources, established in 1980, primarily to regulate the growing krill fishing industry, but with the responsibility for the management of all the activities affecting living resources in the Southern Ocean. At the time it was established, the convention was unique in that it recognized the importance of the ecological relationships between harvested and dependent populations.

Global Impacts

From the International Geophysical Year in 1957–58, Antarctica has been a laboratory for studying both surface and atmospheric processes on a global scale. Since then, there has been a growing recognition of the importance of Antarctica for understanding global change caused by human activity against the background of natural variability. The British Antarctic Survey announced the discovery of the springtime depletion of ozone over Antarctica in 1985; this was among the first indications that human activity is causing environmental impacts at a global scale. At about the same time, scientists became aware of the growing evidence of climatic warming, the phenomenon now known as the greenhouse effect. The study of climate change and greenhouse gases is now perhaps the major focus of Antarctic science. These global changes might also directly harm the environment of Antarctica and its fauna and flora. Increasing ultraviolet radiation caused by ozone depletion could bring about changes in Antarctic phytoplankton that may create effects throughout the food web. Global warming may contribute to the breaking up of ice shelves and the loss of animal habitats. And global change may have effects

Cleaning Up Antarctica
Under the Protocol on Environmental Protection, garbage such as this at abandoned Wilkes Station must be taken away, provided that the site has not been declared historic, or unless removal would cause more damage than leaving it where it is.

Costly Litter
This aircraft wreck on the ice in East Antarctica represents an ongoing problem for the treaty parties: the organization and management of the removal of accumulated refuse and pollutants, especially fuel, from the Antarctic region.

in Antarctica that could have serious consequences elsewhere in the world—for example, changes in the amount of water frozen in the polar ice sheets would radically alter global sea levels.

Managing Environmental Change

During the cold war years, the international community, through the Antarctic Treaty System, acknowledged the value of Antarctica for bridging political chasms. In recent years, political preoccupations have given way to a growing awareness of the effects of industrialization on the global environment, and of Antarctica's unique role in analyzing and containing them. There is no quick or easy solution to the environmental problems that the world faces—but the importance of Antarctica in our understanding of these problems can only increase. SN

The Last Bastion

Antarctica has a symbolic value: if the world cannot protect this most remote of lands, what hope is there for the rest of the planet? In the 1970s, the Antarctic Treaty System (ATS) realized that Antarctica's mineral resources could well attract commercial interests, and that control would be necessary. Throughout the 1980s, the ATS worked toward the establishment of the Convention on the Regulation of Antarctic Mineral Resource Activities, based on the assumption that mineral extraction would occur, but with strong measures to ensure the least possible environmental impact. In 1989, just as the convention was ready for signing, it was set aside when Australia and France declined to sign and sought permanent prohibition of mining in Antarctica and comprehensive protection of the Antarctic environment. The Protocol on Environmental Protection to the ATS was adopted in 1991 and came into force in 1998, committing treaty parties to comprehensive protection of the Antarctic environment, specifically prohibiting commercial mining, and designating Antarctica as a natural reserve devoted to peace and science. However, when the protocol was negotiated, the international community was not ready to lock up the mineral resources of Antarctica forever. The protocol is open for review in 2048, 50 years from the date of its coming into force; at that time, and given the agreement of a majority of the nations engaged in Antarctic activities, it could be amended to permit commercial mining. But if the world decides to turn to Antarctica for its mineral resources, it will mean that no lessons have been learned from the wasteful and short-sighted use of mineral resources that characterized the twentieth century.

Harvesting the Antarctic Region

Antarctica was the last continent discovered, yet the exploitation of its biological resources proceeded so quickly that, by the end of the twentieth century, the populations of many species had been seriously depleted.

Seals, Whales, and Penguins

As early as 1788, a proposal to license sealers and regulate their catches came to nothing. The ensuing exploitation was so intensive that fur seal numbers have only recently recovered. Since the 1950s, however, fur seal populations on some south Atlantic islands have revived so successfully that they may be causing environmental problems. Populations on the southern Indian Ocean islands are still sparse, though the numbers are increasing.

Elephant seal harvesting ceased in 1964. Numbers in the south Atlantic appear to have recovered, but there is evidence that their populations in the southern Indian Ocean have recently declined. There were three attempts at commercial exploitation of pack ice seals— Weddell , leopard, Ross, and crabeater seals—between the 1960s and the 1980s. A few thousand were taken, but these enterprises proved unprofitable.

A Norwegian company established the first whaling station in the region in 1904. In 1986, a moratorium on commercial whaling came into effect, although there has been an annual catch by Japan under a "scientific whaling" exemption. At least 1.5 million whales were harvested from Antarctic waters, and the effects on the ecosystem are still being debated.

Birds have not been commercially important, although penguins were taken for oil: 150,000 kings were taken on Macquarie Island from 1895 to 1919.

Fish

Commercial fishing in Antarctic waters began in 1905, when salt fish from South Georgia were exported to Buenos Aires. Commercial fishing at South Georgia recommenced in the early 1950s, although full-scale commercial trawling did not begin until the late 1960s, off South Georgia and the South Orkneys. Populations of Antarctic rock cod and icefish fell rapidly and have not recovered, even after 30 years of protection. In the south Indian Ocean, commercial fishing began around Iles Kerguelen in the late 1960s and goes on. Icefish and rock cod have recently given way to more profitable longline fishing for Patagonian toothfish. In the Pacific Ocean sector, little fishing took place until the search for new toothfish populations began in the late 1990s. In recent years, pirate longline fishing for toothfish has depleted fish stocks and taken a huge accidental toll on albatrosses and other seabirds.

Krill

In 1901, Erich von Drygalski summarized problems that still plague the krill fishing industry when he reported that krill "tasted quite good, but were rather small and

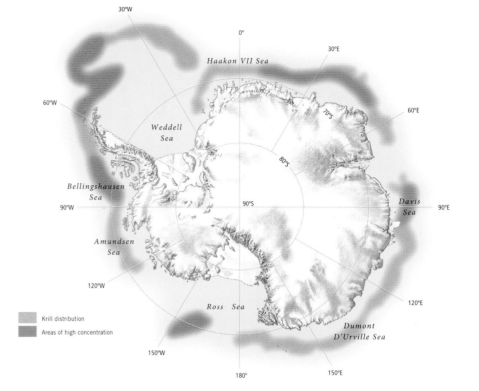

tiresome to peel." Early twentieth-century studies indicated the huge size of the krill population, and during the 1960s interest was further spurred by the suggestion that there existed a "surplus" of krill. The assumption was naive, but did indicate that there was immense commercial potential for Antarctic krill.

The Soviet Union began exploratory fishing for krill in 1961. Soviet vessels made sporadic small catches throughout the 1960s as technologists developed better gear and scientists identified where the biggest krill concentrations lay. In 1972 the Soviet Union set up a permanent Southern Ocean fishery, and by the mid-1970s, the Soviets and Japanese had full-scale commercial operations underway.

Currently, vessels from Japan, Poland, Korea, Argentina, and the Ukraine are fishing for krill. Most krill is caught in the South Atlantic, with fisheries moving south as the ice recedes in summer and north to the ice-free waters of South Georgia in winter. In the Indian Ocean, krill fishing is concentrated along the continental shelf break, but the fishing season is much shorter there because of the abundance of sea ice.

Fishing for the Future
Krill (from the Norwegian *kril,* "young fish") are shrimp-like crustaceans that swarm in immense numbers in Antarctic waters; the map shows the main concentrations. A major krill-fishing industry is a real possibility in the future.

A Versatile Botanist
H. Hamilton, who studied the plant life of Macquarie Island during the 1911–14 Australasian Antarctic Expedition led by Douglas Mawson, systematically quarters an elephant seal for its blubber—an invaluable source of fuel in Antarctica.

Whaling Days

At Grytviken, Frank Hurley wrote, "Slimy waters lapped loathsome foreshores, polluted by offal and refuse—the accumulation of years. ... even the magnificent inland scenery seemed to grow tainted and lose its splendor."

The krill catch peaked in 1982, when over half a million tons were taken—93 percent of this quantity was caught by the Soviet Union. The current low annual krill catch—around 121,000 tons (110,000 tonnes)—reflects a lack of effort rather than any supply difficulties, but there has been renewed interest in krill fishing by a number of countries, notably India, Britain, Norway, the United States, Canada, and Australia.

The projected expansion of the krill fishery is being driven by advances in pharmaceuticals, aquaculture, and food processing. Krill is a near-perfect food for farmed fish, and in the future the fishery may become strongly oriented toward aquaculture. Antarctica could sustain an annual harvest of around 6,600,000 tons (6 million tonnes) of krill; it is probably only a matter of time before a much larger krill fishery develops.

Regulating Resource Exploitation

Initial attempts at regulation were difficult because of conflicting national claims over sections of Antarctica, and because of the high-seas nature of many fisheries. Gradually, however, a number of international treaties and conventions were devised to regulate harvesting in the Antarctic region.

The commercial exploitation of all species is now covered by the International Whaling Commission (IWC), the Convention for the Conservation of Antarctic Seals (CCAS), and the Convention for the Conservation of Antarctic Marine Living Resources (CCAMLR).

The early sealing industry exploited seals so rapidly that by the time it became obvious that there was a problem, there was no industry to regulate. In 1972, the Convention for the Conservation of Antarctic Seals was adopted. It specified the total protection for fur, elephant, and Ross seals south of 60°S, and catch limits on crabeater, leopard, and Weddell seals.

International regulation of whaling did not begin until 1935, with the Convention for the Regulation of Whaling. The convention was largely ineffectual, and the International Convention for the Regulation of

Whaling took over the management of whaling in 1946. This convention established the International Whaling Commission (IWC), which remains responsible for the management of whaling in all oceans. Since the IWC's formation, the vast majority of whales harvested have been in Antarctic waters.

The Convention on the Conservation of Antarctic Marine Living Resources (CCAMLR), which came into force in 1982, regulates the remaining Southern Ocean fisheries. The convention arose from concerns about the expanding krill fishery. CCAMLR was worded to ensure that krill fishing would not harm krill-dependent ecosystems, and in particular that it would not hinder the recovery of baleen whale populations.

At first, CCAMLR developed protective measures for Antarctic fish; in later years, management measures for krill have been introduced. Conservation measures are now in place for all the harvested species within the CCAMLR area, and the convention is now embarking on a much more difficult task: ecosystem management. Its efforts to protect ecosystems are at the leading edge of the development of sustainable fisheries. SN

Abandoned Stations

The once-thriving whaling stations of South Georgia gradually fell apart once abandoned. At Grytviken, these rusted station ruins remained until the early 2000s, but the building shells have now been removed.

Seal Slaughter

In the four decades between 1786 and 1825, some 1.2 million fur seals were slaughtered on South Georgia, bringing them to the verge of extinction. Building up again from small remnant colonies, the rapidly increasing population includes a small proportion of white morphs, like the animal pictured here.

Polar Shipping

Taking Advantage
On the 1911–14 Australasian Antarctic Expedition, the crew of the *Aurora* hoisted blocks of old drift ice aboard and melted them to replenish the ship's supply of freshwater. Sea ice that is more than three years old freshens because the salt leaches back into the sea in the form of brine during the first two or three melt seasons.

Terrifying Collision
To avoid a huge iceberg that emerged out of the gloom in a storm, James Clark Ross's ship, the *Erebus,* turned, only to find her companion ship, the *Terror*, "running down upon us, under her top-sails and foresail." The *Terror* ploughed straight into the *Erebus* and "the concussion when she struck us was such as to throw almost everyone off his feet," wrote Ross later. Locked together, the ships performed a hideous vertical dance that destroyed most of the *Erebus*'s rigging.

I n the sixteenth century, northern polar navigators at least knew that the Arctic Ocean and its ice occupied the high northern latitudes and were surrounded by more-or-less known landmasses. But no one knew if the Antarctic continent even existed, still less what its environment might be like.

The first circumnavigations of the globe, which took place in the sixteenth century by Ferdinand Magellan and Francis Drake, spurred much more daring ocean voyages. Shipbuilders had to allow for the incalculable times at sea, the ports to be served, the weather to be survived, and the cargoes to be carried. But there was almost no provision for sailing through pack ice. Ships were sometimes fitted with hardwood sheathing at the waterline to withstand the abrasion of ice floes, and beams and frames might be stiffened to resist pressure, but no special attention was given to navigation in polar ice until late in the nineteenth century.

Sailing the Polar Seas

Early sailing ships were not designed for ice navigation, for a very simple reason: it was so difficult to make way through ice under sail that the best strategy in the polar regions was to avoid the ice in the first place. Even in the twenty-first century, the most powerful icebreakers stay as clear of the ice as possible, venturing into it only when there is no obvious route through open water.

The first serious scientific attempt to find the Antarctic continent was in 1772, when James Cook set out with two small wooden sailing ships, the *Resolution* and the *Adventure*. Cook crossed the Antarctic Circle on January 17, 1773, but was too far off to sight land and was confronted by pack ice. He prudently turned back into open waters, but continued to probe southeast for two months more, threading his way between masses of icebergs before giving up and making for New Zealand.

The Steam Revolution

Cook's voyages contained lessons for ship construction techniques, but seafarers are notoriously conservative, and ships continued to evolve at a snail's pace.

It was the harnessing of a power source other than the wind—the application of steam power, originally generated by paddle-wheels—that began to make the difference to ice navigation. The totally submerged screw propeller was the answer to propulsion that was independent of the wind. Iron-hulled vessels were built in greater numbers from about 1830, but they were not used in ice until the end of the nineteenth century, probably because the cast-iron plates were riveted and ice would spring the seams and cause damage that was impossible to repair at sea, whereas wooden ships could be repaired at sea, all the materials being to hand to scarf in a new piece of timber and caulk the seams with oakum and tar.

In 1892, the Norwegian Fridtjof Nansen built the small wooden *Fram*, in which he planned to drift across the North Pole in the Arctic ice for three years. She had a rounded hull that would be lifted rather than crushed by the pressure of ice, and internal strengthening of beams, frames, and longitudinal members to withstand pressure, but she was not made for forcing through ice. The ship's steam power was rudimentary; her main propulsion in open water was a full set of sails. She completed her Arctic mission, and was later used by Roald Amundsen in his expedition to the South Pole in 1911, proving that a wooden vessel—if well enough designed—could defeat polar ice.

In the early twentieth century, most explorers set out in converted whaling ships, some of them well past their prime, and all built of wood. The *Endurance*, purpose-built for polar voyages in 1913 and used for Ernest Shackleton's doomed Transantarctic Expedition, still followed the traditional design of the wooden sailing whalers, with only auxiliary steam power. Robert Scott's *Terra Nova* and Shackleton's *Nimrod* were both leaky ex-whalers, sail-driven with auxiliary steam power, not designed for forcing through polar ice. The *Discovery* was purpose-built for Scott's first Antarctic expedition of 1901–04, and it still followed the old wooden whaler design, albeit greatly reinforced.

Steam power was conserved, since coal was needed to warm living quarters ashore. In March 1898, Adrien de Gerlache's expedition became—unintentionally—the first expedition to spend the winter in Antarctica. The steam engine of de Gerlache's ship, the *Belgica*, developed a mere 35 horsepower, but still needed more coal than they could afford when she was trapped in ice for almost a year. Douglas Mawson used the 40-year old *Aurora* for his Australasian Antarctic Expedition in 1911–14. He returned to Antarctica with Scott's *Discovery*, which by then was 30 years' old, for BANZARE, his 1929–31 expedition.

Steam, Steel, and Diesel

The next major effort at scientific discovery in the Antarctic came after World War II. By that time, steel ships had replaced "wooden walls," steam and diesel propulsion had taken over from sail, and oil fuel had eliminated the necessity of carrying vast quantities of coal on board ship. Icebreakers were developed into a specialized class of ship, and, as a spin-off, they doubled as cargo vessels that were capable of navigating the polar ice without needing an icebreaker escort.

The hulls of icebreakers are much stronger than those of most ships, with everything—from frames to plating to stiffeners—made of especially strong steel with greater than standard dimensions. They have a sharply raked stem, which allows them to ride up onto the ice and crush it with their weight. They have no bilge keels or other external hull fittings that might be caught on ice, or ripped off by it. They are powerful for their size, so they can twist and turn to escape from thick pack ice. They have multiple screws, usually twin screws, but larger vessels have triple screws, partly for maneuverability, but also as a fallback if a propeller is damaged by old, hard ice.

Ships strengthened for ice navigation must be certified by national marine classification societies, individual ice classes being rated by a ship's capacity to resist ice damage. But there are as yet no world standards, although for many years the International Maritime Organization (IMO) has been working on a global classification system for polar ice vessels.

Non-polar class vessels may be strengthened to varying degrees to resist ice damage, but none are considered to be icebreakers and their movements in polar ice are strictly confined to the summer and autumn ice navigation seasons. Strengthening does not make a vessel capable of navigating through ice; it merely ensures that she is less likely to be damaged by ice than a non-strengthened ship.   PT

Living in an Icebreaker

In the Russian icebreaker, the *Kapitan Khlebnikov*, living quarters for the crew and passengers are high above the hull, away from the noise of ice grinding along the shell-plates. The ship's bridge is perched well above the ice for the best possible surveillance of the route between the ice floes.

Encountering Pack Ice

Pack ice in the Antarctic Peninsula slows down the progress of the Russian polar research vessel, the *Professor Molchanov,* as she heads toward the ice-edge, visible as a dark line on the horizon. The darker blue of the sky at the ice edge is a sign of open water in that direction, whereas "ice-blink" with a bright sky indicates the presence of ice.

Flying South

Robert Scott became the first to take, somewhat shakily, to the air over the Antarctic continent, albeit secured to the ground by a wire hawser. The place was the Bay of Whales, at the edge of the Ross Ice Shelf, on February 4, 1902. The vessel was a hydrogen observation balloon, from which was suspended a flimsy basket with a nervous passenger. The purpose was to make a reconnaissance flight, and after some 60 minutes aloft, Scott was followed by Ernest Shackleton, with a camera. The balloon was not used again. Nearly two months' later, on the opposite side of the continent, the German explorer, Erich von Drygalski, also made use of a balloon, and managed to reach almost twice the altitude that Scott and Shackleton had done—nearly 1,640 ft (500 m).

In 1911, the fuselage of a Vickers aircraft arrived in Antarctica on the first Australasian expedition. Douglas Mawson had intended to make the first powered flight in the continent, but a training accident en route, in Adelaide, deprived the plane of its wings. Making the best of things, they took the damaged craft with them and used the fuselage and engine as a tractor, which was not very successful.

Aviation's Golden Age

It was not until the late 1920s that the use of heavier-than-air machines in the Antarctic region began to receive serious attention. By that time, the Golden Age of Aviation had begun—and it was little wonder that some aviators turned their attention and ambitions to the frozen south. The late 1920s and early 1930s became an arena for private flights and exploration in the Antarctic area, in most cases combined with some scientific research.

The main players were bankrolled by American backers, including Randolph Hearst, Edsel Ford, and John D. Rockefeller. Australian explorer Hubert Wilkins and American pilot Carl Eielson were the first to venture into Antarctic skies in a powered machine—a Lockheed Vega—from Deception Island, north of the peninsula mainland, on November 16, 1928: their 20-minute flight was curtailed by inclement weather. Subsequently, they made a reconnaissance flight farther south in late December, which lasted for 11 hours, covered 1,350 miles (2,170 km), and overflew extensive areas of the Antarctic Peninsula.

Their achievement was a spur to American Richard Byrd, who claimed to have been the first person to fly to the North Pole. In 1928, at much the same time as Wilkins and Eielson, Byrd set up camp on the Antarctic

mainland. Wilkins pursued a further flying program in the latter part of 1929, but he did not penetrate much farther south than the Antarctic Circle. In a Ford Trimotor piloted by Bernt Balchen, Byrd became the first flier to over-fly the South Pole, early on November 29, 1929, after a flight of just under ten hours.

After 1929, a number of other aircraft were used in Antarctica, mainly for short-distance observation and exploration, and largely restricted to coastal areas. The BANZARE expedition of 1929–31 made extensive use of a DH60 Moth; at around the same time, Norwegian whaling ships began carrying aircraft.

The pouring of American private finance into flights continued with a further Byrd expedition in 1933–35 in which the focus was more on exploration and scientific research. In 1934, the American millionaire, Lincoln Ellsworth, also attempted—and achieved, in December 1935—the first significant transcontinental flight over Antarctica, of around 2,000 miles (3,200 km). Over a period of almost a fortnight, Ellsworth and pilot Hollick-Kenyon flew from Dundee Island to within a short distance of Byrd's then-disused base, Little America, on the Ross Ice Shelf, but lost radio contact with their ship before they could report their location. The resultant search, involving the Australian, New Zealand, and British governments, and private bodies in America, generated a rash of publicity around the world.

Governments Step In

From the mid 1930s, the focus changed from private to government involvement and to territorial claims in Antarctica, many of them competing. At the beginning of 1939, the German government mounted a brief but efficient Dornier flying-boat expedition, using a seaplane tender and taking photographs of some 96,500 sq miles

Deserts of Immensity
An aircraft at rest on a sea-ice runway is massive, but is still dwarfed by the frozen surroundings that stretch as far as the eye can see, from Arrival Heights at McMurdo Sound. Antarctica has been partially tamed by aviation, but it is to be hoped its challenges will never be completely overcome.

Aircraft in Antarctica
By 1939, twin-engined aircraft and metal construction were laying the foundations for future Antarctic aviation. A Barley Grow floatplane was an integral part of Richard Byrd's third Antarctic expedition.

A Craft within a Craft

The immense jaws of a US Galaxy (one of the largest aircraft types in the world) open to regurgitate another, smaller craft—but one that would still have left the first Antarctic aviators gaping with amazement. Unloading takes place on a sea-ice runway 7 ft (2 m) thick at McMurdo Sound.

(250,000 sq km) of Dronning Maud Land, and laying claim to sections of the Antarctic already claimed by Norway. At the end of 1939, the United States returned to the Antarctic in the shape of a government-funded expedition that lasted until early 1941.

World War II put a temporary stop to most of the exploration and aviation in Antarctica. However, it also led to massive advances in aircraft technology and communications, and after the war the importance of aircraft in Antarctic exploration increased dramatically.

In 1946, the United States mounted its Operation Highjump, under the leadership of Byrd (now a rear admiral), from an ice-and-snow airstrip hacked from the Ross Ice Shelf. This huge operation employed United States Navy R4Ds, Martin Mariner flying boats, and ship-borne helicopters. Operation Highjump was a combination of scientific research, exploration (they claimed to have photographed up to one-third of the continent and coastlines), and, not least, an assessment of the operational capacities that might prove useful in a possible Cold War confrontation in the Arctic with the Union of Soviet Socialist Republics.

The 1950s saw the establishment of a proliferation of permanent bases. Flying to Antarctica had become almost routine, and aircraft were increasingly used for local operations and for research. In 1957, the United States Navy started regular flights from Christchurch, New Zealand, to McMurdo Station on the Ross Ice Shelf. The one-way flying time was around 12 hours. Initially, C–124 Globemasters were used, but these were replaced in the 1960s by the Lockheed Hercules. Supply aircraft also began to fly from Argentina and Chile, and from Britain's Falkland Islands, to airstrips established on the Antarctic Peninsula. Aircraft are now integral to certain parts of the Antarctic scene.

Tourist Flights

In 1977, Air New Zealand and the Australian airline, Qantas, began to schedule regular tourist flights over Antarctica. Two years' later, an Air New Zealand DC10 crashed into Mount Erebus. At the time, it was one of the worst air disasters in history, with total loss of life. All tourist flights were suspended until Qantas resumed them in 1994, using Boeing 747s.

Lan Chile also operates sporadic tourist flights, from Punta Arenas. Adventure Network International also runs private flights from Punta Arenas, mainly catering for tourists, adventurers, and expeditioners. PL

Hauling by Helicopter

A helicopter over the Amery Ice Shelf prepares to raise a load in a sling net at a signal from the ice. The helicopter in Antarctica followed the first gyrocopter (the Byrd Expedition of 1933–35), and helicopters were used extensively in Operation Highjump (1946). In ensuing decades, these craft became invaluable for supply work in Antarctica.

Wintering Over

Formidable Wind

Frank Hurley's photograph shows men "getting ice for domestic purposes." Hurley told how he built a shelter to photograph "other members of the party as they struggled about bent—very literally bent—on their duties. On one occasion … I was lifted bodily, carried some fifteen yards [about 14 m] … and dumped on the rocks."

Farewell with Flares

Left behind for winter, a small group of expeditioners gives the last ship of the season out of Horseshoe Harbor a traditional send-off. Not until January or February of the following year will another ship reach Mawson Station to resupply the station.

W hy would any sane person want to spend a year in Antarctica? It is a demanding, even life-threatening, environment. Far from your family and friends, you endure long periods of darkness in temperatures far below freezing. Your companions are not of your choosing, but private life is not an option.

For me, as for most of us who fall under Antarctica's spell, it was a combination of factors. There was the job challenge: I was to be station leader at Australia's Mawson Station, the fourth woman in the role. I was "to create and maintain a productive and harmonious community" of up to 70 people in the summer and 22 in winter. But the main attraction was Antarctica itself: its austere beauty, its enchanting wildlife, the challenge of mastering the skills to live confidently in a hostile environment, and the joy of forging the trust and interdependence that are essential to a long field expedition.

The Antarctic year begins in late spring, when the sea ice has retreated far enough for ships to reach the coast and helicopters to reach the research stations. The days are long and the weather benign. The station swarms with construction and maintenance teams, and with scientists seizing the brief opportunity to study this most extreme of environments. The biologists studied a colony of Adélie penguins on a nearby island, geologists made forays into the mountains, and the glaciologists trekked deep into the ice sheet to measure its depth and ground contours. A field expedition camped in the Prince Charles Mountains, studying its geology and collecting mosses and lichens from its sheltered slopes.

By the end of March, the last ship had borne away most of the scientists and tradespeople, delivered a year's supplies, and removed more than 12 months' garbage, leaving 22 of us behind. But we were not idle. We took the bulldozers and sleds back to the Prince Charles Mountains with fuel for the following summer and to retrieve rubbish and equipment needing repair. We mapped our multimillion-dollar traverse by radar, following a route marked by bamboo canes topped with empty beer cans. I drove one of the bulldozers back, and covered 250 miles (400 km) of featureless ice and snow at a painfully slow 4 mph (7 kph) or less.

In the month we had been away, the sea had frozen over. We visited the nearby islands on foot, on skis, on four-wheeled motor bikes, and in snow vehicles.

At the Auster emperor penguin colony in the grounded icebergs to the east, we saw that the males were huddled together, incubating eggs balanced on their feet under a thick fold of skin. For two months of the harshest winter weather, they fast until the chicks hatch and the females return to feed them—a severe vigil, which is profoundly moving to witness.

Midwinter Blues

The weeks passed and the sun swung lower in the sky, until it disappeared beneath the northern horizon. For a month we worked in a few hours of twilight each day. But there were auroras and flaming sunsets: once we watched the southern sky paling to delicate pastels while a huge full moon floated above the translucent turquoise ice sheet, which was speared by the jagged purple peaks of the inland mountains.

In June, we celebrated winter solstice, the turning point of the Antarctic cycle, with a banquet, a performance of *Cinderella* in drag, some original entertainment—and monumental hangovers.

July and August are the "dog days": the darkness is interminable, morale falls, and tempers get short. Canine company was our salvation. Until 1994, when the Madrid Protocol mandated their removal, the sledge dogs were a feature of life on Mawson Station. In September, we took two dog teams to count the emperor penguins at a rookery 220 miles (350 km) away, carrying everything on sledges and camping each night on the ice. This is a technology so simple that it had remained virtually unchanged since it took Amundsen to the South Pole.

The lengthening days heralded the Antarctic spring and the first ship of the season. Ice had become our only reality, and as we went to collect the supplies, I felt as if I were falling off the edge of the known world into another, unfamiliar medium—water! The vessel carried new expeditioners; our year in Antarctica was over—a year of myriad vivid impressions and deep gratitude for the privilege of such a rare experience. LC

Light Deprivation and Isolation

F rederick Cook, doctor on board the *Belgica*, the first ship to winter over in Antarctica, documented the devastating effects prolonged periods of light deprivation had on humans. In association with NASA and the University of British Columbia, psychologists on the *Sedna IV* Antarctic Mission expedition, led by the ecologist and filmmaker Jean Lemire, have conducted studies on how humans adapt to the continual darkness and the isolation experienced when wintering over. The *Sedna IV* expedition set out for the Antarctic in 2005 on a 430-day voyage, with the purpose of recording the effects of global climate change. Other research is concerned with the effects of ultraviolet B radiation, increased carbon dioxide, and global warming on the biological pump—the process in which carbon dioxide is incorporated into the water column.

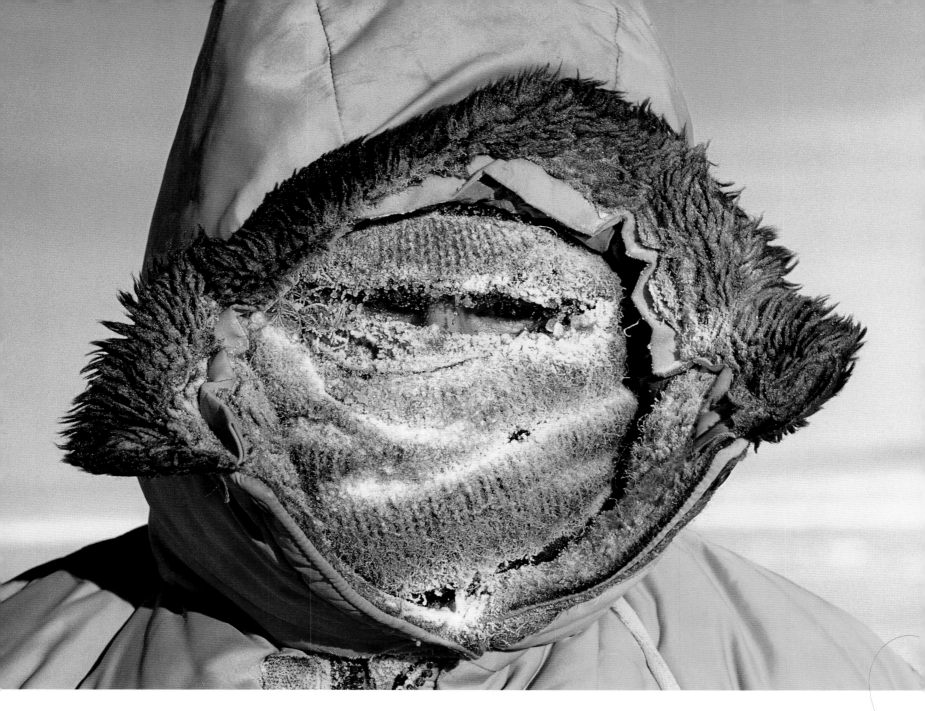

Rugging Up
It is all too easy to be cold in Antarctica, even with the best of modern outdoor clothing. Choosing what to wear is a balance between retaining heat and practicality—extra layers of clothing increase warmth but decrease mobility.

Celestial Curtains
Sweeping celestial arcs reflected in a newly frozen bay wave majestically as the expansive phase of an auroral substorm begins. The individual bands brighten and develop billowing folds, often twisting into spectacular spiral formations as waves of light surge along their entire length. Those wintering over get the opportunity to observe these spectacular skies.

Polar Medicine

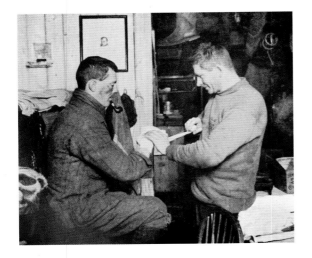

The Dangers of Frostbite
On Scott's last expedition, Petty Officer Evans binds up Dr Edward Atkinson's frostbitten hand after he was lost in a blizzard. Scott noted: "Atkinson had a bad hand today ... blisters on every finger giving them the appearance of sausages ... this bit of experience has done more than all the talking I could ever have accomplished."

In March 1879, an advertisement appeared on the noticeboard at Marischal College, Aberdeen: "Seal fishing—wanted immediately a surgeon for SS *Mazinthien* sailing from Peterhead on Thursday or Friday next (24th or 25th). A junior student not objected to, must be a good shot."

This advertisement epitomizes the challenges faced by polar medicine over the past couple of centuries—and its traditional solution. Whaling and sealing ships from Peterhead operated in waters off Greenland; others sailed to the Southern Ocean, spending years away from home. British government regulations obliged these vessels to carry a qualified doctor, but medical students were often used. The surgeon was comparatively well paid, and it was an excellent way to meet medical school fees. The medical experience gained was minimal because the crews were young, and were healthier than the general population. Occasional wound suturing and bone-setting apart, the days were filled with shooting seals and keeping the log.

Doubling Up
Clearly, the traditional solution to the problem of providing health care in remote, hostile environments was to hire young (and therefore cheap) medical attendants. The main problem—boredom—was averted, and value for money obtained, by giving them some additional tasks. Fortunately for science, nineteenth-century medical courses included much natural history, particularly comparative anatomy and botany, so that the surgeons on polar expeditions often doubled as naturalists.

A classic example was the James Clark Ross *Antarctic* expedition of 1839–43: all four surgeons were also hired as zoologists or botanists, including Joseph Hooker (later director of Kew Gardens) and Robert McCormick, who had been with Charles Darwin on the *Beagle*. Edward Wilson, who perished on Robert Scott's doomed quest for the South Pole in 1912, took the role of surgeon/naturalist to the extreme: although he was medically qualified, his title was "Chief of the Scientific Staff, and Zoologist." Wilson's inability to relieve the medical problems of Edgar Evans and Lawrence Oates on the return journey from the Pole contrasts with the remarkable surgery carried out by the surgeons on Ernest Shackleton's *Endurance*, who aseptically

Standing In for the Bees
Pollinating vegetable flowers in the hydroponics unit at Scott Base is done by hand. In winter, fresh vegetables are a welcome addition to meals.

Pioneering Research
While co-BANZARE expeditioners Mertz, Madigan, Hunter, and Hodgeman occupy themselves, bacteriologist Dr Archibald McLean (second from left) studies a specimen. McLean pioneered medical microbiological studies on microbes that live on humans.

Keeping Out the Cold
Quality materials and layers are combined for maximum warmth—several thinner layers instead of one or two thick ones. Once this man pulls his goggles over his eyes to ride on an unprotected motorized toboggan, every bit of skin will be covered by two or more layers of insulating material.

Emergency Service
The clinic at McMurdo Station is fully equipped, like any modern medical facility. Any group staying in Antarctica over winter needs to be self-sufficient in every way possible. They are a long way from medical—or any other—help.

amputated an expeditioner's gangrenous foot under chloroform on Elephant Island in June 1916.

The brutal and immediate challenges these brave doctors faced were typical of early Antarctic exploration. Frostbite, snow blindness, and inadequate diets, which led to vitamin deficiency and starvation were common on these expeditions, and the presence of doctors did little to prevent them. Key preventive developments—improved equipment and clothing, and advances in nutritional science—were yet to come.

Polar Medicine Today
Antarctic expedition members of today are screened for pre-existing medical problems, so that mortality and morbidity rates in the Antarctic are very low. Most of the serious health risks in modern Antarctica relate not to environmental extremes, which are well understood and well guarded against now, but to hazardous activities, such as light engineering and the use of vehicles and aircraft, and so, in the twenty-first century, Antarctic medical practice is mainly concerned with the treatment of occupational injuries.

The British Antarctic Survey (BAS) offers a good example of current medical practice in Antarctica. Its

contemporary medical unit combines the traditional approach—doctors resident on Antarctic bases—with some modern approaches, such as paramedical training for all staff, electronic access to specialists in Britain, and telemedicine. Until 1962, BAS was called the Falkland Islands Dependencies Survey, and its employees were affectionately referred to as "Fids." Displaying classic Antarctic humor, the BAS medical handbook is called *Kurafid*. Originally written in the 1960s to meet the needs of sledging parties, it has expanded into a comprehensive first-aid manual. Like their nineteenth-century counterparts, BAS doctors need spend only part of their time on medical duties, so they can maintain nearly 200 years of Antarctic tradition by conducting high-quality research as well. HP

Supporting Science in Antarctica

The search for the Magnetic Pole, the race to be first to the South Pole—all the Antarctic ventures that explorers and scientists have undertaken in quest of their Antarctic goals—have succeeded or failed in direct relation to the quality of the logistics that supported their endeavors.

The Antarctic region provides unique opportunities for research, and logistics managers must be able to respond to a wide range of requests from scientists. Researchers may be required to work on glaciers, in the ocean, on mountains, on or under the sea ice, in and above the atmosphere, in every type of weather, at any time of the year, in the highest, driest, coldest, windiest, and loneliest continent on the Earth.

Summer and Winter
Summertime is the most productive season for research scientists, and logistics organizations work at an intense pace throughout this season. At this time, it is relatively easy to reach Antarctica. The sun is shining, the sea ice recedes, animal and bird life around the coast is in abundance, and the weather is comparatively clement. Even so, traveling to and from Antarctica is not simple; despite all the sophisticated systems in place now, it is not unusual for ships to be beset by ice, for aircraft to

be unable to fly, and for planned transit times to be doubled. The distances are huge, and the isolation almost complete. Stores, hospitals, industrial centers, fire brigades, recreation facilities, repair yards—not to mention family and friends—are usually thousands of miles away, so the need for self-sufficiency is high.

But there are also excellent research opportunities in winter, and many nations operate their bases year-round. At most Antarctic bases, winter isolation is absolute; the self-sufficiency required in summer pales into insignificance with that required for wintering over. For most wintering parties, there is no possibility of support from anywhere other than their own resources. Perhaps for this reason, NASA is drawing on the experiences of researchers in Antarctica as a model for those that might occur on extended space flights.

Bases, Camps, and Transport
It is no simple task to conduct research in the region, and many nations have built research stations and bases to provide the infrastructure to support their programs. All have the primary aim of providing a hub from which to conduct scientific programs. Some are the size of small towns, with hundreds of people; others are little more than a collection of tents and

How Thick is the Ice?
Traveling on motor toboggans, known as "ski-doos", a pre-season survey team drills through the sea ice in McMurdo Sound to map its thickness. This survey gauges the suitability of the fast ice for surface travel during the summer months.

Wind-chill Factor

With good, warm clothing, working in temperatures as low as −58°F (−50°C) is no great problem. When blizzards rage, humans simply cannot survive without shelter, no matter what they are wearing.

custom-modified shipping containers that can provide rudimentary living quarters for four to 20 people.

For research at a distance from a base, field camps provide shelter, basic laboratory equipment, and cooking facilities. Travel between field camps and their parent stations can be hazardous and time-consuming; aircraft, helicopters, and all-terrain vehicles of many shapes and sizes are used to make these traverses, and each machine has specialized capabilities that makes it important for supporting Antarctic science. Most of the programs use all types of vehicles at some stage in their work.

New technologies, such as telecommunications, advanced electronics, remote control and monitoring systems, and alternative energy programs, are making the logistics of working in Antarctica much more flexible and have the potential to greatly enhance scientific programs. Today, many data sets are being collected and transmitted in real time from sites in Antarctica to scientists operating in laboratories many thousands of miles away. Systems that are remotely controlled and automated reduce the need for scientists to spend months traveling to and from Antarctica.

Looking after Antarctica

The importance of Antarctica for studying global climate change cannot be overestimated, and this, coupled with the international commitment to protect the Antarctic,

has led many nations to improve their support facilities to ensure that carrying out scientific research does not damage this most precious of environments. Older sites that have outlived their usefulness are being cleaned up, and when new facilities are built they are designed for easy dismantling when no longer required. Leading nations in this field are developing their facilities in such a way that on completion of their research, the station can be removed and the environment left unspoiled and in pristine condition.

Making things happen in Antarctica is a rewarding activity. Those entrusted with the task of supporting scientific research programs never lose sight of their responsibility to protect the Antarctic environment and to ensure the safety of everyone involved. KP

Fit to be Tied

Each summer, colored flags mark the safe routes around Hut Point Peninsula, and over the sea ice and the Ross Ice Shelf. Traditionally, McMurdo Station staff invite Scott Base personnel to a party, with beer, food, live music and lots of bright flags that need tying to the cane poles.

Defying Murphy's Law

"Check: reliable vehicle, fuel, tents, sleeping bags and mats, food, cooking equipment, radio, scientific equipment, survival gear, first-aid kit ..." Every field trip is planned with care. Murphy's Law—if anything can go wrong, it will—is well known to those who live and work in Antarctica.

Current Scientific Research

Antarctica is a vast scientific laboratory, with research initiatives in all the major fields. Since the Antarctic Treaty came into effect in 1961, an increasing number of nations have been establishing research programs to study Antarctica's atmosphere, its ice, its dry land, its freshwater lakes, and its seas.

All the treaty signatories are entitled to send representatives to the Scientific Committee on Antarctic Research (SCAR)—the committee which coordinates international research programs. This committee takes advice from international working groups that have been established to serve the scientific disciplines, with the role of articulating research questions in their fields and coordinating research activities. It also funds workshops and symposiums during the planning and operating phases of the research.

One of the newest advisory groups to be formed is coordinating research into lakes of water that lie below thousands of feet of Antarctic ice. These sub-glacial lakes are very puzzling. How did they form? Do they contain microscopic life forms? What can the sediment of their floors reveal about the history of Antarctica's climate? The focus of international attention is Lake Vostok, the largest of these lakes, which lies underneath about 13,000 ft (4,000 m) of ice in the central East Antarctica area. A phased approach to its exploration is being developed, with the central requirement that no foreign material be introduced into the lake when its icy ceiling is breached. The task requires sophisticated remote sensing and robotic technologies. After the lake is accessed, it will take many years for the findings to be analyzed—but they are likely to be amazing.

The main focuses of Antarctic biological research is the region's biodiversity, including how organisms are able to live, grow, and reproduce in the most hostile environment on the Earth, and an assessment of the consequences of climate change.

A newly established program, called Regional Sensitivity to Climate Change, is using an internationally agreed methodology to monitor the effects of climatic and environmental change on invertebrate animals and vegetation—a long-term project, since the phenomena being studied change very slowly. Project collaborators worldwide are pooling their data, in the typical spirit of openness and cooperation which is a most remarkable hallmark of Antarctic science.

Mud Dipping

Scientists probe for a shallow sediment core through thin sea ice off Mitchell Peninsula. The continual rain of organic matter and the spent calcite shells of tiny marine organisms in the ocean slowly build layers of muddy ooze on the sea floor, thereby preserving a textured history of local climatic conditions.

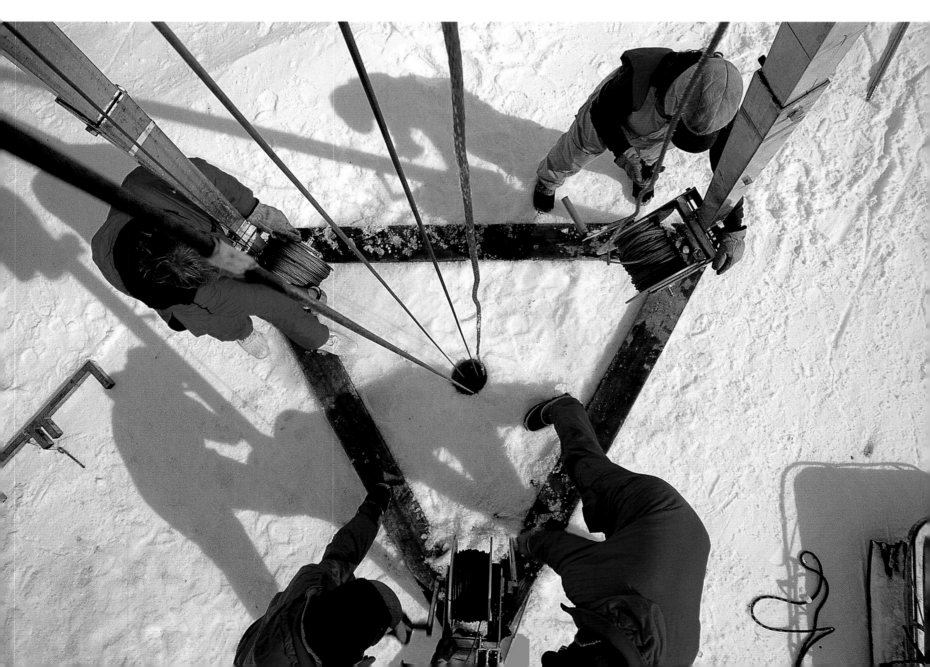

Mobile Laboratory

Scientists on these tractor trains, here parked at an overnight GPS survey stop, are measuring ice-sheet surface velocities in the Lambert Glacier Basin, data vital to calculations of the mass balance of the continent. The spidery web of a radar antennae (in silhouette, right) determines the ice thickness.

Blood Donor

A New Zealand biologist with an Adélie penguin at Cape Bird, near the northern tip of Ross Island. Blood samples from the penguins are being used to study the role of hormones in controlling the length of feeding trips at sea.

Studying the Ice

Considerable effort is being devoted to studying the build-up, movement, and behavior of ice. Contemporary concern about global warming makes this research of great importance, and there are programs to monitor the frequency with which icebergs calve from floating ice sheets. There are also studies to determine whether ice loss at the edge is matched with new ice from precipitation, and it appears that—unlike at the Arctic, where the ice is becoming thinner—the Antarctic ice sheets are roughly in balance at present.

Sea ice is also important climatically. Vast amounts of heat are exchanged annually between the ice and the underlying water as the sea ice waxes and wanes, influencing the temperature of the Southern Ocean. These dynamic interactions have a major effect on the world's climate, and studies are providing valuable information for long-term climate predictions.

Ice cores drilled from the Antarctic ice cap provide an excellent record of the Earth's past climate. The composition of minute gas bubbles trapped in the ice reveals much about the prevailing temperatures, and the amount of methane gives an indication of variations in biological productivity. Traces of sulfur in the ice tell the history of volcanic activity, and the ratio of isotopes of oxygen provides a measure of the temperature when the ice was formed. Ice cores taken from near Casey Station in East Antarctica show that the amount of carbon dioxide in the atmosphere has doubled since the Industrial Revolution. Comparison of ice cores taken from Greenland and Antarctica show many similarities but also many differences between them: this data suggest the Northern and Southern hemispheres have had somewhat different histories, at least since the

break-up of Gondwana, creating the Antarctic Circumpolar Current and causing the mean temperature of the once-luxuriantly vegetated Antarctic continent to fall.

Studying the Seas

In the Southern Ocean, research focuses on the health of the marine ecosystem. Although fishing is permitted in Antarctic waters, a careful watch is kept on how it affects the penguin and seal populations. In 1999, the world's population of pack-ice seals was monitored in one of the largest wildlife surveys ever undertaken, establishing a baseline against which to measure future change. The relationship between penguin numbers and krill stocks has been monitored for more than ten years, and consequently a precautionary limit on krill fishing has been set.

Satellite observations of ocean coloration measure the productivity of the Southern Ocean—the tiny plants that occupy the upper layer of the ocean are green and blue. The sea-ice zone is of particular biological interest, and it appears that variations in the extent of the winter sea ice partly drive the effectiveness of reproduction of plankton and krill. Changes in the composition of the plankton fauna—the tiny, shrimp-like creatures that feed on the microscopic plants—are being monitored by regular surveys carried out with automatic plankton sampling devices towed behind supply vessels.

Studying the Land

SCAR is coordinating several large-scale international studies on dry land. The age and evolution of Antarctica is the subject of a study currently being undertaken by 15 nations. Geologists are locating and defining the ancient boundaries between the building blocks of

Portable Pool

Drilling ice cores has revealed a rich history of retreat and advance of the massive ice sheets. Recent studies on ice shelves have adopted hot-water techniques to melt access boreholes to the ocean cavity below. This camp on the Amery Ice Shelf shows the portable pool of melted snow, ready for drilling, in the foreground.

Test Cases

A marine mammal specialist with his "subjects"—Hooker's sea lions. Earlier, he had glued dive depth and duration recorders and location transmitters onto this nursing mother. This study yielded startling results: dives to more than 1,470 ft (450 m), the deepest ever recorded for sea lions.

Seal Tracking

Scientists regularly fit satellite dive recorders to crabeater seals to track their seasonal movement and foraging behavior. Recorders are attached to an animal's fur with adhesive, preferably to post-molt individuals: the device is likely to remain in place and be active for several months.

Life on Mars?

Compared with meteorites collected elsewhere on the Earth, those found in Antarctica are relatively well preserved. Meteorites thought to originate from Mars (such as the one pictured above) are of particular interest: some NASA scientists believe they have detected some mineral features in them that are characteristic of biological activity.

Antarctica to piece together the continent's history. At the same time, a map of the continent's bedrock is being compiled, complete with data on the magnetic changes that have accompanied its history.

The constant erosion of the ice surface by wind and sun allows meteorites that have landed on Antarctica to be uncovered and collected. Japan maintains a huge collection of these, which is available to scientists from around the world, and they are revealing much about the Earth's neighboring planets. Although this is not strictly Antarctic science, the unique environmental conditions on the ice sheets provide a convenient collection opportunity for these fascinating fragments.

Studying Space

Antarctica has been used for observing astronomical and physical phenomena since exploration began. It is a productive research site for astronomers because the mean elevation is high above the bedrock, and the cold, dry air provides outstanding optical conditions. For upper atmosphere studies, the downward projection of the Earth's protective magnetic field lines at high latitudes provides a window for observation that is undistorted by the deflection induced by the magnetic

shield; cosmic rays can be far more easily observed in Antarctica than in more temperate latitudes, for example. The international Space Ship Earth project is examining how the planet is reacting to the constant bombardment of radiation from outer space. Much international effort is also being expended on a better understanding of space weather—the environment in which satellites, astronauts, and some aircraft must operate—and the practical value of these studies is rising steadily with increasingly sophisticated telecommunications and other satellite applications.

International Cooperation

Antarctic science is remarkable in the degree to which nations cooperate in planning and executing research and publishing the results. Geophysical findings on a range of phenomena from many Antarctic sites are now available on a web database. Antarctica is a laboratory that has no room for those who want to go it alone. This is partly because costs of travel and field support are extremely high. But most of all, the harsh environment and the threat that it poses unite scientists in their common cause, and compel them to support each other in their arduous and often dangerous work. MS

Beneath the Chaos Glacier

Diving in polar waters, such as these in Ranvik Bay, presents specific problems: while salt content keeps the sea liquid below 32°F (0°C), moisture in air and lungs will freeze at this temperature.

Frigid Dip

Glaciologists stand suspended in a cage over the frozen Southern Ocean from the research vessel, the *Aurora Australis* during a winter polynya cruise. Coastal polynyas are regions of perennial open water. In calm weather they are often the site of prodigious sea-ice formation, which is then blown out to sea by fierce katabatic winds. Measurements of the rate of production and types of sea ice yield valuable information about the exchange of heat between atmosphere and ocean.

Resources

Wildlife Conservation Status

The conservation status information from two internationally recognized sources, the World Conservation Union (IUCN) and the Convention on International Trade in Endangered Species of Wild Fauna and Flora (CITES), are included with other statistics at the beginning of species entries in the wildlife chapters of this book. An explanation of the status codes and the organizations that generate them are outlined below.

IUCN

The World Conservation Monitoring Center (WCMC) is part of the United Nations Environment Programme (UNEP), and it provides worldwide biodiversity information for policy and action to conserve the living world. That role includes the publication of the IUCN Red List of Threatened Plants and Animals <www.iucnredlist.org>, an internationally recognized series of lists which categorize the status of species that are globally threatened. The WCMC determines the relative risk of extinction of species, and the main purpose of the Red List is to catalog the species that are regarded as threatened at a global level and are at risk of overall extinction.

Red List Categories and Criteria

EXTINCT

A taxon is Extinct when there is no reasonable doubt that the last individual has died.

EXTINCT IN THE WILD

A taxon is Extinct in the Wild when it is known only to survive in cultivation, in captivity, or as a naturalized population (or populations) well outside its past range.

CRITICALLY ENDANGERED

A taxon is Critically Endangered when it is facing an extremely high risk of extinction in the wild in the immediate future: an 80 percent reduction over ten years or three generations; at least a 50 percent probability of extinction in the wild within ten years.

ENDANGERED

A taxon is Endangered when it is facing a very high risk of extinction in the wild in the near future: a reduction of at least 50 percent over ten years or three generations; severely fragmented or reduced populations; at least a 20 percent probability of extinction in the wild within 20 years.

VULNERABLE

A taxon is Vulnerable when it is facing a high risk of extinction in the wild in the medium-term future: 20 percent probability of extinction in the wild within ten years or three generations; very small or restricted population; at least a 10 percent probability of extinction in the wild within 100 years.

LOWER RISK

The Lower Risk category is separated into three subcategories:

Conservation dependent: Taxa that are the focus of a continuing conservation program, the cessation of which would result in the taxon qualifying for one of the threatened categories (described above) within a period of five years.

Near threatened: Taxa that do not qualify for conservation dependent status, but which are close to qualifying for Vulnerable.

Least concern: Taxa that do not qualify for conservation dependent or near threatened status.

DATA DEFICIENT

A taxon is Data Deficient when there is inadequate information to make an assessment of its risk of extinction based on its distribution and/or population status. Data Deficient is not a category of threat or Lower Risk, and acknowledges the possibility that future research may show that threatened classification is appropriate.

NOT EVALUATED

A taxon is Not Evaluated when it has not yet been assessed against the criteria.

CITES

The international wildlife trade has caused massive declines in the numbers of many species of animals and plants. An international treaty, known as CITES, to which 152 nations subscribe, was introduced in 1975 to ban the commercial international trade in endangered species and to regulate and monitor the trade in other species that might become endangered. See <www.cites.org>.

CITES APPENDIX I

The most endangered species are listed in Appendix I, which includes all species threatened with extinction that are or may be affected by trade.

CITES APPENDIX II

Other species at serious risk are listed in Appendix II: this includes all species that although not necessarily currently threatened with extinction may become so unless trade is subject to strict regulation. Appendix II includes other species which must be subject to regulation in order that trade in endangered species listed in Appendix I may be brought under effective control: for example, species similar in appearance to endangered species.

The Antarctic Treaty FULL TEXT 1961

The Governments of Argentina, Australia, Belgium, Chile, the French Republic, Japan, New Zealand, Norway, the Union of South Africa, the Union of Soviet Socialist Republics, the United Kingdom of Great Britain and Northern Ireland, and the United States of America,

Recognizing that it is in the interest of all mankind that Antarctica shall continue for ever to be used exclusively for peaceful purposes and shall not become the scene or object of international discord;

Acknowledging the substantial contributions to scientific knowledge resulting from international cooperation in scientific investigation in Antarctica;

Convinced that the establishment of a firm foundation for the continuation and development of such cooperation on the basis of freedom of scientific investigation in Antarctica as applied during the International Geophysical Year accords with the interests of science and the progress of all mankind;

Convinced also that a treaty ensuring the use of Antarctica for peaceful purposes only and the continuance of international harmony in Antarctica will further the purposes and principles embodied in the Charter of the United Nations;

Have agreed as follows:

Article I

1. Antarctica shall be used for peaceful purposes only. There shall be prohibited, inter alia, any measure of a military nature, such as the establishment of military bases and fortifications, the carrying out of military manoeuvres, as well as the testing of any type of weapon.
2. The present Treaty shall not prevent the use of military personnel or equipment for scientific research or for any other peaceful purpose.

Article II

Freedom of scientific investigation in Antarctica and cooperation toward that end, as applied during the International Geophysical Year, shall continue, subject to the provisions of the present Treaty.

Article III

1. In order to promote international cooperation in scientific investigation in Antarctica, as provided for in Article II of the present Treaty, the Contracting Parties agree that, to the greatest extent feasible and practicable:
    a. information regarding plans for scientific programs in Antarctica shall be exchanged to permit maximum economy of and efficiency of operations;
    b. scientific personnel shall be exchanged in Antarctica between expeditions and stations;
    c. scientific observations and results from Antarctica shall be exchanged and made freely available.

2. In implementing this Article, every encouragement shall be given to the establishment of cooperative working relations with those Specialized Agencies of the United Nations and other technical organizations having a scientific or technical interest in Antarctica.

Article IV

1. Nothing contained in the present Treaty shall be interpreted as:
    a. a renunciation by any Contracting Party of previously asserted rights of or claims to territorial sovereignty in Antarctica;
    b. a renunciation or diminution by any Contracting Party of any basis of claim to territorial sovereignty in Antarctica which it may have whether as a result of its activities or those of its nationals in Antarctica, or otherwise;
    c. prejudicing the position of any Contracting Party as regards its recognition or non-recognition of any other State's rights of or claim or basis of claim to territorial sovereignty in Antarctica.
2. No acts or activities taking place while the present Treaty is in force shall constitute a basis for asserting, supporting or denying a claim to territorial sovereignty in Antarctica or create any rights of sovereignty in Antarctica. No new claim, or enlargement of an existing claim, to territorial sovereignty in Antarctica shall be asserted while the present Treaty is in force.

Article V

1. Any nuclear explosions in Antarctica and the disposal there of radioactive waste material shall be prohibited.
2. In the event of the conclusion of international agreements concerning the use of nuclear energy, including nuclear explosions and the disposal of radioactive waste material, to which all of the Contracting Parties whose representatives are entitled to participate in the meetings provided for under Article IX are parties, the rules established under such agreements shall apply in Antarctica.

Article VI

The provisions of the present Treaty shall apply to the area south of 60° South Latitude, including all ice shelves, but nothing in the present Treaty shall prejudice or in any way affect the rights, or the exercise of the rights, of any State under international law with regard to the high seas within that area.

Article VII

1. In order to promote the objectives and ensure the observance of the provisions of the present Treaty, each Contracting Party whose representatives are entitled to participate in the meetings referred to in Article IX of the Treaty shall have the right to designate observers to carry out any inspection provided for by the present Article. Observers shall be nationals of the Contracting Parties which designate them. The names of observers shall be communicated to every other Contracting Party having the right to designate observers, and like notice shall be given of the termination of their appointment.
2. Each observer designated in accordance with the provisions of paragraph 1 of this Article shall have complete freedom of access at any time to any or all areas of Antarctica.
3. All areas of Antarctica, including all stations, installations and equipment within those areas, and all ships and aircraft at points of discharging or embarking cargoes or personnel in Antarctica, shall be open at all times to inspection by any observers designated in accordance with paragraph 1 of this Article.
4. Aerial observation may be carried out at any time over any or all areas of Antarctica by any of the Contracting Parties having the right to designate observers.
5. Each Contracting Party shall, at the time when the present Treaty enters into force for it, inform the other Contracting Parties, and thereafter shall give them notice in advance, of
    a. all expeditions to and within Antarctica, on the part of its ships or nationals, and all expeditions to Antarctica organized in or proceeding from its territory;
    b. all stations in Antarctica occupied by its nationals; and
    c. any military personnel or equipment intended to be introduced by it into Antarctica subject to the conditions prescribed in paragraph 2 of Article I of the present Treaty.

Article VIII

1. In order to facilitate the exercise of their functions under the present Treaty, and without prejudice to the respective positions of the Contracting Parties relating to jurisdiction over all other persons in Antarctica, observers designated under paragraph 1 of Article VII and scientific personnel exchanged under sub-paragraph 1(b) of Article III of the Treaty, and members of the staffs accompanying any such persons, shall be subject only to the jurisdiction of the Contracting Party of which they are nationals in respect of all acts or omissions occurring while they are in Antarctica for the purpose of exercising their functions.
2. Without prejudice to the provisions of paragraph 1 of this Article, and pending the adoption of measures in pursuance of subparagraph 1(e) of Article IX, the Contracting Parties concerned in any case of dispute with regard to the exercise of jurisdiction in

Antarctica shall immediately consult together with a view to reaching a mutually acceptable solution.

Article IX

1. Representatives of the Contracting Parties named in the preamble to the present Treaty shall meet at the City of Canberra within two months after the date of entry into force of the Treaty, and thereafter at suitable intervals and places, for the purpose of exchanging information, consulting together on matters of common interest pertaining to Antarctica, and formulating and considering, and recommending to their Governments, measures in furtherance of the principles and objectives of the Treaty, including measures regarding:
   a. use of Antarctica for peaceful purposes only;
   b. facilitation of scientific research in Antarctica;
   c. facilitation of international scientific cooperation in Antarctica;
   d. facilitation of the exercise of the rights of inspection provided for in Article VII of the Treaty;
   e. questions relating to the exercise of jurisdiction in Antarctica;
   f. preservation and conservation of living resources in Antarctica.
2. Each Contracting Party which has become a party to the present Treaty by accession under Article XIII shall be entitled to appoint representatives to participate in the meetings referred to in paragraph 1 of the present Article, during such times as that Contracting Party demonstrates its interest in Antarctica by conducting substantial research activity there, such as the establishment of a scientific station or the despatch of a scientific expedition.
3. Reports from the observers referred to in Article VII of the present Treaty shall be transmitted to the representatives of the Contracting Parties participating in the meetings referred to in paragraph 1 of the present Article.
4. The measures referred to in paragraph 1 of this Article shall become effective when approved by all the Contracting Parties whose representatives were entitled to participate in the meetings held to consider those measures.
5. Any or all of the rights established in the present Treaty may be exercised as from the date of entry into force of the Treaty whether or not any measures facilitating the exercise of such rights have been proposed, considered or approved as provided in this Article.

Article X

Each of the Contracting Parties undertakes to exert appropriate efforts, consistent with the Charter of the United Nations, to the end that no one engages in any activity in Antarctica contrary to the principles or purposes of the present Treaty.

Article XI

1. If any dispute arises between two or more of the Contracting Parties concerning the interpretation or application of the present Treaty, those Contracting Parties shall consult among themselves with a view to having the dispute resolved by negotiation, inquiry, mediation, conciliation, arbitration, judicial settlement or other peaceful means of their own choice.
2. Any dispute of this character not so resolved shall, with the consent, in each case, of all parties to the dispute, be referred to the International Court of Justice for settlement; but failure to reach agreement on reference to the International Court shall not absolve parties to the dispute from the responsibility of continuing to seek to resolve it by any of the various peaceful means referred to in paragraph 1 of this Article.

Article XII

1.
   a. The present Treaty may be modified or amended at any time by unanimous agreement of the Contracting Parties whose representatives are entitled to participate in the meetings provided for under Article IX. Any such modification or amendment shall enter into force when the depositary Government has received notice from all such Contracting Parties that they have ratified it.
   b. Such modification or amendment shall thereafter enter into force as to any other Contracting Party when notice of ratification by it has been received by the depositary Government. Any such Contracting Party from which no notice of ratification is received within a period of two years from the date of entry into force of the modification or amendment in accordance with the provision of subparagraph 1(a) of this Article shall be deemed to have withdrawn from the present Treaty on the date of the expiration of such period.

2.
   a. If after the expiration of thirty years from the date of entry into force of the present Treaty, any of the Contracting Parties whose representatives are entitled to participate in the meetings provided for under Article IX so requests by a communication addressed to the depositary Government, a Conference of all the Contracting Parties shall be held as soon as practicable to review the operation of the Treaty.
   b. Any modification or amendment to the present Treaty which is approved at such a Conference by a majority of the Contracting Parties there represented, including a majority of those whose representatives are entitled to participate in the meetings provided for under Article IX, shall

be communicated by the depositary Government to all Contracting Parties immediately after the termination of the Conference and shall enter into force in accordance with the provisions of paragraph 1 of the present Article
   c. If any such modification or amendment has not entered into force in accordance with the provisions of subparagraph 1(a) of this Article within a period of two years after the date of its communication to all the Contracting Parties, any Contracting Party may at any time after the expiration of that period give notice to the depositary Government of its withdrawal from the present Treaty; and such withdrawal shall take effect two years after the receipt of the notice by the depositary Government.

Article XIII

1. The present Treaty shall be subject to ratification by the signatory States. It shall be open for accession by any State which is a Member of the United Nations, or by any other State which may be invited to accede to the Treaty with the consent of all the Contracting Parties whose representatives are entitled to participate in the meetings provided for under Article IX of the Treaty.
2. Ratification of or accession to the present Treaty shall be effected by each State in accordance with its constitutional processes.
3. Instruments of ratification and instruments of accession shall be deposited with the Government of the United States of America, hereby designated as the depositary Government.
4. The depositary Government shall inform all signatory and acceding States of the date of each deposit of an instrument of ratification or accession, and the date of entry into force of the Treaty and of any modification or amendment thereto.
5. Upon the deposit of instruments of ratification by all the signatory States, the present Treaty shall enter into force for those States and for States which have deposited instruments of accession. Thereafter the Treaty shall enter into force for any acceding State upon the deposit of its instruments of accession.
6. The present Treaty shall be registered by the depositary Government pursuant to Article 102 of the Charter of the United Nations.

Article XIV

The present Treaty, done in the English, French, Russian and Spanish languages, each version being equally authentic, shall be deposited in the archives of the Government of the United States of America, which shall transmit duly certified copies thereof to the Governments of the signatory and acceding States.

Antarctic Links

Every imaginable Antarctic resource can be found on the Internet with a little patience—a recent search under "Antarctica" scored over 61.6 million hits. The sites listed below cover the gamut, from history to glaciology to current events, providing something for every Antarctic taste. Websites have a way of becoming inactive or disappearing outright, but those included here should be around for a while.

Principal Websites

Four of the sites listed stand out because of the quality and depth of their content, their navigational ease, and the fact that they complement one another.

The first is the website of the Scott Polar Research Institute (SPRI), the world's leading academic polar organization <www.spri.cam.ac.uk>. This site is very detailed and comprehensive, and has versions of its homepage in languages other than English. SPRILIB, a searchable database with over 120,000 bibliographic records, and the Picture Library Database are excellent. The Index to Antarctic Expeditions (with links, summaries, and related literature) is useful, as are the worldwide Polar Museums Directory and the Directory of Polar and Cold Regions Organizations. There are even Kid's Pages including Polar Jokes.

For those with a passion for Antarctic history, the Antarctic Philately site <www.south-pole.com> is the place to go. Besides copious information on stamps and postal history, there is voluminous coverage of south polar exploration, from Cook to nearly the present day. The highpoints are the numerous biographical entries and a well-done time line stretching from 1519 to 1959. Some portions are nearly book length. This elegantly designed site also features many seldom-seen photo illustrations.

Gateway Antarctica <www.anta.canterbury.ac.nz> originates at the University of Canterbury in Christchurch, New Zealand. It strives for "increased understanding and more effective management of the Antarctic and the Southern Ocean by being a focal point and a catalyst for Antarctic scholarship, attracting national and international participation in collaborative research, analysis, learning and networking." Gateway to Antarctica is one portion of the Gateway Antarctica site. Once through the Gateway, there are sections on History (including the very lengthy Flight of the Puckered Penguins, focusing on Antarctic aviation in the 1950s), Tourism (the Visitor's Introduction to Antarctica and its Environment is praiseworthy), Logistics, Environment (with lots on ozone depletion), Education, Images, Science, News, and Treaty (a fine background on the Antarctic Treaty).

For keeping up with the news, go to The Antarctican <www.antarctican.com> Produced in Tasmania, Australia, the site describes itself as "delivering the latest news and comment on Antarctic life, South Polar endeavor, the world of the ice, and the Southern Ocean around it." The individual pieces, covering all Antarctic subjects, are original and in a common format, not reprints, or links to newspapers. The Polemic section lets you make your views known on things Antarctic and you can post your Antarctic photos on Ice Picks.

Antarctic Centers

Other sites of the academic centers worth going to include that of the Byrd Polar Research Center at the Ohio State University <http://bprc.osu.edu>, which, among other things, has a biography of Richard Byrd and a notable collections of links entitled Polar Pointers.

The Antarctic Co-operative Research Centre at the University of Tasmania was "established in 1991 and is now one of the largest research organizations in the world concerned with polar regions." Its site, Antarctic CRC at <www.acecrc.org.au> is thoroughly scientific in focus. It gives details on conferences, media releases, and on-line newsletters, and has a listing of commonly used Antarctic acronyms and abbreviations.

Glacier <www.glacier.rice.edu>, administered by Rice University in Texas, has many short sections on all kinds of Antarctic subjects, making it particularly suitable for students. Much of it is presented in question-and-answer format. The coverage of ice and glaciers, life at the bases and polar clothing is first rate.

The Library at CRREL (Cold Regions Research and Engineering Laboratory) in Hanover, New Hampshire, "is recognized as the world's foremost collection of cold regions scientific and technical literature." Its site <www.crrel.usace.army.mil/library> has several searchable databases and many links.

National Programs

Each country that has an Antarctic presence has a website for its program, but most are either sparse in content or of limited interest. But several stand out.

The best starting point is <www.comnap.aq>, the site maintained by COMNAP (Council of Managers of National Antarctic Programs), which "was established in 1988 to bring together those managers of national agencies responsible for the conduct of Antarctic operations in support of science." It lists all research stations, has details on the facilities and activities of each, and links to all the national programs (the "aq" in the web address is Antarctica's own Internet domain).

Another helpful source on national programs is the SCAR (Scientific Committee on Antarctic Research) site at <www.scar.org>. "SCAR is charged with the initiating, developing, and coordination of high quality international scientific research in the Antarctica region, and on the role of the Antarctic rgion in the Earth systems." The SCAR site has an excellent listing of facts (including some Antarctic statistics), as well as information on various scientific working groups and specialists.

National sites of note include:

- the New Zealand Antarctic Institute <www.antarcticanz.govt.nz> (includes some good school resources—frequently asked questions, information sheets, and education database)
- the British Antarctic Survey <www.antarctica.ac.uk> (all about the British bases, news stories)
- the United States National Science Foundation's Office of Polar Programs <www.nsf.gov/dir/index.jsp?org=OPP> (sections on scientific programs by discipline, journal articles, lots of photographic and satellite images)
- the Australian Antarctic Division's site, Antarctica Online <www.aad.gov.au> (maps of station areas, searchable catalog of library holdings, database of polar words and phrases, station webcams, and the Antarctic Artefacts Register).

Another interesting site is The Antarctic Sun at <www.antarcticsun.usap.gov>, which is funded by the US National Science Foundation. The contents are informal and chatty on topical issues; there are even some cartoons.

For a virtual tour of either Amundsen–Scott South Pole Station or McMurdo Station, go to the CARA sites (Center for Astrophysical Research in Antarctica) at: <http://astro.uchicago.edu/cara/vtour/pole> and <http://astro.uchicago.edu/cara/vtour/mcmurdo>.

News

Antarctic news may be found at the 70South site at <www.70south.com> (links to news sources, contributed items, and interesting or important Antarctic anniversaries for the current month).

History

The Antarctic Philately site, noted earlier, is the site when it comes to history. Many webpages will have some coverage of Antarctic history and the better-known explorers. Gateway to Antarctica, also noted earlier, gives good coverage.

Virtual Antarctica, at <www.terraquest.com/antarctica/index.html>, which is maintained by Mountain Travel-Sobek and WorldTravel Partners, includes a chronology of Antarctic exploration from 1772 to 1995, as well as other information aimed at the Antarctic tourist.

Adventure Network, another of the tour operators, has a concise history section at <www.adventure-network.com>.

Those interested in Antarctic historic sites, especially the huts of the explorers, should check out Heritage-Antarctica.org, at <www.heritage-antarctica.org>, which highlights the activities of the New Zealand Antarctic Heritage Trust in maintaining the historic huts in the Ross Sea sector. Included is a listing of historic sites with accompanying maps. The Heritage-Antarctica.org site also serves the United Kingdom Antarctic Heritage Trust.

The efforts to restore Douglas Mawson's huts at Commonwealth Bay in East Antarctica are the focus of the AAP Mawson's Huts Foundation site <www.mawsons-huts.org.au>.

Sites on individual explorers are becoming more common. At the moment Shackleton seems to attract the most attention. The most enthusiastic is Welcome to the Shack! <http://home.nycap.rr.com/gn>. Among its contents are frequently asked questions, Shack Books, Book Review, Shack Links, and News Archive. The Message Board features queries posted by readers with responses from others.

Boston's WGBH public television station has an informative site entitled Shackleton's Antarctic Odyssey at <www.pbs.org/wgbh/nova/shackleton>. There are useful biographies of all Shackleton's men, details on the lesser known Ross Sea party, historic maps, short accounts of other Heroic Age expeditions, educational resources such as lesson plans, and information on clothing, food, etc.

Shackleton's old school, Dulwich College in London, maintains a collection of Shackleton Links at <www.dulwich.org.uk/history/shacklinks>.

The James Caird Society was established in 1994 to honor and perpetuate Shackleton's life and deeds. This very active group has a considerable amount material on its website at <www.jamescairdsociety.com>, including some good photographs, quite a bit of history on Shackleton's *Endurance* expedition, the latest Shackleton news and details on the society itself.

The explorer's cousin and family historian, Jonathan Shackleton, has a site at <http://indigo.ie/~jshack/ernest.html> that notes recent books, videos and exhibitions, and recounts Shackleton's Antarctic expeditions.

Frank Hurley, Australian photographer on the *Endurance* expedition, is the subject of Shane Murphy's site <http://frankhurley.com>.

At present Robert Falcon Scott receives less web attention than Shackleton. His ship, the *Discovery*, is beautifully preserved in Dundee, and the associated web site <www.rrs-discovery.co.uk> has lots on Scott, his men, the ship, and the extensive exhibits on shore at Discovery Point.

The Norwegian site of the Fram Museum in Oslo, <www.fram.museum.no> is based around the Fram, "the strongest vessel in the world" which "has advanced further north and further south than any other vessel." It concentrates on polar exploration, and its most notable explorers—not only Amundsen but also Nansen and Sverdrup.

A wealth of material on Captain Cook, including a very detailed chronology of his second (Antarctic) voyage may be found at <www.captaincooksociety.com>. This has information on stamps, coins, medals, and Cook's ships, and has a long bibliography as well.

Education

In addition to those already mentioned, sites that should appeal to students and teachers include Share the Journey <www.sofweb.vic.edu.au/claypoles>, hosted by the Victoria Department of Education in Australia; it follows the year spent at Commonwealth Bay by Yvonne and Jim Claypole and includes diary entries and questions for students.

Antarctica as an Educational Resource is a site maintained by the School of Biological Sciences at the University of Auckland, New Zealand <www.antarctica.org.nz>. Its aim "is to provide a focus for teachers interested in using Antarctica as an educational resource." There are specific sections (with questions posed) covering History, Environment, Biology, Adaptive Radiation, and Human Involvement.

TEA (Teachers Experiencing Antarctica and the Arctic) <http://tea.rice.edu> is a terrific educational site sponsored by the US National Science Foundation and facilitated by Rice University, the American Museum of Natural History, and CRREL. Teachers participate in research projects in Antarctica and post their journals here, or can e-mail them and learn more.

Tourism

IAATO (International Association of Antarctica Tour Operators) is "a member organization founded in 1991 to advocate, promote and practice safe and environmentally responsible private-sector travel to the Antarctic." <www.iaato.org> has a definitive listing of tour operators with contacts and web addresses, a variety of tourism statistics, a book list, and a section on Guidance for Visitors to the Antarctic.

Lonely Planet, the guidebook publisher, has helpful information such as Facts for the Traveler, When to Go, and Getting There & Away on <www.lonelyplanet.com/worldguide/antarctica/>.

Webcams and Images

Webcams are scattered about Antarctica at various research stations. WebCam Central <www.camcentral.com/search.php?keyword=Antarctica&page=1> is the best place to go for links to those in operation.

AASTO (Automated Astrophysical Site-Testing Observatory) presents from the South Pole live images updated every 10 minutes <http://bat.phys.unsw.edu.au/~aasto>. Also here are researcher diaries and an extensive Photogallery.

For icebergs, the Antarctic Meteorology Research Center's site, Iceberg Images at the AMRC <http://ice.ssec.wisc.edu/iceberg.php>, has a vast number of images and gives the current status and positions of the larger ones.

The US National Ice Center at < www.natice.noaa.gov> has information about ice conditions and icebergs.

For copyright-free satellite aerials, visit NASA's website <http://svs.gsfc.nasa.gov/search/Keyword/Antarctica.html>.

For daily updates on the Antarctic ozone hole, see NASA's < www.nasa.gov/vision/earth/environment/ozone_resource_page>.

Links

Nearly every website has links to other sites. Those of the Byrd Polar Research Center and of CRREL, mentioned above, are particularly extensive. Some other choices such as the Arctic and Antarctic Advice Agency Austria <www.arctic.at> features links by category and country and quite a bit more.

Stephanie Bianchi's Some Polar Websites <www.acs.ucalgary.ca/~tull/polar/polar.htm> is comprehensive, and arranged by subject.

The Polar Web <http://arcticcentre.ulapland.fi/polarweb/> is a collaborative project of the Polar Libraries Colloquy and is managed by the Arctic Centre at the University of Lapland in Finland. Although the emphasis is more Arctic than Antarctic, nonetheless it is a worthwhile compilation with helpful descriptions of each link.

Discoverers Web <www.win.tue.nl/~engels/discovery> is a massive collection of links covering all aspects of travel, discovery, and exploration. Arranged mostly by region and era, it has a more than adequate polar section.

Miscellaneous

Some Antarctic websites defy easy categorization. Here are a few.

A searchable database at <www.scar.org/treaty/> provides probably more than anyone will ever want to know about the Antarctic Treaty.

The *Sedna IV* Antarctic Mission expedition, led by Jean Lemire, has an interesting site at <www.sedna.tv/eng/crew/lemire.php> describing the objectives and progress of the expedition.

Welcome to the Ice <www.theice.org> is the work of Robert Holmes and has a bit of everything. Its Discussion Board includes many e-mail queries, with comments and responses posted by others. It has an informative Did You Know? section and frequently asked questions with answers. Charles Neider's Historic Guide to Ross Island is among the offerings.

The CIA's Factbook on Antarctica may be found at <www.cia.gov/library/publications/the-world-factbook/index.html>. As it employs the standard format used for all countries, the categories are not always Antarctic relevant—percentage of land in permanent crops, for example—but it is still worth a look.

The Tasmanian Polar Network (TPN) <www.tpn.aq> gives Antarctic shipping schedules, information about polar ships calling at Hobart, some news on polar events, and dates of conferences.

Raytheon Polar Services is the logistical arm of the United States Antarctic Program and its site <http://rpsc.raytheon.com> features lots on American activities in the Antarctic region, including the useful Participant Guide and the cruise histories and operating schedules of American polar vessels.

The Antarctic Circle <www.antarctic-circle.org> is a "forum and resource on historical, literary, bibliographical, artistic and cultural aspects of Antarctica and the South Polar regions." There are individual pages covering Antarctic events, books and booksellers, organizations, and historic time lines. Two interesting and lengthy features are The Low-Latitude Antarctic Gazetteer, a compilation of Antarctic sites outside the Antarctic (statues, explorers' houses, etc.), and Tekeli-li, Fauno Cordes's annotated bibliography of Antarctic fiction.

Gazetteer

- The gazetteer contains the names on the following maps: Antarctica (pages 12–13), Antarctic Regions (page 74), Antarctic Peninsula (page 77), Gerlache Strait (page 81), West Antarctica (page 88), Ross Sea and Ross Island (page 93), The Dry Valleys (page 100), Dumont d'Urville (page 104), The Davis Sea Region (page 107), Amery to Enderby Land (page 109), The Haakon VII Sea Region (page 110), The Polar Plateau (page 112), The Weddell Sea Region (page 114), The Sub-Antarctic Islands (pages 116–117).
- The country or ocean is listed for all names outside Antarctica and the sub-Antarctic islands.
- Names that have a symbol (stations and summits) are given a grid reference in the gazetteer according to the location of the symbol. Names without a symbol are referenced in the gazetteer according to the start or center of the name.

A

Abbot Ice Shelf 12 D27, 88
Abbott Peak 93 D8
Aboa Station, Finland 13 D35, 110, 356
Access Point 81 B2
Adams Glacier 13 E12, 107, 117 C4
Adare Peninsula 93 D7
Adelaide Island 12 E30, 77 A4, 117 C10
Adelaide, Australia 117 F5
Adie Inlet 77 D4
Admiralty Mountains 93 D6
Admiralty Sound 77 G3
Alcock Island 81 D2
Alexander Island 12 D29, 117 C10
Alexandra Mountains 93 C10
Alfred Faure Station, France 117 E2
Allison Peninsula 12 D28, 88
Almirante Brown Station (closed), Argentina 12 F30, 77 D3, 81 C2
Amery Ice Shelf 13 E8, 74, 107, 109, 117 C3, 356
Amundsen Glacier 93 A10
Amundsen Sea 12 E26, 74, 88, 117 C9, 356
Amundsen–Scott Station, USA 13 A10, 88, 112, 356
Anare Mountains 13 D17, 93 D6
Andersson Island 77 G2
Andes, South America 12 J29, 116 F10
Andvord Bay 81 C2
Angot Point 81 D1
Antarctic Peninsula 12 E30, 74, 77 C4, 81 C3, 115, 117 C11, 356
Antarctic Sound 77 G2
Antipodes Islands, NZ 117 E7
Anvers Island 12 F30, 77 D3, 81 B2, 117 C10
Arago Glacier 81 C2
Arçtowski Peninsula 81 C2
Arçtowski Station, Poland 12 F31, 77 F2, 356
Argentina 12 J30, 116 F10
Argentine Antarctic Territory 356
Argentine Islands 77 C3, 81 A3
Argo Point 77 E4
Arrowsmith Peninsula 77 B4
Arthur Harbor 81 A2
Artigas Station, Uruguay 12 F31, 77 F2, 356
Arturo Prat Station, Chile 12 F31, 77 F2, 356
Aspland Island 77 G1
Astrolabe Island 77 F2
Asunción, Paraguay 116 G11

Auckland Islands, NZ 117 D6
Auckland, NZ 117 F6
Aurora Glacier 93 D9
Australia 117 G5
Australian Antarctic Territory (East Sector) 356
Australian Antarctic Territory (West Sector) 356
Avery Plateau 77 C4

B

Backdoor Bay 93 C8
Bagshawe Glacier 81 C2
Bahía Grande, Argentina 12 H30
Bailey Peninsula 13 E12, 107
Balleny Islands 13 E17, 104, 117 C6
Bancroft Bay 81 D2
Banzare Coast 13 E13, 104
Barilari Bay 77 C3
Barne Glacier 93 D8
Barne Inlet 93 B6
Bart Bay 81 C2
Bartlett Inlet 93 C10
Bass Strait, Australia 117 F5
Bay of Whales 13 C20, 88, 93 C9
Bear Peninsula 12 D25, 88
Beardmore Glacier 13 B17, 93 B6
Beaufort Island 93 C6
Beaumont Bay 93 B6
Belgica Mountains 13 D3, 109, 110
Belgrano II Station, Argentina 13 C33, 115, 356
Bellingshausen Sea 12 E28, 74, 88, 117 C10, 356
Bellingshausen Station, Russian Federation 12 F31, 77 F2, 356
Beneden Head 81 C2
Berkner Island 12 C32, 115, 117 B11
Bermel Peninsula 77 C5
Berthelot Islands 81 A3
Bird Island 117 D11
Bird Island Station, UK 117 D11
Biscoe Bay 81 B2
Biscoe Islands 12 F30, 77 B3, 117 C10
Bismarck Strait 77 C3, 81 B2
Black Island 93 C6
Blodgett Iceberg Tongue 13 E13
Bluff Island 81 D2
Bob Island 81 B2
Bolivia 116 H10
Bone Bay 77 F2
Booth Island 81 A3
Börgen Bay 81 B2
Botswana 117 G1
Bounty Islands, NZ 117 E7
Bouquet Bay 81 C2
Bourgeois Fjord 77 B4
Bouvetøya, Norway 117 D1
Bowers Mountains 93 D6
Bowman Island 13 E11, 107, 117 C4
Boyle Mountains 77 B4
Brabant Island 12 F30, 77 D3, 81 C2
Bransfield Island 77 G2
Bransfield Strait 12 F30, 77 E2
Brazil 116 H11
Brialmont Cove 81 D2
Britannia Range 93 C5
British Antarctic Territory 356
Brooklyn Island 81 C2
Brown Bluff 77 G2
Bruce Plateau 77 C4
Brunt Ice Shelf 13 D34, 110, 115

Bryde Island 81 B2
Buckle Island 13 E17, 104
Buenos Aires, Argentina 116 F11
Buls Bay 81 C2
Bunger Hills 13 E11, 107
Burke Island 12 D26, 88
Bush Mountains 93 B8
Bussey Glacier 81 A3
Butler Passage 81 B2
Byers Peninsula 77 E2
Byrd Glacier 13 C16, 93 B5
Byth Point 77 E2

C

Cabinet Inlet 77 D4
Cabo de Hornos (Cape Horn), Chile 12 G30
Cabo Vírgenes 12 H30
Caird Coast 13 C34, 110, 115
Campbell Island, NZ 127 D6
Canberra, Australia 117 F5
Cape Adare 13 D18, 93 D7, 117 B6
Cape Agassiz 77 D5
Cape Alexander 12 E30, 77 D4
Cape Andreas 81 E1
Cape Ann 13 E6, 109, 117 C2
Cape Anna 81 C2
Cape Armitage 93 C8
Cape Bellue 77 B4
Cape Bird 93 D8
Cape Boothby 13 E6, 109
Cape Borley 13 E6, 109, 117 C2
Cape Carr 13 E14, 104, 117 C5
Cape Casey 77 D4
Cape Cheetham 13 E17, 93 D6, 104
Cape Cloos 77 C3, 81 A3
Cape Cockburn 81 C2
Cape Colbeck 93 C10
Cape Crozier 93 D10
Cape Darnley 13 E7, 109, 117 C3
Cape Denison 13 E15, 104
Cape Disappointment 77 E3, 81 D3
Cape Ducorps 77 F2
Cape Elliott 13 E11, 107
Cape Errera 81 B2
Cape Evans 93 C8
Cape Fairweather 77 E3, 81 D2
Cape Filchner 13 E10, 107
Cape Flying Fish 12 D26, 88
Cape Framnes 12 E30, 77 E3
Cape Freshfield 13 E16, 104
Cape Garry 77 D2, 81 C1
Cape Goodenough 13 E13, 104, 127 C5
Cape Hallett 93 D7
Cape Herlacher 12 D25, 88, 117 B9
Cape Herschel 77 E3, 81 D2
Cape Hooker 93 D6
Cape Hudson 13 E16, 104, 117 C6
Cape James 77 D2
Cape Juncal 77 G2
Cape Kjellman 77 F2
Cape Knowles 12 D30, 115
Cape Lancaster 81 B2
Cape Longing 77 F3
Cape Lookout 77 H1
Cape MacKay 93 C9
Cape Marsh 77 F3
Cape Mascart 77 A4
Cape Maude 93 B7
Cape Melville 77 G2

Cape Mikhaylov 13 E12, 107
Cape Monaco 81 A2
Cape Moore 13 D17, 93 D6
Cape Murray 93 C6
Cape North 13 D17, 93 D6
Cape Northrop 77 C4
Cape Norvegia 13 D35, 110, 117 B12
Cape of Good Hope, South Africa 117 F1
Cape Peremennyy 13 E11, 107
Cape Poinsett 13 E12, 107, 117 C4
Cape Renard 77 C3, 81 B3
Cape Robert 13 E14
Cape Robinson 77 D4
Cape Roquemaurel 77 F2
Cape Roux 81 C2
Cape Royds 93 D8
Cape Selborne 13 B17, 93 B6
Cape Shirreff 77 E2
Cape Tennyson 93 D9
Cape Town, South Africa 117 F1
Cape Tuxen 81 A3
Cape Valentine 77 H1
Cape Waldron 13 E12, 107, 117 C4
Cape Washington 13 D17, 93 D6
Cape Willems 81 B2
Cape Wollaston 81 E1
Carney Island 12 D24, 88, 117 B9
Casey Station, Australia 13 E12, 107, 356
Chapman Point 77 E3
Charcot Cove 93 C6
Charcot Island 12 E29
Charlotte Bay 81 D2
Chatham Islands, NZ 117 E7
Chile 12 J29, 116 F10
Chilean Antarctic Territory 356
Christchurch, NZ 117 E6
Christiania Islands 77 E2, 81 D1
Churchill Mountains 93 B6
Churchill Peninsula 77 D4
Cierva Cove 81 D2
Clarence Island 12 F31
Clarke Peninsula 13 E12, 107
Claude Point 81 C2
Coats Land 13 C34, 74, 115, 117 B12, 356
Cole Peninsula 77 C4
Collins Bay 81 A3
Comandante Ferraz Station, Brazil 12 F31, 77 F2, 356
Commonwealth Bay 13 E15, 104
Comodoro Rivadavia, Argentina 12 J30, 117 E10
Concordia Station, France & Italy 13 D13, 107, 356
Cook Ice Shelf 13 E16, 104
Copper Peak 81 C2
Coronation Island 12 F32, 117 D11
Coulman Island 13 D18, 93 D7, 117 B6
Crary Ice Rise 13 B20, 93 B8
Crown Prince Olav Coast 13 E5, 109
Crown Princess Martha Coast 13 D35, 110
Cruzen Island 12 D23, 88
Crystal Sound 77 B4
Curtiss Bay 81 E2
Cuverville Island 81 C2

D

Dallmann Bay 77 D3, 81 C2
Dallmeyer Peak 77 D3, 81 C2
Dalton Iceberg Tongue 13 E13, 117 C5
Danco Coast 81 C3

Danco Island 81 C2
Daniell Peninsula 13 D18, 93 D7
Dannebrog Islands 81 A3
Darbel Bay 77 B4
Darwin Glacier 13 C16, 93 C5
David Glacier 13 C16, 93 C6, 100
Davis Coast 81 E2
Davis Island 81 C2
Davis Sea 13 F9, 74, 107, 109, 117 C3, 356
Davis Station, Australia 13 E8, 107, 356
Dean Island 12 D24, 88
Deception Island 77 E2
Delaite Island 81 C2
Dendtler Island 12 D28, 88
Detroit Plateau 77 F3
Deville Glacier 81 C2
Dibble Iceberg Tongue 13 E14, 104, 117 C5
Dion Islands 77 A4
Dixson Island 13 E15
Dolleman Island 12 D30
Dome Argus 13 B8, 107, 112
Dome C 13 C13, 107
Dome Fuji Station, Japan 13 C4, 109, 356
Doumer Island 81 B2
Drake Passage 12 G30, 77 C2, 74, 117 D10
Dronning Maud Land 13 D36, 74, 117 B1, 356
Druzhnaya Station, Russian Federation 13 E8, 107, 109, 356
Dry Valleys 93 C5, 100
Drygalski Glacier 77 E3
Drygalski Ice Tongue 13 C18, 93 C6, 100
Drygalski Island 13 E10, 107, 117 C4
Dumont d'Urville Sea 13 F15, 74, 104, 117 C5, 356
Dumont d'Urville Station, France 13 E15, 104, 356
Dundee Island 12 F31, 77 H2
d'Urville Island 77 H2
Dustin Island 12 D27, 88

E

Eagle Island 77 G2
East Antarctica 13 C9, 74, 107
East Falkland 12 H31, 117 D11
Eckener Point 81 D2
Edward VII Land 13 C22, 74, 88, 93 C11, 117 B8
Edward VIII Bay 13 E6, 109
Ekström Ice Shelf 13 D36, 110
Elephant Island 12 F31, 77 H1, 117 D11
Ellsworth Land 12 C27, 74, 88, 115, 117 B9, 356
Ellsworth Mountains 12 C28, 88, 115
Eltanin Bay 12 D28, 88
Emma Island 81 C2
Enderby Land 13 E5, 74, 109, 117 C2, 356
Enterprise Island 81 C2
Erebus and Terror Gulf 77 G2
Erebus Bay 93 C8
Erebus Glacier Tongue 93 C8
Errera Channel 81 C2
Escudero Station, Chile 12 F31, 77 F2, 356
Esperanza Station, Argentina 12 F31, 77 G2, 356
Eta Island 81 C2
Evans Glacier 77 E3, 81 D3
Evans Peninsula 12 D27, 88
Ewing Island 12 D30
Exasperation Inlet 77 E3, 81 D3

F

Falkland Islands, UK 12 H31, 117 D11
False Cape Renard 81 B3
Farr Bay 13 E10, 107
Faure Islands 77 A5

Filchner Ice Shelf 13 C33, 74, 115
Fimbul Ice Shelf 13 D36, 110
Fisher Bay 13 E15, 104
Fisher Glacier 13 D7, 107, 109
Flandres Bay 81 B2
Flask Glacier 77 D3
Fletcher Peninsula 12 D28, 88
Flood Range 12 C23, 88
Forbes Glacier 77 B4
Forbidden Plateau 77 E3, 81 C2
Fournier Bay 81 B2
Foyn Coast 77 C4
Foyn Point 77 E3, 81 D3
Framnes Mountains 13 E7, 109
Francis Island 77 C4
Franklin Island 13 C18, 93 C6
Freud Passage 81 C2
Fridtjof Island 81 B2

G

Gabriel de Castilla Station, Spain 12 F30, 77 E2, 356
Gand Island 81 C2
Gaussberg 13 E9, 107
Géologie Archipelago 13 E14, 104
George V Land 13 D15, 74, 104, 117 B5
George VI Sound 12 D30
Gerlache Strait 77 D3, 81 C2
Getz Ice Shelf 12 D24, 88, 117 B8
Geysen Glacier 13 D7, 107, 109
Gibbs Island 77 H1
Golfo de San Jorge, Argentina 12 J30
Gonzáles Videla Station (closed), Chile 12 F30, 77 D3, 81 C2
Goodspeed Nunataks 13 D6, 109
Gough Island Station, South Africa 117 E12
Gough Island, UK 117 E12
Gould Bay 12 C32, 115
Gould Coast 93 B10
Gourdon Peninsula 81 B2
Graham Land 12 E30, 74, 77 C4, 81 C3, 117 C10, 356
Granite Harbor 93 C6, 100
Grant Island 12 D23, 88
Graphite Peak 93 B7
Great Australian Bight, Australia 127 F5
Great Wall Station, China 12 F31, 77 F2, 356
Greenwich Island 77 E2
Grosvenor Mountains 93 A7
Grove Mountains 13 D8, 107
Grubb Glacier 81 C2
Guest Peninsula 13 C22, 88
Guyou Bay 81 C2
Guyou Islands 81 B3

H

Haag Inlet 12 C29, 115, 117 B10
Haakon VII Sea 13 E1, 74, 110, 117 C1, 356
Hadley Upland 77 B5
Halley Station, UK 13 C34, 110, 115, 356
Hamburg Bay 81 A2
Hanusse Bay 77 B4
Harold Byrd Mountains 93 A11
Harrison Bluff 93 D8
Haswell Islands 13 E10, 107
Heard and McDonald Islands, Australia 117 D3
Hektoria Glacier 77 E3, 81 D3
Helen Glacier 13 E10, 107
Henkes Islands 77 A4
Henry Bay 13 E13, 107
Henry Ice Rise 12 B30, 115
Hill Bay 81 C2

Hippocrates Glacier 81 C2
Hobart, Australia 117 E5
Hollick-Kenyon Peninsula 77 D5
Hooper Glacier 81 B2
Hope Bay 77 G2
Horlick Mountains 13 A25, 88, 93 A13, 112
Horseshoe Harbor 13 E7, 109
Hoseason Island 77 D2, 81 D1
Hovgaard Island 81 A3
Hughes Bay 77 E3, 81 D2
Hugo Island 77 C3
Hull Bay 12 D23, 88
Hulot Peninsula 81 C2
Hunt Island 81 C2
Hut Point 93 C8
Hydrurga Rocks 81 D2

I

Ile Europa, France 117 G2
Iles Crozet, France 117 E2
Iles Kerguelen, France 117 E3
Inaccessible Island 93 C8
Indian Ocean 117 E3
Inexpressible Island 93 D6
Intercurrence Island 81 D1
Isla de los Estados, Argentina 12 H30
Islas Diego Ramírez, Chile 12 G30
Islas Juan Fernández, Chile 116 F10

J

James Ross Island 12 F31, 77 G3
Jason Peninsula 77 E4
Jelbart Ice Shelf 13 D36, 110
Jenny Island 77 A4
Joerg Peninsula 77 C5
Johannesburg, South Africa 117 G1
Joinville Island 12 F31, 77 H2
Jones Mountains 12 D27, 88, 115
Joubin Islands 81 A2
Juan Carlos Primero Station, Spain 12 F30, 77 E2, 356
Jubany Station, Argentina 12 F31, 77 F2, 356

K

Kainan Bay 93 C9
Kemp Coast 13 E6, 74, 109
Kemp Peninsula 12 D31, 115
Kershaw Peaks 81 B2
Ketley Point 81 C2
King George Island 12 F31, 77 G1, 117 C11
King Peak 13 A28, 88, 115
King Peninsula 12 D26, 88
King Sejong Station, South Korea 12 F31, 77 F2, 356
Kirkwood Islands 77 A5
Knight Island 81 A2
Kohnen Station, Germany 13 C1, 110, 356
Korff Ice Rise 12 C29, 115
Krause Point 13 E10, 107
Krogh Island 77 B4

L

La Gorce Mountains 93 A11
Lady Newnes Bay 13 D18, 93 D6
Laënnec Glacier 81 C2
Lake Vanda 93 C6, 100
Lake Vostok 13 C11, 107, 112
Lambert Glacier 13 D7, 107, 127 B3

Land Glacier 12 C22, 88
Lanusse Bay 81 C2
Larrouy Island 77 C3
Larsemann Hills 13 D8, 107
Larsen Ice Shelf 12 E30, 77 D4, 81 D3
Larsen Inlet 77 F3
Latady Island 12 D29, 117 C10
Laurie Island 12 F32
Lauritzen Bay 13 E16, 104
Laussedat Heights 81 C2
Lavoisier Island 12 E30, 77 B4
Law Dome 13 E12, 107
Law Plateau 13 D8, 107
Law-Racovita Station, Australia & Romania 13 E8, 107
Lazarev Ice Shelf 13 E2, 110
Leay Glacier 81 B3
Lecointe Island 81 C2
Lemaire Channel 81 B3
Lemaire Island 81 C2
Lennox-King Glacier 93 B6
Leonardo Glacier 81 D2
Leppard Glacier 77 D3
Lesotho 117 G1
Leverett Glacier 93 A11
Lewis Bay 93 D9
Liard Island 77 B4
Liège Island 77 D2, 81 C1
Lillie Glacier 93 D6
Lion Island 81 B2
Lister Glacier 81 C2
Liv Glacier 93 A9
Livingston Island 12 F30, 77 E2
Loper Channel 77 H1
Low Island 77 D2, 81 C1
Luitpold Coast 13 C33, 110, 115
Lützow Holm Bay 13 E4, 109
Lyddan Island 13 D34, 110, 115

M

Mac.Robertson Land 13 D7, 74, 109, 117 B3, 356
Macchu Picchu Station, Peru 12 F31, 77 F2, 356
Mackay Glacier 13 C16, 93 C6, 100
Mackenzie Bay, 13 E8, 107, 109
Mackenzie Glacier 81 C2
Macleod Point 81 D2
McMurdo Sound 93 C6, 100
McMurdo Station, USA 13 C17, 93 C8, 100, 356
Macquarie Island Station, Australia 117 D6
Macquarie Island, Australia 117 D6
Madagascar 117 G2
Maitri Station, India 13 D2, 110, 356
Maldonado Station, Ecuador 12 F31, 77 F2, 356
Mamelon Point 77 C4
Marambio Station, Argentina 12 F31, 77 G3, 356
Marguerite Bay 12 E30, 77 A5
Marie Byrd Land 12 C25, 74, 88, 117 B9, 356
Mariner Glacier 93 D6
Mario Zucchelli Station, Italy 13 D17, 93 D6, 100, 356
Marion Island 117 E2
Marion Island Station, South Africa 117 E2
Marshall Mountains 93 B6
Martin Peninsula 12 D25, 88
Masson Island 13 E10, 107, 117 C4
Matha Strait 77 B4
Mawson Coast 13 E7, 109
Mawson Escarpment 13 D7
Mawson Glacier 13 C16, 93 C6, 100
Mawson Peninsula 13 E16, 104
Mawson Station, Australia 13 E7, 109, 356
Mawson's Hut 104

Melbourne, Australia 117 F5
Melchior Islands 81 B2
Mertz Glacier 13 E15, 104
Mertz Glacier Tongue 13 E15, 104
Mikhaylov Island 13 E9, 107, 117 C3
Mikkelsen Harbor 81 E1
Milburn Bay 81 E1
Mill Inlet 77 C4
Mill Island 13 E11, 107, 117 C4
Millerand Island 77 B5
Minna Bluff 13 C18, 93 C6
Mirny Station, Russian Federation 13 E10, 107, 356
Mitchell Point 81 C2
Mobiloil Inlet 77 C5
Molodezhnaya Station, Russian Federation 13 E5, 109, 356
Montagu Island 117 D12
Montevideo, Uruguay 116 F11
Moody Point 12 F31, 77 H2, 117 C11
Moser Glacier 81 C2
Moubray Bay 13 D18, 93 D7
Mozambique 127 G2
Mozambique Channel, Indian Ocean 117 G2
Mt Albert Markham 93 B5
Mt Bird 93 D8
Mt Borland 13 D7
Mt Bulcke 81 C2
Mt Coman 12 D30, 115
Mt Dudley 77 B5
Mt Erebus 13 C17, 93 D9, 100
Mt Ford 93 D6
Mt Frakes 12 C25, 88
Mt Français 77 D3, 81 B2
Mt Frazier 93 C10
Mt Gaudry 77 A4
Mt Haddington 77 G3
Mt Joyce 93 C6
Mt Larsen 93 D6
Mt Longhurst 93 C5
Mt Markham 13 B17, 93 B6, 117 A6
Mt McClintock 13 B16, 93 B5
Mt Melbourne 13 D17, 93 D6, 100
Mt Menzies 13 D7, 109, 117 B3
Mt Miller 93 B6
Mt Minto 13 D17, 93 D6, 117 B6
Mt Moberly 81 B2
Mt Murchison 93 D6
Mt Northampton 93 D6
Mt Reeves 77 A4
Mt Sandow 13 E11, 107
Mt Saunders 93 A6
Mt Scott 81 A3
Mt Shackleton 81 B3
Mt Sidley 12 C24, 88, 117 B8
Mt Siple 12 D24, 88
Mt Southard 13 D16, 93 D5, 104
Mt Takahe 12 C25, 88
Mt Tennant 81 C2
Mt Terra Nova 93 D9
Mt Terror 93 D9
Mt Walker 81 C2
Mt William 81 B2
Mt Zeppelin 81 D2
Mulock Glacier 93 C6, 100
Mulock Inlet 13 C17, 93 C6

N

Namibia 117 G1
Nansen Island 81 C2
Napier Mountains 13 E6, 109
Neko Harbor 81 C2
Nelson Island 77 F2
Nemo Peak 81 B2

Neny Fjord 77 B5
Neptune's Bellows 77 E2
Neumayer Channel 81 B2
Neumayer Station, Germany 13 D36, 110, 356
New Bedford Inlet 12 D30, 115
New Zealand 117 E6
Newman Island 12 C22, 88
Neyt Point 81 D1
Nilsen Plateau 93 A9
Nimrod Glacier 93 B6
Nimrod Passage 81 A2
Ninnis Glacier 13 E15, 104
Ninnis Glacier Tongue 13 E15, 104
Nobile Glacier 81 D2
Nordenskjöld Coast 81 E2
Northcliffe Glacier 13 E10, 107
Norwegian Antarctic Territory 356
Noville Peninsula 12 D27, 88
Novolazarevskaya Station, Russian Federation 13 D2, 110, 356

O

O'Higgins Station, Chile 12 F31, 77 G2, 356
Oates Land 13 D16, 74, 93 D6, 104, 117 B6
Ob' Bay 13 D17, 93 D6, 104
Observation Hill 93 C8
Ohridiski Station, Bulgaria 12 F30, 77 E2, 356
Okuma Bay 93 C10
Omega Island 81 C2
Orcadas Station, Argentina 12 F32, 117 C11
Orléans Strait 81 E1
Orne Harbor 81 C2
Oscar II Coast 77 E3, 81 D3

P

Palmer Archipelago 12 F30, 77 D3, 81 A2
Palmer Land 12 D30, 74, 117 B10, 356
Palmer Station, USA 12 F30, 77 C3, 81 A2, 356
Paradise Harbor 81 C2
Paraguay 116 G11
Pasteur Peninsula 81 C2
Patagonia, Argentina/Chile 116 E10
Paulding Bay 13 E13, 104, 107
Paulet Island 77 H2
Pavlov Peak 81 D2
Pelseneer Island 81 C2
Pendleton Strait 77 B3
Península Valdés, Argentina 12 K30
Penola Strait 81 A3
Pensacola Mountains 13 B31, 115
Periphery Point 77 C5
Perrier Bay 81 B2
Peter I Øy 12 E27, 88, 117 C9
Petermann Island 81 A3
Petermann Ranges 13 D2, 110
Philippi Glacier 13 E9, 107
Piccard Cove 81 C2
Pine Island Bay 12 D26, 88
Pinel Point 81 C2
Pitt Islands 77 C3
Pléneau Island 81 A3
Point Wild 77 H1
Polar Plateau 13 A4, 74
Porpoise Bay 13 E13, 104, 117 C5
Port-aux-Français Station, France 117 E3
Port Elizabeth, South Africa 117 F1
Port Foster 77 E2
Port Lockroy 81 B2
Port Stephens, Falkland Islands 12 H30
Portal Pt 81 D2
Possession Islands 93 D7
Pourquois Pas Island 77 B4

Powell Island 12 F32
Pram Point 93 C8
Presidente Eduardo Frei Station, Chile 12 F31, 77 F2, 356
Prestrud Inlet 93 C10
Pretoria, South Africa 117 G1
Priestly Glacier 93 D6
Prince Albert Mountains 93 C6, 100
Prince Charles Mountains 13 D7, 109
Prince Edward Island 117 E2
Prince Edward Islands, South Africa 117 E2
Prince Gustav Channel 77 F2
Prince Olav Mountains 93 B8
Princess Astrid Coast 13 D2, 110
Princess Elizabeth Land 13 D9, 74, 107, 117 B3, 356
Progress II Station, Russian Federation 13 E8, 107
Prospect Pt 77 C4
Prydz Bay 13 E8, 107
Puerto Deseado, Argentina 12 J30
Puerto Natales, Chile 12 H29, 117 D10
Punta Arenas, Chile 12 H29, 117 D10
Puzzle Islands 81 B2
Py Point 81 B2

Q

Queen Alexandra Range 93 B6
Queen Elizabeth Range 93 B6
Queen Fabiola Mountains 13 D4, 109, 110
Queen Mary Land 13 D10, 74, 107, 117 C4, 356
Queen Maud Mountains 13 A19, 93 A10, 112, 117 A7

R

Rabot Island 77 B3
Rawson Mountains 93 A9
Rayner Peak 13 E6, 109
Recess Cove 81 D2
Reclus Peninsula 81 D2
Reedy Glacier 13 A23, 88, 93 A12
Renard Glacier 81 D2
Renaud Island 12 F30, 77 B3
Rennick Bay 93 D6
Rennick Glacier 13 D17, 93 D6, 104
Revelle Inlet 77 D5
Richards Inlet 93 B7
Ridley Island 77 F1
Riiser-Larsen Ice Shelf 13 D35, 110, 115
Riiser-Larsen Peninsula 13 E4, 109, 110, 117 C2
Rio de Janeiro, Brazil 116 G11
Río Gallegos, Argentina 12 H30, 117 D10
Ritscher Uplands 13 D36, 110
Robert Glacier 13 E6, 109
Robert Island 77 F2
Robertson Bay 13 D18, 93 D7
Robertson Island 12 E31, 77 F3
Rockefeller Mountains 93 C10
Rockefeller Plateau 13 B23, 88, 93 B12
Rongé Island 81 C2
Ronne Ice Shelf 12 C30, 74, 115, 117 B11, 356
Roosevelt Island 13 C20, 93 C9, 117 B7
Ross Dependency, New Zealand 356
Ross Ice Shelf 13 B19, 74, 88, 93 B8, 100, 112, 117 A7, 356
Ross Island 13 C17, 93 C6, 100, 117 B6
Ross Sea 13 C20, 74, 88, 93 D8, 100, 117 B7, 356
Rothera Station, UK 12 E30, 77 A4, 356
Rothschild Island 12 E29

Roundel Dome 77 D3
Rudolph Glacier 81 C2
Rydberg Peninsula 12 D28, 88
Rymill Bay 77 B5

S

Sabrina Coast 13 E12, 107
Salvesen Cove 81 D2
San Martin Station, Argentina 12 E30, 77 B5, 356
SANAE IV Station, South Africa 13 D36, 110, 356
Santiago, Chile 116 F10
Saunders Island 12 H30
Scar Inlet 77 D3
Scotia Sea 117 D11
Scott Glacier, East Antarctica 13 E11, 107
Scott Glacier, West Antarctica 13 A21, 88, 93 A10
Scott Island 13 E18, 117 C7
Scott Mountains 13 E5, 109
Scott Station, NZ 13 C17, 93 C9, 100, 356
Scott's Discovery Hut 93 C8
Scott's Terra Nova Hut 93 C8
Scullin Monolith 13 E7, 109
Seal Nunataks 77 E3
Seaplane Point 81 E2
Seligman Inlet 77 C4
Seymour Island 77 G3
Shackleton Glacier 13 A18, 93 A8, 112
Shackleton Ice Shelf 13 E10, 107, 117 C4
Shackleton Inlet 13 B18, 93 B6, 112
Shackleton Range 13 B34, 115
Shackleton's Hut 93 D8
Shirase Coast 93 C10
Signy Station, UK 12 F32, 117 C11
Siple Coast 93 B10
Siple Island 12 D24, 88, 117 B8
Skelton Glacier 13 C16, 93 C6, 100
Skytrain Ice Rise 12 C29, 115
Slessor Glacier 13 C34, 115
Slessor Peak 77 C4
Small Island 81 D2
Smith Island 12 F30, 77 D2
Smith Peninsula 12 D30, 115
Smyley Island 12 D29
Snares Islands, NZ 117 E6
Snow Hill Island 77 G3
Snow Island 12 F30, 77 E2
Solberg Inlet 77 C5
Solvay Mountains 81 C2
Sør-Rondane Mountains 13 D3, 110
South Africa 117 G1
South Atlantic Ocean 117 F12
South Georgia, UK 117 D11
South Magnetic Pole 13 F14, 74, 104
South Orkney Islands, UK 12 F32, 117 C11
South Pacific Ocean 12 H27, 117 E8
South Pole 13 A28, 74, 88, 112, 117 A4, 356
South Sandwich Islands, UK 117 D12
South Shetland Islands, UK 12 F30, 77 E1, 117 C10
Southern Ocean 117 D1
Spaatz Island 12 D29
Spallanzani Point 81 C2
Spert Island 81 D1
Spigot Peak 81 C2
Spring Point 81 D2
Spur Point 77 D4
Stanley Island 77 D4
Stanley, Falkland Islands 12 H31, 117 D11
Stefansson Bay 13 E6, 109
Stewart Island, NZ 117 E6
Stopford Peak 81 D1
Strait of Magellan, Argentina/Chile 12 H30, 117 D10
Strath Point 81 C2

Stratton Inlet 77 E4
Sturge Island 13 E17, 104, 117 C6
Sulzberger Bay 93 C10
Sulzberger Ice Shelf 13 C21, 88, 93 C11, 117 B8
Swaziland 117 G1
Sydney, Australia 117 F6
Syowa Station, Japan 13 E4, 109, 356

T

Tabarin Peninsula 77 G2
Tange Promontory 13 E5, 109, 117 C2
Tasman Sea, Pacific Ocean 117 F6
Taylor Valley 93 C6, 100
Terra Firma Islands 77 B5
Terra Nova Bay 13 C17, 93 C6
Terre Adélie 13 E14, 74, 104, 117 C5, 356
Terror Glacier 93 D9
Thiel Mountains 13 A28, 88, 112, 115
Thompson Peninsula 81 B2
Thurston Island 12 D27, 88, 117 B9
Thwaites Glacier 12 C26, 88
Thwaites Glacier Tongue 12 D26, 88
Tierra del Fuego, Argentina/Chile 12 H30
Tonkin Island 77 C4
Tor Station, Norway 13 D1, 110, 356
Totten Glacier 13 E12, 107
Tournachon Peak 81 D2
Tower Hill 81 E1
Tower Island 77 E2
Trail Inlet 77 C5

Transantarctic Mountains 13 A16, 74, 93 A4, 112, 117 A5, 356
  See also Dry Valleys
Trelew, Argentina 12 K30
Trinity Island 77 E2, 81 E1
Trinity Peninsula 77 F2
Tristan de Cunha, UK 117 F12
Troll Station, Norway 13 D36, 110, 356
Trooz Glacier 81 B3
Truant Island 81 B2
Tucker Glacier 13 D18, 93 D6
Tula Mountains 13 E6, 109
Tuna Bay 93 C10
Two Hummock Island 81 D2

U

Uruguay 116 F11
Usarp Mountains 93 D5
Ushuaia, Argentina 12 H30, 117 D10

V

Vahsel Bay 13 C33, 115
Valdivia Point 81 D2
Valkyrie Dome 13 C4, 110
Vanderford Glacier 13 E12, 107
Vedel Islands 81 A3
Vega Island 77 G2
Veier Head 77 E4
Venable Ice Shelf 12 D28, 88

Vernadsky Station, Ukraine (previously Faraday Station, UK) 12 E30, 77 C3, 81 A3, 356
Vestfold Hills 13 E8, 107
Vicente Station, Eucador 12 F31, 77 F2, 356
Victoria Land 13 D16, 93 D6, 117 B6, 356
Victoria Valley 93 C6, 100
Victory Mountains 93 D6
Vincennes Bay 13 E11, 107, 117 C4
Vinson Massif 12 C28, 88, 115, 117 B10
Vostok Station, Russian Federation 13 C11, 107, 112, 356
Voyeykov Ice Shelf 13 E13, 104

W

Wasa Station, Sweden 13 D35, 110, 356
Waterboat Point 81 C2
Watkins Island 77 B4
Watson Escarpment 93 A12
Watt Bay 13 E15, 104
Wauwermans Islands 81 B2
Weddell Sea 12 D32, 74, 77 F4, 115, 117 B11
Wellington, NZ 117 E6
Wellman Glacier 81 D2
West Antarctica 12 C27, 74, 88
West Falkland 12 H30, 117 E10
West Ice Shelf 13 E9, 107, 117 C3
West Point 13 E9, 107, 117 C3
West Point Island 12 H30
Whirlwind Inlet 77 C4
White Island 13 E5, 109, 117 C2
White Island, Ross Sea 93 C6

Whitmore Mountains 13 B26, 88, 115
Wiencke Island 81 B2
Wiggins Glacier 81 A3
Wilhelm Archipelago 77 C3, 81 A3
Wilhelm II Land 13 D10, 74, 107, 117 B4
Wilhelmina Bay 81 C2
Wilkes Land 13 D13, 74, 107, 117 B5, 356
Wilkins Sound 12 D29
William Glacier 81 B2
Williamson Head 13 E16, 104
Wilma Glacier 13 E6, 109
Wilson Hills 13 D16, 104
Windless Bight 93 C9
Windmill Islands 13 E11, 107
Wohlschlag Bay 93 D8
Wood Bay 13 D17, 93 D6
Wordie Ice Shelf 12 E29
Wright Island 12 D25, 88
Wright Valley 93 C6, 100
Wylie Bay 81 A2

Y

Yalour Islands 81 A3
Young Island 13 E17, 104, 117 C6
Yule Bay 93 D6

Z

Zavadovskiy Island 13 E9, 107
Zhongshan Station, China 13 E8, 107, 356

Index

Plain numbers indicate references in the body text.
Italicized numbers indicate references in image
captions or maps, while bold numbers indicate
references in feature boxes.

A

"A" icebergs 45–46
A Voyage Towards the South Pole 275
Abbott Ice Shelf 89
Abbott, J. L. 89
Academician Vernadsky station 86–87
accidents and disasters
    Air New Zealand flight 92, 371
    Deception Island volcanic eruption 128–129
    Erebus–Terror collision 361, *368*
    Vostok station fire 112–113
acidification of seawater 69
Adams Island *130*
Adams, Jameson Boyd 308, 310
Adare Peninsula 102
Adelaide Island 276
Adélie penguins *86, 147, 246–247, 246, 247*
    changes in number of 68
    feeding chicks *152*
    lichens and *144*
    on Cape Royds 98, *98*
    on Coronation Island *122*
    on Paulet Island 78
    on Petermann Island 86
    on Possession Islands 102
    on South Shetland Islands *125*
    reproductive behavior **246**
    scientific study *379*
    swimming speed *236*
Adenita grandis **138**
Admiralty Bay 125
Adventure (ship) 266, *368*
African penguins 258–259
Air New Zealand crash 92, 371
air pressure, changes in 60
air transport 370–371
    aircraft wrecks *365*
    AN-2 (plane) *106*
    balloon flights 296, *296*, 300, 370
    BAS Otters *126*
    Byrd expeditions 338–341
    Dornier flying boats 370–371
    Fox Moth biplane *344, 345*
    Globemaster 350
    helicopters *106, 371*
    Northrop Gamma 346–347, *347*
    sea-ice airstrips *370–371*
    Skytrain 350
    to Mawson station 109
    to McMurdo station 97
    tourism and 361
    US Galaxy 371
    Wilkins expeditions 336–337
    Winfly flights 97
airgun array *70*
Aitcho Islands 125–126, *126*
Akademik Ioffe (ship) *127*
Aladdin's Cave 322, 324
albatrosses 131, *146, 204–205*, 206–211, *299*
albedo 38
Albert P. Crary Science and Engineering Center
    97–98
Alcock Island 80
Alexander Island 75, 272, 345

Alfred–Faure station 135
algae, *see also* plankton
    growing under ice 148, *148*
    in Continental Antarctica 142
    in Dry Valleys 101
    snow algae *142–143*, 144
Allan Hills *91*
allopreening *210*, 236, 252
Almirante Brown base 83
ambergris 200
American sheathbills 230
Amery Ice Shelf 46, 107
ammonites *34*
Amsterdam Island fur seals 165
Amundsen, Roald *312*
    comments on Charcot 307
    comments on Ross 285
    comments on Ross Ice Shelf 94
    comments on Shackleton 311
    on Axel Heiberg Glacier 90
    on de Gerlache expedition 292–293
    polar expedition 312–315
    uses *Fram* 368
    winters in Antarctic 88
Amundsen–Scott station 59, 112, *113*, 350, *358*
Amundsen Sea 88–89, *88–89*
Anas acuta eatoni 227
Anas flavirostris 227
Anas georgica 120–121, *226*, 227, *227*
Anas superciliosa 227
Anatidae 226–227
Andean Orogeny 27
Andersson, Gunnar 302–303, *303*
Andes mountains 23
Andvord Bay 82, *82*
Anous stolidus 232
Antarctic bottlenose whales 203
Antarctic Bottom Water 88, 149
Antarctic Circle 87, 266, 270
Antarctic Circumpolar Current 30
Antarctic cod 153, *153*
Antarctic Continental Margin Drilling Program 71
Antarctic Convergence (Polar Front) 20, 36–37,
    36–37, 120, 146–147
Antarctic Development Project 348
Antarctic fulmars 214
Antarctic fur seals 121, *121, 157*, 164–165, *165*
Antarctic hair grass *30*, 66, 82, 126, 141, *145*
Antarctic herring 153
Antarctic krill *151, see also* krill
Antarctic limpets 231
Antarctic penguins, *see* chinstrap penguins
Antarctic Peninsula *24–25*, 75
    as mountain system 23
    early expeditions *273*
    Gerlache Strait *81*
    Northern region 76–79
    Southern region 84–89
Antarctic petrels 110, 214
Antarctic prions 216
Antarctic rock cod 366
Antarctic sea-urchins *155*
Antarctic shags 224, *224*
Antarctic (ship) 78, 102, 290, *290–291*,
    302–303, 374
Antarctic skuas 228, 229
Antarctic Sound *302–303*
Antarctic terns 232, *232, 233*
Antarctic toothfish 153
Antarctic Treaty 352, 357–359, **359, 385–386**
Antarctic zones 138

Antarctica 74–75
    Arctic regions compared with **18**
    climate change evidence 59–69
    defined 20
    evolution in 30–35
    formation of 24–29
    law in 358–359
    ozone depletion 54–57
    plants and invertebrates 138, 142–144
    territorial claims 356–357
antarcticite 101
Anvers Island 68, 83
Aptenodytes 238–243
Aptenodytes forsteri, see emperor penguins
Aptenodytes patagonicus, see king penguins
Arctic regions **18**, 37–39
Arctic terns 233, *233*
Arctocephalus australis 164
Arctocephalus forsteri 160, *162*, 164
Arctocephalus gazella, see Antarctic fur seals
Arctocephalus pusillus 165
Arctocephalus tropicalis 165
Arçtowski, Henryk 292–293, *293*
Arçtowski station 125
Argentina 121, 356–357
Argentine Islands 86–87
Aristarchus 262
Aristotle 262
Arktos 262
Armitage, Albert *297*, 299
Arnoux's beaked whale 203
arterio-venous anastomoses 171
Arthur Harbor 83
Artigas station 125
Astrolabe Island 79
Astrolabe (ship) 280, *281*
astronomy 380
Atkinson, Edward *374*
atmospheric changes 60, *see also* carbon dioxide
Auckland Island teals 226
Auckland Islands 130–131, *130*
Aurora Australis (ship) *381*
Aurora (ship) 322–331, *322, 368, 368*
auroras 52–53, *53, 373*
Auster penguin colony *108, 372*
Australasian Antarctic Expedition 322–325
Australia 342–343, 356
Australian fur seals 165
Australophocaena dioptrica 199
aviation, *see* air transport
Axel Heiberg Glacier 40, 90, 314
Azimuth Hill 105
Azorella spp. 141

B

"B" icebergs 45–46, 94–95
Backstairs Passage 310
bacteria 150, *150*, 155
Bagshawe, Thomas 82, **333**
Bahia Inutil 80
Bahia Paraiso (ship) 83, 121
Baily Head *127, 128*
Balaenidae 183, 186, 188–189
Balaenoptera bonaerensis 182, *185*, 186, 191,
    191, 361
Balaenoptera borealis 191, *191*
Balaenoptera edeni 191
Balaenoptera musculus 119, *182*, 192
Balaenoptera physalus 182, 191–192

Balaenopterae 190–193
Balboa, Vaso Núñez de 264, *265*
Balchen, Bernt 340, 346, *347*, 370
baleen whales 182, 186–193, *187*
Balleny Islands 130, *130*, 278
Balleny, John 278, 284
Balloon Bight 296
balloon flights 296, *296*, 300, 370
Banks, Joseph 266
BANZARE expedition 342–343, 369, 370, *375*
Barley Grow floatplane *340–341*, 341, 370
barnacles *68*
Barne Glacier *42–43*
Barne, Michael *297*
Barrientos Island 126
BAS 129, 375
BAS Otters *126*
basal sliding 40
Base G 125
Base H 122
Bay of Whales 95, 296, 308, 312, *313*
Beacon Supergroup **27**
beaked whales *182*, 203
bearded penguins, *see* chinstrap penguins
bearded seals *157*
Beardmore Glacier 40, 90, 309–310, 317–318
Beardmore Orogeny 27
Beardmore, William 308
Beauchene Island *276*
Beaufoy (ship) 274–275, *274*
Bechervaise, John 345
Belgica expedition 88
Belgica (ship) 88, 292–293, *368*
Belgrano II station 115
Bellingshausen Sea 88–89, *89*
Bellingshausen station 125, *125*
Bellingshausen, Thaddeus von 119, 132,
    270–272, *270*
beluga *182*
Bennett, Suzanne *337*
benthic communities *71*, 154–155
Berardius arnouxii 203
bergy bits 47
Berkner Island 114
Bernacchi, Louis 130, 294–296, *294, 297*,
    299
Bertrab nunatak 115
Big Ben 132
big-eyed seals 178
Bingham, Edward *345*
biological habitats 100–101, 146–147
birds 120–121, 204–259
Biscoe House 128
Biscoe, John 108, 275–276
Bismarck Strait 84
black ducks 227
black-bellied storm-petrels 222, *222*
Blackborow, Perce 326
black-browed albatrosses *123*, 210
black-faced sheathbills 230
blackfish (orca) *195*, 197–198, *197*
blubber 170, *185*
blue icebergs *46*
blue penguins 258, *258*
blue petrels *217*
blue whales 119, *182*, 192
blue-eyed shags 224
Bluff Depot 130, 310
Bond, Charles 348
Bonney Laboratory 87
Book of Isaiah 263

Borchgrevink, Carsten 102, 290–291, 294–295, 295
Borradaile Island 130
bottlenose dolphins 195
Bouvet de Lozier, Jean-Baptiste-Charles 135
Bouvetøya 135
Bowers, "Birdie" 99, 317
bowhead whales 183, 188
Bransfield, Edward 124
brash ice 50
breaching 188
Bristow, Abraham 130
Britain 121, 342, 356–357
British Antarctic Expedition 308–311
British Antarctic Survey (BAS) 129, 375
British Graham Land Expedition 344–345, 362
British Imperial Expedition 333
British National Antarctic Expedition 296–299, 298
broad-billed prion 216, 216
Brown Bluff 76, 78–79
brown ducks 227
Brown Glacier 65
brown noddies 232
Brown Peak 130
brown petrels 219
brown skuas 228, 229, 229
Browning Peninsula 75
Bruce, William Spiers 122, 304
Brunt Ice Shelf 43, 64
brush-tailed penguins 244–251
Bryde's whales 191
bryophytes 141–142, see also moss beds
Bryum argenteum **138**
Bryum pseudotriquetrum 144
bubble rills 47
Buckle Island 130
Budretski, A. 38
Bull, Henryk Johan 273, 290–291
Bull Island 291
bull kelp 140–141
Bull Pass 101
Bunger, David 348
Bunger Hills 348
Burmeister's porpoise 199
Butcher's Shop 314
Byrd, Richard 338–341, 338, 348, 348, 350
  claims North Pole flight 370
  in Bay of Whales 95
Byrd station 90, 112, 350
Byrd Subglacial Basin 90

C

"C" icebergs 46
Cabral, Pedro 264
calcium carbonate formation 69
callosities 188
CAML 67–68, 71
Campbell, George 288
Campbell Island 130–131
Campbell Island shags 133
Campbell, Victor **319**
Cape Adare 102, 102–103, 272, 294–295
Cape Circumcision 266
Cape Crozier 98, 99, 296, 317
Cape Denison 323, 324, 343
  hut at 322, 324
  wind speeds **38**
Cape Disappointment 118, 268
Cape Evans 98, 318, 331
  hut at 316, 317, 318, 330, 330
Cape Fold Belt 28
Cape fur seals 165
Cape Hallett 102, 102, 350

cape hens 219
Cape Hudson 283
Cape Lookout 124
Cape Norvegia 110–111
cape petrels 212, 214, 215
cape pigeons, see cape petrels
Cape Prud'homme 105
Cape Renard 84, 84, 292
Cape Royds 25, 48, 98–99, 98, 310
  hut at 99, 308
Cape Seymour 79
Cape Valentine 327
Cape Well-met 303
Cape Wild 124, 327
Caperea marginata 188, 189
carbon dioxide 58, 62, **148**
Carcass Island 123
cartography, see maps
Casey station 66, 106
castellated icebergs 47
Castle Rock 96
Casuarina cunninghamiana 33
catchalots 200–202
Catharacta antarctica 228, 229, 229
Catharacta genus 228
Catharacta maccormicki 228, 228–229, 229
Cecilia (ship) 273
celestial poles 17
Cenozoic era 25, 33, 35
Census of Antarctic Marine Life 67–68, 71
Cephalorhynchus commersonii 198
Cetacea 182–203, see also whaling
CFCs 54–55
Challenger expedition 288–289, 289
Champsocephalus gunnari 153
Charadriiformes 205
Charcot, Jean-Baptiste 85, 86, 306–307, 307
Charcot Land 307
Cherry-Garrard, Apsley 92, 99, 215, 316–319
Chile 83, 356
chinstrap penguins 129, 248–249, 248, 248–249, 249
  on Deception Island 127, 128, 129
Chionidae 230
Chionis alba 83, 230, 230
Chionis minor 230
chlorofluorocarbons 54–55
Christchurch Cathedral 123
Circumpolar Current 29, 36, **60–61**, 146, 152
cirque glaciers 43
Claudius Ptolemy 263
climate change evidence 38–39, 365
  Antarctic ecosystem 66–69
  ecological changes 144–145
  melting ice 62–65
  Polar Front 36–37
  Southern Ocean and 148
clothing 373, 375
coaming 186
coastal and continental shelf zone 147
Coats brothers 304
Coats Land 304
cod 153
Colbeck, William 294–295, 297
cold deserts 27, 59, 91, 91, 100–101, 100
Coleridge, Samuel Taylor 44, 269
Col–Lyall Saddle 131
Colobanthus spp. 82, 126, 141–142, 141
Comandante Ferraz station 125
commercial fishing, see fishing
Commerson's dolphin 198
common diving petrels 223, 223
common rorquals 191–192
common seals 157
Commonwealth Bay 22, 104–105, 342

Commonwealth Transantarctic Expedition 350, 351–352
Concordia station 113
condominium sovereignty 357
conifers 31–32
Convention for the Conservation of Antarctic Marine Living Resources 365, 367
Convention for the Conservation of Antarctic Seals 359, 367
Convention on the Regulation of Antarctic Mineral Resource Activities 359, **365**
Cook, Frederick 292–293
Cook Ice Shelf 63
Cook Island 272
Cook, James 266
  Antarctic Circle crossings 368
  explorations 266–269
  furthest southern point reached 88
  narrative of voyage 268
  on Iles Kerguelen 134
  on South Georgia 118
Copacabana station 125
Cope, John 333
copepods 152, 154
Coquille (ship) 280, 281
cormorants 224
Cornice Channel 87
coronal point 53
Coronation Island 122, 122
Correll, Percy 325
Coulman Island 102, 321
crabeater seals 87, 156, 172–173, 172–173, 380
Crean, Tom 119, 318, 326, 328–329
crested penguins 254–255
Cretaceous period 31, 32–33
crevasses 97
Crozet Islands 135
Crozet, Jules Marie 135
Crozier, Francis 284
Cruise of the Antarctic, The 290–291, 291
crustaceans 68, 152
Cruzen, Richard 348
Crystal Sound 86, 87
Cumberland Bay 335
Curtiss Bay 80
Curtiss Condor biplane 340–341
cushion plants 66
Cuverville Island 19, 82
Cuvier's beaked whale 203
cyanobacteria 144
cyclones 60

D

da Gama, Vasco 264
Dakshin Gangotri station 110
Dall's porpoise 196
Dampier, William 275, 276
Danco, Emile 293
Danco Island 82
Daption capense 212, 214, 215
Dark Ages, exploration during 263
dark-mantled sooty albatrosses 211
"Dash Patrol" 321
data collection 377
Davis, John King 273, 311, 322, 329, 331, 342
Davis Sea region 106–107
Davis station 106
de Fresne, Marion 135
de Gerlache, Adrien 80, 84, **292–293**, 368
de Kerguelen-Trémarec, Yves-Joseph 133–134
de la Ferrière, Laurence 112–113
de la Roché, Antoine 118
de Palma. Emilio Marcos 76
de Palma. Silvia Morello 76

de Solfo, Juan 275
Deacon Peak 126
Debenham, Frank 333
Decepciòn station 129
Deception Island 29, 126–128, 126, 127, 271, 273, 336, 336
  Ellsworth expedition 346
  volcanic caldera 29
Deep Freeze Range 103
Deepfreeze 350–351
Deloncie Bay 85
Delphinidae 183, 196–199
Deschampsia antarctica 30, 82, 126, 141, 145
Detaille Island station 87, 352
Deutschland (ship) 320
devil's fish 183
diatoms 146, 149
Diaz, Bartolomeu 263–264
dinosaurs 31, 32
Diomedea epomophora 206, 208–209
Diomedea exulans 120, 206, 208, 208
Diomedeidae 131, 146, 204–205, 206–211, 299
Dion Islands 87
Dirty Ice 50–51
disasters, see accidents and disasters
Discovery II (ship) 335, 347
Discovery investigations 334–335
Discovery (ship) 268–269, 296–299, 334–335, 335, 342, 368–369
Dissosfichus eleginoides 67, 153, 366
diving petrels 223
Dobson spectrophotometer 54
dogs and dog sledges 108, 362
  Amundsen expedition 315
  at Mawson station 372
  remains of 331
dolphins 183, 196–199
Dome A 71, 112
Dome Fuji 112
Dome, The 112
Dominican gulls 231
Don Juan Pond 26, 101
Dornier flying boats 370–371
dragonfish 153
Drake Passage 33, 63, 265
Drake, Sir Francis 263–264
Dronning Maud Land 110
Dry Valleys 27, 59, 91, 91, 100–101, 100
Drygalski, Erich von 106, 132, 300–301, 300, 370
Drygalski Fiord 121, 320
Drygalski Ice Tongue 64–65, 300–301
du Toit, Alex 28
Dufek complex 28
Dufek, George 348, 350, 350
Dufek Massif 40–41, 114
Dufek Ventifact 29
Dumont d'Urville Coast 104–105
Dumont d'Urville, Jules Sébastien 79, 280–281, 280, 281
Dumont d'Urville station 105
Dundee Island 346
Durvillaea antarctica 140–141
Duse, Samuel 302–303, 303
dusky dolphins 195, 196, 198, 199
Dutch East India Company 265
dwarf sperm whales 202

E

eared seals 156, 160–167
Earth, average temperature of 62
East Antarctic Ice Sheet 40, 65
East Falkland 123
echolocation by whales 186, 190, 194–195, 200–201

ecological changes 144–145, 365
  Antarctic ecosystem 66–69
  climate change evidence 59–69
  food web 151
  marine ecosystem 379
  melting ice 62–65
  Polar Front 36–37
  Southern Ocean and 148
Edgeworth David, T. W. 99, 310–311, 311
Eduardo Frei Montalvo station 125, 358
Eendracht (ship) 265
Eielson, Carl 336–337, 370
Eisenhower Range 103
Ekström Ice Shelf 111
El Niño/Southern Oscillation events 169
Elephant Island 124, 124, 329
elephant seals 120, 121, 134–135, 156,
  179–181, 366
Eliza Scott (ship) 278
Ellsworth Land 89
Ellsworth, Lincoln 79, 335, 337, 346–347, 346,
  347, 370
Ellsworth Mountains 23, 28, 89, 114, 114–115
Ellsworth station 350
emperor penguins 98, 240, 241
  Auster rookery 108
  changes in number of 68–69
  chicks 239
  huddling behavior 236–237, **240**
  Laurie Island 305
  Scott's observations 296
Enderby Brothers 275–276
Enderby, Charles 131
Enderby Land 275, 342
Endurance (ship) 326–329, 327, 328, 368
energy technologies 364
England, see Britain
ENSO 169
environmental issues 362–365, see also
  ecological changes; human activity,
  impact of; tourism
Epimeria 68
equinoxes 17
equipment 351–352, 373, 375, see also
  air transport
  nutrition for cold climates 315
  vehicles 108
Eratosthenes of Cyrene 262–263
Erebus Bay 50
Erebus (ship) 284–287, 286, 368
erect-crested penguins 258
Errera Channel 82
Escudero, Julio 357
Esperanza station 76
Eubalaena australis 183, 184–185, 185, 188,
  188
Eudyptes chrysocome 252, 254–255, 254, 255
Eudyptes chrysolophus 120, 252, 253, 256,
  256
Eudyptes pachyrhynchus 258
Eudyptes robustus 258
Eudyptes schlegeli 257
Eudyptes sclateri 258
Eudyptes spp. 252–258
Eudyptula minor 258, 258
Euphausia crystallorophias 147, 150
Euphausia spp. 150–153, 151, see also krill
Eusirus spp. 68
Evans, Edgar 299, 318, 374
Evans, Edward "Teddy" 98, 316–319, 318
Evans, Hugh Blackwall 294
evolution 30–35, 183–185
Exasperation Inlet 82
exploration 261–353
Explorer (ship) 361
extinctions 33

F
fairy penguins 258, 258
fairy prion 216
Falkland fur seals 164
Falkland Islands 118, 123, 334, 375
Falklands War 121
False Cape Renard 84
Faraday station 86
fast ice 48
feldmark 140
Ferrar, Hartley 297
field camps 377
Filchner Ice Shelf 114, 320
Filchner, Wilhelm 320
Fimbul Ice Shelf 46, 110
finback whales (finners) 182, 191–192
Finch Creek 134
finnesko 311
Fiordland penguins 258, 258
fire oak 33
First on the Antarctic Continent 295
fish 153
Fish Islands 87
fishing 207, 290–291, 364–367
flat-headed whales 203
Fletcher, Francis 265
flowering plants 32
Flower's whales 203
Flying Fish (ship) 282–283
food web 151
Ford, Edsel 338, 340
Ford Trimotor 339
Forster, George 269
Forster, John Reinhold 266, 269
fossils 30–31, 35
  ammonites 34
  Cenozoic 33
  Glossopteris 23, 27, 29, 30, 91
  marsupials 29
  moss beds 67
  Nothofagus 29
  Paleozoic 27
  penguins 79, 302
  seabirds 206, 212
Foster, Henry 126
Fox Moth biplane 344, 345
Foyn Land 290
Foyn, Svend 290, 291
Fram (ship) 312–315, 313, 315, 321, 368
Framheim 312, 312
Framnes Mountains 23
Français (ship) 306–307, 306
France, territorial claims 355
Franklin Island 102
frazil 48
Fregetta tropica 222, 222
French Antarctic Expeditions 306–307
Fuchs, Vivian 351–352, 351
fuel handling 362
fulmar prions 216–217, 217, 217
fulmarine petrels 212
fulmars 212–221
Fulmarus glacialoides 212, 214
fur seals 156, 158, 164–165, 164, 170
  diving by **163**
  exploitation of 67
  on South Georgia 276, 367

G
Gabriel de Castilla station 129
gadfly petrels 212, 219
Galapagos penguins 234, 259
Gamburtsev Subglacial Mountains 22, 71

Garrodia nereis 222
gastroliths 158
Gauss expedition 106
Gauss (ship) 300–301, 301
Gaussberg 300, 325
gentoo penguins 83, 244, 250–251, 250–251,
  251
  at Port Lockroy 363
  nomenclature **250**
  on Antarctic Peninsula 250
  on South Georgia 120
geographic poles 17
geology
  Challenger expedition 289
  Ellsworth Land 89
  formation of Antarctica 24–29
  polar landscape 22–23
  scientific study 379–380
  Transantarctic Mountains 90–91
geomagnetic poles 18
Gerlache Strait 80–83, 80, 292
German Antarctic Expeditions 300–301, 320,
  347
giant floes 50
giant sea star 289
ginkgo 32
glaciation 42–43
  climate change evidence 63
  Dry Valleys formed by 100
  glacial erratics 41
  glacial sediments 68
  global 41
  history of 33, 62–63
  retreating 60
global environmental impacts **365**
Globemaster 350
Globicephala melas 199
Glossopteris fossils 23, 27, 29, 30, 91
glycolyis, in seals 171
glycoproteins 153
Gobernador Bories (ship) 127
Godfroy, E. 307
Godthul whaling base 119
Gondwana 25–26, 25, 28, 91
Gondwanan Orogeny 28
González Videla base 82
goose-backed whales 203
Gore, Al 61
Gospel of St. Mark 263
Gourdon, E. 307
Graham Land 75, 75
grampuses (orcas) 195, 197–198, 197
Grand Chasm 42
Granite Harbor 143
gray ducks 227
gray petrels 219
gray whales 183
gray-backed storm-petrels 222
gray-faced petrels 218
gray-headed albatrosses 211, 211
Gray's beaked whales 203
grease ice 48
Great Britain, see Britain
Great Lakes 65
Great Ocean Conveyor Belt 58
great shearwater 220–221
Great Wall station 107, 125
greater sheathbills 230
great-winged petrels 218, 218
greenhouse gases 58, 62, see also climate
  change evidence
Greenland whales 183, 188
Greenwich Island 126
growlers 47
growth rings 32
Grunden, Toralf 302–303, 303

Grytviken whaling station 118–119, 119, 120,
  334, 335, 367
gulls 231, **232**
gulp feeding 192

H
Haakon VII Sea region 110–111
Haggitts Pillar 130, 299
Hagglund vehicles 108
hair grass 30, 66, 82, 126, 141, 145
hair seals 156, 168–181
Half Moon Island 126, 127
Halley V station 59, 114–115, 115
Halobaena caerula 217
Hamilton, H. 366
Hannah Point 126
Hannam, Walter 325
Hanson, Nicolai 102, 294–295
harbor seals (common seals) 157
hardheads (gray whales) 183
Hardy, Alistair 335
Hasselburg, Frederick 130, 132
Haswell Islands 106
Hawaiian Islands 269
Hawaiian monk seals 156
Heard Island 132, 145, 343
  altitude 138
  climate change evidence 65
  ecological changes 66
Heard, John 132
hearing underwater **187**
Hearst, William Randolph 336
Hector's beaked whales 203
Hektoria (ship) 336
helicopters 106, 371, see also air transport
hemispheres 74
Henry the Navigator, Prince of Portugal 263,
  263
Hero (ship) 273
High Nutrient–Low Chlorophyll (ocean
  environment) 149
Highjump 348
Hillary, Sir Edmund 112, 351–352, 351
Hodgeman, Alfred 105, 325, 375
Hodges, William 267
Hodgson, Thomas Vere 297
Hollick-Kenyon, Herbert 335, 346, 346, 370
Home of the Blizzard, The 39
Hooker, Joseph Dalton 130, 284, 374
Hooker's sea lions 131, 132, 161, 166, 166,
  167, 380
Hoorn (ship) 265
Hope Bay 28–29, 76, 302–303
Hope Point 332
Horn Bluff 322
horses on expeditions 311, 316–317, 316
Horseshoe Harbor 19, 372
hourglass dolphins 199, 199
Hovgaard Island 85
Hughes Bay 273
Hughes Doctrine 356
human activity, impact of 145, 362–364,
  see also ecological changes; fishing;
  seals; tourism; whaling
Humboldt penguins 259
humpback whales 147, 183, 186–187, 190,
  193, 193
hunting, see also fishing; seals; whaling
  harvesting 366–367
  impact of 364–365
  of king penguins 243
Hurley, Frank 124, **325**, 325, 326, 327, 372
huskies, see dogs and dog sledges
Husvik whaling base 119

Hut Point *95*, 96, *97*, *297*, 330–331
huts *295*, *298*, 310, *318*, *324*
Hydrobatidae 222
Hydrographers Cove 125
hydroponics *375*
Hydrurga leptonyx *156*, *158*, *174*, *175*
Hyperoodon planifrons 203, *203*

I

IAATO 361
ice arches *361*
ice cores *61*, 379, *380*
Ice Islands with Ice Blink *267*
ice mass 62–63
ice sheets 40–41, *41*
ice shelves 42, 64
Iceberg Bay 122
icebergs 44–47, *45*, *47*, *84*
  ecological changes due to 155
  from Ross Ice Shelf 94–95
  grounded 85
  ice arches *361*
  near South Georgia *118*
  scientific study 379
icebreakers 368
icefish *153*, 366
Ile aux Cochons 135
Ile de la Possession 135
Ile de l'Est 135
Iles Crozet 135
Iles des Apôtres 135
Iles des Pingouins 124–125, 135
Iles Kerguelen 65, 133–134, *135*, 269, *343*
imperial shags *204*, 224, *225*
Imperial Transantarctic Expedition (1914–17)
  326–329
India 110
Inexpressible Island 102, *319*
interglacial stage 41
Intergovernmental Panel on Climate Change 59,
  61
International Association of Antarctica Tour
  Operators 361
International Convention for the Regulation of
  Whaling 367
International Council for Science 70
International Council of Scientific Unions 350
International Geophysical Year 59, 112, 350–352,
  357
International Polar Year 70–71
International Space Ship Earth 380
International Whaling Commission 367
introduced species 145, 362–363, *362*
invertebrates 138–145
  coastal *68*
  in Continental Antarctica 143
  in Southern Ocean *146*
  numbers of species **141**
IPCC 59, 61
Irizar, Julián 303

J

jackass penguins 258–259
jade icebergs *44*
James Caird (lifeboat) 328
Jane (ship) 274–275
Japan 109, *321*
Japan finner whales 191
Jensen, Bernhard 294–295
johnny penguins, *see* gentoo penguins
Jonassen, Ole 303
Jones, Evan *325*

Joyce, Ernest 330–331
Juan Carlos I station 126
Juan Fernandez Islands 275

K

Kainan Maru (ship) 321, *321*
Kap Nor (ship) 290
Kapitan Khlebnikov (ship) 95, *369*
katabatic winds *39*, *104*
kelp gulls 231, *231*
Kemp Coast 278
Kemp, Peter 108, 132, 278
Kemp, Stanley Wells 334–335
Kerguelen cabbage 135
Kerguelen fur seals, *see* Antarctic fur seals
Kerguelen Islands 65, 133–134, *135*, 269, *343*
Kerguelen petrels 218
Kerguelen pintails 227
Kerguelen Plateau 132
Kerguelen shags 224
Kerguelen terns 233
killer whales (orcas) *195*, 197–198, *197*
Ki-moon, Ban *71*
King Edward Point 118
King George Island 125
king penguins *234*, 242–244, *242*, **243**, *243*
  chicks *235*, *238*
  expanding populations 365
  on Iles Crozet 135
  on South Georgia 69, *116*, 120, *268–269*
King Sejong station 125
kleptoparasitism 228
Koettlitz, Reginald *297*, 299
Kogia breviceps 202
Kogia simus 202
Kogiidae 202
Komsomolskaya station 112
Koonya (ship) 308–311
krill 68, 150–153, 366–367, *366*, 379, *see also*
  Euphausia spp.
Kristensen, Leonard 290–291
Kurafid *375*

L

Lagenorhynchus crociger 199, *199*
Lagenorhynchus obscurus *195*, *196*, *198*, 199
Lake Vanda *101*
Lake Vostok 40, 113, 378
lakes 143–144, *145*
Lambert Glacier 42–43
lanternfish 153
lanugo 170
Laridae 231, 232
Lars Christensen Peak 89
Larsen, Carl Anton 78, 119, 290, *290*, 302
Larsen Ice Shelf 42, **62**, 67–68
Larsøya Island 135
Larus dominicanus 231, *231*
Laseron, Charles *325*
Lashly, William 299, 318
Laurasia 25
Laurie Island 122, *122*, 304, *305*
Law Dome 106
law in Antarctica 358–359
Law, Phillip 106, 108–109
Lazarev Ice Shelf 270
le Maire, Isaac 265
Leaellynasaura *31*, 32
Lecointe, George 292
Lemaire Channel 84–85, *87*
Lemire, Jean **372**, 388
leopard seals *156*, *158*, *174*, *175*

Leptonychotes weddellii 69, 79, *152*, *156*,
  158, *168*, *169*, *170*, 176–177, *176*, *177*,
  274–275
Lesser Antarctica 74
lesser rorquals 191
lesser sheathbill 230
lesser sperm whales 202
Lester, Maxim 82
Lester, Michael **333**
lichens *30–31*, *138*, 143, *144*
Lieutenant Juan Ruperto Elichiribehety station 76
light, *see* sunlight
light-mantled sooty albatrosses 211, *211*
Lillie, D. G. *154*
Lindblad, Lars-Eric 360
Lindsay, Captain 135
Lissodelphis peronii 199, *199*
Little America base 338, 340
Little America II base *346*
Little America IV base 348, *348*
Little America V base 350–351
little penguins 258, *258*
little shearwater 221
Lively (ship) 275–276
Livingston Island 126, *129*
Lobodon carcinophagus *87*, *156*, 172–173,
  172–173, *380*
local environmental impacts 362–364
long-finned pilot whale 199
longitude 20–21
Long-Term Ecological Research sites 59
Lymeburner, Jack 347

M

macaroni penguins 120, *252*, *253*, 256, *256*
McCormick, Robert 374
McCormick's skua 228, *228–229*, 229
McDonald Island 132
McDonald, William 132
Mackay, Alistair 310–311, *311*
Mackenzie, K. N. 342
Mackintosh, Aeneas 330–331
Macklin, Alexander 326, 332
McMurdo Sound *48*, 92, *154*, 376
McMurdo station 92, 97–98, 350, *352*
  clinic *374*
  safe route marking *377*
  warming at 59–60
McNeish, Harry 328
Macquarie Island 131–132, *134*
  altitude 138
  climate change evidence 61
  radio station at 322
  sealers' huts *323*
Macquarie Island cormorant 132
Macquarie Island station 132
Mac.Robertson Land 108
Macronectes giganteus *205*, 213
Macronectes halli *205*, 213, *213*
Madigan, Cecil *325*, *375*
Madrid Protocol 359–364
Magellan, Ferdinand 264, *264*
magellanic penguins 259, *259*
Magnet (ship) 278
magnetic poles 17–18
magnetic reconnection 52
magnetograph house 104–105
Maitri station 110
Malvinas (Falkland Islands) *118*, 121, 123,
  334, *375*
maps
  albatross distribution *208*, *209*, *210*, *211*
  Amundsen expedition *314*

Antarctic Peninsula 77
Antarctic Peninsula expeditions *273*
Antarctica 74, 265, *314*
Australasian Antarctic Expedition *322*
BANZARE expedition *342*
Bellingshausen expedition *270*
biological zones *150*
Borchgrevink expedition *294*
British Antarctic Expedition *308*
Byrd expeditions *338*
Challenger expedition *288*
Commonwealth Transantarctic Expedition *350*
Cook's voyages *266*, *268*
cross-sectional profile *40*
Davis Sea region *107*
de Gerlache expedition *292*
Dry Valleys *100*
Dumont d'Urville Coast *104*
Dumont d'Urville's explorations *280*
Enderby and Mac.Robertson Land *109*
fossil sites *35*
French Antarctic Expeditions *307*
fulmar distribution *214*
Gerlache Strait region *81*
German Antarctic Expedition *300*
Haakon VII Sea region *110*
ice shelves *43*
Imperial Transantarctic Expedition *326*
kelp gull distribution *231*
krill distribution *366*
Larsen expeditions *290*
penguin distribution *240*, *242*, *246*, *248*, *250*,
  254, *256*, *257*
petrel distribution *213*, *214*, *215*, *217*
Polar Plateau *112*
prion distribution *215*, *216*, *217*
Ross expeditions *284*
Ross Sea region *93*
Scott expeditions *296*, *317*
Scottish National Expedition *304*
sea ice *18*, *50*
seal distribution *164*, *165*, *166*, *173*, *174*,
  176, *178*, *179*
skua distribution *229*
snowy sheathbill distribution *230*
sooty shearwater distribution *220*
South America *263*
Southern Ocean currents *36*
sub-Antarctic islands *117*
Swedish South Polar Expedition *303*
tern distribution *232*, *233*
territorial claims *356*
time zones *18*
Weddell Sea region *115*
West Antarctica *88*
Wilkes's expeditions *282*
world *262*, *263*
marbled rockcod 153, *153*
Marguerite Bay 345
Marie Byrd Land 90
Marie Byrd Seamount 88
marine ecosystem 379
marine plants 138
Marion Island 61, 65, 66
Maritime Antarctica 138, 141–142
markers *113*
Markham, Clements 292, 296, 326
Marshall, Eric 308, 310
Marston, George 326
marsupials 29, 32
Maudheimia wilsoni **138**
Mawson, Douglas 310–311, *311*, *322*, *325*, *342*
  Australasian Antarctic Expedition (1911–14)
    322–325
  BANZARE expedition 342–343
  main hut *105*

names Mac.Robertson Land 108
on Commonwealth Bay 104
on Heard Island 132
on Macquarie Island 132
ships used by 368–369
Mawson Peak 132
Mawson station 52–53, 108–109, 372
Maxwell Bay 125
measurement of Antarctic ice 21
meatuses 168
medical treatment 374–375
Megadyptes antipodes 131, 258
megaherbs 140
Megalestris Hill 86, 306
Megaptera novaengliae 147, 183, 186–187, 190, 193, 193
melon (oil gland) 194
Mertz Glacier 105
Mertz Ice Tongue 105
Mertz, Xavier 322–323, 325, 375
Mesoplodon spp. 203
Mesozoic era 27, 28
meteorites 322, 380, 380
meteorology 38–39, 58–61, *see also* climate change evidence
microbial loop 150
microoganisms **148**
microplankton 149
migration behavior 190, **193, 235**
Mikkelsen Harbor 79
Mikkelsen, Klarius and Karoline 106
Milankovitch, Milutin 41
Mill, Hugh Robert 294
mineral resources **365**
minke whales 182, 185, 186, 191, 191, 361
Mirny station 106
Mirnyi (ship) 270–272, 271
Mirounga leonina 120, 121, 134–135, 156, 158–159, 169, 171, 179–181, 179, 180, 181, 278, 279, 366
mixotrophy 149
Mizhuo station 112
mollymawks (albatrosses) 131, 146, 204–205, 206–211, 299
molting
in penguins 235–236, 252
in seals 161–162
monotremes 32
Montreal Protocol on Substances that Deplete the Ozone Layer 57
moraines 43
Morning (ship) 297, 299, 299
Morrell, Benjamin 135
moss beds 30, 66, 67, 125, 139, 143
Mount Barr-Smith 325
Mount Discovery 92
Mount Erebus 92, 96, 99, 285, 285, 309, 309
Mount Flora 28–29
Mount Gardner 114
Mount Jackson 75
Mount Kirkpatrick 90
Mount Melbourne 25, 102, 103
Mount Minto 294–295
Mount Nivea 122
Mount Paget 118
Mount Ross 134
Mount Scott 86
Mount Shackleton 84, 86
Mount Shinn 114
Mount Terror 285
Mount Tyree 114
mountains 23–25, 40
mud dipping 378
Murphy's Law 377
mussel digger whales 183

muttonbirds 220
Myctophids 153
Myrtaceae 32
Mysticeti 182, 186, 186–193, 187

N
Nacella concinna 231
nacreous clouds 52, 331
nanoplankton 149
Nansen, Fridtjof 368
Nares, George Strong 288–289
narrow-billed prions 215
narwhals 182
navigation by penguins **235**
Neko Harbor 82
Neko (ship) 82
Nelson Island 125
nemertean worm 154
Neobalaenidae 183, 186, 188–189
Neptune's Bellows 126
Neptune's Window 127
nesting stones **244**
Neumayer Glacier 360
Neumayer station 111, 111
New Island 123
New Zealand 97, 130–131, 356
New Zealand beaked whale 203
New Zealand fur seal 160, 162, 164
New Zealand sea lion 131, 132, 161, 166, 166, 167, 380
Newnes, Sir George 294
nilas 48
Nilsen, Thorvald 312
Nimrod (ship) 308–311, 368
Ninnis, Belgrave 322–323, 325
noctilucent clouds 60
Nordenskjöld, Otto 79, 302–303, 302
Norris, George 135
North Head 134
North Magnetic Pole, shifts in 18
northern giant petrels/fulmars 205, 213, 213
northern lights, *see* auroras
northern royal albatrosses 209
Northrop Gamma 346–347, 347
Northwest Passage 269
Northwind (ship) 348
Norvegia expedition 89
Norway 356
Nothofagus spp. 32, 34, 91
fossils 29, 33, 34–35
Notothenioidei 153, 153
Novolazarevskaya station 110
nunataks 22
nutrition for cold climates 315

O
oases 142
Oates, Lawrence "Titus" 316–319, 316, 374
O'Brien Bay 349
Observation Hill 97, 317, 319
Ocean Camp 327
Ocean Harbor whaling base 119
Oceanites oceanicus 222, 222
oceanography 88–89
acidification and 69
ecological changes 67–68
International Polar Year 70–71
Polar Front 36–37
sea ice and 50–51
sea level changes 62–64
Odobenidae 156, 157
Odontaster validus 154–155

Odontoceti 182, 185, 194–195
oil spills 362
Olavtoppen 135
Olstad, Ola 89
Ommatophoca rossi 156, 178, 178
One Ton Camp 317
Onyx River 101
Operation Deepfreeze 350–351
Operation Highjump 348, 371
Orcadas station 122, 305
orcas 195, 197–198, 197
Ormond House 122
Orne Harbor 80
Otaria flavescens 166
Otariidae 156, 160–167
Owen's pygmy sperm whales 202
ozone depletion 54–57, **54**, 61, 66, 114–115, 145

P
Pachyptila belcheri 215
Pachyptila crassirostris 216–217, 217, 217
Pachyptila desolata 216
Pachyptila turtur 216
Pachyptila vittata 216, 216
Pacific beaked whales 203
pack ice 48–49, 368, 369
Pagodroma nivea 212, 215, 215
pale-faced sheathbills 230
Palmer Land 75
Palmer, Nathaniel 83, 122, 127, 272–273
Palmer station 83
Panagrolaimus davidi **138**
pancake ice 48, 51, 87
Pangaea 25
Paradise Harbor 83
Parborlasia corrugatus 154
Parmenides 262
passenger ships 361
Patagonian toothfish 67, 153, 366
Patience Camp 327, 327
Patric (ship) 292
Paulet Island 78, 79
Peacock (ship) 282–283
Peale's dolphin 195, 196, 198, 199
pearlwort 82, 126, 141–142, 141
pedidunkers 219
Pegasus airstrip 97
pelagic fish 153
Pelagodroma marina 222
Pelecaniformes 204–205
Pelecanoides georgicus 223
Pelecanoides urinatrix 223, 223
Pelecanoididae 223
Pendulum Cove 129
Penguin Island 124–125, 135
Penguin Point 324
penguins 95, 234–259
as meat 292
fossils of 302
harvesting of 364–366
on Heard Island 132
on Macquarie Island 132
on Saunders Island 123
Penola (ship) 344–345, 344
Penola Strait 345
permanently open ocean zone 147
Peruvian penguins 259
Peter I Øy 88, 88, 89, 90, 272
Petermann, August 86
Petermann Island 28, 84, 86, 145
Petermann Ranges 86
Peterson Island 349

Petrel Island 105
petrels (fulmars) 212–221
petrified tree 32
Phaeocystis algae 147
Phalacrocoracidae 224
Phalacrocorax atriceps 204, 224, 225
Phalacrocorax bransfieldensis 224, 224
Phalacrocorax verrucosus 224
Philodina gregaria **138**
Phocarctos hookeri 131, 132, 161, 166, 166, 167, 380
Phocidae 156, 168–181
Phocoena spinipinnis 199
Phocoenidae 183, 196–199
Phoebetria fusca 211
Phoebetria palpebrata 211, 211
photosynthesis 149, *see also* sunlight
Phyctenactis anemone 149
Physeteridae 182, 185, 194, 195, 200–202, 200–201, 203
Physeter macrocephalus 200–202
phytoplankton 68, 147, 149, 151
Pic Marion-Dufresne 135
picoplankton 149
piked (minke) whales 182, 185, 186, 191, 191, 361
pilot whales 182–183, 196, 199
Pine Island Glacier 63
Pinnipedia, *see* seals
pintado petrels 212, 214, 215
Pionerskaya station 112
plankton 68, 147, 149, 151, 379
plants 138–145
evolution of 30–35
numbers of species **141**
on South Shetland Islands 125, 126
plate tectonics 24–29
Plateau station 112
Pléneau Island 85
Pléneau, Paul 306–307
Pleuragramma antarcticum 153
Pleurophyllum hookeri 134
Pleurophyllum speciosum 132
Pliocene epoch 33
Poa folosia 134
poikilohydrous plants 143
Point Thomas 125
Polar Front 20, 36–37, 36–37, 120, 146–147
polar ice vessels 369
polar medicine 374–375
Polar Plateau 74, 75, 112–113, 339
polar regions 16–23
Polar Sea (ship) 71
Polaris (ship) 326
poles of inaccessibility 18, 350
poles of rotation 17
pollack whales 191
pollen fossils 33
Pollux (ship) 294
polynas 50, 298–299, 381
Pomponius Mela 262–263
Ponting, Herbert 228, 316
Pony Lake 98–99, 99
Porosira pseudodenticulata 149
Porpoise (ship) 281–283
porpoises 183, 196–199
Port Charcot 306
Port Circumcision 86, 307
Port Egmont 123
Port Foster 126
Port Lockroy 83, 306, 363
Port Ross 131
Portal Point 80
Port-aux-Français 134
Poseidonius 263
Possession Islands 102, 135, 285

Pourquoi-Pas? (ship) 307
Powell, George 122
Powell Island 122
Precambrian Era 26
precipitation 39, 66
predators 68–69, 155
preening, in penguins 236
Presidente Gabriel González Videla station *333*
Presidente Pedro Aguirre Cerda station 129
Priestley, Sir Raymond 128
Prime Meridian 20–21
Prince Olaf Harbor whaling base 119
Pringlea antiscorbutica 135
prions 212
proboscis worm *154*
Procellaria aequinoctialis 219, *219*
Procellaria cinerea 219
Procellariiformes 204
Procellariidae 212–221
Professor Molchanov (ship) *369*
prokaryotes 144
Proteaceae 32
protists 149–150
Protocol on Environmental Protection 359,
    360–361, 362–364
protozoa 149
Prydz Bay *106–107*, 343
Pterodroma brevirostris 218
Pterodroma lessoni 218, *218–219*
Pterodroma macroptera 218, *218*
Pterodroma mollis 218, *219*
Ptolemy, Claudius 263
Puffinus assimilis 221
Puffinus gravis 220–221
Puffinus griseus 220, *220*, *220–221*
pygmy right whales 188, 189
pygmy sperm whales 202
Pygoscelis adeliae, see Adélie penguins
Pygoscelis antarctica, see chinstrap penguins
Pygoscelis papua, see gentoo penguins
Pygoscelis spp. 244–251
Pythagoras 262

Q

Queen Maud Land 110
Queen Maud Mountains 340, *340*
Quest expedition 332

R

Racovitza, Emile 292
radio use 322, *325*
radiolarians *288*
Ranvik Bay *381*
rat porpoises 202
rata forests 131, *131*
Ravn Rock 127
razorback whales *182*, 191–192
reaction wood 35
Regional Sensitivity to Climate Change 378
reindeer *362*
Relief (ship) 282–283
reproductive behavior **189, 246**
Resolution (ship) 266, **267**, 268–269, 368
resource regulation 367
resurrection plants 143
retia mirabilia 184
Rhone Glacier *74*
Richards, Dick 331
Ridley Beach 102, *102–103*
Ridley Camp 294, *295*
right whales *183*, 186, 188–189
ringed penguins, *see* chinstrap penguins

Ringgold, Cadwalader 283
Ritscher, Alfred **347**
river dolphins *183*
Robertson, Thomas 304
rock cod 153, *153*, 366
Rockefeller, John D. 338
rockhopper penguins *252*, 254–255, *254*, **255**, *255*
Rogers, Erasmus Darwin 132
Rongé Island 82
Ronne–Filcher Ice Shelf 45, 114
rorquals *182*, 186–193, *187*
Ross Ice Shelf *42*, 46, 92, *92*, 94–95, *94*, *285*,
    286, *286–287*
Ross Island *16*, 22–23, 92, *92*, 96–99, *96–97*
Ross, James Clark 284–287, *284*
    Antarctic expedition 374
    discovers Ross Sea 92, 284
    discovers Transantarctic Mountains 90
    names Paulet Island 78
    on James Weddell 275
    on Wilkes's charts 283
Ross Orogeny 27
Ross Sea expedition 330–331
Ross Sea region 92, *93*, 102–103
    discovery of 92, 284
    geology of 23
    volcanoes 28
Ross seals *156*, 178, *178*
Rossbank observatory 284
Rothera station 87
Row Island 130
royal albatrosses *206*, 208–209
Royal Bay station 119
royal penguins 257
Royds, Charles 296, *297*
Russia 350, 356–357, 366
Rymill, John 85, 344–345, *345*

S

Sabrina Coast 278
Sabrina Island 130
Sabrina (ship) 278
sagittal otoliths 170
salinity, *see* seawater
salps 68, 152
Salvesen Range *121*
SANAE IV 110, *111*
sandstone boulders *107*
Sandwich Land 268, *272*
sardine whale 191
sastrugi 39, *110–111*
satellite imaging 64, 71
Saunders Island 123
scamperdown whales 203
SCAR 378
Schirmacher Oasis 110
Schmidt-Nielsen, Knut **205**
Schwabenland Expedition (1938–39) 340, **347**
Scientific Committee on Antarctic Research 378
scientific research 376–380
    into seal diets 170
    into stress **249**
    population estimation **157**
scientific study
    ozone depletion *56–57*
Scotia Sea 29
Scotia (ship) 304
Scott base 92, 96, 352, *352, 375*
Scott Island 130
Scott, Robert Falcon 296–299, *297*, 316, *318*
    first Antarctic balloon flight 370
    on Beardmore Glacier 90
    on Cape Royds 99
    polar expedition 316–319

Scottish National Expedition 304
sea caves 65
sea floor communities *71*, 154–155
Sea Gull (ship) 282–283
sea ice *18*, *38*, 48–51, *48–49*
    core samples *61*
    ecology of 148
    receding 60
    scientific research *60*, 379
    zone of 88, *89*
sea leopards *156*, *158*, 174, *175*
sea level changes 62–64
sea lions 156, 160–167
sea spiders 68, *155*
sea squirts (salps) 68, 152
seabirds 204–259
seals (Pinnipedia) 156–181
    baby fur **170**
    expanding populations 365, 366
    hunting of 275–277, *277*, 364–365, 366
    on Macquarie Island 132
    on South Georgia 118–119
seasonal ice zone 147
seasonal variations *17*, *39*, 48–49, 60–61
seawater
    ecological changes in 68–69
    freezing point of 50–51, 147–148
    processing by seabirds **205**
    processing by seals **160**
seaweeds 154–155, *see also* algae
sei whales 191, *191*
Seymour Island 79
Shackleton, Ernest *297*, *326*, *327*
    British Antarctic Expedition 308–311
    British National Antarctic Expedition 296
    Imperial Transantarctic Expedition 326–329
    names Bay of Whales 95
    on Cape Royds 99
    on South Georgia 119–120
    Quest expedition 332
    Ross Sea expedition 330–331
    sledging expedition 297
Shackleton Ice Shelf 283, 324–325
Shag Islands 132
shags 224
shearwaters 212–213
sheathbills 230
Shingle Cove 122, *122*
shipping **207**, *334*, 360, 368–369, *see also*
    names of ships e.g., *Discovery*
Shirase Glacier 43, 110
Shirase, Nobu 321
shoemaker petrels 219
short-headed sperm whales 191, *191*
short-headed whales 202
shy albatrosses 210
Signy Island 66, 122
silverfish 153
silver-gray fulmars 214
singing seals *156*, **177**, 178, *178*
Siple, Paul *338*, *348*, 350, *350*
Sirius Formation flora 33, 34–35
Skelton, Reginald 296, *297*
ski-doos *376*
skimming feeding 191
Skottsberg, Carl 303
skuas 228–229
Skytrain 350
slender-billed prions 215
small whales 202
Smith, William 124, 273
Snares crested penguins 258
snow algae *142–143*, 144
Snow Hill Island 79, 302, 303
snow petrels 212, 215, *215*
snowy sheathbills *83*, 230, *230*

social behavior in cetaceans **194**, 197
soft-plumaged petrels 218, *219*
solar corona 52
sooty albatrosses 211
sooty shearwaters 220, *220*, *220–221*
Sørlle, Thoralf 329
sound underwater **187**
    echolocation 186, 190, 194–195,
        200–201
South Africa 110
South African fur seals 165
South American fur seals 164
South American sea lions 166
South Georgia 20–21, 118–121, *119*, *267*,
    329
    Bellingshausen expedition 270
    death of Shackleton in 332
    discovery of 268
    ecological changes 65, 66
    fur seals *276*
    Imperial Transantarctic Expedition (1914–17)
        328–329
    king penguin colonies 69, 116, *268–269*
    penguin feeding at 235
    sealing industry 275
    whaling relics *335*
South Georgia diving petrels 223
South Georgia pintail 120–121, 227
South Georgia pipit 121
South Ice 351
South Magnetic Pole 18, 284–287, 310–311
South Orkney Islands 122
south polar skuas 228, *228–229*, 229
South Polar Times 296
South Pole 20, *315*, *see also* Amundsen–Scott
    station
    Amundsen expedition 312–315
    cooling at 59
    expeditions to 112
    flights over 338–341
    IGY expeditions 350–351
South Pole, The 315
South Sandwich Islands 25, 29, 118
South Shetland Islands 124–129, 235, 273
southern beech, *see Nothofagus* spp.
southern black-backed gulls 231, *231*
southern bottlenose whales 203, *203*
Southern Cross (ship) 294–295
southern dolphin 198
southern elephant seals 120, 121, *134–135*,
    156, *158–159*, *169*, *171*, 179–181, *179*,
    180, *181*, *278*, *279*, 366
southern fulmars *212*, 214
southern fur seals 164
southern giant petrels/fulmars *205*, 213
Southern Hunter (ship) 127
southern lights, *see* auroras
Southern Ocean 20, *see also* oceanography
    climate change evidence 61
    ecology of 67, 146–155
    marine plants 138
    Polar Front 36–37
Southern Oscillation 169
southern right whale dolphins 199, *199*
southern right whales *183*, *184–185*, *185*,
    188, *188*
southern royal albatrosses 209
southern sea elephants, *see* southern elephant
    seals
southern sea lions 166
Sovetskaya station 112
Soviet Union (Russia) 350, 356–357, 366
Sparrman, Anders **267**
speckled teal 227
spectacled porpoise 199
Spencer-Smith, Arnold 330

sperm whales *182*, *185*, *194*, 195, 200–202, *200–201*, **201**, *203*
spermaceti organ 200
Sphenicus demersus 258–259
Sphenicus humboldti 259
Sphenicus magellanicus 259, *259*
Sphenicus spp. 258–259
Spheniscidae, *see* penguins
Spheniscus mendiculus 234, 259
Spigot Peak 80, *82*
sponges *154*, 155
squid *123*, 153–154, 201
Stanley 123
Station B 128
Station N 83
steel ships 369
Stenhouse, John 335
Stercorariidae 228–229
Stercorarius genus 228
Sterna paradisaea 233, *233*
Sterna virgata 233
Sterna vittata 232, *232*, *233*
Sternidae 232–233
Stevens, Fred 330
Stilbocarpa polaris *134*, *139*, 140–141
Stillwell, Frank *325*
stone-cracker penguins, *see* chinstrap
  penguins
storm-petrels 222
Strait of Magellan 264
strap-toothed whales 203
stratosphere, cooling of 60
Stromness whaling base 119
Sturge Island 130
Subantarctic fur seals 165
sub-Antarctic islands 116–117, *116*, *117*
  climate change evidence 61, 65
  ecological changes 69
  plants and invertebrates 138–141
subglacial lakes 378
sulfur compounds **148**
sulfurbottom whales 192
sunlight 16, 41, **144**
sunrise *16*
superswarms of krill 151
sustainable krill harvesting 150–151
Svarthamaren 110
Swedish South Polar Expedition 79, 302–303
Syowa station 109, *109*

T

tabular icebergs *44*, *46–47*
Tasman, Abel Janszoon 265
Tasmanian fur seals 165
Taylor Glacier 100, *100–101*
Taylor, Griffith 316
Taylor Valley *74*, 100
technological change 377
Teille Island, *see* Deception Island
temperature
  climate change evidence 62
  records for 38
  seasonal variations *39*
  thermoregulation in animals 171, 183–184,
    235, **240**
Teniente Camára station 126
Teniente Jubany station 125
Teniente Rodolfo Marsh Martin station 125
terns 232–233, **232**
Terra Australis Incognita 262
Terra Nova Bay *39*, 102, *102*
Terra Nova expedition 99
Terra Nova (ship) 98, 299, 312, 316, *316*,
  319, 368

Terre Adélie 68–69, *104*, 280, 311
territorial claims 356–357
Terror (ship) 284–287, *286*, *368*
Thalassarche cauta *132*, *206*, *209*, 210, *210*
Thalassarche chlororhynchos 209, *209*
Thalassarche chrysostoma 211, *211*
Thalassarche melanophris *123*, 210
Thalassoica antarctica 110, 214
The Cruise of the Antarctic 290–291, *291*
The Dome 112
The Home of the Blizzard 39
The South Pole 315
The Voyage of the Discovery 299
The Voyages of the Morning 299
The Worst Journey in the World 317
Thermohaline Current 58
thermoregulation in animals 171, 183–184,
  235, **240**
thin-billed prions 215
Thomson, Wyville *288*, 289
Thule Island *272*
Thurston Island 89
Thwaites Glacier 63
Tierra del Fuego 142
time zones *18*, 20–21
Titanic (ship) 47
toothed whales 182, *185*, *194–195*
Torgersen Island *142*
Total Ozone Mapping Spectrometer *56*
Totten Glacier 63
tourism 145, 360–361, 371
tractor trains *379*
Transantarctic Mountains 23, 27, 40, 90–91,
  103
transportation *334*, 361, 368–369, *377*,
    see also air transport; *names of ships*;
    vehicles
tree line 20
Trinity House *76*
Trinity Island 79
Trinity Peninsula 273
true diver penguins 252–257
true seals 156, 168–181
tufted penguins *252*, 254–255, *254*, *255*
Tula (ship) 275–276
Tunzelman, A. H. F. von 291
Turret Point 125
tussock grass 140

U

Ukraine 86–87
ultraviolet radiation 54, **144**
Umbilicaria aprina **138**
United Kingdom, *see* Britain
United States 97
  IGY expeditions 350
  naval involvement 348
  Operation Highjump 348
  territorial claims 356
United States Exploring Expedition 282–283
United States National Ice Center 45
upper atmosphere studies 380
Upper Glacier Depot 318
Upper Wright Glacier *26*
Uruguay (ship) 303, 304
US Galaxy (plane) *371*
USSR (Russia) 350, 356–357, 366

V

Vahsel Bay 351
valley glaciers 42–43
Vancouver, George **267**

Vanderford Glacier *43*, *107*
vast floes 50
vehicles, *see also* air transport
  Commonwealth Transantarctic Expedition
    351–352
  Hagglund vehicles *108*
  ski-doos *376*
  tractor trains *379*
Venable Ice Shelf 89
Verkhoyansk 38
Vernadsky base 60, *76*
Vesleskarvet Cliff *111*
Vesleskarvet nunatak 110
Vespucci, Amerigo 264
Vestfold Hills *106*, 143
vibrissae 171
Vicecomodoro Marambio station 79
Victoria Land 28, 142, 287, *287*, 310
Victoria (ship) 264
Vienna Convention for the Protection of the
    Ozone Layer 56
Vince, George 296
Vincennes (ship) 282–283
Vincent, John 328
Vinson Massif 23, 74, 114, 360
viruses 150
vocalization by seals **177**
volcanoes 25–26, *28*, 29, 74
von Bellingshausen, Thaddeus 119, *132*,
    270–272, *270*
von Drygalski, Erich *106*, 300–301, *300*,
    370
von Tunzelman, A. H. F. 291
Vostok (ship) 270–272, *271*
Vostok station 38, 112–113
Voyage of the Discovery, The 299
Voyage Towards the South Pole 275
Voyages of the Morning, The 299

W

Wales, William 269
Walkabout Rocks *337*
Walker Bay *129*
walruses 156, *157*
wandering albatrosses 120, 206, 208,
    208
waratahs *32–33*
waste recycling *364*, *365*
water, *see also* seawater
  precipitation 39, 66
  processing by plants **143**
  processing by seabirds **205**
  processing by seals **160**
Waterboat Point 82–83, 333, *333*
waterfowl 226–227
weather patterns 38–39, 58–61, *see also*
    climate change evidence
Weddell, James 122, 125, **274–275**, *274*
Weddell Sea region 23, 78, 114–115, 275
Weddell seals 69, 79, *152*, 156, *158*, *168*,
    169, *170*, 176–177, *176*, *177*, 274–275
Weddellian Biogeographic province **33**
Wennersgaard, Ole 78
West Antarctic Ice Sheet 40–41, 75
West Antarctic Rift System 26, 29
West Antarctica 24–25, 60, 88–91
West Falkland 123
West Point Island 123
Whalers Bay 127–128
whales (Cetacea) 182–203
  migration by **193**
  reproductive behavior **189**
  social behavior **194**
  "tantrums" **190**

whaling 275–277, *278–279*, 365, 366
  exploration due to 290, 334–335
  history of 366
  on South Georgia 119–120
  sperm whales **201**
whiskers in seals 171
white-capped albatrosses *132*, *206*, *209*,
    210, *210*
white-chinned petrels 219, *219*
white-faced storm-petrels 222
white-headed petrels 218, *218–219*
whitesided dolphins 195, *196*, *198*, 199,
    199
Wiencke, Auguste-Karl 293
Wild, Ernest 331
Wild, Frank 124, 296, 308, 310, 322–326,
    332
Wilhelm II Land 300
Wilhelm II Plateau 135
Wilhelmina Bay 80
Wilkes, Charles 106, 281, **282–283**, *282*
Wilkes station 106, 350, *352*, *357*, 365
Wilkins, Hubert 333, 336–337, *336*, *337*,
    347, 370
Wilkins Ice Shelf 87
William Scoresby (ship) 335
Williams Field 97
Wilson, Edward "Bill" *52*, 99, 297, *297*, 317,
    374
Wilson's storm-petrels, *see Oceanites*
    oceanicus
Windmill Islands 143, *244–245*
winds 37, 38–39
  at Cape Denison **38**
  changes over time 60
  katabatic *39*, *104*
  Polar Front and 36
  seasonal variations *39*
Winfly flights 97
wingless diver penguins 238–243
wintering in Antarctica 372
woody plants 33
Wordie House 87
Wordie Ice Shelf 87
Wordie, James 87
World Heritage listing 358
World Meteorological Organization 70
Worsley, Frank 46, 119, 326–327, 329,
    332
Worst Journey in the World, The 317
wreathed terns 232, *232*, *233*
Wyatt Earp (ship) 346–347

Y

Yalour Islands 86
Yalour, Jorge 86
Yankee Harbor 126
Yelcho (ship) 329
yellow-billed pintail 120–121, *226*, 227,
    227
yellow-billed sheathbills *83*, 230, *230*
yellow-billed teals 227
yellow-eyed penguins 131, 258
yellow-nosed albatrosses 209, *209*
Young Island 130

Z

Zelée ship) 280
Zhongshan station 107
Ziphius cavirostris 203
Zipiidae *182*, 203
zooplankton 152, 165

Acknowledgments

The authors would like to thank the many people who have contributed to this remarkable project. Most thanks go to all the contributors named within the author list who must be acknowledged here for their exceptional assistance far beyond their written work. Thanks, too, to the dedicated production team and all at Global Publishing. We owe a lot to those we have traveled with, and those who have taken us to the polar regions. Among the many Antarcticans we would like to thank, in no particular order, are Chris Downes, Margot Morrell, Andrew Prossin, the Wikander family, Carla Santos, Greg Mortimer, Andie Smithies, Graeme Watt, Emily Slatten, Ingrid Visser, Dava Sobel, Ben Hodges, Lee Belbin, Peter Gill, Nigel de Winser, Francis Herbert, Sir Edmund Hillary, Phil Law, Glenn A. Baker, Dean Peterson, Rob McNaught, Rinie Van Meurs, Chris McKay, Esteban de Salas, John Borthwick, Dennis Collaton, Chris Doughty, Tony Press, Kerry Lorimer, Bernhard Lettau, Glenn Browning, Pat Quilty, Aaron and Cathy Lawton, Rob Harcourt, Austin Simpson, Pamela Wright, Scott McPhail, Andrew Fountain, Michael Bryden, Zaz Shackleton, Jayne Paramor, Bill Davis, Des Cooper, Jack Sayers, Darrel Schoeling, Barry Griffiths, Dave Briscoe, Denise Landau, Heather Jeffery, Dick Filby, Mark Eldridge, Tim Bowden, Ken Morton, Andy Beatty, Lincoln Hall, Stephen Martin, Warren Papworth, Adelie Hurley, Toni Hurley-Mooy, Craig Bowen, Barry Boyce, Bob Finch, Tony Harrington, John Palmer and Ross Brewer. There are many others in the Antarctic community who helped a lot, not least of all our Russian captains and their crews. Several institutes were of inestimable assistance, including the Scott Polar Research Institute, Oficina Antártida Ushuaia, Australian Geographic Society, Royal Geographical Society, Cambridge University Library, State Library of NSW, Library of the Australian Antarctic Division, US Naval Library.

The publishers would like to thank the following people for their assistance in the production of this book: Dannielle Doggett, Paul Hopler, Paula Kelly, Robert B. Stephenson, Tracy Tucker, and Kim Westerkov.

Photographers

Kim Westerskov, David McGonigal, Grant Dixon, Mike Craven, Tony Palliser, Luke Saffigna, Malcolm Ludgate, Craig Potton, Stewart Campbell, Kelvin Aitken, Kevin Deacon, Debra Glasgow, Peter Gill, Albert Kuhnigk, Robyn Stewart, Garry Phillips, Harvey Marchant, Steve Nicol, André Martin, Graham Robertson, Patrick Toomey, Jane Francis, J. Howe, R. Hunt, Geoff Longford, Glen A. Baker, David Keith Jones, and Barry Griffiths, Quest Nature Tours.
Cover photograph: Colin Monteath

Global Book Publishing would be pleased to hear from photographers wishing to supply photographs.

Photography credits

All photographs are from the Global Photo Archive except where credited below:

CB = Corbis Australia
GI = Getty Images
NASA = National Aeronautical and Space Administration
NLA = National Library of Australia
PL = Photolibrary
RGS = Royal Geographical Society
USAP = US Antarctic Program

Cover GI/Hegdgehog House/Colin Monteath, **20**tl USAP/ National Science Foundation/Josh Landis, **28**tr USAP/ National Science Foundation/Zee Evans, **28**bl USAP/ National Science Foundation /Josh Landis **29**tl CB, **29**br USAP/National Science Foundation/Bill Meurer, **34**b GI/Ralph Lee Hopkins, **35**t PL, **40–41** USAP/ National Science Foundation/Bill Meurer, **43**cr PL, **52**tl USAP/National Science Foundation/Chad Carpenter, **55**t NASA, **56**b NASA, **56**bcl NASA, **56**bc NASA, **56**bcr NASA, **56**br NASA, **59**tr CB, **60**bl PL, **60**cr USAP/ National Science Foundation/David Vaughan, **61**bl PL, **62**bl NASA, **62**bcl NASA, **62**bc NASA, **62**bcr NASA, **62**br NASA, **67**cl PL, **67**br PL, **68**tl GI/AFP/Stringer, **68**bl PL, **69**bl PL, **70**b PL, **71**tl PL, **71**tr CB, **88**tc GI/Maria Stenzel, **89**b GI/Maria Stenzel, **110–11** CB, **112–13** CB, **113**cr USAP/National Science Foundation/Glenn Grant, **113**b USAP/National Science Foundation/ Dwight Bohnet, **114**cl USAP/National Science Foundation/Bill Meurer, **114**bl GI/Hedgehog House/Colin Monteath, **114–15** GI/National Geographic, **136**br USAP/National Science Foundation/Glenn Grant, **142**cl USAP/National Science Foundation/Kurtis Burmeister, **186–87** GI/Doug Allan, **188**cr GI/Shaen Adey, **193**t GI/Foto Nora/Flip De Nooyer, **195**tr GI/Norbert Wu, **196**tr GI/Tui De Roy, **200–01** GI/Ken Usami, **260–61** Mitchell Library, State Library of New South Wales, **262**bc RGS, **263**bc CB, **264**tl NLA, **264–65** CB , **265**tr CB, **267**cr Mitchell Library, State Library of New South Wales, **271**b Hakluyt Society, **277**cl CB, **278**tl NLA, **278–79** NLA, **281**tr NLA, **281**br NLA, **288**tl State Library of Victoria, **288**br RGS, **289**t NLA, **327**br NLA, **328**tc NLA, **328**cr NLA, **329**tr NLA, **329**cr NLA, **337**br CB, **338**bl NLA, **339**t CB, **340**tl GI/National Geographic, **340–41** CB, **343**t NLA, **344**tr NLA, **344**bl NLA, **345**tr NLA, **345**br NLA, **346**tl NLA, **346**b NLA, **348**tr CB, **348**cr CB, **350**cl CB, **351**t CB, **351**bc CB, **358**bl USAP/National Science Foundation/ Liesl Schernthanner, **362**cl NLA, **370**br CB, **374**b USAP/ National Science Foundation/Peter Rejcek

The publisher believes that permission for use of the historical and satellite images in this publication, listed above, has been correctly obtained; however if any errors or omissions have occurred, Global Book Publishing would be pleased to hear from copyright owners.

Illustration credits

All illustrations are from the Global Illustration Archive except where credited below:
31br Pat Vickers/Peter Schouten

Captions for cover photos, preliminary pages and section openers

Front cover Adélie penguin on iceberg, South Shetland Islands.

Back cover photographs

Main photograph Barne Glacier, Ross Island.

Insets (from top) Weddell seal pup with mother; ice mask; blue icebergs; wandering albatrosses.

Cover spine photograph Adélie penguin.

Preliminary pages and section openers

1 Law Dome on the Polar Plateau.

2–3 Emperor penguins, Ross Ice Shelf.

4–5 Nacreous clouds, Ross Sea region.

8–9 Admiralty Range, Victoria Land.

10–11 Fur seals at play.

14–15 (clockwise from left) Sea ice off the Antarctic coast; hoar crystals on sea-ice surface, Erebus Bay; ancient sandstone, Transantarctic Mountains; full moon over Castle Rock, Hut Point Peninsula.

72–73 (clockwise from left) Andvord Bay, Antarctic Peninsula; Airdevronsix Icefall, Transantarctic Mountains; blue icebergs, Curtiss Bay; Shingle Cove, Coronation Island in the South Orkneys.

136–37 (clockwise from left) Penguins on ice; young wandering albatrosses; *Colabanthus* (pearlwort) in the sub-Antarctic islands; southern elephant seals.

260–61 (clockwise from left) The *Endurance* in ice, 1915, photographed by Frank Hurley; remains of hut built in 1903 by shipwrecked sailors from the Antarctica on Paulet Island; the Aurora in pack ice off East Antarctica, 1912; BANZARE expeditioners, 1929–31.

354–55 (clockwise from left) Sunset, Paradise Harbor, Antarctic Peninsula; McMurdo Station; a white-capped albatross preens its mate; negotiating heavy polar ice conditions.

382–83 (clockwise from left) Icebreaker widening channel through ice; Hercules LC–130 on ice runway in front of the Royal Society Range; spring at Hut Point Peninsula; snow plough, McMurdo Sound.

385 Chinstrap penguins.